AF509019

Biological and Pharmacological Activity of Plant Natural Compounds II

Biological and Pharmacological Activity of Plant Natural Compounds II

Editors

Raffaele Pezzani
Sara Vitalini

MDPI • Basel • Beijing • Wuhan • Barcelona • Belgrade • Manchester • Tokyo • Cluj • Tianjin

Editors

Raffaele Pezzani
Phytotherapy Lab
Endocrinology Unit
Dept. of Medicine
University of Padova
Padova
Italy

Sara Vitalini
Department of Agricultural and
Environmental Sciences
Milan State University
Milan
Italy

Editorial Office
MDPI
St. Alban-Anlage 66
4052 Basel, Switzerland

This is a reprint of articles from the Special Issue published online in the open access journal *Molecules* (ISSN 1420-3049) (available at: www.mdpi.com/journal/molecules/special_issues/ Natural_Compounds_II).

For citation purposes, cite each article independently as indicated on the article page online and as indicated below:

LastName, A.A.; LastName, B.B.; LastName, C.C. Article Title. *Journal Name* **Year**, *Volume Number*, Page Range.

ISBN 978-3-0365-3506-7 (Hbk)
ISBN 978-3-0365-3505-0 (PDF)

Cover image courtesy of Raffaele Pezzani

Contents

About the Editors

Raffaele Pezzani

Raffaele Pezzani is both a biologist and a physician with a PhD in Neuroscience at the University of Padova, Italy. He has over 70 scientific publications that have more than 1000 citations. He has been serving as an editorial board member and as an editor of reputed journals with major impacts. He is the principal investigator of his research group in Padova, with long-lasting and effective experience (more than 20 years) in biology and medicine.

Field of expertise: phytotherapy; endocrinology; oncology; nutraceutics.

Sara Vitalini

Sara Vitalini is a biologist with a PhD in Plant Biology and Crop Production at the University of Milano, Italy. She is the author of more than 60 peer-reviewed papers and serves as a regular referee, editorial board member, and guest editor of some international journals. Her research group is mainly concerned with ethnopharmacology, pharmaceutical biology, and phytochemistry.

Field of expertise: bioactivity; medicinal plants; natural products; phytochemicals.

Review

In Vitro and In Vivo Antidiabetic Potential of Monoterpenoids: An Update

Lina T. Al Kury [1,*] , Aya Abdoh [2], Kamel Ikbariah [2], Bassem Sadek [3] and Mohamed Mahgoub [4]

[1] Department of Health Sciences, College of Natural and Health Sciences, Zayed University, Abu Dhabi P.O. Box 144534, United Arab Emirates

[2] School of Medicine, Royal College of Surgeons in Ireland—Bahrain, Muharraq P.O. Box 15503, Bahrain; 18253687@rcsi.com (A.A.); 18206441@rcsi.com (K.I.)

[3] Zayed Center for Health Sciences, College of Medicine and Health Sciences, United Arab Emirates University, Abu Dhabi P.O. Box 144534, United Arab Emirates; bassem.sadek@uaeu.ac.ae

[4] Pharmacy Department, SEHA, Abu Dhabi Health Services, Abu Dhabi P.O. Box 144534, United Arab Emirates; Momahgoub@seha.ae

* Correspondence: Lina.AlKury@zu.ac.ae; Tel.: +971-50-6623975

Abstract: Diabetes mellitus (DM) is a chronic metabolic condition characterized by persistent hyperglycemia due to insufficient insulin levels or insulin resistance. Despite the availability of several oral and injectable hypoglycemic agents, their use is associated with a wide range of side effects. Monoterpenes are compounds extracted from different plants including herbs, vegetables, and fruits and they contribute to their aroma and flavor. Based on their chemical structure, monoterpenes are classified into acyclic, monocyclic, and bicyclic monoterpenes. They have been found to exhibit numerous biological and medicinal effects such as antipruritic, antioxidant, anti-inflammatory, and analgesic activities. Therefore, monoterpenes emerged as promising molecules that can be used therapeutically to treat a vast range of diseases. Additionally, monoterpenes were found to modulate enzymes and proteins that contribute to insulin resistance and other pathological events caused by DM. In this review, we highlight the different mechanisms by which monoterpenes can be used in the pharmacological intervention of DM via the alteration of certain enzymes, proteins, and pathways involved in the pathophysiology of DM. Based on the fact that monoterpenes have multiple mechanisms of action on different targets in in vitro and in vivo studies, they can be considered as lead compounds for developing effective hypoglycemic agents. Incorporating these compounds in clinical trials is needed to investigate their actions in diabetic patients in order to confirm their ability in controlling hyperglycemia.

Keywords: diabetes mellitus; anti-diabetic drugs; monoterpenes

Citation: Al Kury, L.T.; Abdoh, A.; Ikbariah, K.; Sadek, B.; Mahgoub, M. In Vitro and In Vivo Antidiabetic Potential of Monoterpenoids: An Update. *Molecules* **2022**, *27*, 182. https://doi.org/10.3390/molecules27010182

Academic Editors: Raffaele Pezzani and Sara Vitalini

Received: 14 November 2021
Accepted: 25 December 2021
Published: 29 December 2021

Publisher's Note: MDPI stays neutral with regard to jurisdictional claims in published maps and institutional affiliations.

1. Introduction

Diabetes mellitus (DM) is a chronic metabolic condition characterized by endocrine abnormalities and persistent hyperglycemia [1–3]. DM can be classified into several types based on the etiology, clinical manifestations, and management; however, persistent high levels of glucose and hyperlipidemia are the major common aspects between all the major types of DM [4–7]. Due to its complexity, DM and its complications remain a substantial medical problem. Most of the available conventional drugs, despite their therapeutic benefits, can produce some undesirable side effects and are expensive. Therefore, the search for antidiabetic drugs, specifically plant-based medicine, gains importance due to their potential therapeutic effects. Recently, several phytochemicals have been shown to possess antidiabetic properties, and many efforts have been carried out to elucidate their possible antidiabetic mechanisms. Monoterpenes are a group of secondary plant metabolites that are widespread in nature and have significant hypoglycemic effect, which has been well-documented in several experimental studies [8–11]. The aim of this review is to overview the activities and the underlying mechanisms by which monoterpenes exhibit their antidiabetic effects against DM. The novelty of this study stems from the fact that it highlights the most recent findings on the mechanisms of monoterpenes in in vitro and in vivo studies using animal models, which in turn provides a window of opportunity for future research in this field.

2. Diabetes Mellitus and Its Pathogenesis

DM is classified into four main subtypes including type 1 diabetes mellitus (T1DM), type 2 diabetes mellitus (T2DM), gestational diabetes mellitus [12], and maturity-onset diabetes of the young (MODY) [13]. T1DM, also known as insulin-dependent DM, occurs due to the destruction of insulin-producing β-cells in the pancreas via autoimmune mechanisms. Consequently, this leads to the scantiness of insulin levels and hence patients require exogenous insulin supply [14–17]. T2DM, however, is characterized by what is known as insulin resistance (IR) [18,19]. On the contrary, gestational diabetes is an acute form of DM affecting pregnant women as a result of perturbations in the levels of different hormones such as estrogen, progesterone, and cortisol [4,20]. MODY, the rarest type of DM, results from mutations in the genes involved in glucose metabolism [5,21].

Under normal conditions, the molecular events involved in insulin signaling are initiated by glucose oxidation and its facilitated diffusion into β-cell by glucose transporter 2 (GLUT2), the main transporter of glucose in the intestine, pancreas, liver, and kidney. Following the entry of glucose, it is phosphorylated by glucokinase enzyme into glucose-6-phospahte (G6P) which is considered the sensor for glucose in the pancreatic β-cell and plays a central function in insulin secretion. Further metabolism of G6P produces ATP, which inhibits ATP-sensitive K^+ channels and results in membrane depolarization and calcium influx through L-type voltage-dependent calcium channels. The rise in intracellular calcium stimulates insulin release into the bloodstream [22].

Unlike T1DM, pancreatic production of insulin in T2DM may remain intact. However, the action of insulin on various body organs is the cardinal pathological condition which occurs due to IR, causing impaired glucose uptake by muscle tissue, inhibition of hepatic glucose synthesis, and increased lipolysis (Figure 1) [23,24]. Typically, pancreatic β-cells counteract for the diminished effect of insulin through increasing the release of insulin to reverse hyperglycemia; however, as IR worsens, this compensatory mechanism becomes less effective. Consequently, the insulin-producing capacity of the pancreas progressively diminishes, leading to the eventual loss of pancreatic β-cells mass, apoptosis, and complete loss of insulin production [25–28]. It is important to mention that insulin sensitivity and/or activity is physiologically regulated by various factors such as circulating hormone levels, plasma lipids, adipokines, and their respective signaling pathways [29–31]. The interaction between those pathways and the insulin pathway tunes the sensitivity and activity of insulin.

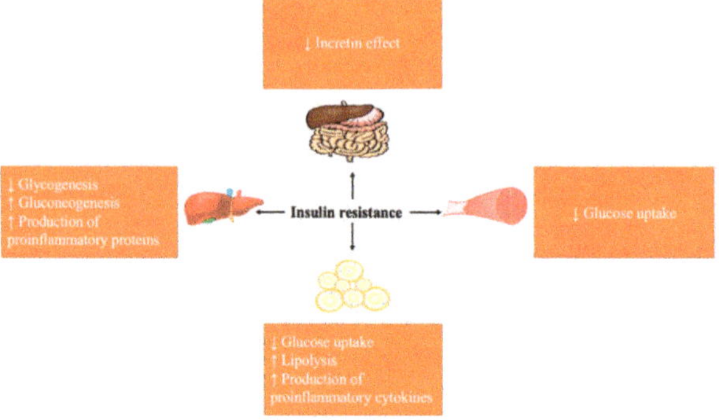

Figure 1. Effects of insulin resistance on body organs and tissues.

After a meal, approximately two-thirds of the ingested glucose is utilized by skeletal muscles through an insulin-dependent mechanism. Following its binding to its receptor, insulin enhances the migration of the glucose transporter 4 (GLUT4) from the intracellular compartment to the plasma membrane, where it facilitates the uptake of glucose [32,33]. Insulin binds to the α-subunit of the insulin receptor (INSR) and causes phosphorylation of tyrosine residues in the β-subunit, which is followed by the recruitment of different substrates such as insulin receptor substrate-1 (IRS-1), insulin receptor substrate-2 (IRS-2), and phosphoinositide 3-kinase (PI3K) [34]. In addition to the utilization by skeletal muscle, a large portion of glucose is absorbed from the intestines and taken up by hepatocytes to be converted into glycogen via the action of insulin [35]. Upon binding to its receptor, insulin causes a cascade of phosphorylation for several downstream proteins that regulate various metabolic pathways such as gluconeogenesis, glycogen synthesis, glycogenolysis, and lipid synthesis [36]. These metabolic processes are finely tuned by the actions of insulin and glucagon, where insulin promotes glucose storage and glycogen synthesis, while glucagon promotes hepatic glucose production and glycogen breakdown [35,37,38]. It is important to mention that development of hepatic IR impairs insulin response in the hepatocytes, which results in the inhibition of glycogen synthesis and the increase in hepatic gluconeogenesis, lipogenesis, and synthesis of proinflammatory proteins such as C-reactive protein (CRP). This can lead to an ongoing inflammatory state in the liver that consequently exacerbates IR [39,40].

Postprandially, insulin binding to its receptor in adipose tissue facilitates the uptake of glucose by GLUT4. This subsequently activates glycolysis, from which glycerol-3-phosphate (G3P) is produced and esterified with other fatty acid- forming triacylglycerols that act as a source of energy in the fasting state [41]. Adipose IR impairs the actions of insulin and can therefore lead to impaired uptake of free fatty acids from the blood, enhanced lipolysis, and impaired glucose uptake [42]. At the molecular level, it was found that adipose IR causes activation of a defective form of AKT that impairs the translocation of GLUT4 to the membrane and activates lipolytic enzymes, which consequently worsens hyperglycemia. On the contrary, high levels of free fatty acids in the bloodstream can lead to their accumulation in other organs such as the liver, which eventually affects insulin sensitivity and hepatic gluconeogenesis and worsens T2DM [39,41].

Adipose tissue has a dynamic endocrine role and releases different proteins known as adipokines [43,44]. It has been reported that an increase in adipose tissue size and/or mass is associated with fibrosis, hypoxia, macrophage-mediated inflammation, and pathologic vascularization [45]. High-fat diet can stimulate mitochondrial proteins and transcription factors that cause adipose tissue inflammation and dysfunction [46]. The changes in the size of adipocytes and the infiltration of immune cells induce the production of proinflammatory cytokines such as tumor necrosis factor-α (TNF-α) and interleukins (IL-6 and IL-1β). This causes a chronic state of inflammation known as metabolic inflammation which plays a significant part in IR and T2DM, consequently [47].

In addition to the above-mentioned events, two types of incretins, namely glucagon-like peptide 1 (GLP-1) and glucose-dependent insulinotropic peptide (GIP) are released from the intestine after meals to stimulate pancreatic insulin secretion [14,48,49]. These peptides have a short duration of action due to their deactivation via the dipeptidyl peptidase-4 (DPP-4) enzyme [50]. While both GLP-1 and GIP share the same effect on insulin secretion [51–53], only GLP-1 can suppress the secretion of glucagon [54,55] and exhibit growth-factor-like effects on pancreatic β-cells, stimulating insulin gene expression and insulin biosynthesis [56,57]. For this reason, GLP-1 arose as an important pharmacological target in the formulation of antidiabetic therapies via mimicking its effect [58,59]. In T2DM, the action and the level of incretins are adversely affected [60], and the glucose-dependent secretion of insulin is reduced in the fed state [61,62]. The pancreas becomes less responsive to GIP, while it remains responsive to GLP-1 [63]. This could be justified by either an uprise in the expression of DPP-4 or a reduction in the expression of GIP and GLP-1 receptors [64,65].

3. Conventional Hypoglycemic Agents

Up to this day, different pharmacologic agents have been used to limit the effects of hyperglycemia in diabetes. The mechanisms by which hypoglycemia is achieved include stimulation of insulin secretion by sulfonylureas and meglitinides, stimulation of peripheral glucose absorption by thiazolidinediones and biguanides, delay of carbohydrate absorption from the intestine by alpha-glucosidase, and reduction of hepatic gluconeogenesis by biguanides. Combining lifestyle modifications (such as diet and exercise) and using hypoglycemic agents is important to achieve long-term metabolic control and to protect against health complications caused by DM. Several studies investigated this treatment modality and showed the superiority of combining both lifestyle changes and pharmacological agents in the management of T2DM over using antidiabetic agents alone [66–72]. Various injectable and oral therapeutic agents have been developed and used clinically in the management of T2DM, each of which has a unique mechanism of action that targets different pathological events occurring in T2DM [18,73,74] (Figure 2). For example, metformin exhibits its effects by inhibiting hepatic gluconeogenesis [75–77], reducing insulin resistance in skeletal muscle and adipose tissue and promoting the release of GLP-1 [78]. Furthermore, metformin lowers plasma lipid levels by acting on the peroxisome proliferator-activated receptor (PPAR-α) pathway.

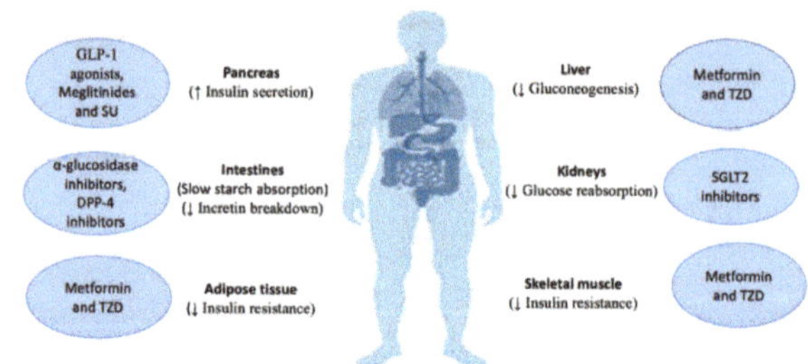

Figure 2. Mechanisms of action of hypoglycemic agents: dipeptidyl peptidase-4 (DPP-4); glucagon-like peptide 1 (GLP-1); sodium-glucose co-transporter-2 (SGLT2); sulfonylureas (SU); thiazolidinediones (TZD).

Sulfonylureas (SU) are insulin secretagogues that exert their action directly on the pancreas by inhibiting ATP-dependent potassium channels on the pancreatic β-cells, which causes cell depolarization and increases intracellular Ca^{2+} levels, resulting in insulin secretion [74]. Additionally, they inhibit the breakdown of lipids in the liver and decrease insulin clearance [79]. Although SU are associated with weight gain and hypoglycemic attacks, they remain one of the most widely used agents in the management of T2DM due to their high efficacy in reducing blood glucose levels [80]. Another group of insulin secretagogues are meglitinides, which work through a mechanism similar to that of SU [81]. However, they cause less weight gain and hypoglycemic attacks in comparison to SU, which makes them an ideal alternative for patients complaining of these side effects [74]. Thiazolidinediones (TZD) are a group of drugs that exert their effects by acting on the liver, skeletal muscle, and adipose tissue where they reduce insulin resistance and improve tissue sensitivity to insulin through the activation of PPAR-γ [82]. Moreover, TZD can also act on another isoform of PPAR-α which accounts for its lipid-lowering properties. TZD administration results in multiple actions such as maintaining pancreatic β-cell integrity, decreasing the levels of inflammatory cytokines, and increasing the levels of a protein known as adiponectin that is released from adipose tissue, causing an overall improvement in insulin sensitivity [27,83]. Alpha-glucosidase inhibitors such as acarbose, work by inhibiting the enzyme α-glucosidase, which functions via the conversion of oligosaccharides into monosaccharides in the small intestines [84]. Acarbose has a similar structure to that of oligosaccharides, which allows it to compete for the binding site in the enzyme. As a result, a delay in the postprandial absorption of glucose is achieved along with a reduction in hyperglycemia. The enzyme DPP-4 is responsible for the breakdown of incretin. Due to its physiological function, it arose as a target for the management of T2DM [85]. In 2007, sitagliptin was approved by the Food and Drug Administration (FDA), making it the first DPP-4 inhibitor. By inhibiting DPP-4, the action of incretins is prolonged, which in turn improves insulin secretion, reduces glucagon secretion, and decreases the rate of nutrient absorption into the bloodstream [86,87]. As mentioned previously, GLP-1 agonists became available for use in the management of T2DM in 2005 when the first GLP-1 agonist was approved by the FDA [88,89]. GLP-1 and GLP-1 agonists bind to the GLP-1 receptor on pancreatic β-cells and inhibit ATP-activated K^+ channels through activation of protein kinase A (PKA)-dependent pathway [90,91]. Sodium glucose co-transporter-2 inhibitors are the newest class of oral hypoglycemics that exert their action on renal tubules by suppressing the sodium glucose co-transporter-2, which reduces the reabsorption of glucose and enhances its excretion [28,92–96].

4. Monoterpenes in Diabetes

Despite the management of diabetes via the use of conventional pharmacological agents, DM and its complications remain a substantial medical problem. The majority of synthetic oral glucose-lowering drugs exhibit significant side effects and are expensive. Therefore, there has recently been a shift of interest toward exploring natural plant products for their pharmacological effects, including the treatment of diabetes. Monoterpenes are an important group of secondary metabolites that belong to the terpenoids family of natural products and have been recognized for their wide range of cellular and molecular activities that could potentially underlie their positive therapeutic index. Furthermore, their low cost, availability, low undesirable side effects, and better safety profile mark them as promising source for synthesizing new and effective agents to treat DM. For example, monoterpenes such as thymol and carvacrol are common ingredients of food and therefore, not expected to have undesirable effects. Monoterpenes are composed of two isoprene units with a general molecular formula of $C_{10}H_{16}$ and frequently contain one double bond in their structures [11]. Monoterpenes exist in over 30 known skeletons and can be classified into three subgroups: acyclic, monocyclic, and bicyclic monoterpenes [97] (Figure 3). Common examples of the acyclic form include linalool, citral, and geraniol, while important representatives of monocyclic monoterpenes include limonene, carveol, and menthol.

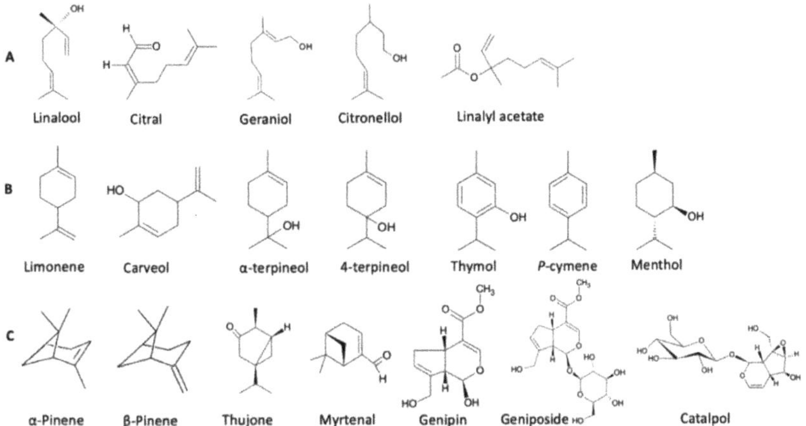

Figure 3. (**A**) Acyclic monoterpenes, (**B**) monocyclic monoterpenes, (**C**) bicyclic monoterpenes.

According to the size of their second ring, bicyclic monoterpenes can be classified into three classes. The first ring in each class is a six-membered ring while the second can be either a three (e.g., thujone), four (e.g., α- and β-pinene), or five (e.g., borneol and camphor)-membered ring. Their hydrophobic property along with their small molecular weight makes them the major components found in nearly all essential oils. Studies have reported that both natural monoterpenes and their synthetic derivatives have a vast array of pharmacological actions including anti-diabetic, hypocholesterolemic, antioxidant, antibacterial, anti-inflammatory, anti-cancer, antihistaminic, and analgesic actions [98–100]. This review highlights the potential therapeutic effects of monoterpenes in DM.

4.1. Acyclic Monoterpenes

4.1.1. Linalool

Linalool (3,7-dimethyl-1,6-octadiene-3-ol) is one of the main monoterpenoids found in herbal essential oils of many plants such as lavender (*Lavandula* spp.), which is known for its antiarrhythmic effect. Furthermore, linalool is a main component of rose (*Rosa* spp.), basil (*Ocimum basilicum*), neroli oil (*Citrus aurantium*) [101] and found in both green and black tea. Linalool has been implicated in aroma and flavoring [102]. Previous studies have reported potent antioxidant and antidiabetic activity of linalool [103,104]. Linalool was found to have favorable effects on glucose metabolism in animal models of diabetes [105]. Garba et al., 2020 investigated the antidiabetic action of lemongrass tea in T2DM model of rats. The findings of this study have shown that consumption of lemongrass reduced blood glucose levels by 60.3% [106]. Linalool, one of the main active ingredients of lemongrass, was shown to attenuate hyperglycemia and its associated complications [105]. The results were supported by higher glucose tolerance in lemongrass-treated diabetic rats in comparison to control diabetic rats which could be associated with the high content of linalool [106].

The enzymes α-amylase and α-glucosidase are accountable for the breakdown of carbohydrates and for the hydrolysis of starch into glucose pre-absorption. A reduction in hyperglycemia postprandially is due to the inhibition of α-amylase, which retards carbohydrate digestion and decreases glucose levels in the blood [107]. Therefore, inhibition of carbohydrate digestion in the gastrointestinal tract by α-amylase is one of the approaches to treat diabetes. Previous studies have demonstrated that lemongrass could effectively inhibit α-amylase and α-glucosidase activity [108]. For example, α-amylase inhibitory activity of the essential oil of lemon grass, for which linalool is the main active constituent, was found to be fifteen times higher compared to the currently used glucose lowering drug acarbose [109], while the inhibitory activity of methanol extract of lemon grass on α-glucosidase was more than 50% [108].

The uptake of glucose using rat diaphragm is a commonly used method to measure peripheral utilization of glucose in in vitro studies [110]. Linalool demonstrated dose-dependent uptake of glucose. At a concentration of 3 mM, linalool causes an increased uptake of glucose that is almost equivalent to two units of insulin. Furthermore, linalool was found to reduce oxidative stress and stimulate the activity of the antioxidant enzymes, catalase, and superoxide dismutase [105].

4.1.2. Citral

Citral (3,7-dimethylocta-2,6-dienal) is a combination of the *cis* and *trans* isomers geranial and neral, and can be found in all citrus fruits and lemon grass (*Cymbopogon citratus*) [111]. *Cymbopogon citratus* has been used over the years in Indian traditional medicine as a sedative and to treat headaches and fever [111]. Citral was shown to reduce hyperglycemia and attenuate diabetes-associated complications in earlier studies [112]. A study has reported that citral exhibits a 45.7% inhibitory effect on α-amylase at a concentration of 10 mM [98]. In streptozotocin-treated rats, citral inhibited mammalian α-amylase, with an IC_{50} of 120 μM, and reduced α-amylase levels in vivo. In addition, citral treatment caused a moderate decrease in postprandial glucose and normalized blood lipid profile [112]. Due to their direct influence on the control of energy balance via glucose uptake, lipogenesis, and lipolysis, 3T3-L1 adipocytes are among the most commonly used cell culture models to study obesity and T2DM. In 3T3-L1 adipocytes, 1 μM of citral was found to suppress the proliferation by 29.2% [98]. The results of these studies suggest that citral could be a potential antihyperlipidemic agent in diabetes. It is worth noting that several antihyperlipidemic agents such as bile acid sequestrants exhibited a promising glucose lowering activity. Such agents target bile acid receptors, which play a crucial role in metabolic diseases [113,114]. In fact, colesevelam, a bile acid sequestrant, caused a significant reduction in HbA1c and fasting plasma glucose levels. Additionally, it resulted in an increase in the levels of circulating incretins when used by patients with T2DM [115,116]. Furthermore, other types of lipid lowering agents such as fibrates [117] and cholesterol absorption inhibitors such as ezetimibe [118] have also been reported to improve glycemic control and insulin activity through unknown mechanisms.

Citral inhibits the retinaldehyde dehydrogenase enzyme and therefore raises adipose tissue retinaldehyde levels, leading to the inhibition of adipogenesis, increase in metabolic rate, reduction of weight gain, and enhanced tolerance to glucose. Treating 6-week-old male Sprague–Dawley rats with citral (10, 15, and 20 mg/kg bodyweight for 28 days) caused a noticeable reduction in the increase of body weight. Additionally, citral-treated rats had lower fasting glucose levels, enhanced glucose tolerance and metabolic rate, and lower abdominal fat accumulation [119].

Supporting the above findings, a study recently conducted by Mishra et al., 2019 revealed that citral has antidiabetic as well as dyslipidemic activities. In streptozotocin-induced diabetic rats on a high-fat diet, citral application significantly diminished glucose levels in the blood and increased insulin levels in the plasma. Moreover, citral ameliorated oxidative markers along with anti-oxidative enzymes of the pancreas, liver, and adipose tissue, and regulated the activity of the glucose-metabolic enzymes in the liver [120].

4.1.3. Geraniol

Geraniol (3,7-dimethylocta-trans-2,6-dien-1-ol) is an acyclic monoterpene alcohol found in many aromatic plants including *Cinnamomum tenuipilum* and *Valeriana officinalis*. In traditional medicine, geraniol has been used to treat many ailments including diabetes [121]. In streptozotocin-induced diabetic rats, application of geraniol for 45 days led to a significant dose-dependent increase in insulin levels and reduction in glycated hemoglobin, HbA1c. Furthermore, geraniol was found to ameliorate the function of the enzymes responsible for the metabolism and utilization of glucose. Geraniol additionally improved glycogen content in hepatocytes and preserved the histology of hepatic and pancreatic β-cells in streptozotocin-induced diabetic rats [122].

A recent work conducted by Kamble et al., 2020 demonstrated for the first time the efficacy of geraniol in inhibiting GLUT2 [123]. Inhibition of GLUT2 in the intestine, liver, and kidney plays a critical role in lowering glucose levels in the blood. Moreover, the inhibition of GLUT2 on pancreatic β-cells is anticipated to guard β-cells from glucotoxicity.

Prolonged treatment with geraniol (29.37 mm/kg body weight twice a day for 60 days) enhanced the lipid profile and HbA1c levels [123]. In another study, 1 μM of geraniol resulted in the suppression of 3T3-L1 pre-adipocyte proliferation by 19.9% [98]. It is clear from these findings that geraniol could be a novel drug in treatment of DM due to the fact that it is effective in lowering blood glucose and improving lipid profile.

4.1.4. Citronellol

Citronellol (3,7-dimethyl-6-octen-1-ol) is a linear monoterpene alcohol naturally found in about 70 essential oils, with abundance in *Cymbopogon nardus* (L.) and citrus oil [124,125]. *Cymbopogon nardus* was previously used in Chinese medicine to treat rheumatism, fever, and digestive problems [126]. Although citronellol has been reported to possess strong antioxidant, anti-inflammatory, anti-cancer, and cardioprotective properties [127,128], its role in diabetes is not well-investigated.

Oral administration of citronellol (25, 50, and 100 mg/kg bodyweight for 30 days) attenuated the hyperglycemia in streptozotocin-induced diabetic rats. Citronellol improved insulin, hemoglobin, and hepatic glycogen levels and decreased HbA1c concentration. Furthermore, there was a near to normal restoration of the altered activity of carbohydrate metabolic enzymes as well as hepatic and kidney markers. Citronellol supplement preserved the histology of hepatic cells and pancreatic β-cells in streptozotocin-treated rats [124].

Glucose uptake plays an important role in the control of plasma glucose level, thus directly influencing glucose tolerance. Treating 3T3-L1 adipocytes with 1 μM of citronellol exerted about 16% enhancement in glucose uptake [98].

4.1.5. Linalyl Acetate

Linalyl acetate (3,7-dimethylocta-1,6-dien-3-yl acetate) is the primary constituent of lavender (*Lavandula angustifolia*) which is known in folk medicine for its sedative effect [129]. It is also a main component of *Salvia sclarea* oil [130]. It has been shown that linalyl acetate possesses an anti-inflammatory effect and can restore endothelial function in rats after oxidative stress [104,131]. To date, the reported therapeutic effects of linalyl acetate in hyperglycemia are scarce. Treatment with 100 mg/kg linalyl acetate was more efficient in correcting serum glucose than the antidiabetic drug metformin in streptozotocin-induced diabetic rats. In addition, the observed cardiovascular protective and metabolic stabilization effects of linalyl acetate could be attributed to its antioxidative and anti-inflammatory properties, its increase in AMP-activated protein kinase expression, and its suppression of excess serum NO [132]. The antidiabetic effects of acyclic monoterpenes are summarized in Table 1.

Table 1. Antidiabetic effects of acyclic monoterpenes.

Compound	Model	Concentration	Antidiabetic Activities	References
Linalool	T2DM rat model	Tea preparation (0.25 g/ 100 mL and 0.5 g/100 mL for 4 weeks)	Lowered serum glucose and lipids; increased insulin sensitivity and levels of serum insulin; improved β-cell function, increased liver glycogen	[106]
	Diaphragm of streptozotocin-induced diabetic rat	3 mM	Decreased oxidative stress, increased the activity of the antioxidant enzymes catalase and superoxide dismutase.	[105]
Citral	Hemi diaphragm of Albino rat	3 mM	Increased glucose uptake	[105]
	Streptozotocin-induced diabetic rats	2, 8, 16 or 32 mg/kg body weight	Inhibited adipogenesis; increased metabolic rate, reduced weight gain; enhanced glucose tolerance.	[112]
	Streptozotocin-induced diabetic rats	2, 8, 16 or 32 mg/kg body weight	Inhibition of α-amylase.	[112]
	3T3-L1 adipocytes	1 μM	Suppression of adipocyte proliferation of by 29.2%.	[98]
	6-week-old male Sprague–Dawley rats	10, 15, and 20 mg/kg body weight for 28 days	Increased energy dissipation; reduced lipid accumulation; prevention of diet-induced obesity; improved insulin sensitivity and glucose tolerance.	[119]
	Streptozotocin-induced diabetic rats fed with high-fat diet	45 mg/kg/body weight for 28 days	Decreased blood glucose and increased plasma insulin; increased anti-oxidative enzymes of the liver, adipose tissue, and pancreas; regulated enzyme activity of glycolysis and gluconeogenesis in the liver.	[120]

Table 1. *Cont.*

Compound	Model	Concentration	Antidiabetic Activities	References
Geraniol	Streptozotocin-induced diabetic rats	100, 200, 400 mg/kg body weight for 45 days	Increased the levels of insulin and hemoglobin; decreased plasma glucose HbA1c; ameliorated carbohydrate metabolism; preserved normal histological appearance of hepatic and pancreatic β-cells.	[122]
		648.34 μM	Inhibited GLUT2 transporter.	[123]
		60 days with 29.37 mm/kg B.W. twice a day	Improved lipid profile, HbA1c levels and renal parameters.	[123]
Citronellol	Streptozotocin-induced diabetic rats	Oral administration of 25, 50, and 100 mg/kg body weight for 30 days	Improved levels of insulin, hemoglobin, and hepatic glycogen; decreased levels of HbA1c; restored altered activities of carbohydrate metabolic enzymes, hepatic and kidney markers; preserved normal histological appearance of hepatic cells and insulin-positive β-cells	[124]
	3T3-L1 adipocytes	1 μM	Enhanced glucose uptake	[98]
Linalyl acetate	Streptozotocin-induced diabetic rats	100 mg/kg	Decreased serum glucose; reduced oxidative stress and inflammation	[132]

4.2. Monocyclic Monoterpene

4.2.1. Limonene

Limonene [1-methyl-4-(1-methylethenyl)-cyclohexene] is the main constituent of oils extracted from orange, lemon, grapefruit, and other citrus plants. It is also frequently used as a food additive, and a constituent of soaps and perfumes. As per the Code of Federal Regulations, D-limonene is classified as a safe flavoring compound [133].

Limonene was shown to reduce hyperglycemia and attenuate diabetes-associated complications in earlier studies [105,134]. Inhibition of protein glycation is known to improve secondary complications in diabetes. In streptozotocin-induced diabetic rats, limonene (100 μM) revealed 85.61% reduction in protein glycation [105]. In a study conducted by Joglekar et al., 2013, limonene was shown to inhibit protein glycation by 56.3% at a concentration of 50 μM. Furthermore, BSA was used as a model protein in PatchDock studies, which have shown that limonene has the ability to bind to the key glycation sites IB, IIA, and IIB sub domains. It was concluded that limonene is a powerful inhibitor of protein glycation that exhibits its effects by a novel mechanism of stabilization of protein structure through hydrophobic interactions [135]. In 3T3-L1 adipocytes, 1 μM of (R)-(+)-limonene stimulated both the uptake of glucose and breakdown of fats. It also upregulated glucose transporter 1 (GLUT1) expression and suppressed adipose triglyceride lipase (ATGL). (R)-(+)-limonene (at mM range) also suppressed both α-amylase and α-glucosidase; however, such outcome was weak [98].

In oral streptozotocin-induced diabetic rats, administration of D-limonene (50, 100 and 200 mg/kg body weight) for 45 days resulted in a significant drop in plasma glucose and HbA1c levels. Furthermore, it resulted in a decrease in the activity of the enzymes involved in gluconeogenesis, including glucose 6-phosphatase (G6Pase) as well as fructose 1,6-bisphosphatase. On the contrary, D-limonene inhibited liver glycogen as well as the activity of the glycolytic enzyme glucokinase in diabetic rats. Such antidiabetic effects were proportional with glibenclamide [136]. These findings support the potential antihyperglycemic activity of D-limonene reported in the literature.

Limonene, alone and in combination with linalool, was found to reduce oxidative stress and intensify the activity of the antioxidant enzymes catalase and superoxide dismutase [105]. The shielding role of D-limonene against diabetes and its complications was demonstrated by Bacanlı et al., 2017 [134]. In streptozotocin-induced diabetic rats, D-limonene treatment (50 mg/kg body weight for 28 days) caused a remarkable reduction in DNA damage, glutathione reductase enzyme activity, and malondialdehyde (MDA) levels in the plasma. In addition, it caused a significant increase in the levels of glutathione and the activities of catalase, superoxide dismutase, and glutathione peroxidase. Overall, lipid levels and liver enzymes were adjusted in diabetic rats [134].

4.2.2. Carveol

The monoterpene carveol [2-methyl-5-(prop-1-en-2-yl)cyclohex-2-en-1-ol] is a component of the essential oils of *Cymbopogon giganteus* [137], *Illicium pachyphyllum* [138], and *Carum carvi* [139]. It is also present in orange peel, caraway seeds, and dill. Carveol is broadly used in perfumes, soap, and shampoos [140] and has several pharmacological activities including antioxidant, anticancer [141], antimicrobial [99], and anti-inflammatory [142] effects. In addition, carveol has a low toxicity profile [143].

Recently, the antidiabetic capacity of carveol was evaluated in in vivo, in vitro, and in silico studies. In alloxan-induced diabetic rats, carveol caused concentration- and time-dependent decrease in the level of glucose in the blood. Carveol (394.1 μM/kg) amended oral glucose tolerance surplus in rats and attenuated the HbA1c level and mediated hepatoprotective and anti-hyperlipidemic effects [8]. In in vitro assay, carveol inhibited α-amylase activity in a dose-dependent manner. In addition, carveol revealed binding affinity toward different targets associated with diabetes. In silico evaluation showed that carveol had maximum binding affinity (lowest energy value) toward the sodium-glucose co-transporter, intermediate binding affinity against fructose-1,6-bisphosphatase, and lowest affinity toward phosphoenolpyruvate carboxykinase (PEPCK) and glycogen synthase kinase-3β (PEPCK) [142]. The results of this study support the antidiabetic potential of carveol.

4.2.3. Terpineol

Terpineol [2-(4-methyl-3-cyclohexen-1-yl)-2-propanol] is a main constituent of Marjoram (*Origanum majorana*) and Maritime pine (*Pinus pinaster*) [144]. Terpineol is widely used in food and household products. Although the antioxidant and anti-inflammatory effects of terpineol have been documented previously, studies highlighting its direct antidiabetic effects are very limited. In a recent study, in vitro α-amylase enzymatic assay has shown that both α-terpineol and its structural isomer 4-terpineol caused an inhibition in its enzymatic activity by 33% (IC_{50} 1.01 ± 0.0221 mg/mL) and 40% (IC_{50} 0.838 ± 0.0335 mg/mL) respectively, when tested individually at a concentration of 0.670 mg/mL [145]. Furthermore, terpineol was recently reported to upregulate insulin sensitivity and lessen serum levels of pro-inflammatory cytokines in rats fed with high fat diet [146].

4.2.4. Thymol

Thymol (2-isopropyl-5-methylphenol), a natural phenolic monoterpenoid obtained mainly from the Thymus species (*Trachyspermum ammi* L. Sprague) [145], has been used in folk medicine to treat various ailments such as diabetes and respiratory disorders [147]. Thymol is a potent antioxidant and scavenger for hydroxyl radicals and superoxide anions [148]. Earlier studies on thymol have reported antimicrobial [149], anti-inflammatory [150], as well as anticancer potential [151].

In obese murine model fed with high fat diet, thymol treatment decreased body weight gain as well as visceral fat-pad weight. Additionally, an overall reduction in the levels of lipids was observed. The enzymes alanine aminotransferase, aspartate aminotransaminase, and lactate dehydrogenase were also reduced. Furthermore, thymol decreased the levels of glucose and leptin, decreased serum lipid peroxidation, and improved the levels of antioxidants [152]. Similarly, in mice fed with high-fat diet, thymol treatment (20, 40 mg/kg daily) significantly reversed body weight gain and peripheral insulin resistance [153]. Saravanan and Pari, (2015) tested the antihyperglycemic and antihyperlipidemic effects of thymol in diabetic C57BL/6J mice fed with high-fat diet. Daily intragastric application of thymol (40 mg/kg body weight) for 5 weeks caused a significant decline in plasma glucose, HbA1c, insulin resistance, and leptin. Moreover, it lowered the levels of plasma triglycerides, total cholesterol, free fatty acids, and low-density lipoprotein. On the other hand, thymol increased high density lipoprotein cholesterol. In addition, thymol significantly decreased hepatic lipid content including triglycerides, free fatty acids, total cholesterol, and phospholipids [154]. More recently, Saravanan and Pari [155] have shown that thymol possesses a protective role against diabetic nephropathy in C57BL/6J mice. Thymol hindered the activation of transforming growth factor-β1 (TGF-β1) and vascular endothelial growth factor (VEGF). In addition, it caused a substantial increase in the antioxidants, inhibited lipid peroxidation markers in erythrocytes and kidney tissue and reduced the lipid accumulation in kidney [156].

Supporting these results, a more recent study has shown that in streptozotocin-treated diabetic rats, 20 and 40 mg/kg thymol significantly reduced the levels of creatinine, low-density lipoprotein cholesterol, and hepatic enzymes including aspartate aminotransferase and alanine aminotransferase. Furthermore, the antioxidant enzyme status was also modulated after treatment with thymol [157]. Such findings indicate that thymol may possess promising protective and anti-diabetic activity.

The antidiabetic and antioxidant properties of *Thymus quinquecostatus* Celak, of which thymol is the main active constituent, were investigated. High level of thymol in *T. quinquecostatus* shows the potential of this plant as a crude drug and dietary health supplement. The ethyl acetate fraction of the methanol crude extract of *T. quinquecostatus* possessed a strong antioxidant activity. In hexane fraction, α-glucosidase inhibitory activity was positively correlated with the amount of thymol, indicating that thymol is the primary source for antioxidant and antidiabetic activity of *T. quinquecostatus* [158].

The inhibitory activity of thymol (5.0 mg/mL) and its synergistic effect with *p*-cymene (2.5 mg/mL) were linked to their antioxidant property by reducing the formation of advanced glycation end products. Based on spectroscopic and electrochemical methods, in combination with molecular docking study, it was found that the binding affinity of thymol with bovine serum albumin is greater than glucose. Furthermore, thymol had a protective effect toward arginine or lysine modification, indicating that it has an anti-glycation property [9].

4.2.5. *p*-Cymene

p-Cymene [1-methyl-4-(1-methylethyl) benzene] is an essential oil component found in over 100 plants, including *Cuminum cyminum* and thyme. Due to its use as an intermediate in the industrial manufacturing of food flavoring, fragrances, herbicides, and medications, *p*-Cymene possesses a significant commercial role [159,160]. *p*-Cymene is the biological precursor of carvacrol and has a structure that is similar to thymol [161]. Earlier studies have reported antioxidant [162] and anti-inflammatory [160] activity of *p*-cymene. In high fat diet-treated adult NMRI mice, *p*-cymene (20 mg/kg) led to an apparent drop in blood glucose levels as well as alanine aminotransferase and alkaline phosphatase. Additionally, a slight alteration was detected in lipid profile. Interestingly, the effects of *p*-cymene were comparable with metformin [163]. Similar findings were also observed with thymol [152].

In streptozotocin-induced diabetic rats, administration of *p*-cymene (20 mg/kg body weight for 60 days) was found to lower HbA1c. Biophysical studies showed that *p*-cymene can inhibit glycation-mediated conversion of α-helix to β-pleated sheet structure of bovine serum albumin. Interestingly, it produced antiglycation effects when used in concentrations that were 10–20 times less than the known protein glycation inhibitors, without exhibiting any toxic effects [164].

4.2.6. Menthol

Menthol [5-methyl-2-(propan-2-yl) cyclohexan-1-ol], is a component of essential oils such as eucalyptus and lemongrass and is responsible for the characteristic smell and flavor of *Mentha longifloia* that has been used traditionally in Asia for the treatment of respiratory illnesses. Menthol occurs in four isomers namely, (+)- and (−)-menthol, (+)- and (−)-neomenthol, (+)- and (−)-neoisomenthol, and (+)- and (−)-isomenthol; however, (−)-

menthol (L-menthol) is the major form that exists in nature [165]. Menthol is used to treat several conditions including the common cold and other respiratory conditions, gastrointestinal disorders, as well as musculoskeletal pain [166]. In streptozotocin-nicotinamide induced diabetic rats, application of menthol (25, 50, and 100 mg/kg/body weight) and glibenclamide (600 μg/kg/body weight) for 45 days caused a significant reduction in the overall levels of blood glucose and HbA1c. It also resulted in an increase in the level of plasma insulin, liver glycogen, and total hemoglobin. Furthermore, menthol ameliorated glucose-metabolizing enzymes, protected hepatic and pancreatic islets, and suppressed pancreatic β-cells apoptosis in diabetic rats. The later effect was coupled with a rise in anti-apoptotic Bcl-2 expression and a fall in pro-apoptotic Bax expression [167]. In a more recent study, acute oral (200 mg/kg) and topical administration (10% w/v) of menthol to high fat-fed diabetic mice were found to increase serum glucagon concentration 2 h after administration. Furthermore, chronic oral administration of menthol (50 and 100 mg/kg/day) for 12 weeks and topical application (10% w/v) prevented high fat diet-induced weight gain, adipose tissue hypertrophy, liver triacylglycerol depletion, and insulin resistance. The consequent metabolic changes of menthol in the liver and adipose tissue imitated the role of glucagon. In the liver, an increase in glycogenolysis and gluconeogenesis was observed. Additionally, the thermogenic activity of adipose tissue was boosted. Interestingly, in mature 3T3L1 adipocytes, treatment with the serum of menthol-treated mice improved the markers of energy expenditure, which was blocked following the administration of the non-competitive glucagon receptor antagonist, L-168,049. This effect shows that the increase in serum glucagon induced by menthol administration is responsible for the rise in energy expenditure [168]. The antidiabetic effects of monocyclic monoterpenes are summarized in Table 2.

Table 2. Antidiabetic effects of monocyclic monoterpenes.

Compound	Model	Concentration	Antidiabetic Activities	References
Limonene	Streptozotocin-induced diabetic rats	50 μM and 100 μM	Inhibited protein glycation.	[105,135]
	Streptozotocin-induced diabetic rat	100 μM	Increased activity of catalase and superoxide dismutase.	[105]
	3T3-L1 adipocytes	1 μM	Increased glucose uptake and lipolysis; upregulated mRNA expression GLUT1 and suppressed ATGL.	[98]
		mM range	Inhibited α-amylase and α-glucosidase	[98]
		50 mg/kg body weight	Decreased DNA damage, decreased glutathione reductase enzyme activity, decreased the levels of MDA in the plasma; increased total glutathione levels, catalase, superoxide dismutase and glutathione peroxidase activities	[134]
		50, 100 and 200 mg/kg body weight and for 45 days	Increased plasma glucose, HbA1c levels, and activities of gluconeogenic enzymes; decreased the activity of glucokinase.	[136]
Carveol	Alloxan-induced diabetic rat	394.1 μM/kg	Improved oral glucose tolerance overload in; decreased the level of HbA1c; inhibited α-amylase activity.	[8]
Terpineol	α-amylase enzymatic assay	α-terpineol 0.670 mg/mL 4-terpineol 0.670 mg/mL	Inhibited α-amylase activity Inhibited α-amylase activity	[145]
Thymol	High-fat diet induced T2DM in C57BL/6J mice	Intragastric administration of 40 mg/kg body weight daily for 5 weeks.	Decreased plasma glucose, insulin resistance, HbA1c, leptin and adiponectin; lowered the levels of plasma triglyceride, total cholesterol, free fatty acids, low density lipoprotein; increased high density lipoprotein cholesterol; decreased in hepatic lipid content.	[154]

Table 2. *Cont.*

Compound	Model	Concentration	Antidiabetic Activities	References
	C57BL/6J mice	40 mg/kg body weight daily for 5 weeks	Protected against diabetic nephropathy; inhibited the activation of transforming growth factor-β1 (TGF-β1) and vascular endothelial growth factor (VEGF), elevated antioxidants, inhibited lipid peroxidation markers in erythrocytes and kidney tissue, reduced the lipid accumulation in kidney	[156]
	High-fat diet-induced obesity in murine model	14 mg/kg orally twice a day to 4 weeks	Decreased body weight gain, visceral fat-pad weights, lipids, alanine aminotransferase, aspartate aminotransaminase, lactate dehydrogenase, glucose, insulin, and leptin levels	[152]
	Streptozotocin-induced diabetic rats	20 and 40 mg/kg thymol	Reduced creatinine, low-density lipoprotein cholesterol, and liver function-related enzymes, aspartate aminotransferase and alanine aminotransferase	[157]
	1,1-dephenyl-2-picryl-hydrazyl free radical scavenging and a reducing power assay		Increased radical scavenging activity	[158]
	In vitro α-glucosidase assay		Decreased α-glucosidase activity	[158]
p-Cymene	High-fat diet fed adult NMRI mice	20 mg/kg body weight for 6 weeks	Decreased levels of blood glucose, alanine aminotransferase and alkaline phosphatase; altered lipid profile.	[163]
	Streptozotocin-induced diabetic rat	20 mg/kg body weight for 60 days	Lowered HbA1c, prevented glycation-mediated transition of α-helix to β-pleated sheet structure of bovine serum albumin.	[164]
Menthol	High-fat diet fed mice	Acute oral (200 mg/kg) and topical administration (10% *w/v*)	Increased serum glucagon concentration;	[168]
		Chronic oral administration (50 and 100 mg/kg/day for 12 weeks) and topical Application (10% *w/v*)	Prevented high fat diet-induced weight gain, insulin resistance, adipose tissue hypertrophy and triacylglycerol deposition in liver.	[168]
	Mature 3T3L1 adipocytes treated with serum of menthol-treated mice in	0.3 μM	Improved energy expenditure markers, which was blocked in the presence of non-competitive glucagon receptor antagonist, L-168,049.	[168]
	Streptozotocin-nicotinamide -induced diabetic rats	25, 50, and 100 mg/kg/body weight for 45 days	Reduced the level of blood glucose and HbA1c; increased the level of total hemoglobin, plasma insulin, and liver glycogen.	[167]

4.3. Bicyclic Monoterpenes

4.3.1. α- and β-Pinene

α-pinene [(1S,5S)-2,6,6-trimethylbicyclo[3.1.1]hept-2-ene ((−)-α-Pinene)], is a major component of the volatile oil extract of the herb *Foeniculum vulgare* (fennel). Earlier studies have reported anti-inflammatory, hypoglycemic, and hepatoprotective effects of fennel [169]. In alloxan-induced diabetic mice, α-pinene evoked hypoglycemia at the 2nd and 24th hours of treatment. In addition, it was reported that α-pinene possesses a strong anti-inflammatory effect at a concentration of 0.50 mL/kg [169].

β-Pinene [6,6-dimethyl-2-methylidenebicyclo[3.1.1]heptane Pin-2(10)-ene] is found in numerous essential oils which possess antioxidant potential. It is one of the key constituents of the hexanic extract of *Eryngium carlinae*, commonly referred to as the "frog herb", which has been shown to reduce hyperglycemia and hyperlipidemia and exert antioxidant activity in diabetic rats [170,171].

Pistacia atlantica has been proposed to have a protective effect against conditions associated with oxidative stress [172]. A- and β-Pinene are the main constituents of gum essential oil of *P. atlantica*. Administration of the essential oil to diabetic rats caused a significant decrease in MDA and increase in glutathione, glutathione peroxidase, superoxide dismutase, and catalase [173]. In a recent study, in vitro α-amylase enzymatic assay has shown that both α-pinene (IC_{50} 1.05 ± 0.0252 mg mL^{-1}) and β-pinene (IC_{50} 1.17 ± 0.0233 mg mL^{-1}) resulted in a 32% and 29% drop in enzymatic activity respectively [145].

4.3.2. Thujone

Thujone [(1S,4R,5R)-4-methyl-1-propan-2-yl)bicyclo[3.1.0]hexan-3-one] occurs mainly as a mixture of α and β diastereoisomers in many plants including *Salvia officinalis* L. (sage), *Artemisia absinthium* L., and *Thuja occidantalis* L. Traditionally, it was used by native Americans as a remedy for several ailments such as headache, constipation, wounds, and birthmarks. This monoterpene is commonly used as a flavoring substance in food and beverages [174]. Interestingly, sage tea is known for its metformin-like effect, in particular for the essential oil fraction which contains thujone. Therefore, thujone could possibly exhibit some sort of an antidiabetic effect [175]. Nevertheless, animal studies that have pointed to the potential antidiabetic activity of thujone are limited. For example, in soleus muscles, palmitate-induced insulin resistance was assessed in the presence of thujone (0.01 mg/mL). Initially, insulin resistance was induced with high concentrations of palmitate [176]. Subsequently, the ability of thujone to restore sensitivity to insulin while preserving high palmitate concentrations was tested. The findings of this study indicated that thujone can ameliorate palmitate oxidation and prevent palmitate-induced insulin resistance via AMP-activated protein kinase (AMPK)-dependent pathway that involves partial restoration of insulin-stimulated translocation of GLUT4 [177]. Al-Haj Baddar, et al., 2011 demonstrated that oral administration of 5 mg/kg body weight of thujone in diabetic rats over 28 days can restore the normal levels of cholesterol and triglycerides [175]. While this finding is promising, the adverse effects of thujone necessitates careful analysis of the results. The narrow therapeutic window of thujone is evident in 2-year studies in rats and mice due to the dose-dependent incidence of seizures [178].

4.3.3. Myrtenal

Myrtenal [6,6-dimethylbicyclo[3.1.1]hept-2-ene-2-carbaldehyde] is a natural monoterpene present in plants such as pepper, mint, cumin, and eucalyptus and used as a food additive. It has various biological effects and acts as an antioxidant, anticancer agent, cyclooxygenase-inhibitor, and immunostimulant [179,180]. Recently, it was found that myrtenal exhibits antihyperglycemic, antihyperlipidemic, hepatoprotective, and β-cell protective effects [181,182].

Oral treatment with myrtenal (20, 40, and 80 mg/kg body weight) resulted in a significant depletion in plasma glucose and HbA1c in diabetic rats treated with streptozotocin. Additionally, there was a rise in insulin, hemoglobin (Hb), and glycogen levels in the liver and muscles. An enhancement of the main enzymes involved in carbohydrate metabolism (hexokinase, glucose-6-phosphatase, fructose-1,6-bisphosphatase, and glucose-6-phosphate dehydrogenase) was observed. Furthermore, myrtenal enhanced hepatic enzyme function and restored islet cells and liver histology [182].

In parallel to the above findings, another study has shown that myrtenal-treated diabetic rats displayed a reduction in plasma glucose and a simultaneous rise in plasma insulin. Additionally, myrtenal caused an upregulation in the expression of proteins involved in insulin signaling such as IRS2 (insulin receptor substrate 2), Akt, and GLUT2 in hepatocytes as well as IRS2, Akt, and GLUT4 in skeletal muscle [183].

Recently, the influence of myrtenal on oxidative stress, inflammation, and lipid peroxidation was tested on diabetic rats treated with streptozotocin. Oral administration of 80 mg/kg body weight of myrtenal for four weeks significantly decreased the diabetes-associated alterations in hepatic and pancreatic cells. This includes antioxidant levels, lipid peroxidation, and proinflammatory cytokines such as TNF-α, IL-6, and the p65 subunit of nuclear factor-kappa B (NF-kB p65). The findings of this work indicated that myrtenal can potentially act as an antioxidant and anti-inflammatory compound against oxidative stress and inflammation associated with diabetes [184].

4.3.4. Genipin and Geniposide

The iridoids genipin [methyl-1-hydroxy-7-(hydrozymethyl)-1,4a,5,7 tetrahydrocyclope nta[c]pyran-4-carboxylate] and geniposide [methyl (1S,4aS,7aS)-7-(hydroxymethyl)-1-[(2S, 3R,4S,5S,6R)-3,4,5-trihydroxy-6-(hydroxymethyl)oxan-2-yl]oxy-1,4a,5,7a tetrahydrocyclope nta[c]pyran-4-carboxylate] exist in many plants as secondary metabolites. The basic structural skeleton of iridoids is a cyclopentane-[C]-pyran ring fused with a six-membered heterocycle oxygenate [185]. At C1 position of the pyran ring, the hydroxyl group can be replaced with a sugar moiety to form the genipin glycoside, geniposide. Genipin is found in unripe *Genipa americana* L. (genipa) fruits, while geniposide is found in the fruits of *Gardenia jasminoides* J. (gardenia, Rubiaceae family) that has been used in traditional Chinese medicine for its choleretic and hepatoprotective activity. Earlier studies have shown that geniposide is converted to genipin by the intestinal microflora enzymes, which indicates that genipin is the main form of geniposide in circulating blood [186].

Genipin was shown to have anticancer, anti-inflammatory, hepatoprotective as well as antioxidative activity [187]. Geniposide exhibits many biological effects including antioxidative stress [188], anti-inflammatory [189] and antiapoptosis [190]. In addition, studies have shown that it exerts a promising anti-diabetic activity. For example, in C(2)C(12) myotubes, genipin (10 μM) stimulated glucose uptake in a time- and concentration-dependent manner. It also enhanced GLUT4 translocation to the cell surface and increased the phosphorylation of IRS-1, AKT, and GSK3β. Genipin also caused a rise in ATP levels, which inhibited ATP-dependent K^+ channels and resulted in elevated cytoplasmic Ca^{2+} content [191].

Administration of 25 mg/kg of genipin per day for 12 days to aged rats ameliorated systemic as well as hepatic insulin resistance. It also alleviated hyperinsulinemia, hyperglyceridemia, and hepatic steatosis. Furthermore, genepin reduced hepatic oxidative stress as well as mitochondrial dysfunction. It also improved insulin sensitivity, suppressed cellular ROS overproduction, and alleviated the reduction in mitochondrial membrane potential (MMP) and ATP levels [192]. Guan et al., 2018 studied the effect of genipin on obesity and lipid metabolism in diet-induced obese rats. The findings of this study demonstrated that genipin caused an overall drop in body weight and total fat. Additionally, it reversed insulin and glucose intolerance, dyslipidemia, adipocyte hypertrophy, and hepatic steatosis. It also caused a reduction in serum TNF-α levels [193]. Similar results were reported by Zhong et al., 2018, where genipin alleviated hyperlipidemia and hepatic steatosis in high-fat diet fed mice [194].

Earlier study has shown that geniposide exhibits anti-obesity, anti-oxidant, and insulin resistance-alleviating effects. Additionally, it was shown to adjust abnormal lipid metabolism. In spontaneously obese T2DM TSOD mice, geniposide caused a reduction in visceral fat and body weight and improved lipid metabolism. Furthermore, geniposide had a positive therapeutic impact on glucose tolerance and hyperinsulinemia. Interestingly, geniposide had a direct effect on the liver. In mice treated with free fatty acids, genipin not only inhibited lipid accumulation hepatocytes, but also improved the expression of PPARα [195].

Emerging body of evidence revealed that lipotoxicity may be a leading cause of pancreatic β-cell apoptosis and oxidative stress in diabetes. Increased levels of plasma-free fatty acids not only induce cytotoxicity in pancreatic β-cells leading to apoptosis, but also promote mitochondrial perturbation, resulting in oxidative stress. In pancreatic INS-1 cells, application of geniposide (1 or 10 μM) for 7 h alleviated β-cell apoptosis induced by palmitate and activated caspase-3 expression. Furthermore, geniposide improved glucose-induced insulin secretion via the activation of GLP-1 receptor [196]. Another study has demonstrated that when INS-1 cells are chronically exposed to elevated glucose concentrations, insulin secretion was impaired and cell apoptosis was observed. This change was reversed by the application of geniposide [197]. However, the effects of geniposide on insulin secretion after acute exposure to glucose was dependent on glucose concentration. When INS-1 cells were acutely stimulated with high glucose concentrations, the protective effect of geniposide was diminished. This could be attributed to the capability of geniposide to protect the cells from damage resulting from prolonged release of insulin and glucotoxicity under high glucose load [198].

An earlier study has assessed the direct effect of geniposide on β-cell function using both rat pancreatic islets and dispersed single islet cells [199]. Geniposide was found to mediate insulin release via the activation of GLP-1R and adenylyl cyclase (AC)/cAMP signaling pathway. In general, the effect of GLP-1R agonists is linked to cAMP signaling [200]. In this study, PKA suppression inhibited geniposide-mediated secretion of insulin, implying that geniposide exhibited its actions mainly via the activation of cAMP-dependent PKA [199]. It is well known that activation of pancreatic voltage-gated K^+ channels repolarizes cells and suppresses insulin release. Therefore, inhibition of these channels could prolong the duration of the action potential and promote glucose-dependent insulin secretion [201]. Interestingly, Zhang et al., 2016 stated that geniposide can inhibit voltage-gated K^+ channels in a concentration-dependent manner. This was diminished upon treating β-cells with GLP-1R and PKA inhibitors. Collectively, the findings of this study suggest that inhibition of voltage-gated K^+ channels is coupled to geniposide-induced insulin release by activating the downstream of GLP-1/cAMP/PKA signaling pathway [199].

4.3.5. Catalpol

Catalpol[(2S,3R,4S,5S,6R)-2-[[(1S,2S,4S,5S,6R,10S)-5-hydroxy-2-2(hydroxymethyl)-3,9-dioxatricyclo[4.4.0.0^{2,4}]dec-7-en-10-yl]oxy]-6-(hydroxymethyl) oxane-3,4,5-triol, is an iridoid glucoside isolated from the root of *Rehmannia glutinosa*, which has previously been used in traditional Chinese medicine to manage hyperglycemia for decades. Earlier studies have reported that catalpol exhibits an antidiabetic potential, which is attributed to its antioxidant property. In animal models, the oral dose of catalpol that caused a significant antidiabetic effect ranged from 2.5 to 200 mg/kg and 10 to 200 mg/kg in rats and in mice, respectively [202].

Catalpol acts through several mechanisms that affect insulin-sensitive organs like the liver, skeletal muscle, adipose tissue, and pancreas. Furthermore, catalpol adjusts several genes and proteins in the pancreas, skeletal muscle, and adipose tissue that have a crucial role in the management of diabetes [202].

In high-fat and streptozotocin-treated diabetic C57BL/6J mice, administration of 100 and 200 mg/kg catalpol over four weeks decreased the p (Ser 307)-IRS-1 and increased the p (Ser 347)-AKT and p (Ser 9)-GSK3 β. Such effect adjusted the impaired insulin pathway in the liver through PI3K/AKT pathway. Furthermore, catalpol prevented gluconeogenesis by enhancing the activity of AMPK and inhibiting PEPCK and G6Pase protein expression [203]. In spontaneous diabetic db/db mice treated with 80 or 160 mg/kg catalpol for four weeks, p-AMPK and GLUT expression were significantly enhanced in liver, skeletal muscle, as well as adipose tissue, which promoted the uptake of glucose into the cells [204].

In spontaneous diabetic db/db mice, the lowered expression of IRS-1 resulted in negative regulation of insulin signaling cascades, as IRS-1 is an important ligand in activating the PI3K/AKT pathway. Furthermore, decreased activity of isocitrate dehydrogenase 2 (IDH2), an enzyme that catalyzes the citrate cycle, attenuates glucose metabolism and ATP production. It is well-known that glucose-6-phosphate 1-dehydrogenase (G6PD2) catalyzes the pentose phosphate pathway that utilizes glucose to produce NADPH and ribose-5-phosphate. The downregulation of G6PD2 enzyme decreases the glucose metabolism. On the other hand, upregulation of suppressor of cytokine signaling 3 (SOCS3) enzyme can inhibit the tyrosine phosphorylation of the insulin receptor, leading to the suppression of insulin signaling pathway [205–207]. Liu et al., 2018 reported that oral treatment with catalpol (25, 50, 100, and 200 mg/kg) upregulated IRS-1, IDH2, and G6PD2 expression, and downregulated SOCS3. Collectively, the findings indicate that catalpol can increase glucose metabolism through accelerating the citrate cycle and pentose phosphate pathway and promoting insulin signaling pathway [204].

The antidiabetic effects of bicyclic monoterpenes are summarized in Table 3. The mechanisms of action of the above-mentioned monoterpenes are summarized in Figure 4.

Table 3. Antidiabetic effects of bicyclic monoterpenes.

Compound	Model	Concentration	Antidiabetic Activities	References
α-Pinene	Alloxan-induced diabetic mice	i.p. injection of 0.25 mL/kg α-pinene	Evoked hypoglycemia activity at the 2nd and 24th hours.	[10]
	α-amylase enzymatic assay	0.670 mg/mL	Inhibited α-amylase activity.	[145]
β-Pinene	Streptozotocin-induced diabetic rat	Oral administration of 30 mg/kg of hexanic extract (17.53% β-pinene) daily for 7 weeks	Ameliorated hyperglycemia and oxidative damage.	[170]
	α-amylase enzymatic assay	0.670 mg/mL	Inhibited α-amylase activity.	[145]
Thujone	Palmitate-induced insulin resistance in soleus muscles of male Sprague-Dawley rats	0.01 mg/mL (incubation for 6 h in presence of palmitate)	Restored insulin sensitivity; ameliorated palmitate oxidation and rescued palmitate-induced insulin resistance via AMPK-dependent mechanism involving partial restoration of insulin-stimulated GLUT4 translocation.	[177]
	Alloxan monohydrate-induced diabetic rats	5 mg/kg thujone for 28 days	Adjusted cholesterol and triglyceride levels to normal levels.	[175]
Myrtenal	Streptozotocin-induced diabetic rat	80 mg/kg body weight (orally)	Adjusted antioxidant levels, lipid peroxidation, and proinflammatory cytokines (TNF-α, IL-6, NF-kB p65).	[184]
	Streptozotocin-induced diabetic rat	80 mg/kg body weight (orally)	Reduced plasma glucose; increased plasma insulin; upregulated IRS2, Akt, and GLUT2 in hepatocytes; upregulated IRS2, Akt, and GLUT4 in skeletal muscle.	[183]
	Streptozotocin-induced diabetic rat	20, 40, and 80 mg/kg body weight (orally)	Depleted plasma glucose and HbA1c; increased insulin, Hb, and hepatic and muscle glycogen; enhanced carbohydrate metabolic enzymes and hepatic enzyme function; restored islet cells and liver histology.	[182]
Genipin	C2C12 myotubes	10 μM	Promoted GLUT4 translocation to the cell surface; increased the phosphorylation of IRS-1, AKT, and GSK3β; increased ATP levels which inhibited ATP-dependent potassium channels; increased cytoplasmic calcium.	[191]
	Aging rats	25 mg/kg genipin or vehicle once daily for 12 days	Adjusted insulin resistance; ameliorated systemic and hepatic insulin resistance; alleviated hyperinsulinemia, hyperglyceridemia, and hepatic steatosis; reduced hepatic oxidative stress and mitochondrial dysfunction; improved insulin sensitivity; inhibited cellular ROS overproduction; alleviated the reduction of levels of MMP and ATP.	[192]
	Diet-induced obese rats		Reduced body fat; Reversed dyslipidemia, glucose and insulin intolerance, adipocyte hypertrophy, and hepatic steatosis. Reduced serum tumor necrosis factor-α levels.	[193]
	Diet-induced obese mice	5 or 20 mg/kg/day	Alleviated high-fat diet induced hyperlipidemia and hepatic steatosis.	[194]
Geniposide	Spontaneously obese T2DM TSOD mice		Caused a reduction in body weight and visceral fat accumulation, improved lipid metabolism and intrahepatic lipid accumulation, adjusted hyperinsulinemia glucose tolerance, inhibited the accumulation of lipid in hepatocytes of free fatty acid treated rats, improved the expression of PPARα.	[195]

Table 3. *Cont.*

Compound	Model	Concentration	Antidiabetic Activities	References
	Pancreatic INS-1 cells	1 or 10 µM for 7 h	Alleviated β-cell apoptosis induced by palmitate, activated caspase-3 expression, improved glucose stimulated insulin secretion by activating GLP-1R	[198]
	Pancreatic INS-1 cells	1 or 10 µM for 5 days	Increased insulin secretion in β-cells and decreased apoptosis	[197]
	Pancreatic islets and dispersed single islet cells from Male Sprague-Dawley (SD) rat	1 and 10 µM	Inhibition of voltage-dependent potassium, activated GLP-1/cAMP/PKA signaling pathway and insulin secretion.	[199]
Catalpol	High-fat diet and streptozotocin-induced diabetic C57BL/6J mice	100 or 200 mg/kg, p.o., four weeks	Adjusted the impaired insulin pathway in the liver through PI3K/AKT pathway (decreased p (Ser 307)-IRS-1 and increased the p (Ser 347)-AKT and p (Ser 9)-GSK3 β), prevented gluconeogenesis by enhancing the activity of AMPK and inhibiting PEPCK and glucose G6Pase protein expression.	[203]
	db/db mice	25, 50, 100, and 200 mg/kg (orally)	Upregulated the expression of IRS-1, IDH2, and G6PD2, and downregulated the expression of the SOCS3.	[205]
	High fat diet and streptozotocin-induced diabetic mice	100 or 200 mg/kg for four weeks (orally)	Upregulated SOD2 and GSH-Px, suppressed the serum level of MDA and NOX4.	[203]
	Glucosamine-treated HepG2 cells	20–80 µM	Increased the levels of SOD and GSH-Px, decreased the MDA level and NOX4 protein expression.	[203]
	C57BL6/J mice fed with high fat diet	200 mg/kg for 4–8 weeks	Increased skeletal muscle insulin sensitivity by activating IRS-1/AKT/GLUT4.	[203]
	db/db mice	200 mg/kg for 8 weeks	Augmented myogenesis by increasing expression of MyoD, MyoG and MHC expressions	[204]
	High glucose treated C2C12 cells	10, 30, 100 µM for 24 h	Increased MyoD and MyoG mRNA/protein levels.	[203]
	Skeletal muscle of db/db mice	200 mg/kg/day for 8 weeks (orally)	Increased number of mitochondria, mitochondrial DNA levels, and expression of genes involved in mitochondrial biogenesis.	[205]

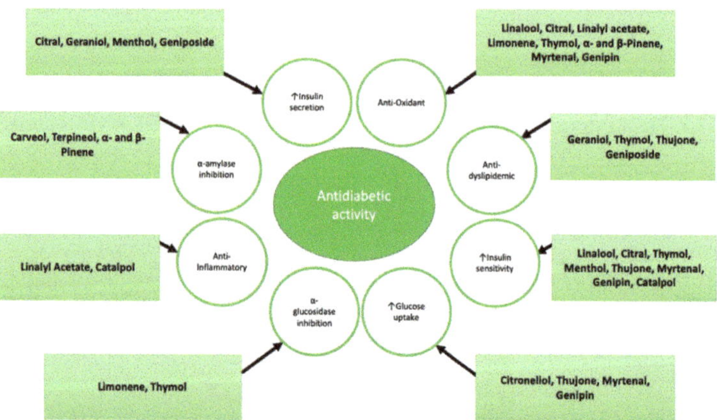

Figure 4. Mechanisms of action of different monoterpenes.

5. Structure–Activity Relationship

Although monoterpenes possess multiple pharmacological and molecular mechanisms of action, their structure–activity relationship has not been fully elucidated yet. In vitro and in vivo data summarized in this review demonstrate that there is a wide range of mechanisms of action by which monoterpenes exhibit their antidiabetic effects. These include (1) inhibition of α-amylase and α-glucosidase, (2) stimulation of insulin release, (3) stimulation of glucose uptake, (4) increase in insulin sensitivity, (5) inhibition of gluconeogenesis, (6) reduction in cellular oxidative stress, (7) reversal of dyslipidemia, (8) increase in anti-inflammatory activity, and (9) inhibition of pancreatic β-cell apoptosis. The current review discusses the antidiabetic effect of different monoterpenes using in vitro, as well as in vivo models, in which oxidative metabolism is an essential factor to consider. For example, *p*-cymene could be hydroxylated as a result of oxidative metabolism at a position comparable to the hydroxyl group position in α-terpineol. Hydroxylation of *p*-cymene also leads to the biosynthesis of an entirely different monoterpene, namely thymol, in which the antioxidant and antidiabetic properties are attributed to the pharmacophore of the phenolic hydroxyl group in its chemical structure. Therefore, it could be highly anticipated that structural modification of the parent molecule (*p*-cymene), such as the introduction of hydroxyl group, enhances its antioxidant activity. This is also applicable to other compounds, such as citral, which contains an aldehyde group. It is well-known that aldehydes are highly resistant to oxidative deterioration [208]. Citral has a high tendency to be oxidized and therefore, the aldehyde group could be easily converted to a carboxylic acid group. Such potential metabolism of the aldehyde group is also applicable to the compound myrtenal. Moreover, limonene is a precursor for carveol. Considering the carbon numbering relative to limonene, the presence of an oxygenated group at carbon-6 conjugated to a double bond at carbon-1 and an isopropenyl group at carbon-4 were found to be the major chemical features relevant for activity and potency of carveol. For example, compared to limonene and other limonene derivatives, carveol significantly decreased lipopolysacharide (LPS)-induced nitric oxide (NO) production in murine macrophages. This anti-inflammatory activity was credited to the chemical features that are absent in other compounds [209]. Earlier studies have attributed the effect of monoterpenes to their volatility [11], hydrophobicity [210], and non-specific [211] and non-competitive [212] mechanisms of action. The lipophilic characteristic of the monoterpene skeleton combined with the nature of the functional group is essential for its activity. It has been proposed that the rank of activity is the greatest for aldehydes (e.g., citral), followed by alcohols (e.g., linalool and geraniol), followed by hydrocarbons (e.g., p-cymene and limonene). It should also be noted that some monoterpenes (e.g., catalpol) that exist in glycosylated form are very polar, which also affects their biological activity [213]. Compounds that contain phenolic groups are known to confer protection against the deleterious effects of free radicals both by absorbing or neutralizing free radicals and by augmenting endogenous antioxidants [214]. Additionally, studies have shown that the presence of a phenolic structural moiety displays potent antioxidant effects and/or direct radical scavenging that can account for the antidiabetic activity of monoterpenes. Thymol and 4-terpineol are typical examples that have been reported for their antihyperglycemic effects [145,154]. Supporting these findings, Zunino and Zygadlo (2004) concluded that most potent monoterpenes are those that are alcohols and phenols [215]. A study conducted by Javan and Javan (2014) evaluated the structure-radical scavenging activity of thymol derivatives. It was concluded that the presence of an unsaturated double bond is the main factor that determines the antioxidant and radical scavenging activity of the monoterpene derivatives [216]. Interestingly, it was shown that the incorporation of monoterpenes into other groups such as flavonoids augments their antioxidant effect [217]. Whether a monoterpene is a simple hydrocarbon (e.g., *p*-cymene and limonene), hydroxy derivative, or phenolic, a potential antidiabetic effect has been reported at low doses. However, due to the wide range of variations in experimental settings (e.g., range of concentrations tested, modes of drug administration, cell type, and animal models used), in addition to controversial in vitro and in vivo findings and their species dependency, direct comparison of in vitro and in vivo potency between the various subtypes of monoterpenes is difficult. In fact, more in vivo studies should be undertaken to confirm in vitro findings. Furthermore, a full-scale pharmacokinetic profiling is needed to interpret the inconsistency between results observed in in vitro and in vivo preclinical studies.

Based on the above, structure–activity relationship among monoterpenes can be made only when the effect of each compound (acyclic, monocyclic, and bicyclic) is investigated using a single target in vitro, in which pharmacokinetic profile (absorption, distribution, metabolism, and elimination) is excluded. In addition, an in silico molecular docking approach must be used to predict the molecular mechanism of action of each monoterpene on its potential target related to diabetes. Determination of the order of potency of the monoterpenes under standardized conditions, will help in correlating the activity with structural features to identify the relevant structural determinants of antidiabetic activity.

6. Summary and Conclusions

DM is a disease associated with high rates of morbidity and mortality and one of the leading causes of death in the world. The major complications associated with diabetes mellitus are classified as microvascular (including retinopathy, neuropathy, and nephropathy) and macrovascular (including cardiovascular myopathy and cerebrovascular diseases) [218,219]. Hyperglycemia plays an important role in the onset and development of these complications, mainly by generating reactive oxygen species (ROS) which causes lipid peroxidation and membrane damage. Cardiovascular (CV) risk factors such as obesity, hypertension, and dyslipidemia are common in patients with DM, placing them at increased risk for cardiac events. DM can be controlled by targeting multiple components like glucose transport, insulin signaling, insulin secretion, lipid regulation, inflammation, and oxidation. Despite the availability of different classes of antidiabetic agents, side effects like weight gain and hypoglycemia affect patients' adherence to therapy. Novel medicinal compounds can be synthesized and designed for the treatment of several diseases based on the chemical structure of these molecules. Monoterpenes are the main components of essential oils and have been recognized for their wide range of cellular and molecular activities that could potentially underlie their positive therapeutic index. Due to their abundance in occurrence, various biological activities, and high safety profile, monoterpenes became central for research and development around the globe. In this article, the pathogenesis of DM and the classes of antidiabetic agents used for the management of the disease were discussed. Moreover, we summarized the effects of selected acyclic, monocyclic, and bicyclic monoterpenes that are supposed to possess a potential role in the management of DM. Based on the fact that monoterpenes show structural complexity and diversity, comparison of the net antidiabetic effect between the three subcategories of monoterpenes cannot be made due to inconsistency in dose, duration, mode of drug administration, target tissue, and animal model used. To accurately determine which category of monoterpenes (acyclic, monocyclic, bicyclic) can exhibit the greatest antidiabetic effect, a comparison must be made using the exact same experimental conditions (concentration used, cell and tissue type targeted, etc.). However, based on extensive review of experimental studies, it has been proposed that the rank of activity is the greatest for aldehydes (e.g., the acyclic monoterpene citral), followed by alcohols (e.g., the acyclic monoterpenes linalool and geraniol), followed by hydrocarbons (e.g., the monocyclic monoterpenes p-cymene and limonene) [213]. Due to the fact that monoterpenes provide a promising area of research, further studies with regards to their structure-activity relationship as well as structural modification are crucial to maximize their therapeutic effects. Their use in combination with other monoterpenes or natural compounds should be carried out in the future to fill in the gaps. Additionally, more research is still needed to investigate the actions of these molecules on diabetic patients in order to confirm their therapeutic ability in controlling hyperglycemia and dyslipidemia caused by the disease.

Author Contributions: L.T.A.K. and M.M.; writing—original draft preparation, L.T.A.K. and M.M.; writing—review and editing, A.A.; writing—review and editing, K.I.; writing—review and editing, B.S.; writing—review and editing. All authors have read and agreed to the published version of the manuscript.

Funding: This research received no external funding.

Institutional Review Board Statement: Not applicable.

Conflicts of Interest: The authors declare no conflict of interest.

References

1. American Diabetes Association. 4. Lifestyle Management: Standards of Medical Care in Diabetes—2018. *Diabetes Care* **2018**, *41* (Suppl. 1), S38–S50. [CrossRef]
2. American Diabetes Association. Standards of medical care in diabetes–2014. *Diabetes Care* **2014**, *37* (Suppl. 1), S14–S80. [CrossRef]
3. Adeghate, E. Diabetes mellitus-multifactorial in aetiology and global in prevalence. *Arch. Physiol. Biochem.* **2001**, *109*, 197–199. [CrossRef]
4. Chiefari, E.; Arcidiacono, B.; Foti, D.; Brunetti, A. Gestational diabetes mellitus: An updated overview. *J. Endocrinol. Investig.* **2017**, *40*, 899–909. [CrossRef] [PubMed]
5. Anik, A.; Catli, G.; Abaci, A.; Bober, E. Maturity-onset diabetes of the young (MODY): An update. *J. Pediatr. Endocrinol. Metab.* **2015**, *28*, 251–263. [CrossRef] [PubMed]
6. American Diabetes Association. Diagnosis and Classification of Diabetes Mellitus. *Diabetes Care* **2010**, *33* (Suppl. 1), S62–S69.
7. Diabetes Control and Complications Trial Research Group. Effect of intensive diabetes treatment on the development and progression of long-term complications in adolescents with insulin-dependent diabetes mellitus: Diabetes Control and Complications Trial. *J. Pediatr.* **1994**, *125*, 177–188.
8. Ahmed, M.S.; Khan, A.U.; Kury, L.T.A.; Shah, F.A. Computational and Pharmacological Evaluation of Carveol for Antidiabetic Potential. *Front. Pharmacol.* **2020**, *11*, 919. [CrossRef] [PubMed]
9. Zielińska-Błajet, M.; Feder-Kubis, J. Monoterpenes and Their Derivatives-Recent Development in Biological and Medical Applications. *Int. J. Mol. Sci.* **2020**, *21*, 7078. [CrossRef]
10. Özbek, H.; Yilmaz, B. Anti-inflammatory and hypoglycemic activities of alpha-pinene. *ACTA Pharm. Sci.* **2017**, *55*, 7. [CrossRef]
11. Habtemariam, S. Antidiabetic Potential of Monoterpenes: A Case of Small Molecules Punching above Their Weight. *Int. J. Mol. Sci.* **2017**, *19*, 4. [CrossRef] [PubMed]
12. Salsali, A.; Nathan, M. A Review of Types 1 and 2 Diabetes Mellitus and Their Treatment with Insulin. *Am. J. Ther.* **2006**, *13*, 349–361. [CrossRef] [PubMed]
13. Urakami, T. Maturity-onset diabetes of the young (MODY): Current perspectives on diagnosis and treatment. *Diabetes Metab. Syndr. Obes.* **2019**, *12*, 1047–1056. [CrossRef]
14. Lin, X.; Xu, Y.; Pan, X.; Xu, J.; Ding, Y.; Sun, X.; Song, X.; Ren, Y.; Shan, P.F. Global, regional, and national burden and trend of diabetes in 195 countries and territories: An analysis from 1990 to 2025. *Sci. Rep.* **2020**, *10*, 14790. [CrossRef]
15. van Belle, T.L.; Coppieters, K.T.; von Herrath, M.G. Type 1 diabetes: Etiology, immunology, and therapeutic strategies. *Physiol. Rev.* **2011**, *91*, 79–118. [CrossRef] [PubMed]
16. Gillespie, K.M. Type 1 diabetes: Pathogenesis and prevention. *CMAJ Can. Med. Assoc. J. J. L'association Med. Can.* **2006**, *175*, 165–170. [CrossRef]
17. Gerich, J.E. Insulin-dependent diabetes mellitus: Pathophysiology. *Mayo Clin. Proc.* **1986**, *61*, 787–791. [CrossRef]
18. Chaudhury, A.; Duvoor, C.; Reddy Dendi, V.S.; Kraleti, S.; Chada, A.; Ravilla, R.; Marco, A.; Shekhawat, N.S.; Montales, M.T.; Kuriakose, K.; et al. Clinical Review of Antidiabetic Drugs: Implications for Type 2 Diabetes Mellitus Management. *Front. Endocrinol.* **2017**, *8*, 6. [CrossRef]
19. Goran, M.I.; Ball, G.D.; Cruz, M.L. Obesity and risk of type 2 diabetes and cardiovascular disease in children and adolescents. *J. Clin. Endocrinol. Metab.* **2003**, *88*, 1417–1427. [CrossRef]
20. Gilmartin, A.B.H.; Ural, S.H.; Repke, J.T. Gestational Diabetes Mellitus. *Rev. Obstet. Gynecol.* **2008**, *1*, 129–134.
21. Amed, S.; Oram, R. Maturity-Onset Diabetes of the Young (MODY): Making the Right Diagnosis to Optimize Treatment. *Can. J. Diabetes* **2016**, *40*, 449–454. [CrossRef] [PubMed]
22. Czech, M.P. Insulin action and resistance in obesity and type 2 diabetes. *Nat. Med.* **2017**, *23*, 804–814. [CrossRef] [PubMed]
23. Weyer, C.; Bogardus, C.; Mott, D.M.; Pratley, R.E. The natural history of insulin secretory dysfunction and insulin resistance in the pathogenesis of type 2 diabetes mellitus. *J. Clin. Investig.* **1999**, *104*, 787–794. [CrossRef] [PubMed]
24. Stumvoll, M.; Goldstein, B.J.; van Haeften, T.W. Type 2 diabetes: Principles of pathogenesis and therapy. *Lancet* **2005**, *365*, 1333–1346. [CrossRef]
25. Chatterjee, S.; Khunti, K.; Davies, M.J. Type 2 diabetes. *Lancet* **2017**, *389*, 2239–2251. [CrossRef]
26. Nyenwe, E.A.; Kitabchi, A.E. The evolution of diabetic ketoacidosis: An update of its etiology, pathogenesis and management. *Metab. Clin. Exp.* **2016**, *65*, 507–521. [CrossRef]
27. Marín-Peñalver, J.J.; Martín-Timón, I.; Sevillano-Collantes, C.; del Cañizo-Gómez, F.J. Update on the treatment of type 2 diabetes mellitus. *World J. Diabetes* **2016**, *7*, 354–395. [CrossRef]
28. Weir, M.R. The kidney and type 2 diabetes mellitus: Therapeutic implications of SGLT2 inhibitors. *Postgrad. Med.* **2016**, *128*, 290–298. [CrossRef]
29. Cai, C.; Qian, L.; Jiang, S.; Sun, Y.; Wang, Q.; Ma, D.; Xiao, G.; Li, B.; Xie, S.; Gao, T.; et al. Loss-of-function myostatin mutation increases insulin sensitivity and browning of white fat in Meishan pigs. *Oncotarget* **2017**, *8*, 34911–34922. [CrossRef]
30. Nigro, E.; Scudiero, O.; Ludovica Monaco, M.; Polito, R.; Schettino, P.; Grandone, A.; Perrone, L.; Miraglia Del Giudice, E.; Daniele, A. Adiponectin profile and Irisin expression in Italian obese children: Association with insulin-resistance. *Cytokine* **2017**, *94*, 8–13. [CrossRef]

31. Qiu, S.; Cai, X.; Yin, H.; Zugel, M.; Sun, Z.; Steinacker, J.M.; Schumann, U. Association between circulating irisin and insulin resistance in non-diabetic adults: A meta-analysis. *Metab. Clin. Exp.* **2016**, *65*, 825–834. [CrossRef]

32. Petersen, K.F.; Shulman, G.I. Cellular mechanism of insulin resistance in skeletal muscle. *J. R. Soc. Med.* **2002**, *95* (Suppl. 42), 8–13. [PubMed]

33. Petersen, K.F.; Shulman, G.I. Pathogenesis of skeletal muscle insulin resistance in type 2 diabetes mellitus. *Am. J. Cardiol.* **2002**, *90*, 11g–18g. [CrossRef]

34. Satoh, T. Molecular mechanisms for the regulation of insulin-stimulated glucose uptake by small guanosine triphosphatases in skeletal muscle and adipocytes. *Int. J. Mol. Sci.* **2014**, *15*, 18677–18692. [CrossRef]

35. Galicia-Garcia, U.; Benito-Vicente, A.; Jebari, S.; Larrea-Sebal, A.; Siddiqi, H.; Uribe, K.B.; Ostolaza, H.; Martín, C. Pathophysiology of Type 2 Diabetes Mellitus. *Int. J. Mol. Sci.* **2020**, *21*, 6275. [CrossRef] [PubMed]

36. Titchenell, P.M.; Lazar, M.A.; Birnbaum, M.J. Unraveling the Regulation of Hepatic Metabolism by Insulin. *Trends Endocrinol. Metab.* **2017**, *28*, 497–505. [CrossRef]

37. Cherrington, A.D.; Moore, M.C.; Sindelar, D.K.; Edgerton, D.S. Insulin action on the liver in vivo. *Biochem. Soc. Trans.* **2007**, *35*, 1171–1174. [CrossRef]

38. Edgerton, D.S.; Lautz, M.; Scott, M.; Everett, C.A.; Stettler, K.M.; Neal, D.W.; Chu, C.A.; Cherrington, A.D. Insulin's direct effects on the liver dominate the control of hepatic glucose production. *J. Clin. Investig.* **2006**, *116*, 521–527. [CrossRef]

39. Meshkani, R.; Adeli, K. Hepatic insulin resistance, metabolic syndrome and cardiovascular disease. *Clin. Biochem.* **2009**, *42*, 1331–1346. [CrossRef]

40. Leclercq, I.A.; Da Silva Morais, A.; Schroyen, B.; Van Hul, N.; Geerts, A. Insulin resistance in hepatocytes and sinusoidal liver cells: Mechanisms and consequences. *J. Hepatol.* **2007**, *47*, 142–156. [CrossRef]

41. Gastaldelli, A.; Gaggini, M.; DeFronzo, R.A. Role of Adipose Tissue Insulin Resistance in the Natural History of Type 2 Diabetes: Results From the San Antonio Metabolism Study. *Diabetes* **2017**, *66*, 815–822. [CrossRef]

42. Czech, M.P. Mechanisms of insulin resistance related to white, beige, and brown adipocytes. *Mol. Metab.* **2020**, *34*, 27–42. [CrossRef] [PubMed]

43. Coelho, M.; Oliveira, T.; Fernandes, R. Biochemistry of adipose tissue: An endocrine organ. *Arch. Med. Sci.* **2013**, *9*, 191–200. [CrossRef] [PubMed]

44. Rosen, E.D.; Spiegelman, B.M. Adipocytes as regulators of energy balance and glucose homeostasis. *Nature* **2006**, *444*, 847–853. [CrossRef]

45. Scherer, P.E. The many secret lives of adipocytes: Implications for diabetes. *Diabetologia* **2019**, *62*, 223–232. [CrossRef] [PubMed]

46. Roden, M.; Shulman, G.I. The integrative biology of type 2 diabetes. *Nature* **2019**, *576*, 51–60. [CrossRef] [PubMed]

47. Maki, K.C.; Kelley, K.M.; Lawless, A.L.; Hubacher, R.L.; Schild, A.L.; Dicklin, M.R.; Rains, T.M. Validation of insulin sensitivity and secretion indices derived from the liquid meal tolerance test. *Diabetes Technol. Ther.* **2011**, *13*, 661–666. [CrossRef]

48. Meier, J.J.; Nauck, M.A. Is the diminished incretin effect in type 2 diabetes just an epi-phenomenon of impaired beta-cell function? *Diabetes* **2010**, *59*, 1117–1125. [CrossRef]

49. Surampudi, P.N.; John-Kalarickal, J.; Fonseca, V.A. Emerging concepts in the pathophysiology of type 2 diabetes mellitus. *Mt. Sinai. J. Med.* **2009**, *76*, 216–226. [CrossRef] [PubMed]

50. Baggio, L.L.; Drucker, D.J. Biology of incretins: GLP-1 and GIP. *Gastroenterology* **2007**, *132*, 2131–2157. [CrossRef]

51. Inzucchi, S.E.; Bergenstal, R.M.; Buse, J.B.; Diamant, M.; Ferrannini, E.; Nauck, M.; Peters, A.L.; Tsapas, A.; Wender, R.; Matthews, D.R. Management of hyperglycemia in type 2 diabetes, 2015: A patient-centered approach: Update to a position statement of the American Diabetes Association and the European Association for the Study of Diabetes. *Diabetes Care* **2015**, *38*, 140–149. [CrossRef] [PubMed]

52. Nauck, M.A.; Homberger, E.; Siegel, E.G.; Allen, R.C.; Eaton, R.P.; Ebert, R.; Creutzfeldt, W. Incretin effects of increasing glucose loads in man calculated from venous insulin and C-peptide responses. *J. Clin. Endocrinol. Metab.* **1986**, *63*, 492–498. [CrossRef]

53. Nauck, M.; Stöckmann, F.; Ebert, R.; Creutzfeldt, W. Reduced incretin effect in type 2 (non-insulin-dependent) diabetes. *Diabetologia* **1986**, *29*, 46–52. [CrossRef] [PubMed]

54. Drucker, D.J. Enhancing incretin action for the treatment of type 2 diabetes. *Diabetes Care* **2003**, *26*, 2929–2940. [CrossRef] [PubMed]

55. Drucker, D.J. Glucagon-like peptide-1 and the islet beta-cell: Augmentation of cell proliferation and inhibition of apoptosis. *Endocrinology* **2003**, *144*, 5145–5148. [CrossRef] [PubMed]

56. Buteau, J. GLP-1 receptor signaling: Effects on pancreatic beta-cell proliferation and survival. *Diabetes Metab.* **2008**, *34* (Suppl. 2), S73–S77. [CrossRef]

57. Gautier, J.F.; Fetita, S.; Sobngwi, E.; Salaun-Martin, C. Biological actions of the incretins GIP and GLP-1 and therapeutic perspectives in patients with type 2 diabetes. *Diabetes Metab.* **2005**, *31*, 233–242. [CrossRef]

58. Irwin, N.; Hunter, K.; Frizzell, N.; Flatt, P.R. Antidiabetic effects of sub-chronic activation of the GIP receptor alone and in combination with background exendin-4 therapy in high fat fed mice. *Regul. Pept.* **2009**, *153*, 70–76. [CrossRef]

59. Cao, L.; Li, D.; Feng, P.; Li, L.; Xue, G.F.; Li, G.; Holscher, C. A novel dual GLP-1 and GIP incretin receptor agonist is neuroprotective in a mouse model of Parkinson's disease by reducing chronic inflammation in the brain. *Neuroreport* **2016**, *27*, 384–391. [CrossRef]

60. Boer, G.A.; Holst, J.J. Incretin Hormones and Type 2 Diabetes—Mechanistic Insights and Therapeutic Approaches. *Biology* **2020**, *9*, 473. [CrossRef]

61. Toft-Nielsen, M.B.; Damholt, M.B.; Madsbad, S.; Hilsted, L.M.; Hughes, T.E.; Michelsen, B.K.; Holst, J.J. Determinants of the impaired secretion of glucagon-like peptide-1 in type 2 diabetic patients. *J. Clin. Endocrinol. Metab.* **2001**, *86*, 3717–3723. [CrossRef] [PubMed]
62. Faerch, K.; Vaag, A.; Holst, J.J.; Glümer, C.; Pedersen, O.; Borch-Johnsen, K. Impaired fasting glycaemia vs impaired glucose tolerance: Similar impairment of pancreatic alpha and beta cell function but differential roles of incretin hormones and insulin action. *Diabetologia* **2008**, *51*, 853–861. [CrossRef] [PubMed]
63. Nauck, M.A.; Meier, J.J. The incretin effect in healthy individuals and those with type 2 diabetes: Physiology, pathophysiology, and response to therapeutic interventions. *Lancet Diabetes Endocrinol.* **2016**, *4*, 525–536. [CrossRef]
64. Lynn, F.C.; Thompson, S.A.; Pospisilik, J.A.; Ehses, J.A.; Hinke, S.A.; Pamir, N.; McIntosh, C.H.; Pederson, R.A. A novel pathway for regulation of glucose-dependent insulinotropic polypeptide (GIP) receptor expression in beta cells. *FASEB J. Off. Publ. Fed. Am. Soc. Exp. Biol.* **2003**, *17*, 91–93.
65. Lynn, F.C.; Pamir, N.; Ng, E.H.; McIntosh, C.H.; Kieffer, T.J.; Pederson, R.A. Defective glucose-dependent insulinotropic polypeptide receptor expression in diabetic fatty Zucker rats. *Diabetes* **2001**, *50*, 1004–1011. [CrossRef] [PubMed]
66. Balakumar, P.; Maung, U.K.; Jagadeesh, G. Prevalence and prevention of cardiovascular disease and diabetes mellitus. *Pharmacol. Res.* **2016**, *113*, 600–609. [CrossRef]
67. Von Ah Morano, A.E.; Dorneles, G.P.; Peres, A.; Lira, F.S. The role of glucose homeostasis on immune function in response to exercise: The impact of low or higher energetic conditions. *J. Cell Physiol.* **2020**, *235*, 3169–3188. [CrossRef]
68. Deed, G.; Barlow, J.; Kawol, D.; Kilov, G.; Sharma, A.; Hwa, L.Y. Diet and diabetes. *Aust. Fam. Phys.* **2015**, *44*, 192–196.
69. Tuso, P. Prediabetes and lifestyle modification: Time to prevent a preventable disease. *Perm. J.* **2014**, *18*, 88–93. [CrossRef]
70. Thent, Z.C.; Das, S.; Henry, L.J. Role of Exercise in the Management of Diabetes Mellitus: The Global Scenario. *PLoS ONE* **2013**, *8*, e80436.
71. Knowler, W.C.; Barrett-Connor, E.; Fowler, S.E.; Hamman, R.F.; Lachin, J.M.; Walker, E.A.; Nathan, D.M. Reduction in the incidence of type 2 diabetes with lifestyle intervention or metformin. *N. Engl. J. Med.* **2002**, *346*, 393–403. [PubMed]
72. Pan, X.R.; Li, G.W.; Hu, Y.H.; Wang, J.X.; Yang, W.Y.; An, Z.X.; Hu, Z.X.; Lin, J.; Xiao, J.Z.; Cao, H.B.; et al. Effects of diet and exercise in preventing NIDDM in people with impaired glucose tolerance. The Da Qing IGT and Diabetes Study. *Diabetes Care* **1997**, *20*, 537–544. [CrossRef] [PubMed]
73. Olokoba, A.B.; Obateru, O.A.; Olokoba, L.B. Type 2 Diabetes Mellitus: A Review of Current Trends. *Oman. Med. J.* **2012**, *27*, 269–273. [CrossRef] [PubMed]
74. Tran, L.; Zielinski, A.; Roach, A.H.; Jende, J.A.; Householder, A.M.; Cole, E.E.; Atway, S.A.; Amornyard, M.; Accursi, M.L.; Shieh, S.W.; et al. Pharmacologic treatment of type 2 diabetes: Oral medications. *Ann. Pharmacother.* **2015**, *49*, 540–556. [CrossRef] [PubMed]
75. Song, R. Mechanism of Metformin: A Tale of Two Sites. *Diabetes Care* **2016**, *39*, 187–189. [CrossRef] [PubMed]
76. Shin, N.R.; Lee, J.C.; Lee, H.Y.; Kim, M.S.; Whon, T.W.; Lee, M.S.; Bae, J.W. An increase in the Akkermansia spp. population induced by metformin treatment improves glucose homeostasis in diet-induced obese mice. *Gut* **2014**, *63*, 727–735. [CrossRef]
77. Wen, S.; Wang, C.; Gong, M.; Zhou, L. An overview of energy and metabolic regulation. *Sci. China. Life Sci.* **2019**, *62*, 771–790. [CrossRef]
78. Viollet, B.; Guigas, B.; Sanz Garcia, N.; Leclerc, J.; Foretz, M.; Andreelli, F. Cellular and molecular mechanisms of metformin: An overview. *Clin. Sci. (Lond.)* **2012**, *122*, 253–270. [CrossRef]
79. Proks, P.; Reimann, F.; Green, N.; Gribble, F.; Ashcroft, F. Sulfonylurea stimulation of insulin secretion. *Diabetes* **2002**, *51* (Suppl. 3), S368–S376. [CrossRef]
80. Holman, R.R.; Paul, S.K.; Bethel, M.A.; Matthews, D.R.; Neil, H.A. 10-year follow-up of intensive glucose control in type 2 diabetes. *N. Engl. J. Med.* **2008**, *359*, 1577–1589. [CrossRef]
81. Becker, M.; Galler, A.; Raile, K. Meglitinide analogues in adolescent patients with HNF1A-MODY (MODY 3). *Pediatrics* **2014**, *133*, e775–e779. [CrossRef] [PubMed]
82. Skliros, N.P.; Vlachopoulos, C.; Tousoulis, D. Treatment of diabetes: Crossing to the other side. *Hell. J. Cardiol. HJC Hell. Kardiol. Ep.* **2016**, *57*, 304–310. [CrossRef]
83. Jung, Y.; Cao, Y.; Paudel, S.; Yoon, G.; Cheon, S.H.; Bae, G.U.; Jin, L.T.; Kim, Y.K.; Kim, S.N. Antidiabetic effect of SN158 through PPARalpha/gamma dual activation in ob/ob mice. *Chem.-Biol. Interact.* **2017**, *268*, 24–30. [CrossRef] [PubMed]
84. Gopal, S.S.; Lakshmi, M.J.; Sharavana, G.; Sathaiah, G.; Sreerama, Y.N.; Baskaran, V. Lactucaxanthin-a potential anti-diabetic carotenoid from lettuce (*Lactuca sativa*) inhibits alpha-amylase and alpha-glucosidase activity in vitro and in diabetic rats. *Food Funct.* **2017**, *8*, 1124–1131. [CrossRef] [PubMed]
85. Ishii, H.; Hayashino, Y.; Akai, Y.; Yabuta, M.; Tsujii, S. Dipeptidyl peptidase-4 inhibitors as preferable oral hypoglycemic agents in terms of treatment satisfaction: Results from a multicenter, 12-week, open label, randomized controlled study in Japan. *J. Diabetes Investig.* **2017**, *9*, 137–145. [CrossRef]
86. Kobayashi, K.; Yokoh, H.; Sato, Y.; Takemoto, M.; Uchida, D.; Kanatsuka, A.; Kuribayashi, N.; Terano, T.; Hashimoto, N.; Sakurai, K.; et al. Efficacy and safety of the dipeptidyl peptidase-4 inhibitor sitagliptin compared with alpha-glucosidase inhibitor in Japanese patients with type 2 diabetes inadequately controlled on sulfonylurea alone (SUCCESS-2): A multicenter, randomized, open-label, non-inferiority trial. *Diabetes Obes. Metab.* **2014**, *16*, 761–765. [PubMed]

87. Deacon, C.F.; Mannucci, E.; Ahren, B. Glycaemic efficacy of glucagon-like peptide-1 receptor agonists and dipeptidyl peptidase-4 inhibitors as add-on therapy to metformin in subjects with type 2 diabetes-a review and meta analysis. *Diabetes Obes. Metab.* **2012**, *14*, 762–767. [CrossRef]

88. Zhou, M.; Mok, M.T.; Sun, H.; Chan, A.W.; Huang, Y.; Cheng, A.S.; Xu, G. The anti-diabetic drug exenatide, a glucagon-like peptide-1 receptor agonist, counteracts hepatocarcinogenesis through cAMP-PKA-EGFR-STAT3 axis. *Oncogene* **2017**, *36*, 4135–4149. [CrossRef]

89. Liu, J.; Hu, Y.; Zhang, H.; Xu, Y.; Wang, G. Exenatide treatment increases serum irisin levels in patients with obesity and newly diagnosed type 2 diabetes. *J. Diabetes Its Complicat.* **2016**, *30*, 1555–1559. [CrossRef]

90. Seino, Y.; Fukushima, M.; Yabe, D. GIP and GLP-1, the two incretin hormones: Similarities and differences. *J. Diabetes Investig.* **2010**, *1*, 8–23. [CrossRef]

91. Ojha, A.; Ojha, U.; Mohammed, R.; Chandrashekar, A.; Ojha, H. Current perspective on the role of insulin and glucagon in the pathogenesis and treatment of type 2 diabetes mellitus. *Clin. Pharmacol.* **2019**, *11*, 57–65. [CrossRef]

92. Adeghate, E.; Mohsin, S.; Adi, F.; Ahmed, F.; Yahya, A.; Kalász, H.; Tekes, K.; Adeghate, E.A. An update of SGLT1 and SGLT2 inhibitors in early phase diabetes-type 2 clinical trials. *Expert Opin. Investig. Drugs* **2019**, *28*, 811–820. [CrossRef] [PubMed]

93. Steen, O.; Goldenberg, R.M. The Role of Sodium-Glucose Cotransporter 2 Inhibitors in the Management of Type 2 Diabetes. *Can. J. Diabetes* **2017**, *41*, 517–523. [CrossRef] [PubMed]

94. Nauck, M.A. Update on developments with SGLT2 inhibitors in the management of type 2 diabetes. *Drug Des. Dev. Ther.* **2014**, *8*, 1335–1380. [CrossRef] [PubMed]

95. Kosiborod, M.; Gause-Nilsson, I.; Xu, J.; Sonesson, C.; Johnsson, E. Efficacy and safety of dapagliflozin in patients with type 2 diabetes and concomitant heart failure. *J. Diabetes Its Complicat.* **2017**, *31*, 1215–1221. [CrossRef]

96. Bailey, C.J.; Gross, J.L.; Hennicken, D.; Iqbal, N.; Mansfield, T.A.; List, J.F. Dapagliflozin add-on to metformin in type 2 diabetes inadequately controlled with metformin: A randomized, double-blind, placebo-controlled 102-week trial. *BMC Med.* **2013**, *11*, 43. [CrossRef]

97. Brahmkshatriya, P.P.; Brahmkshatriya, P.S. Terpenes: Chemistry, Biological Role, and Therapeutic Applications. In *Natural Products: Phytochemistry, Botany and Metabolism of Alkaloids, Phenolics and Terpenes*; Ramawat, K.G., Mérillon, J.-M., Eds.; Springer: Berlin/Heidelberg, Germany, 2013; pp. 2665–2691.

98. Tan, X.C.; Chua, K.H.; Ravishankar Ram, M.; Kuppusamy, U.R. Monoterpenes: Novel insights into their biological effects and roles on glucose uptake and lipid metabolism in 3T3-L1 adipocytes. *Food Chem.* **2016**, *196*, 242–250. [CrossRef]

99. Guimarães, A.C.; Meireles, L.M.; Lemos, M.F.; Guimarães, M.C.C.; Endringer, D.C.; Fronza, M.; Scherer, R. Antibacterial Activity of Terpenes and Terpenoids Present in Essential Oils. *Molecules* **2019**, *24*, 2471. [CrossRef]

100. Kozioł, A.; Stryjewska, A.; Librowski, T.; Sałat, K.; Gaweł, M.; Moniczewski, A.; Lochyński, S. An overview of the pharmacological properties and potential applications of natural monoterpenes. *Mini. Rev. Med. Chem.* **2014**, *14*, 1156–1168. [CrossRef]

101. Ke, J.; Zhu, C.; Zhang, Y.; Zhang, W. Anti-Arrhythmic Effects of Linalool via Cx43 Expression in a Rat Model of Myocardial Infarction. *Front. Pharmacol.* **2020**, *11*, 926. [CrossRef]

102. Pripdeevech, P.; Machan, T. Fingerprint of volatile flavour constituents and antioxidant activities of teas from Thailand. *Food Chem.* **2011**, *125*, 797–802. [CrossRef]

103. Sajid, M.; Khan, M.R.; Ismail, H.; Latif, S.; Rahim, A.A.; Mehboob, R.; Shah, S.A. Antidiabetic and antioxidant potential of Alnus nitida leaves in alloxan induced diabetic rats. *J. Ethnopharmacol.* **2020**, *251*, 112544. [CrossRef] [PubMed]

104. Peana, A.T.; D'Aquila, P.S.; Panin, F.; Serra, G.; Pippia, P.; Moretti, M.D. Anti-inflammatory activity of linalool and linalyl acetate constituents of essential oils. *Phytomed. Int. J. Phytother. Phytopharm.* **2002**, *9*, 721–726. [CrossRef] [PubMed]

105. More, T.; Kulkarni, B.; Nalawade, M.; Arvindekar, A. Antidiabetic activity of linalool and limonene in streptozotocin-induced diabetic rat: A combinatorial therapy approach. *Int. J. Pharm. Pharm. Sci.* **2014**, *6*, 159–163.

106. Garba, H.A.; Mohammed, A.; Ibrahim, M.A.; Shuaibu, M.N. Effect of lemongrass (*Cymbopogon citratus* Stapf) tea in a type 2 diabetes rat model. *Clin. Phytosci.* **2020**, *6*, 19. [CrossRef]

107. Kwon, Y.I.; Apostolidis, E.; Shetty, K. Evaluation of pepper (*Capsicum annuum*) for management of diabetes and hypertension. *J. Food Biochem.* **2007**, *31*, 370–385. [CrossRef]

108. Boaduo, N.K.; Katerere, D.; Eloff, J.N.; Naidoo, V. Evaluation of six plant species used traditionally in the treatment and control of diabetes mellitus in South Africa using in vitro methods. *Pharm. Biol.* **2014**, *52*, 756–761. [CrossRef] [PubMed]

109. Jumepaeng, T.; Prachakool, S.; Luthria, D.L.; Chanthai, S. Determination of antioxidant capacity and α-amylase inhibitory activity of the essential oils from citronella grass and lemongrass. *Int. Food Res. J.* **2013**, *20*, 481–485.

110. Deepa, B.; Venkataraman, A. Linalool, a plant derived monoterpene alcohol, rescues kidney from diabetes-induced nephropathic changes via blood glucose reduction. *Diabetol. Croat.* **2011**, *40*, 121–137.

111. Shah, G.; Shri, R.; Panchal, V.; Sharma, N.; Singh, B.; Mann, A.S. Scientific basis for the therapeutic use of Cymbopogon citratus, stapf (*Lemon grass*). *J. Adv. Pharm. Technol. Res.* **2011**, *2*, 3–8. [CrossRef]

112. Najafian, M.; Ebrahim-Habibi, A.; Yaghmaei, P.; Parivar, K.; Larijani, B. Citral as a potential antihyperlipidemic medicine in diabetes: A study on streptozotocin-induced diabetic rats. *Iran. J. Diabetes Lipid Disord.* **2011**, *10*, 3.

113. Molinaro, A.; Wahlström, A.; Marschall, H.U. Role of Bile Acids in Metabolic Control. *Trends Endocrinol. Metab.* **2018**, *29*, 31–41. [CrossRef] [PubMed]

114. Schaap, F.G.; Trauner, M.; Jansen, P.L. Bile acid receptors as targets for drug development. *Nat. Rev. Gastroenterol. Hepatol.* **2014**, *11*, 55–67. [CrossRef] [PubMed]

115. Ooi, C.P.; Loke, S.C. Colesevelam for type 2 diabetes mellitus. *Cochrane Database Syst. Rev.* **2012**, *12*, Cd009361. [CrossRef] [PubMed]

116. Zieve, F.J.; Kalin, M.F.; Schwartz, S.L.; Jones, M.R.; Bailey, W.L. Results of the glucose-lowering effect of WelChol study (GLOWS): A randomized, double-blind, placebo-controlled pilot study evaluating the effect of colesevelam hydrochloride on glycemic control in subjects with type 2 diabetes. *Clin. Ther.* **2007**, *29*, 74–83. [CrossRef]

117. Damci, T.; Tatliagac, S.; Osar, Z.; Ilkova, H. Fenofibrate treatment is associated with better glycemic control and lower serum leptin and insulin levels in type 2 diabetic patients with hypertriglyceridemia. *Eur. J. Intern. Med.* **2003**, *14*, 357–360. [CrossRef]

118. Nozue, T.; Michishita, I.; Mizuguchi, I. Effects of ezetimibe on glucose metabolism in patients with type 2 diabetes: A 12-week, open-label, uncontrolled, pilot study. *Curr. Ther. Res. Clin. Exp.* **2010**, *71*, 252–258. [CrossRef]

119. Modak, T.; Mukhopadhaya, A. Effects of citral, a naturally occurring antiadipogenic molecule, on an energy-intense diet model of obesity. *Indian J. Pharmacol.* **2011**, *43*, 300–305. [CrossRef] [PubMed]

120. Mishra, C.; Khalid, M.A.; Fatima, N.; Singh, B.; Tripathi, D.; Waseem, M.; Mahdi, A.A. Effects of citral on oxidative stress and hepatic key enzymes of glucose metabolism in streptozotocin/high-fat-diet induced diabetic dyslipidemic rats. *Iran. J. Basic Med. Sci.* **2019**, *22*, 49–57.

121. Lei, Y.; Fu, P.; Jun, X.; Cheng, P. Pharmacological Properties of Geraniol-A Review. *Planta Med.* **2019**, *85*, 48–55. [CrossRef]

122. Babukumar, S.; Vinothkumar, V.; Sankaranarayanan, C.; Srinivasan, S. Geraniol, a natural monoterpene, ameliorates hyperglycemia by attenuating the key enzymes of carbohydrate metabolism in streptozotocin-induced diabetic rats. *Pharm. Biol.* **2017**, *55*, 1442–1449. [CrossRef]

123. Kamble, S.P.; Ghadyale, V.A.; Patil, R.S.; Haldavnekar, V.S.; Arvindekar, A.U. Inhibition of GLUT2 transporter by geraniol from Cymbopogon martinii: A novel treatment for diabetes mellitus in streptozotocin-induced diabetic rats. *J. Pharm. Pharmacol.* **2020**, *72*, 294–304. [CrossRef] [PubMed]

124. Srinivasan, S.; Muruganathan, U. Antidiabetic efficacy of citronellol, a citrus monoterpene by ameliorating the hepatic key enzymes of carbohydrate metabolism in streptozotocin-induced diabetic rats. *Chem.-Biol. Interact.* **2016**, *250*, 38–46. [CrossRef] [PubMed]

125. De Toledo, L.G.; Ramos, M.A.; Spósito, L.; Castilho, E.M.; Pavan, F.R.; Lopes Éde, O.; Zocolo, G.J.; Silva, F.A.; Soares, T.H.; Dos Santos, A.G.; et al. Essential Oil of *Cymbopogon nardus* (L.) Rendle: A Strategy to Combat Fungal Infections Caused by Candida Species. *Int. J. Mol. Sci.* **2016**, *17*, 1252. [CrossRef] [PubMed]

126. Abena, A.A.; Gbenou, J.D.; Yayi, E.; Moudachirou, M.; Ongoka, R.P.; Ouamba, J.M.; Silou, T. Comparative chemical and analgesic properties of essential oils of *Cymbopogon nardus* (L) Rendle of Benin and Congo. *Afr. J. Tradit. Complement. Altern. Med.* **2007**, *4*, 267–272. [CrossRef]

127. Boukhris, M.; Bouaziz, M.; Feki, I.; Jemai, H.; El Feki, A.; Sayadi, S. Hypoglycemic and antioxidant effects of leaf essential oil of Pelargonium graveolens L'Hér. in alloxan induced diabetic rats. *Lipids Health Dis.* **2012**, *11*, 81. [CrossRef] [PubMed]

128. Santos, M.; Moreira, F.; Fraga, B.; Sousa, D.; Bonjardim, L.; Quintans-Júnior, L. Cardiovascular effects of monoterpenes: A review. *Rev. Bras. Farmacogn.* **2011**, *21*, 764–771. [CrossRef]

129. Buchbauer, G.; Jirovetz, L.; Jäger, W.; Dietrich, H.; Plank, C. Aromatherapy: Evidence for sedative effects of the essential oil of lavender after inhalation. *Z. Für Nat. C* **1991**, *46*, 1067–1072. [CrossRef]

130. Dzumayev, K.; Tsibulskaya, I.; Zenkevich, I.; Tkachenko, K.; Satzyperova, I. Essential Oils of Salvia sclarea L. Produced from Plants Grown in Southern Uzbekistan. *J. Essent. Oil Res.* **1995**, *7*, 597–604. [CrossRef]

131. Yang, H.J.; Kim, K.Y.; Kang, P.; Lee, H.S.; Seol, G.H. Effects of Salvia sclarea on chronic immobilization stress induced endothelial dysfunction in rats. *BMC Complement. Altern. Med.* **2014**, *14*, 396. [CrossRef]

132. Shin, Y.K.; Hsieh, Y.S.; Kwon, S.; Lee, H.S.; Seol, G.H. Linalyl acetate restores endothelial dysfunction and hemodynamic alterations in diabetic rats exposed to chronic immobilization stress. *J. Appl. Physiol. (Bethesda Md. 1985)* **2018**, *124*, 1274–1283. [CrossRef] [PubMed]

133. Sun, J. D-Limonene: Safety and clinical applications. *Altern. Med. Rev.* **2007**, *12*, 259–264. [PubMed]

134. Bacanlı, M.; Anlar, H.G.; Aydın, S.; Çal, T.; Arı, N.; Ündeğer Bucurgat, Ü.; Başaran, A.A.; Başaran, N. D-limonene ameliorates diabetes and its complications in streptozotocin-induced diabetic rats. *Food Chem. Toxicol. Int. J. Publ. Br. Ind. Biol. Res. Assoc.* **2017**, *110*, 434–442. [CrossRef] [PubMed]

135. Joglekar, M.M.; Panaskar, S.N.; Chougale, A.D.; Kulkarni, M.J.; Arvindekar, A.U. A novel mechanism for antiglycative action of limonene through stabilization of protein conformation. *Mol. Biosyst.* **2013**, *9*, 2463–2472. [CrossRef]

136. Murali, R.; Saravanan, R. Antidiabetic effect of d-limonene, a monoterpene in streptozotocin-induced diabetic rats. *Biomed. Prev. Nutr.* **2012**, *2*, 269–275. [CrossRef]

137. Bossou, A.D.; Mangelinckx, S.; Yedomonhan, H.; Boko, P.M.; Akogbeto, M.C.; De Kimpe, N.; Avlessi, F.; Sohounhloue, D.C. Chemical composition and insecticidal activity of plant essential oils from Benin against Anopheles gambiae (Giles). *Parasit Vectors* **2013**, *6*, 337. [CrossRef]

138. Liu, P.; Liu, X.C.; Dong, H.W.; Liu, Z.L.; Du, S.S.; Deng, Z.W. Chemical composition and insecticidal activity of the essential oil of Illicium pachyphyllum fruits against two grain storage insects. *Molecules* **2012**, *17*, 14870–14881. [CrossRef]

139. Fang, R.; Jiang, C.H.; Wang, X.Y.; Zhang, H.M.; Liu, Z.L.; Zhou, L.; Du, S.S.; Deng, Z.W. Insecticidal activity of essential oil of Carum Carvi fruits from China and its main components against two grain storage insects. *Molecules* **2010**, *15*, 9391–9402. [CrossRef]
140. Bhatia, S.P.; McGinty, D.; Letizia, C.S.; Api, A.M. Fragrance material review on carveol. *Food Chem. Toxicol. Int. J. Publ. Br. Ind. Biol. Res. Assoc.* **2008**, *46* (Suppl. 11), S85–S87. [CrossRef]
141. Wagner, K.H.; Elmadfa, I. Biological relevance of terpenoids. Overview focusing on mono-, di- and tetraterpenes. *Ann. Nutr. Metab.* **2003**, *47*, 95–106. [CrossRef]
142. Marques, F.M.; Figueira, M.M.; Schmitt, E.F.P.; Kondratyuk, T.P.; Endringer, D.C.; Scherer, R.; Fronza, M. In vitro anti-inflammatory activity of terpenes via suppression of superoxide and nitric oxide generation and the NF-κB signalling pathway. *Inflammopharmacology* **2019**, *27*, 281–289. [CrossRef] [PubMed]
143. Rossi, Y.E.; Palacios, S.M. Fumigant toxicity of Citrus sinensis essential oil on Musca domestica L. adults in the absence and presence of a P450 inhibitor. *Acta Trop.* **2013**, *127*, 33–37. [CrossRef] [PubMed]
144. Sales, A.; Felipe, L.; Bicas, J. Production, Properties, and Applications of α-Terpineol. *Food Bioprocess Technol.* **2020**, *13*, 1261–1279. [CrossRef]
145. Capetti, F.; Cagliero, C.; Marengo, A.; Bicchi, C.; Rubiolo, P.; Sgorbini, B. Bio-Guided Fractionation Driven by In Vitro α-Amylase Inhibition Assays of Essential Oils Bearing Specialized Metabolites with Potential Hypoglycemic Activity. *Plants* **2020**, *9*, 1242. [CrossRef]
146. Sousa, G.M.; Cazarin, C.B.B.; Maróstica Junior, M.R.; Lamas, C.A.; Quitete, V.; Pastore, G.M.; Bicas, J.L. The effect of α-terpineol enantiomers on biomarkers of rats fed a high-fat diet. *Heliyon* **2020**, *6*, e03752. [CrossRef]
147. Bouyahya, A.; Chamkhi, I.; Guaouguaou, F.E.; Benali, T.; Balahbib, A.; El Omari, N.; Taha, D.; El-Shazly, M.; El Menyiy, N. Ethnomedicinal use, phytochemistry, pharmacology, and food benefits of Thymus capitatus. *J. Ethnopharmacol.* **2020**, *259*, 112925. [CrossRef]
148. Abbasi, S.; Gharaghani, S.; Benvidi, A.; Rezaeinasab, M. New insights into the efficiency of thymol synergistic effect with p-cymene in inhibiting advanced glycation end products: A multi-way analysis based on spectroscopic and electrochemical methods in combination with molecular docking study. *J. Pharm. Biomed. Anal.* **2018**, *150*, 436–451. [CrossRef]
149. Miladi, H.; Zmantar, T.; Kouidhi, B.; Chaabouni, Y.; Mahdouani, K.; Bakhrouf, A.; Chaieb, K. Use of carvacrol, thymol, and eugenol for biofilm eradication and resistance modifying susceptibility of Salmonella enterica serovar Typhimurium strains to nalidixic acid. *Microb. Pathog.* **2017**, *104*, 56–63. [CrossRef]
150. Veras, H.N.; Araruna, M.K.; Costa, J.G.; Coutinho, H.D.; Kerntopf, M.R.; Botelho, M.A.; Menezes, I.R. Topical antiinflammatory activity of essential oil of Lippia sidoides cham: Possible mechanism of action. *Phytother. Res. PTR* **2013**, *27*, 179–185. [CrossRef]
151. Kang, S.H.; Kim, Y.S.; Kim, E.K.; Hwang, J.W.; Jeong, J.H.; Dong, X.; Lee, J.W.; Moon, S.H.; Jeon, B.T.; Park, P.J. Anticancer Effect of Thymol on AGS Human Gastric Carcinoma Cells. *J. Microbiol. Biotechnol.* **2016**, *26*, 28–37. [CrossRef]
152. Haque, M.R.; Ansari, S.H.; Najmi, A.K.; Ahmad, M.A. Monoterpene phenolic compound thymol prevents high fat diet induced obesity in murine model. *Toxicol. Mech. Methods* **2014**, *24*, 116–123. [CrossRef] [PubMed]
153. Fang, F.; Li, H.; Qin, T.; Li, M.; Ma, S. Thymol improves high-fat diet-induced cognitive deficits in mice via ameliorating brain insulin resistance and upregulating NRF2/HO-1 pathway. *Metab. Brain Dis.* **2017**, *32*, 385–393. [CrossRef] [PubMed]
154. Saravanan, S.; Pari, L. Role of thymol on hyperglycemia and hyperlipidemia in high fat diet-induced type 2 diabetic C57BL/6J mice. *Eur. J. Pharmacol.* **2015**, *761*, 279–287. [CrossRef] [PubMed]
155. Homocysteine Lowering Trialists Collaboration. Dose-dependent effects of folic acid on blood concentrations of homocysteine: A meta-analysis of the randomized trials. *Am. J. Clin. Nutr.* **2005**, *82*, 806–812. [CrossRef]
156. Saravanan, S.; Pari, L. Protective effect of thymol on high fat diet induced diabetic nephropathy in C57BL/6J mice. *Chem.-Biol. Interact.* **2016**, *245*, 1–11. [CrossRef]
157. Oskouei, B.G.; Abbaspour-Ravasjani, S.; Jamal Musavinejad, S.; Ahmad Salehzadeh, S.; Abdolhosseinzadeh, A.; Hamishehkar, H.; Ghahremanzadeh, K.; Shokouhi, B. In vivo Evaluation of Anti-Hyperglycemic, Anti-hyperlipidemic and Anti-Oxidant Status of Liver and Kidney of Thymol in STZ-Induced Diabetic Rats. *Drug Res.* **2019**, *69*, 46–52. [CrossRef]
158. Hyun, T.K.; Kim, H.-C.; Kim, J.-S. Antioxidant and antidiabetic activity of Thymus quinquecostatus Celak. *Ind. Crops Prod.* **2014**, *52*, 611–616. [CrossRef]
159. Behera, G.C.; Parida, K.M.; Das, D.P. Facile fabrication of aluminum-promoted vanadium phosphate: A highly active heterogeneous catalyst for isopropylation of toluene to cymene. *J. Catal.* **2012**, *289*, 190–198. [CrossRef]
160. Bonjardim, L.R.; Cunha, E.S.; Guimarães, A.G.; Santana, M.F.; Oliveira, M.G.; Serafini, M.R.; Araújo, A.A.; Antoniolli, A.R.; Cavalcanti, S.C.; Santos, M.R.; et al. Evaluation of the anti-inflammatory and antinociceptive properties of p-cymene in mice. *Z. Für Nat. C* **2012**, *67*, 15–21. [CrossRef]
161. Nabavi, S.M.; Marchese, A.; Izadi, M.; Curti, V.; Daglia, M.; Nabavi, S.F. Plants belonging to the genus Thymus as antibacterial agents: From farm to pharmacy. *Food Chem.* **2015**, *173*, 339–347. [CrossRef]
162. Nickavar, B.; Adeli, A.; Nickavar, A. TLC-Bioautography and GC-MS Analyses for Detection and Identification of Antioxidant Constituents of Trachyspermum copticum Essential Oil. *Iran. J. Pharm. Res.* **2014**, *13*, 127–133.
163. Lotfi, P.; Yaghmaei, P.; Ebrahim-Habibi, A. Cymene and Metformin treatment effect on biochemical parameters of male NMRI mice fed with high fat diet. *J. Diabetes Metab. Disord.* **2015**, *14*, 52. [CrossRef]

164. Joglekar, M.; Panaskar, S.; Arvindekar, A. Inhibition of advanced glycation end product formation by cymene-A common food constituent. *J. Funct. Foods* **2013**, *6*, 107–115. [CrossRef]

165. Oz, M.; El Nebrisi, E.G.; Yang, K.S.; Howarth, F.C.; Al Kury, L.T. Cellular and Molecular Targets of Menthol Actions. *Front. Pharmacol.* **2017**, *8*, 472. [CrossRef] [PubMed]

166. Patel, T.; Ishiuji, Y.; Yosipovitch, G. Menthol: A refreshing look at this ancient compound. *J. Am. Acad. Dermatol.* **2007**, *57*, 873–878. [CrossRef] [PubMed]

167. Muruganathan, U.; Srinivasan, S.; Vinothkumar, V. Antidiabetogenic efficiency of menthol, improves glucose homeostasis and attenuates pancreatic β-cell apoptosis in streptozotocin-nicotinamide induced experimental rats through ameliorating glucose metabolic enzymes. *Biomed. Pharmacother. Biomed. Pharmacother.* **2017**, *92*, 229–239. [CrossRef]

168. Khare, P.; Mangal, P.; Baboota, R.K.; Jagtap, S.; Kumar, V.; Singh, D.P.; Boparai, R.K.; Sharma, S.S.; Khardori, R.; Bhadada, S.K.; et al. Involvement of Glucagon in Preventive Effect of Menthol Against High Fat Diet Induced Obesity in Mice. *Front. Pharmacol.* **2018**, *9*, 1244. [CrossRef]

169. Ceylan, E.; Özbek, H.; Aðaoðlu, Z. Investigation of The Level of The Median Lethal Dose (LD 50) and The Hypoglycemic Effect of *Cuminum cyminum* L. Fruit Essential Oil Extract in Healthy and Diabetic Mice. *Van Tıp. Derg. (Van Med. J.)* **2003**, *10*, 29–35.

170. Peña-Montes, D.J.; Huerta-Cervantes, M.; Ríos-Silva, M.; Trujillo, X.; Huerta, M.; Noriega-Cisneros, R.; Salgado-Garciglia, R.; Saavedra-Molina, A. Protective Effect of the Hexanic Extract of Eryngium carlinae Inflorescences In Vitro, in Yeast, and in Streptozotocin-Induced Diabetic Male Rats. *Antioxidants* **2019**, *8*, 73. [CrossRef] [PubMed]

171. Noriega-Cisneros, R.; Ortiz-Ávila, O.; Esquivel-Gutiérrez, E.; Clemente-Guerrero, M.; Manzo-Avalos, S.; Salgado-Garciglia, R.; Cortés-Rojo, C.; Boldogh, I.; Saavedra-Molina, A. Hypolipidemic Activity of Eryngium carlinae on Streptozotocin-Induced Diabetic Rats. *Biochem. Res. Int.* **2012**, *2012*, 603501. [CrossRef]

172. Nuzzo, D.; Galizzi, G.; Amato, A.; Terzo, S.; Picone, P.; Cristaldi, L.; Mulè, F.; Di Carlo, M. Regular Intake of Pistachio Mitigates the Deleterious Effects of a High Fat-Diet in the Brain of Obese Mice. *Antioxidants* **2020**, *9*, 317. [CrossRef] [PubMed]

173. Bagheri, S.; Sarabi, M.M.; Khosravi, P.; Khorramabadi, R.M.; Veiskarami, S.; Ahmadvand, H.; Keshvari, M. Effects of Pistacia atlantica on Oxidative Stress Markers and Antioxidant Enzymes Expression in Diabetic Rats. *J. Am. Coll. Nutr.* **2019**, *38*, 267–274. [CrossRef] [PubMed]

174. Höld, K.M.; Sirisoma, N.S.; Ikeda, T.; Narahashi, T.; Casida, J.E. Alpha-thujone (the active component of absinthe): Gamma-aminobutyric acid type A receptor modulation and metabolic detoxification. *Proc. Natl. Acad. Sci. USA* **2000**, *97*, 3826–3831. [CrossRef]

175. Baddar, N.W.; Aburjai, T.A.; Taha, M.O.; Disi, A.M. Thujone corrects cholesterol and triglyceride profiles in diabetic rat model. *Nat. Prod. Res.* **2011**, *25*, 1180–1184. [CrossRef]

176. Alkhateeb, H.; Chabowski, A.; Glatz, J.F.; Luiken, J.F.; Bonen, A. Two phases of palmitate-induced insulin resistance in skeletal muscle: Impaired GLUT4 translocation is followed by a reduced GLUT4 intrinsic activity. *Am. J. Physiol. Endocrinol. Metab.* **2007**, *293*, E783–E793. [CrossRef] [PubMed]

177. Alkhateeb, H.; Bonen, A. Thujone, a component of medicinal herbs, rescues palmitate-induced insulin resistance in skeletal muscle. *Am. J. Physiol. Regul. Integr. Comp. Physiol.* **2010**, *299*, R804–R812. [CrossRef] [PubMed]

178. Lachenmeier, D.W.; Walch, S.G. The choice of thujone as drug for diabetes. *Nat. Prod. Res.* **2011**, *25*, 1890–1892. [CrossRef]

179. Lindmark-Henriksson, M.; Isaksson, D.; Vanek, T.; Valterová, I.; Högberg, H.E.; Sjödin, K. Transformation of terpenes using a Picea abies suspension culture. *J. Biotechnol.* **2004**, *107*, 173–184. [CrossRef]

180. Vibha, J.; Choudhary, K.; Singh, M.; Rathore, M.; Shekhawat, N. A Study on Pharmacokinetics and Therapeutic Efficacy of Glycyrrhiza glabra: A Miracle Medicinal Herb. *Bot. Res. Int.* **2009**, *2*, 157–163.

181. Ayyasamy, R.; Leelavinothan, P. Myrtenal alleviates hyperglycaemia, hyperlipidaemia and improves pancreatic insulin level in STZ-induced diabetic rats. *Pharm. Biol.* **2016**, *54*, 2521–2527. [CrossRef]

182. Rathinam, A.; Pari, L.; Chandramohan, R.; Sheikh, B.A. Histopathological findings of the pancreas, liver, and carbohydrate metabolizing enzymes in STZ-induced diabetic rats improved by administration of myrtenal. *J. Physiol. Biochem.* **2014**, *70*, 935–946. [CrossRef]

183. Rathinam, A.; Pari, L. Myrtenal ameliorates hyperglycemia by enhancing GLUT2 through Akt in the skeletal muscle and liver of diabetic rats. *Chem.-Biol. Interact.* **2016**, *256*, 161–166. [CrossRef] [PubMed]

184. Rathinam, A.; Pari, L.; Venkatesan, M.; Munusamy, S. Myrtenal attenuates oxidative stress and inflammation in a rat model of streptozotocin-induced diabetes. *Arch. Physiol. Biochem.* **2019**, 1–9. [CrossRef] [PubMed]

185. Neves, G.N.; Nogueira, G.; Vardanega, R.; Meireles, M.A. Identification and quantification of genipin and geniposide from Genipa americana L. by HPLC-DAD using a fused-core column. *Food Sci. Technol.* **2018**, *38*, 116–122. [CrossRef]

186. Wang, S.C.; Liao, H.J.; Lee, W.C.; Huang, C.M.; Tsai, T.H. Using orthogonal array to obtain gradient liquid chromatography conditions of enhanced peak intensity to determine geniposide and genipin with electrospray tandem mass spectrometry. *J. Chromatogr. A* **2008**, *1212*, 68–75. [CrossRef]

187. Shanmugam, M.K.; Shen, H.; Tang, F.R.; Arfuso, F.; Rajesh, M.; Wang, L.; Kumar, A.P.; Bian, J.; Goh, B.C.; Bishayee, A.; et al. Potential role of genipin in cancer therapy. *Pharmacol. Res.* **2018**, *133*, 195–200. [CrossRef] [PubMed]

188. Wang, J.; Zhang, Y.; Liu, R.; Li, X.; Cui, Y.; Qu, L. Geniposide protects against acute alcohol-induced liver injury in mice via up-regulating the expression of the main antioxidant enzymes. *Can. J. Physiol. Pharmacol.* **2015**, *93*, 261–267. [CrossRef]

189. Koo, H.J.; Lim, K.H.; Jung, H.J.; Park, E.H. Anti-inflammatory evaluation of gardenia extract, geniposide and genipin. *J. Ethnopharmacol.* **2006**, *103*, 496–500. [CrossRef]

190. Jiang, Y.Q.; Chang, G.L.; Wang, Y.; Zhang, D.Y.; Cao, L.; Liu, J. Geniposide Prevents Hypoxia/Reoxygenation-Induced Apoptosis in H9c2 Cells: Improvement of Mitochondrial Dysfunction and Activation of GLP-1R and the PI3K/AKT Signaling Pathway. *Cell Physiol. Biochem.* **2016**, *39*, 407–421. [CrossRef]

191. Ma, C.J.; Nie, A.F.; Zhang, Z.J.; Zhang, Z.G.; Du, L.; Li, X.Y.; Ning, G. Genipin stimulates glucose transport in C2C12 myotubes via an IRS-1 and calcium-dependent mechanism. *J. Endocrinol.* **2013**, *216*, 353–362. [CrossRef]

192. Guan, L.; Feng, H.; Gong, D.; Zhao, X.; Cai, L.; Wu, Q.; Yuan, B.; Yang, M.; Zhao, J.; Zou, Y. Genipin ameliorates age-related insulin resistance through inhibiting hepatic oxidative stress and mitochondrial dysfunction. *Exp. Gerontol.* **2013**, *48*, 1387–1394. [CrossRef]

193. Guan, L.; Gong, D.; Yang, S.; Shen, N.; Zhang, S.; Li, Y.; Wu, Q.; Yuan, B.; Sun, Y.; Dai, N.; et al. Genipin ameliorates diet-induced obesity via promoting lipid mobilization and browning of white adipose tissue in rats. *Phytother. Res. PTR* **2018**, *32*, 723–732. [CrossRef] [PubMed]

194. Zhong, H.; Chen, K.; Feng, M.; Shao, W.; Wu, J.; Liang, T.; Liu, C. Genipin alleviates high-fat diet-induced hyperlipidemia and hepatic lipid accumulation in mice via miR-142a-5p/SREBP-1c axis. *Febs J.* **2018**, *285*, 501–517. [CrossRef] [PubMed]

195. Kojima, K.; Shimada, T.; Nagareda, Y.; Watanabe, M.; Ishizaki, J.; Sai, Y.; Miyamoto, K.; Aburada, M. Preventive effect of geniposide on metabolic disease status in spontaneously obese type 2 diabetic mice and free fatty acid-treated HepG2 cells. *Biol. Pharm. Bull.* **2011**, *34*, 1613–1618. [CrossRef]

196. Liu, J.; Yin, F.; Xiao, H.; Guo, L.; Gao, X. Glucagon-like peptide 1 receptor plays an essential role in geniposide attenuating lipotoxicity-induced β-cell apoptosis. *Toxicol Vitr.* **2012**, *26*, 1093–1097. [CrossRef] [PubMed]

197. Guo, L.X.; Liu, J.H.; Zheng, X.X.; Yin, Z.Y.; Kosaraju, J.; Tam, K.Y. Geniposide improves insulin production and reduces apoptosis in high glucose-induced glucotoxic insulinoma cells. *Eur. J. Pharm. Sci.* **2017**, *110*, 70–76. [CrossRef]

198. Liu, J.; Guo, L.; Yin, F.; Zhang, Y.; Liu, Z.; Wang, Y. Geniposide regulates glucose-stimulated insulin secretion possibly through controlling glucose metabolism in INS-1 cells. *PLoS ONE* **2013**, *8*, e78315. [CrossRef] [PubMed]

199. Zhang, Y.; Ding, Y.; Zhong, X.; Guo, Q.; Wang, H.; Gao, J.; Bai, T.; Ren, L.; Guo, Y.; Jiao, X.; et al. Geniposide acutely stimulates insulin secretion in pancreatic β-cells by regulating GLP-1 receptor/cAMP signaling and ion channels. *Mol. Cell Endocrinol.* **2016**, *430*, 89–96. [CrossRef]

200. Hodson, D.J.; Tarasov, A.I.; Gimeno Brias, S.; Mitchell, R.K.; Johnston, N.R.; Haghollahi, S.; Cane, M.C.; Bugliani, M.; Marchetti, P.; Bosco, D.; et al. Incretin-modulated beta cell energetics in intact islets of Langerhans. *Mol. Endocrinol.* **2014**, *28*, 860–871. [CrossRef] [PubMed]

201. Jacobson, D.A.; Kuznetsov, A.; Lopez, J.P.; Kash, S.; Ammälä, C.E.; Philipson, L.H. Kv2.1 ablation alters glucose-induced islet electrical activity, enhancing insulin secretion. *Cell Metab.* **2007**, *6*, 229–235. [CrossRef]

202. Bai, Y.; Zhu, R.; Tian, Y.; Li, R.; Chen, B.; Zhang, H.; Xia, B.; Zhao, D.; Mo, F.; Zhang, D.; et al. Catalpol in Diabetes and its Complications: A Review of Pharmacology, Pharmacokinetics, and Safety. *Molecules* **2019**, *24*, 3302. [CrossRef] [PubMed]

203. Yan, J.; Wang, C.; Jin, Y.; Meng, Q.; Liu, Q.; Liu, Z.; Liu, K.; Sun, H. Catalpol ameliorates hepatic insulin resistance in type 2 diabetes through acting on AMPK/NOX4/PI3K/AKT pathway. *Pharmacol. Res.* **2018**, *130*, 466–480. [CrossRef] [PubMed]

204. Bao, Q.; Shen, X.; Qian, L.; Gong, C.; Nie, M.; Dong, Y. Anti-diabetic activities of catalpol in db/db mice. *Korean J. Physiol. Pharmacol.* **2016**, *20*, 153–160. [CrossRef] [PubMed]

205. Liu, J.; Zhang, H.R.; Hou, Y.B.; Jing, X.L.; Song, X.Y.; Shen, X.P. Global gene expression analysis in liver of db/db mice treated with catalpol. *Chin. J. Nat. Med.* **2018**, *16*, 590–598. [CrossRef]

206. Trembath, D.G. Chapter 26-Molecular Testing for Glioblastoma. In *Diagnostic Molecular Pathology*; Coleman, W.B., Tsongalis, G.J., Eds.; Academic Press: Cambridge, MA, USA, 2017; pp. 339–347.

207. Kim, Y.; Kim, E.Y.; Seo, Y.M.; Yoon, T.K.; Lee, W.S.; Lee, K.A. Function of the pentose phosphate pathway and its key enzyme, transketolase, in the regulation of the meiotic cell cycle in oocytes. *Clin. Exp. Reprod. Med.* **2012**, *39*, 58–67. [CrossRef]

208. Wang, Y.; Jiang, Z.-T.; Li, R. Antioxidant Activity, Free Radical Scavenging Potential and Chemical Composition of Litsea cubeba Essential Oil. *J. Essent. Oil-Bear. Plants JEOP* **2012**, *15*, 134–143. [CrossRef]

209. Sousa, C.; Leitão, A.J.; Neves, B.M.; Judas, F.; Cavaleiro, C.; Mendes, A.F. Standardised comparison of limonene-derived monoterpenes identifies structural determinants of anti-inflammatory activity. *Sci. Rep.* **2020**, *10*, 7199. [CrossRef]

210. Nazzaro, F.; Fratianni, F.; De Martino, L.; Coppola, R.; De Feo, V. Effect of essential oils on pathogenic bacteria. *Pharmaceuticals* **2013**, *6*, 1451–1474. [CrossRef]

211. Boskabady, M.H.; Jandaghi, P. Relaxant effects of carvacrol on guinea pig tracheal chains and its possible mechanisms. *Pharmazie* **2003**, *58*, 661–663.

212. Brum, L.F.; Elisabetsky, E.; Souza, D. Effects of linalool on [(3)H]MK801 and [(3)H] muscimol binding in mouse cortical membranes. *Phytother. Res.* **2001**, *15*, 422–425. [CrossRef]

213. Koroch, A.R.; Rodolfo Juliani, H.; Zygadlo, J.A. Bioactivity of Essential Oils and Their Components. In *Flavours and Fragrances: Chemistry, Bioprocessing and Sustainability*; Berger, R.G., Ed.; Springer: Berlin/Heidelberg, Germany, 2007; pp. 87–115.

214. Wojdyło, A.; Oszmiański, J.; Czemerys, R. Antioxidant activity and phenolic compounds in 32 selected herbs. *Food Chem.* **2007**, *105*, 940–949. [CrossRef]

215. Zunino, M.P.; Zygadlo, J.A. Effect of monoterpenes on lipid oxidation in maize. *Planta* **2004**, *219*, 303–309.

216. Javan, A.J.; Javan, M.J. Electronic structure of some thymol derivatives correlated with the radical scavenging activity: Theoretical study. *Food Chem.* **2014**, *165*, 451–459. [CrossRef] [PubMed]
217. Malmir, M.; Gohari, A.R.; Saeidnia, S.; Silva, O. A new bioactive monoterpene-flavonoid from Satureja khuzistanica. *Fitoterapia* **2015**, *105*, 107–112. [CrossRef]
218. Forbes, J.M.; Cooper, M.E. Mechanisms of diabetic complications. *Physiol. Rev.* **2013**, *93*, 137–188. [CrossRef] [PubMed]
219. Patel, D.K.; Kumar, R.; Laloo, D.; Hemalatha, S. Diabetes mellitus: An overview on its pharmacological aspects and reported medicinal plants having antidiabetic activity. *Asian Pac. J. Trop. Biomed.* **2012**, *2*, 411–420. [CrossRef]

Article

Rosmanol and Carnosol Synergistically Alleviate Rheumatoid Arthritis through Inhibiting TLR4/NF-κB/MAPK Pathway

Lianchun Li , Zhenghong Pan *, Desheng Ning and Yuxia Fu

Guangxi Key Laboratory of Functional Phytochemicals Research and Utilization, Guangxi Institute of Botany, Chinese Academy of Sciences, Guilin 541006, China; llc@gxib.cn (L.L.); ndsh@gxib.cn (D.N.); fyx@gxib.cn (Y.F.)
* Correspondence: pan7260@126.com

Abstract: *Callicarpa longissima* has been used as a Yao folk medicine to treat arthritis for years in China, although its active anti-arthritic moieties have not been clarified so far. In this study, two natural phenolic diterpenoids with anti-rheumatoid arthritis (RA) effects, rosmanol and carnosol, isolated from the medicinal plant were reported on for the first time. In type II collagen-induced arthritis DBA/1 mice, both rosmanol (40 mg/kg/d) and carnosol (40 mg/kg/d) alone alleviated the RA symptoms, such as swelling, redness, and synovitis; decreased the arthritis index score; and downregulated the serum pro-inflammatory cytokine levels of interleukin 6 (IL-6), monocyte chemotactic protein 1 (MCP-1), and tumor necrosis factor α (TNF-α). Additionally, they blocked the activation of the Toll-like receptor 4 (TLR4)/nuclear factor κB (NF-κB)/c-Jun N-terminal kinase (JNK) and p38 mitogen-activated protein kinase (MAPK) pathways. Of particular interest was that when they were used in combination (20 mg/kg/d each), the anti-RA effect and inhibitory activity on the TLR4/NF-κB/MAPK pathway were significantly enhanced. The results demonstrated that rosmanol and carnosol synergistically alleviated RA by inhibiting inflammation through regulating the TLR4/NF-κB/MAPK pathway, meaning they have the potential to be developed into novel, safe natural combinations for the treatment of RA.

Keywords: rheumatoid arthritis; rosmanol; carnosol; *Callicarpa longissima*; TLR4/NF-κB/MAPK; synergistic effect

Citation: Li, L.; Pan, Z.; Ning, D.; Fu, Y. Rosmanol and Carnosol Synergistically Alleviate Rheumatoid Arthritis through Inhibiting TLR4/NF-κB/MAPK Pathway. *Molecules* **2022**, *27*, 78. https://doi.org/10.3390/molecules27010078

Academic Editors: Raffaele Pezzani and Sara Vitalini

Received: 7 November 2021
Accepted: 20 December 2021
Published: 23 December 2021

Publisher's Note: MDPI stays neutral with regard to jurisdictional claims in published maps and institutional affiliations.

1. Introduction

Rheumatoid arthritis (RA) is a chronic autoimmune disease that is characterized by the presence of autoantibodies, lasting synovitis, and systemic inflammation. It can cause joint destruction and disability with a prevalence of 0.5–1%, leading to considerable costs to both the individual and the community [1]. Nonsteroidal anti-inflammation drugs, glucocorticoids, disease-modifying anti-rheumatic drugs, and biological agents are the currently used therapeutic drugs for RA. However, they have different side effects, including gastrointestinal, liver, pulmonary, hematological, neurological, and renal toxicities, as well as risks of serious infections [2]. Therefore, it is urgently needed to develop novel alternative drugs with less adverse effects for the treatment of RA. To meet this requirement, more anti-inflammatory medicinal plants traditionally used to treat RA have been considered, and many medicinal plant-derived anti-RA drugs have been successfully developed, including glucosides of *Tripterygium wilfordii* and total glucosides of paeony and sinomenine [3–5]. These successful stories demonstrate that undertaking an ethnopharmacology-based investigation is an effective strategy for the development of anti-RA drugs.

Callicarpa longissima (Hemsl.) Merr. is a shrub that widely grows in southern China, which has been used as a Yao folk medicine to treat arthritis, common cold, cough, bleeding, and abdominal pain [6]. The reported phytochemical composition of *C. longissima* include glycosides [6], terpenoids [7,8], phenylpropanoids, lignans, steroids, essential oils, and flavonoids [9]. Some of these compounds exhibit considerable physiological functions, such as anti-cancer [7], skin-whitening [8], anti-inflammatory [7,9], and antioxidant effects [10]. Unfortunately, no study on the anti-arthritic effect of this ethnomedicine has been reported yet.

Recently, we investigated the phytochemicals of *C. longissima* and isolated two phenolic diterpenoids, rosmanol and carnosol. Rosmanol and carnosol are abundant in this medicinal plant, with contents of no less than 0.46 mg/g and 2.37 mg/g (Figure S1), respectively, and could be regarded as its characteristic constituents. Interestingly, they are also present in *Rosmarinus officinalis* (rosemary), a medicinal and edible plant native to the Mediterranean region [11]. As major constituents of rosemary extracts, they are commonly used as flavoring agents in cooking and as antioxidants in food preservation, which have been approved by the European Union [12]. Additionally, rosmanol and carnosol in the dose range of 50–200 mg/kg/d did not exhibit any signs of acute toxicity in male Swiss mice [13,14]. Therefore, they are safe for human consumption.

The anti-inflammatory effects of rosmanol and carnosol have been investigated mainly in vitro with cell experiments. A previous study reported that rosmanol potently decreased the expression of inducible nitric oxide synthase (iNOS) and cyclooxygenase 2 in lipopolysaccharide-stimulated RAW 264.7 cells, which were mediated by inhibiting the activation of nuclear factor κB (NF-κB), signal transducer and activator of transcription-3, CCAAT/enhancer binding protein (C/EBP), and mitogen-activated protein kinase (MAPK) signaling pathways [15]. Chang et al. found that rosmanol inhibited not only the migration and proliferation of rat-fibroblast-like synoviocytes, but also the endothelial tube formation of human umbilical vein endothelial cells, which were mediated via regulating the C/EBP δ signaling pathway [16]. Schwager et al. disclosed that carnosol strongly inhibited the production of nitric oxide and prostaglandin E_2 and significantly downregulated the gene expression of iNOS, cytokines, and chemokines in both macrophages and chondrocytes, primarily by regulating the NF-κB signaling pathway [17]. Using a carrageen-induced paw edema mouse model, we found that the combination of rosmanol and carnosol exhibited more powerful anti-inflammatory effects than either compound alone (Figure S2), indicating that they synergistically suppressed inflammation.

Toll-like receptor 4 (TLR4) is a member of the Toll-like receptor family. Its downstream signaling pathways includes NF-κB and MAPKs [18]. The NF-κB is an important regulator of pro-inflammatory genes expression, such as tumor necrosis factor α (TNF-α), interleukin 6 (IL-6), IL-1β, and cyclooxygenase 2 (Cox-2), and plays a key role in the regulation of inflammation [19]. MAPK family proteins, including the extracellular-signal-regulated kinases, c-Jun N-terminal kinase (JNK), and p38 in mammals, are tightly associated with RA pathogenesis [20]. JNK regulates the expression of matrix metalloproteinases (MMPs), as well as the proliferation, migration, and invasion of synoviocytes and the destruction of joints [21,22]. P38 is critical for RA pathogenesis, as its activation involves almost all aspects of RA-related pathologies, including the expression of pro-inflammatory cytokines, synovitis, cartilage degradation, bone destruction, and angiogenesis [23,24]. Therefore, the TLR4/NF-κB/MAPK pathway is an important target for the treatment of RA.

Altogether, based on the reported studies, it is reasonable to speculate that rosmanol and carnosol are anti-arthritic constituents of *C. longissima*. Inspired by their synergistic effect in carrageen-induced paw edema mice, we assume that rosmanol and carnosol may synergistically alleviate RA. To confirm our hypothesis, the pharmacodynamics and effects on TLR4/NF-κB/MAPK pathway of rosmanol, carnosol, and the combination of both were investigated in this study with a collagen-induced arthritis (CIA) DBA/1 mouse model.

2. Materials and Methods

2.1. Plant Materials

The branches and leaves of *C. longissima* were collected from the Botanic Garden of Guangxi Institute of Botany, Chinese Academy of Sciences (CAS) in October 2019, and were identified by Associate Professor Yu-song Huang (Guangxi Institute of Botany, CAS). A voucher specimen (CTM201916) was deposited at Guangxi Key Laboratory of Functional Phytochemicals Research and Utilization, Guangxi Institute of Botany, CAS.

2.2. Preparation of Rosmanol and Carnosol

The dried branches and leaves of *C. longissima* (5.0 kg) were extracted with 95% ethanol at room temperature three times. The ethanol extract was concentrated under reduced pressure at 50 °C to give a crude extract (557 g), which was dissolved in water and partitioned successively with petroleum ether (PE) and ethyl acetate. The ethyl acetate extract (251.3 g) was subjected to a 100–200 mesh silica gel column (Qingdao Marine Chemical Factory, Qingdao, China) and eluted with PE-Me$_2$CO (from 20:1 to 1:1, *v/v*) to give five fractions (Fr.1–Fr.5). Fr.5 (18.3 g) was performed on a silica gel column (PE-Me$_2$CO, from 20:1 to 2:1, *v/v*) and subsequently purified on semi-preparative HPLC column (Agilent Technologies Inc., Santa Clara, CA, USA) with CH$_3$CN/H$_2$O (50:50) to give **1** (421.2 mg, purity >95% (Figure S3)). Fr.2 (36.2 g) was purified repeatedly on a silica gel column (PE-Me$_2$CO, 5:1, *v/v*) to yield **2** (2.1 g, purity > 95% (Figure S4)). HR-ESI-MS data was measured on an LC/MS-IT-TOF mass spectrometer (Shimadzu Co., Ltd., Kyoto, Japan) and NMR spectra were recorded with an AVANCE III HD 500 spectrometer at 25 °C (Bruker Co., Ltd., Ettlingen, Germany). Compounds **1** and **2** were identified as rosmanol and carnosol (Figure 1), respectively, by direct comparison of their MS and NMR spectral data (Figures S5–S10) with the literature [25].

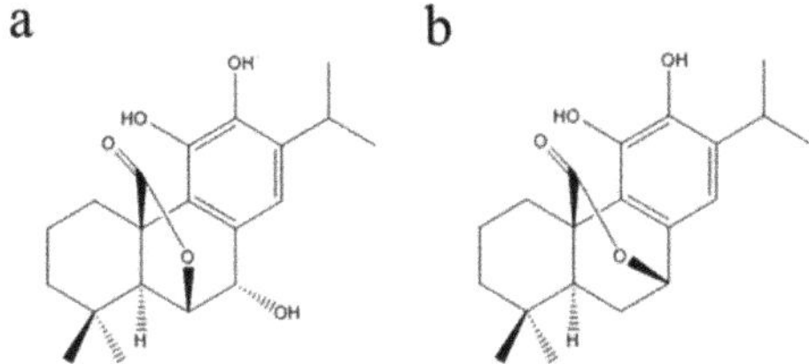

Figure 1. Chemical structures of rosmanol (**a**) and carnosol (**b**).

2.3. Chemicals and Antibodies

Bovine type II collagen (BIIC, 20022), Complete Freund's Adjuvant (CFA, 7009), and Incomplete Freund's Adjuvant (IFA, 7002) were acquired from Chondrex (Redmond, WA, USA). IL-6 (E-EL-M0044c), monocyte chemotactic protein 1 (MCP-1, E-EL-M3001), and TNF-α (E-EL-M0049c) enzyme-linked immunosorbent assay (ELISA) kits were obtained from Elabscience Biotechnology (Wuhan, China). The total RNA extraction kit (CW0560S) and cDNA synthesis kit (CW0744M) were obtained from CoWin Biosciences (Beijing, China). The quantitative real-time polymerase chain reaction (PCR) kit (FP205) was acquired from Tiangen Biotech (Beijing, China). RIPA lysis buffer (P0013B), phenylmethylsulphonyl fluoride (PMSF, ST506), sodium orthovanadate (S1873), the BCA protein quantification kit (P0012S), enhanced chemiluminescence reagents (ECL, P0018S), horse radish peroxidase (HRP)-labeled goat anti-mouse (A0286), and anti-rabbit (A0277) secondary antibodies were acquired from Beyotime (Nantong, China). Primary antibodies against glyceraldehyde-3-phosphate dehydrogenase (GAPDH, 60004-1-Ig) and TLR4 (66350-1-Ig) were obtained from Proteintech (Wuhan, China). Primary antibodies against MyD88 (4283S), NF-κB p65 (8242T), p-p65 (3033T), JNK (9252T), p-JNK (4668T), p38 (8690T), and p-p38 (4511T) were purchased from Cell Signaling Technology (Boston, MA, USA).

2.4. Animals

Thirty male DBA/1 mice aged 5−6 weeks were obtained from Changzhou Cavens Laboratory Animals Limited Company (Changzhou, China, certification No. SCXK (Su) 2016−0010). The mice were given free access to food and water. All animal care procedures followed the Guidelines for the Care and Use of Laboratory Animals from the Ministry of

Science and Technology of China. The animal experimental protocols used in this research were approved by the Laboratory Animal Management and Ethics Committee of Guangxi Institute of Botany, CAS (Guilin, China).

2.5. Establishment of CIA DBA/1 Mouse Model

After 7 days of adaption to laboratory conditions, all mice were randomly divided into 5 groups (6 mice/group), namely a drug-untreated control group (normal), BIIC-immunized group (CIA model), BIIC combined with rosmanol-treated (40 mg/kg) group, BIIC combined with carnosol-treated (40 mg/kg) group, and BIIC combined with the combination-treated (20 mg/kg rosmanol and 20 mg/kg carnosol) group. The protocol used for establishment of CIA DBA/1 mice was described by Miyoshi and Liu [26]. Briefly, on day 0, 100 μL of emulsion of BIIC and CFA (containing 100 μg BIIC) was subcutaneously injected into the base of each mouse tail, except for the normal group. A booster injection of the emulsion of BIIC and IFA was performed 21 days after the initial immunization. After the booster immunization, rosmanol, carnosol, or their combination was administered by intragastric gavage once daily on days 21–42, respectively, whereas the normal and model group mice were fed with normal saline instead (Figure 2a).

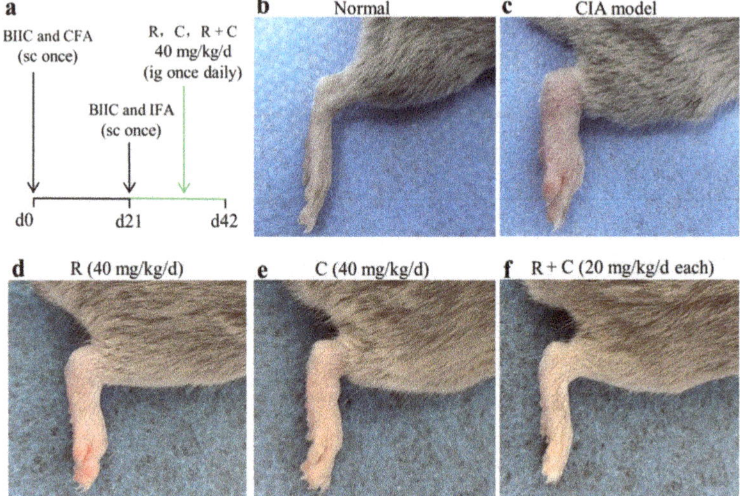

Figure 2. Rosmanol, carnosol, and their combination alleviated RA severity in CIA mice. Schematic of establishment of CIA DBA/1 mouse model and the treatments of rosmanol (R), carnosol (C), and their combination (R + C) in the CIA mice (**a**). Representative pictures from the right hind limbs of the normal group (**b**), model group (**c**), rosmanol-treated group (**d**), carnosol-treated group (**e**), and combination-treated group (**f**) on day 42.

2.6. Evaluation of Arthritis Severity

The severity of arthritis was evaluated on days 28, 35, and 42 using an established macroscopic scoring system of 0–4 per paw as described previously [27]. The severity of arthritis in each paw was scored as follows: 0 = normal joint; 1 = swelling of one joint (toe/wrist/ankle/footpath); 2 = swelling of more than one joint; 3 = swelling of all joints; 4 = bursting of the skin, dysfunction, or distortion of the joint. The cumulative score for all four paws in each mouse was used as an arthritis index to represent the overall disease severity and progression.

2.7. Hematoxylin and Eosin (H&E) Staining

Paraffin sections of synovium tissues were stained with H&E. The right hind limb specimens were fixed with 10% (*v/v*) neutral formalin for 24 h, then embedded in paraffin and sliced into 4-μm-thick tissue sections. H&E staining was performed according to protocols described previously [28].

2.8. ELISA

On day 42, all mice were euthanized by CO_2 asphyxiation. Blood samples from each mouse were collected immediately, coagulated naturally at room temperature for 1 h, and then centrifuged by $3000 \times g$ for 10 min at 4 °C. The IL-6, MCP-1, and TNF-α levels of the supernatant (serum) were determined with ELISA kits according to the manufacturer's instructions. The absorbance at the wavelength of 450 nm was determined on a SPARK microplate reader (Tecan, Männedorf, Switzerland).

2.9. Quantitative Real-Time PCR

Total RNA was prepared from mice synovial tissues and used as a template for first strand cDNA synthesis. Quantitative real-time PCR analysis was performed with the QuantStudio 6 System (ABI, Foster, CA, USA). Primer sequences used in real-time PCR were as follows: GAPDH forward and reverse primers, 5′-ATGGGTGTGAACCACGAGA-3′ and 5′-CAGGGATGATGTTCTGGGCA-3′; TLR4 forward and reverse primers, 5′-GCCCTACCA AGTCTCAGCTA-3′ and 5′-CTGCAGCTCTTCTAGACCCA-3′; MyD88 forward and reverse primers, 5′-CCCACTCGCAGTTTGTTG-3′ and 5′- CACCTGTAAAGGCTTCTCG-3′; p65 forward and reverse primers, 5′-CACCGGATTGAAGAGAAGCG-3′ and 5′- AAGTTGATG GTGCTGAGGGA-3′. The reaction mixtures contained 10 μL of SYBR Green Master Mix, 0.4 μL of ROX Reference Dye (50×), 1 μL of cDNA, 0.4 μL of forward and reverse primers (10 μM), and 7.8 μL of RNase-free water. The amplification protocols were as follows: a 10 min initial denaturation step at 95 °C, followed by 40 three-step cycles, including a denaturation step (95 °C, 15 s), an annealing step (60 °C, 60 s), and an extension step (72 °C, 15 s). The experimental results were calculated using the $2^{-\Delta\Delta Ct}$ method.

2.10. Western Blot

Frozen synovial tissues were homogenized in cold RIPA lysis buffer containing 1 mM sodium orthovanadate and 1 mM PMSF on ice. The homogenates were centrifuged at $13,000 \times g$ for 10 min at 4 °C to remove debris. The total protein concentrations of the supernatants were quantified with BCA assay. Equal amounts of proteins (40 μg) were separated by 12% SDS–polyacrylamide gel electrophoresis and transferred to polyvinylidene difluoride (PVDF) membranes. The membranes were blocked with 5% (w/v) skimmed milk powder in Tris-buffered saline Tween 20 (TBST, 10 mM Tris, 150 mM NaCl, pH 7.4, 0.5% Tween 20) for 1.5 h, then were incubated with primary antibodies against mice, including GAPDH (1:10,000), TLR4 (1:2000), MyD88 (1:1000), p38 (1:1000), p-p38 (1:800), JNK (1:1000), p-JNK (1:1000), NF-κB p65 (1:1000), and p-p65 (1:500), at 4 °C overnight. After washing with TBST (3 × 10 min), the membranes were incubated with HRP-labeled goat anti-mouse or anti-rabbit secondary antibodies (1:1000) for 1.5 h at room temperature. The protein bands were developed with an ECL detection reagent according to the manufacturer's instructions and were exposed to X-ray film. The intensity levels of antibody-reactive bands were analyzed with Image Lab 3.0 (Bio-Rad, Hercules, CA, USA).

2.11. Statistical Analysis

All data used were expressed as means ± standard deviation (SD). Statistical differences were analyzed by unpaired *t*-test with GraphPad Prism 5.0 software (GraphPad Software Inc., San Diego, CA, USA), where $p < 0.05$ was considered significantly different.

3. Results

3.1. Rosmanol and Carnosol Synergistically Alleviated RA Pathologies in CIA Mice

To examine the effects of rosmanol and carnosol on RA in vivo, a CIA DBA/1 mouse model was established. It was observed clearly that the development of arthritis was induced by BIIC on day 42, with the symptoms including swelling and redness (Figure 2b,c).

Notably, the arthritis severity was alleviated in mice treated by rosmanol, carnosol, or the combination of both (Figure 2d–f). To compare the arthritis severity quantitatively, the arthritis index scores for each group were calculated on days 28, 35, and 42, respectively. The arthritis index score for the normal group was 0 on each day. On day 28, there were no significant differences among BIIC-induced groups. On day 35, compared to the model group, though the arthritis severity of the rosmanol-treated and carnosol-treated mice was attenuated slightly, for the combination-treated mice it was attenuated significantly ($p < 0.05$). On day 42, the arthritis severity was markedly attenuated in the rosmanol-treated and carnosol-treated group compared to the model group ($p < 0.05$), with arthritis index scores of 8.7, 8.5, and 10.0, respectively. In particular, the arthritis severity was further significantly attenuated in the combination-treated group with an arthritis index score of 6.8 ($p < 0.05$) (Figure 3). Taken together, these data clearly demonstrated that rosmanol and carnosol synergistically alleviated RA in CIA DBA/1 mice.

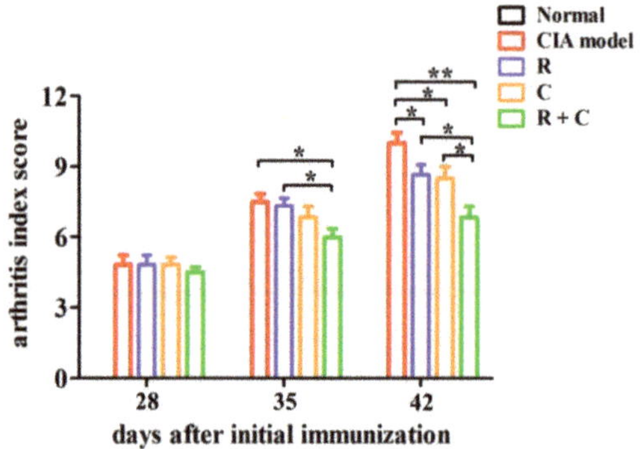

Figure 3. Effects of rosmanol (R), carnosol (C), and their combination (R + C) on arthritis index scores in CIA mice. Arthritis index scores were the cumulative scores for all four paws, which represented the overall disease severity. The arthritis index score for each group is presented as the mean $\pm$ SD ($n = 6$). The significance of differences was analyzed by unpaired t-test. Note: * $p < 0.05$; ** $p < 0.01$.

3.2. Rosmanol and Carnosol Synergistically Alleviated Synovitis in CIA Mice

To study the effects of rosmanol, carnosol, and their combination on synovitis, an H&E staining experiment was performed on mice synovial tissue sections. Representative results are shown in Figure 4. Compared to the normal group, severely proliferated synovial cells, hyperplastic fibrous tissues, and infiltrated inflammatory cells were observed in the model group (Figure 4a,b). These pathology changes were alleviated by daily gavage feeding with 40 mg/kg of rosmanol, carnosol, or a combination of both; the combination-treated mice in particular retained nearly normal architecture of synovial tissues (Figure 4c–e), suggesting that rosmanol and carnosol could synergistically alleviate synovitis in CIA DBA/1 mice.

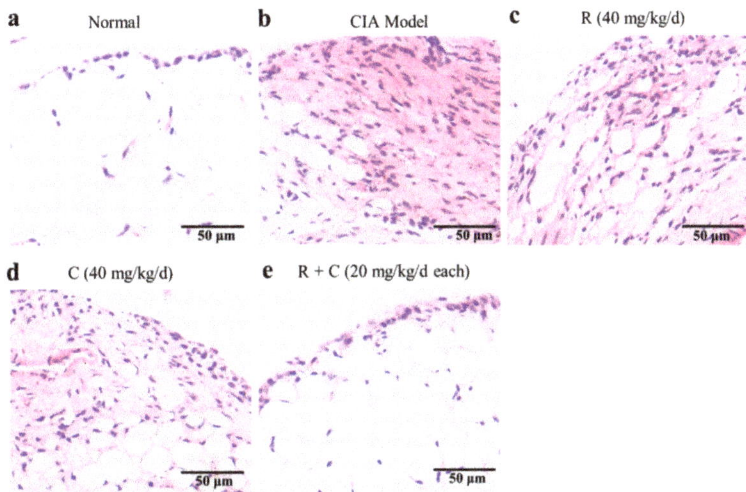

Figure 4. Rosmanol (R), carnosol (C), and their combination (R + C) alleviated synovitis in CIA mice. Representative H&E staining results from synovial tissue samples of the normal group (**a**), model group (**b**), rosmanol-treated group (**c**), carnosol-treated group (**d**), and combination-treated group (**e**) mice are shown ($200\times$, $n = 3$).

3.3. Rosmanol and Carnosol Synergistically Decreased Pro-Inflammatory Cytokines in Serum

Persistent inflammation is maintained during RA pathogenesis [1]. To study the effects of rosmanol, carnosol, and their combination on the inflammation response of CIA mice, the serum levels of pro-inflammatory cytokines such as TNF-α, MCP-1, and IL-6 were quantified by ELISA on day 42 (Figure 5). These pro-inflammatory cytokines increased dramatically after stimulation by BIIC and decreased slightly in the rosmanol-treated group but decreased significantly in the carnosol-treated group ($p < 0.01$), indicating that rosmanol and carnosol exerted anti-inflammatory roles in CIA mice. Of note, the levels of TNF-α, MCP-1, and IL-6 further decreased in the combination-treated group ($p < 0.05$), indicating that rosmanol and carnosol synergistically inhibited the inflammation response in CIA DBA/1 mice.

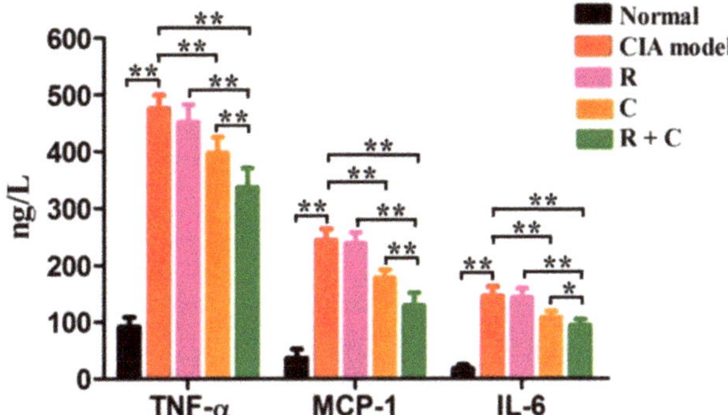

Figure 5. The effects of rosmanol (R), carnosol (C), and their combination (R + C) on the production of TNF-α, MCP-1, and IL-6. The serum levels of TNF-α, MCP-1, and IL-6 of each group were determined by ELISA on day 42. Data are presented as the means $\pm$ SD ($n = 6$). The significance of differences was analyzed by unpaired *t*-test. Note: * $p < 0.05$; ** $p < 0.01$.

3.4. Rosmanol and Carnosol Synergistically Inhibited the TLR4/NF-κB Pathway

TLR4/NF-κB pathway is an important regulator of pro-inflammatory gene expression, such as that of TNF-α, IL-6, IL-1β, and Cox-2. It plays a key role in the regulation of inflammation [18,19] and has been proven to be a therapeutic target for RA treatment [29]. To study the influences of rosmanol, carnosol, and their combination on the TLR4/NF-κB pathway, the transcription levels of TLR4, MyD88, and NF-κB p65 in synovial tissue were tested using real-time PCR on day 42. Compared to the normal group, the TLR4, MyD88, and NF-κB p65 mRNA levels of the model group were upregulated remarkably ($p < 0.01$). After treatment with rosmanol or carnosol, the transcription levels of TLR4, MyD88, and NF-κB p65 were downregulated to different degrees and the downregulation was strengthened when treated with rosmanol and carnosol together ($p < 0.05$) (Figure 6a–c). Similar results were obtained at the protein level, as examined by Western blot, whereby rosmanol and carnosol alone inhibited not only the translation of TLR4 and MyD88, but also the phosphorylation of p65. The inhibition was strengthened when mice were treated with their combination ($p < 0.05$) (Figure 6d–f). Taken together, these data demonstrate that rosmanol and carnosol synergistically inhibited the activation of the TLR4/NF-κB pathway.

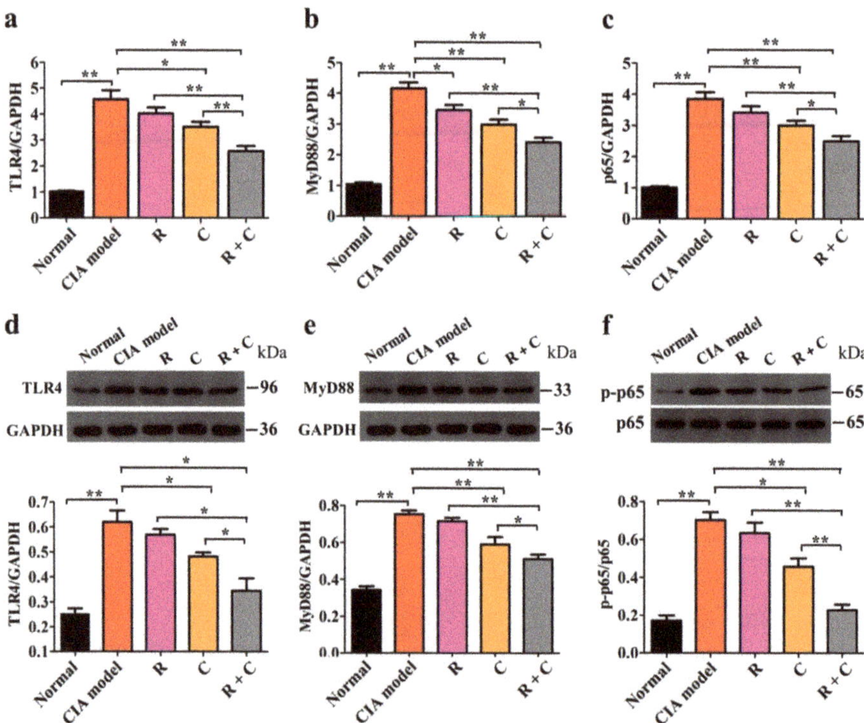

Figure 6. Rosmanol (R), carnosol (C), and their combination (R + C) inhibited the TLR4/NF-κB pathway. The relative transcription levels of TLR4 (**a**), MyD88 (**b**), and NF-κB p65 (**c**) in synovial tissue samples of each group were quantified by real-time PCR on day 42. The representative Western blot analysis results and protein relative expression levels for TLR4 (**d**), MyD88 (**e**), and NF-κB p-p65 (**f**) in synovial tissue samples of each group on day 42 are shown. GAPDH was used as an internal reference to quantify the expression levels of TLR4 and MyD88. The phosphorylation of p65 was expressed as p-p65/p65. Data are presented as the mean ± SD ($n = 3$). The significance of differences was analyzed by unpaired *t*-test. Note: *, $p < 0.05$; **, $p < 0.01$.

3.5. Rosmanol and Carnosol Synergistically Inhibited MAPK Activation

MAPKs are downstream of TLR4 and play important roles in the pathogenesis of RA, as the abnormal activation of JNK and p38 almost participate in all aspects of RA-related pathologies [21–24]. To study the effects of rosmanol, carnosol, and their combination on the MAPKs, the expression levels of JNK, p-JNK, p38, and p-p38 in synovial tissue were examined by Western blot on day 42 (Figure 7). Compared to the normal group, though the expression levels of JNK and p38 were not influenced, the phosphorylation rates of JNK and p38 were upregulated remarkably in the model group ($p < 0.01$), indicating the activation of JNK and p38 MAPK pathways. The phosphorylation of JNK and p38 was inhibited by both rosmanol and carnosol alone ($p < 0.05$), while the inhibitory effect was enhanced when rosmanol and carnosol were used jointly ($p < 0.05$), indicating that rosmanol and carnosol synergistically inhibited the activation of JNK and p38 MAPK pathways.

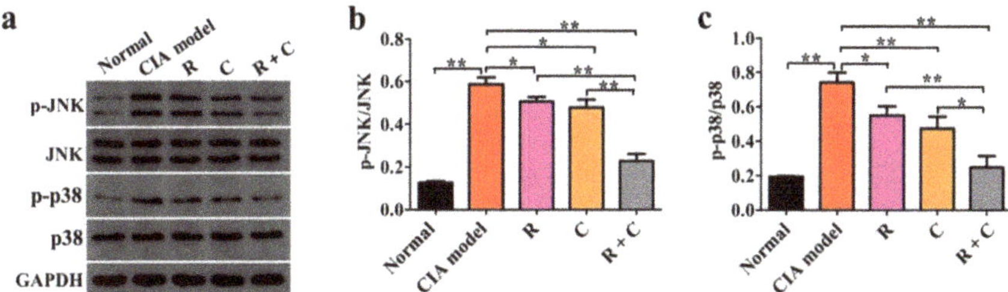

Figure 7. Rosmanol (R), carnosol (C), and their combination (R + C) inhibited the JNK and p38 MAPK pathways. The protein expression levels of JNK, p-JNK, p38, and p-p38 in synovial tissue samples of each group were examined by Western blot on day 42 (**a**) and the relative phosphorylation rates of JNK (**b**) and p38 (**c**) were calculated. GAPDH was used as the internal reference to quantify the expression levels of JNK, p-JNK, p38, and p-p38. The phosphorylation rates of JNK and p38 were expressed as p-JNK/JNK and p-p38/p38, respectively. Data are presented as the means $\pm$ SD (n = 3). The significance of differences was analyzed by unpaired t-test. Note: * $p < 0.05$; ** $p < 0.01$.

4. Discussion

Rheumatoid arthritis is considered to be a multi-factorial autoimmune disease. Its pathogenesis involves epigenetic alterations, post-translational modifications, autophagy, T-cells, and other factors [30]. Natural products are abundant in quantity and diverse in structure, so they may target various biological processes; for example, curcumin and resveratrol are modulators of NF-κB [31,32], while Akone et al. reviewed the advances in natural products that could modulate DNA methylation and histone deacetylation [33], indicating that natural products are important sources for the discovery of drugs for the treatment of multi-factorial diseases such as RA.

Callicarpa longissima has been used as a Yao folk medicine to treat arthritis for a long time in China [6], although its anti-arthritic active moieties have not been clarified so far. In this study, we reported on anti-RA constituents and their active mechanism of this ethnomedicine for the first time. Rosmanol and carnosol, two major phenolic diterpenoids from the branches and leaves of *C. longissima*, were found to alleviate swelling, redness, and synovitis in CIA DBA/1 mice (Figures 2–4), demonstrating that they were the constituents responsible for the anti-RA effect of *C. longissima*. The safety of rosmanol and carnosol has been proven in previous studies [13,14]. In this study, though rosmanol and carnosol exhibited a slight ability to alleviate body weight loss in CIA mice, their combination significantly alleviated body weight loss after day 35 (Table S1), indicating the safety of rosmanol, carnosol, and their combination. Therefore, they have the potential to be developed into novel, safe natural agents for the treatment of RA.

A synergistic effect is an interaction of two or more ingredients that results in a greater effect than the total of their separate effects. Drug combinations with synergistic effects can often alleviate drug resistance and improve the therapeutic efficacy [34], so it is highly significant to continue to research such combinations. Rosmanol and carnosol synergistically inhibited paw swelling (Figure S2); inspired by this discovery, the anti-RA effect of their combination was also assessed in this study. Compared to rosmanol or carnosol alone, their combination alleviated RA symptoms such as swelling, redness, and synovitis and reduced the arthritis index score more significantly (Figures 2–4). In addition, on day 35, the RA-related pathologies in the combination-treated mice were significantly alleviated, although the inhibitory effect was not obvious in the rosmanol-treated or carnosol-treated mice (Figure 3). These data powerfully demonstrated that rosmanol and carnosol synergistically alleviated RA in CIA DBA/1 mice.

It is well-known that the inflammation response, which is involved in many pro-inflammatory mediators, takes part in every phase of RA pathogenesis [1]. Compared to rosmanol-treated and carnosol-treated groups, the serum levels of TNF-α, IL-6, and MCP-1 in the combination-treated mice were all significantly reduced (Figure 5), indicating that rosmanol and carnosol could synergistically alleviate RA via inhibiting the inflammation response. This effect was similar to those of their respective analogs rosmarinic acid and carnosic acid, whose anti-RA activity also via inflammation suppression [35,36].

Earlier studies have demonstrated that the TLR4/NF-κB pathway is a plausible therapeutic target for RA treatment [29,37]. The combination of rosmanol and carnosol suppressed both the expression levels of TLR4 and MyD88 and the phosphorylation rates of NF-κB p65 more significantly than either compound alone (Figure 6), indicating that they synergistically inhibited the activation of the TLR4/NF-κB pathway in CIA DBA/1 mice. The combination of rosmanol and carnosol also blocked the phosphorylation rates of both JNK and p38 more powerfully than either single compound did (Figure 7), indicating that they synergistically inhibited the activation of both JNK and p38 MAPK pathways. Rosmanol and carnosol could possibly inhibit angiogenesis, cartilage degradation, and bone destruction, as the activation of JNK and p38 involves these RA-related pathologies [21–24], although this speculation requires further validation.

In conclusion, the data presented in this study clearly demonstrate that rosmanol and carnosol synergistically alleviated RA via inhibiting inflammation through regulating the TLR4/NF-κB/MAPK pathway (Figure 8), and they have the potential to be developed into novel, safe natural combinations for the treatment of RA. The results also proved that *Callicarpa longissima* can be used as a source of natural anti-inflammatory drugs.

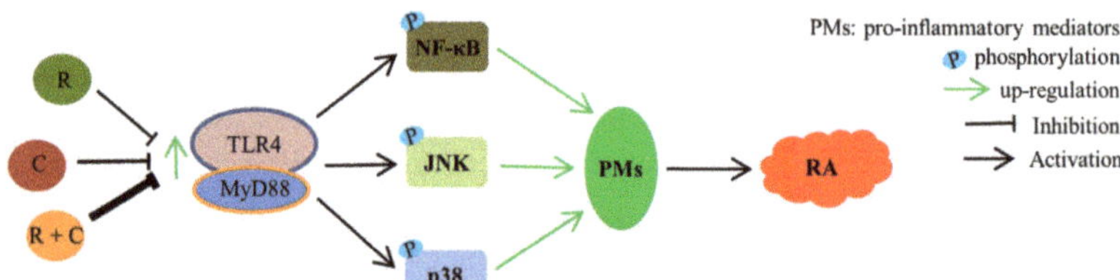

Figure 8. The signaling pathways that rosmanol (R), carnosol (C), and their combination (R + C) target in RA. The intensity of inhibition is proportional to the line width of the inhibition symbol.

Supplementary Materials: The following are available online. Figure S1: Chromatogram of standard reference materials and methanol extract of *C. longissima*; Figure S2: Rosmanol and carnosol synergistically inhibited paw edema in carrageenan-induced Kunming mouse model; Figure S3: HPLC spectrum of rosmanol (1); Figure S4: HPLC spectrum of carnosol (2); Figure S5: HR-ESI-MS spectrum of rosmanol (1); Figure S6: ^{1}H-NMR spectrum of rosmanol (1) (500 MHz in CD$_3$OD); Figure S7: ^{13}C-NMR spectrum of rosmanol (1) (125 MHz in CD$_3$OD); Figure S8: HR-ESI-MS spectrum of carnosol (2); Figure S9: ^{1}H-NMR spectrum of carnosol (2) (500 MHz in CD$_3$OD); Figure S10: Figure S10 ^{13}C-NMR spectrum of carnosol (2) (125 MHz in CD$_3$OD); Table S1: Effects of rosmanol, carnosol and their combination on body weight in CIA DBA/1 mice.

Author Contributions: L.L., D.N. and Y.F. performed the experiments. L.L. and D.N. wrote the manuscript. Z.P. designed the experiments and revised the manuscript. All authors have read and agreed to the published version of the manuscript.

Funding: This research was financially supported by the Natural Science Foundation of Guangxi (2018GXNSFDA050016, 2018GXNSFBA281094), the Special Funds for Local Science and Technology Development Guided by the Central Committee (ZY20111010), and the Guangxi Innovation-Driven Development Special Project (Guike AA18118015).

Institutional Review Board Statement: The animal experiments were approved by the Laboratory Animal Management and Ethics Committee of Guangxi Institute of Botany, CAS (Approval ID: 2020-09-16).

Informed Consent Statement: Not applicable.

Data Availability Statement: Data are provided in the manuscript and Supplementary Materials.

Conflicts of Interest: The authors declare no conflict of interest.

Sample Availability: Samples of rosmanol and carnosol are not available from the authors.

Abbreviations

BIIC	bovine type II collagen
CIA	collagen-induced arthritis
H&E	hematoxylin and eosin
IL-6	interleukin 6
JNK	c-Jun N-terminal kinase
MAPK	mitogen-activated protein kinase
MCP-1	monocyte chemotactic protein 1
MyD88	myeloid differentiation factor 88
NF-κB	nuclear factor κB
RA	rheumatoid arthritis
TLR4	Toll-like receptor 4
TNF-α	tumor necrosis factor α

References

1. Scott, D.L.; Wolfe, F.; Huizing, T.W. Rheumatoid arthritis. *Lancet* **2010**, *376*, 1094–1108. [CrossRef]
2. Kour, G.; Haq, S.A.; Bajaj, B.K.; Gupta, P.N.; Ahmed, Z. Phytochemical add-on therapy to DMARDs therapy in rheumatoid arthritis: In vitro and in vivo bases, clinical evidence and future trends. *Pharmacol. Res.* **2021**, *169*, 105618. [CrossRef] [PubMed]
3. Zhang, D.; Lyu, J.T.; Zhang, B.; Zhang, X.M.; Jiang, H.; Lin, Z.J. Comparative efficacy, safety and cost of oral Chinese patent medicines for rheumatoid arthritis: A Bayesian network meta-analysis. *BMC Complement. Med. Ther.* **2020**, *20*, 210. [CrossRef]
4. Jiang, H.; Li, J.; Wang, L.; Wang, S.; Nie, X.; Chen, Y.; Fu, Q.; Jiang, M.; Fu, C.; He, Y. Total glucosides of paeony: A review of its phytochemistry, role in autoimmune diseases, and mechanisms of action. *J. Ethnopharmacol.* **2020**, *258*, 112913. [CrossRef] [PubMed]
5. Liu, W.; Zhang, Y.; Zhu, W.; Ma, C.; Ruan, J.; Long, H.; Wang, Y. Sinomenine inhibits the progression of rheumatoid arthritis by regulating the secretion of inflammatory cytokines and monocyte/macrophage subsets. *Front. Immunol.* **2018**, *9*, 2228. [CrossRef] [PubMed]
6. Yuan, J.Q.; Qiu, L.; Zou, L.H.; Wei, Q.; Miao, J.H.; Yao, X.S. Two new phenylethanoid glycosides from *Callicarpa longissima*. *Helv. Chim. Acta* **2015**, *98*, 482–489. [CrossRef]
7. Liu, Y.W.; Cheng, Y.B.; Liaw, C.C.; Chen, C.H.; Guh, J.H.; Hwang, T.L.; Tsai, J.S.; Wang, W.B.; Shen, Y.C. Bioactive diterpenes from *Callicarpa longissima*. *J. Nat. Prod.* **2012**, *75*, 689–693. [CrossRef]
8. Yamahara, M.; Sugimura, K.; Kumagai, A.; Fuchino, H.; Kuroi, A.; Kagawa, M.; Itoh, Y.; Kawahara, H.; Nagaoka, Y.; Iida, O.; et al. *Callicarpa longissima* extract, carnosol-rich, potently inhibits melanogenesis in B16F10 melanoma cells. *J. Nat. Med.* **2016**, *70*, 28–35. [CrossRef] [PubMed]
9. Yuan, J.Q. Studies on the Constituents of *Callicarpa longissima*. Ph.D. Thesis, Jinan University, Guangzhou, China, 2015.
10. Kawamura, T.; Momozane, T.; Sanosaka, M.; Sugimura, K.; Iida, O.; Fuchino, H.; Funaki, S.; Shintani, Y.; Inoue, M.; Minami, M.; et al. Carnosol is a potent lung protective agent: Experimental study on mice. *Transplant. Proc.* **2015**, *47*, 1657–1661. [CrossRef] [PubMed]
11. Haraguchi, H.; Saito, T.; Okamura, N.; Yagi, A. Inhibition of lipid peroxidation and superoxide generation by diterpenoids from *Rosmarinus officinalis*. *Planta Med.* **1995**, *61*, 333–336. [CrossRef] [PubMed]

12. Andrade, J.M.; Faustino, C.; Garcia, C.; Ladeiras, D.; Reis, C.P.; Rijo, P. *Rosmarinus officinalis* L.: An update review of its phytochemistry and biological activity. *Future Sci. OA* **2018**, *4*, FSO283. [CrossRef] [PubMed]
13. Abdelhalim, A.; Karim, N.; Chebib, M.; Aburjai, T.; Khan, I.; Johnston, G.A.; Hanrahan, J. Antidepressant, anxiolytic and antinociceptive activities of constituents from *Rosmarinus Officinalis*. *J. Pharm. Pharm. Sci.* **2015**, *18*, 448–459. [CrossRef] [PubMed]
14. Khan, I.; Karim, N.; Ahmad, W.; Abdelhalim, A.; Chebib, M. GABA-A receptor modulation and anticonvulsant, anxiolytic, and antidepressant activities of constituents from *Artemisia indica* Linn. *Evid.-Based Complement. Alternat. Med.* **2016**, *2016*, 1215393. [CrossRef] [PubMed]
15. Lai, C.S.; Lee, J.H.; Ho, C.T.; Liu, C.B.; Wang, J.M.; Wang, Y.J.; Pan, M.H. Rosmanol potently inhibits lipopolysaccharide-induced iNOS and COX-2 expression through downregulating MAPK, NF-kappaB, STAT3 and C/EBP signaling pathways. *J. Agric. Food Chem.* **2009**, *57*, 10990–10998. [CrossRef]
16. Chang, L.H.; Huang, H.S.; Wu, P.T.; Jou, I.M.; Pan, M.H.; Chang, W.C.; Wang, D.D.; Wang, J.M. Role of macrophage CCAAT/enhancer binding protein delta in the pathogenesis of rheumatoid arthritis in collagen-induced arthritic mice. *PLoS ONE* **2012**, *7*, e45378. [CrossRef]
17. Schwager, J.; Richard, N.; Fowler, A.; Seifert, N.; Raederstorff, D. Carnosol and related substances modulate chemokine and cytokine production in macrophages and chondrocytes. *Molecules* **2016**, *21*, 465. [CrossRef]
18. Chen, J.Q.; Szodoray, P.; Zeher, M. Toll-like receptor pathways in autoimmune diseases. *Clin. Rev. Allergy Immunol.* **2016**, *50*, 1–17. [CrossRef] [PubMed]
19. Tak, P.P.; Firestein, G.S. NF-kappaB: A key role in inflammatory diseases. *J. Clin. Investig.* **2001**, *107*, 7–11. [CrossRef]
20. Thalhamer, T.; McGrath, M.A.; Harnett, M.M. MAPKs and their relevance to arthritis and inflammation. *Rheumatology* **2008**, *47*, 409–414. [CrossRef]
21. Han, Z.; Boyle, D.L.; Chang, L.; Bennett, B.; Karin, M.; Yang, L.; Manning, A.M.; Firestein, G.S. C-Jun N-terminal kinase is required for metalloproteinase expression and joint destruction in inflammatory arthritis. *J. Clin. Investig.* **2001**, *108*, 73–81. [CrossRef]
22. Zhu, S.; Ye, Y.; Shi, Y.; Dang, J.; Feng, X.; Chen, Y.; Liu, F.; Olsen, N.; Huang, J.; Zheng, S.G. Sonic hedgehog regulates proliferation, migration and invasion of synoviocytes in rheumatoid arthritis via JNK signaling. *Front. Immunol.* **2020**, *11*, 1300. [CrossRef] [PubMed]
23. Cuenda, A.; Rousseau, S. P38 MAP-kinases pathway regulation, function and role in human diseases. *Biochim. Biophys. Acta* **2007**, *1773*, 1358–1375. [CrossRef]
24. Schett, G.; Zwerina, J.; Firestein, G. The p38 mitogen-activated protein kinase (MAPK) pathway in rheumatoid arthritis. *Ann. Rheum. Dis.* **2008**, *67*, 909–916. [CrossRef]
25. Liu, X.; Du, J.; Ou, Y.; Xu, H.; Chen, X.; Zhou, A.; He, L.; Cao, Y. Degradation pathway of carnosic acid in methanol solution through isolation and structural identification of its degradation products. *Eur. Food Res. Technol.* **2013**, *237*, 617–626. [CrossRef]
26. Miyoshi, M.; Liu, S. Collagen-induced arthritis models. *Methods Mol. Biol.* **2018**, *1868*, 3–7.
27. Yeremenko, N.; Härle, P.; Cantaert, T.; van Tok, M.; van Duivenvoorde, L.M.; Bosserhoff, A.; Baeten, D. The cartilage protein melanoma inhibitory activity contributes to inflammatory arthritis. *Rheumatology* **2014**, *53*, 438–447. [CrossRef] [PubMed]
28. Cardiff, R.D.; Miller, C.H.; Munn, R.J. Manual hematoxylin and eosin staining of mouse tissue sections. *Cold Spring Harb. Protoc.* **2014**, *2014*, 655–658. [CrossRef] [PubMed]
29. Arjumand, S.; Shahzad, M.; Shabbir, A.; Yousaf, M.Z. Thymoquinone attenuates rheumatoid arthritis by downregulating TLR2, TLR4, TNF-α, IL-1, and NFκB expression levels. *Biomed. Pharmacother.* **2019**, *111*, 958–963. [CrossRef]
30. Mueller, A.L.; Payandeh, Z.; Mohammadkhani, N.; Mubarak, S.M.H.; Zakeri, A.; Alagheband Bahrami, A.; Brockmueller, A.; Shakibaei, M. Recent advances in understanding the pathogenesis of rheumatoid arthritis: New treatment strategies. *Cells* **2021**, *10*, 3017. [CrossRef] [PubMed]
31. Buhrmann, C.; Brockmueller, A.; Mueller, A.L.; Shayan, P.; Shakibaei, M. Curcumin attenuates environment-derived osteoarthritis by Sox9/NF-kB signaling axis. *Int. J. Mol. Sci.* **2021**, *22*, 7645. [CrossRef] [PubMed]
32. Shakibaei, M.; Csaki, C.; Nebrich, S.; Mobasheri, A. Resveratrol suppresses interleukin-1beta-induced inflammatory signaling and apoptosis in human articular chondrocytes: Potential for use as a novel nutraceutical for the treatment of osteoarthritis. *Biochem. Pharmacol.* **2008**, *76*, 1426–1439. [CrossRef] [PubMed]
33. Akone, S.H.; Ntie-Kang, F.; Stuhldreier, F.; Ewonkem, M.B.; Noah, A.M.; Mouelle, S.E.M.; Müller, R. Natural products impacting DNA methyltransferases and histone deacetylases. *Front. Pharmacol.* **2020**, *11*, 992. [CrossRef] [PubMed]
34. Liu, Q.; Xie, L. TranSynergy: Mechanism-driven interpretable deep neural network for the synergistic prediction and pathway deconvolution of drug combinations. *PLoS Comput. Biol.* **2021**, *17*, e1008653. [CrossRef] [PubMed]
35. Xia, G.; Wang, X.; Sun, H.; Qin, Y.; Fu, M. Carnosic acid (CA) attenuates collagen-induced arthritis in db/db mice via inflammation suppression by regulating ROS-dependent p38 pathway. *Free Radic. Biol. Med.* **2017**, *108*, 418–432. [CrossRef]
36. Youn, J.; Lee, K.H.; Won, J.; Huh, S.J.; Yun, H.S.; Cho, W.G.; Paik, D.J. Beneficial effects of rosmarinic acid on suppression of collagen induced arthritis. *J. Rheumatol.* **2003**, *30*, 1203–1207. [PubMed]
37. Wang, Y.; Zheng, F.; Gao, G.; Yan, S.; Zhang, L.; Wang, L.; Cai, X.; Wang, X.; Xu, D.; Wang, J. MiR-548a-3p regulates inflammatory response via TLR4/NF-κB signaling pathway in rheumatoid arthritis. *J. Cell. Biochem.* **2019**, *120*, 1133–1140. [CrossRef] [PubMed]

 *molecules*

Article

Antioxidant and Anti-Inflammatory Effects of *Peganum harmala* Extracts: An In Vitro and In Vivo Study

Malik Waseem Abbas [1], Mazhar Hussain [1,*], Muhammad Qamar [2], Sajed Ali [3], Zahid Shafiq [1], Polrat Wilairatana [4,*] and Mohammad S. Mubarak [5,*]

[1] Institute of Chemical Sciences, Bahauddin Zakariya University, Multan 60800, Pakistan; wasimchemist229@gmail.com (M.W.A.); Zahidshafiq@bzu.edu.pk (Z.S.)

[2] Institute of Food Science and Nutrition, Bahauddin Zakariya University, Multan 60800, Pakistan; Muhammad.qamar44@gmail.com

[3] Department of Biotechnology, University of Management and Technology, Sialkot 51041, Pakistan; sajed.ali@skt.umt.edu.pk

[4] Department of Clinical Tropical Medicine, Faculty of Tropical Medicine, Mahidol University, Bangkok 10400, Thailand

[5] Department of Chemistry, The University of Jordan, Amman 11942, Jordan

* Correspondence: mazharhussain@bzu.edu.pk (M.H.); polrat.wil@mahidol.ac.th (P.W.); mmubarak@ju.edu.jo (M.S.M); Tel.: +962-791016126 (M.S.M.)

Citation: Abbas, M.W.; Hussain, M.; Qamar, M.; Ali, S.; Shafiq, Z.; Wilairatana, P.; Mubarak, M.S. Antioxidant and Anti-Inflammatory Effects of *Peganum harmala* Extracts: An In Vitro and In Vivo Study. *Molecules* **2021**, *26*, 6084. https://doi.org/10.3390/molecules26196084

Academic Editors: Raffaele Pezzani and Sara Vitalini

Received: 12 September 2021
Accepted: 6 October 2021
Published: 8 October 2021

Publisher's Note: MDPI stays neutral with regard to jurisdictional claims in published maps and institutional affiliations.

Abstract: *Peganum harmala* (*P. harmala*) belongs to the family *Zygophyllaceae*, and is utilized in the traditional medicinal systems of Pakistan, China, Morocco, Algeria, and Spain to treat several chronic health disorders. The aim of the present study was to identify the chemical constituents and to evaluate the antioxidant, anti-inflammatory, and toxicity effects of *P. harmala* extracts both in vitro and in vivo. Sequential crude extracts including 100% dichloromethane, 100% methanol, and 70% aqueous methanol were obtained and their antioxidant and anti-inflammatory effects evaluated both in vitro and in vivo. The anti-inflammatory effect of the extract was investigated using the carrageenan-induced paw edema method in mice, whereas the toxicity of the most active extract was evaluated using an acute and subacute toxicity rat model. In addition, we have used the bioassay-guided approach to obtain potent fractions, using solvent–solvent partitioning and reversed phase high performance liquid chromatography from active crude extracts; identification and quantification of compounds from the active fractions was achieved using electrospray ionization mass spectrometry and high performance liquid chromatography techniques. Results revealed that the 100% methanol extract of *P. harmala* exhibits significant in vitro antioxidant activity in DPPH assay with an IC_{50} of 49 µg/mL as compared to the standard quercetin with an IC_{50} of 25.4 µg/mL. The same extract exhibited 63.0% inhibition against serum albumin denaturation as compared to 97% inhibition by the standard diclofenac sodium in an in vitro anti-inflammatory assay, and in vivo anti-inflammatory against carrageenan-induced paw edema (75.14% inhibition) as compared to 86.1% inhibition caused by the standard indomethacin. Furthermore, this extract was not toxic during a 14 day trial of acute toxicity when given at a dose of 3 g/kg, indicating that the lethal dose (LD_{50}) of *P. harmala* methanol extract was greater than 3 g/kg. *P. harmala* methanolic fraction 2 obtained using bioassay-guided fractionation showed the presence of quinic acid, peganine, harmol, harmaline, and harmine, confirmed by electrospray ionization mass spectrometry and quantified using external standards on high performance liquid chromatography. Taken all together, the current investigation further confirms the antioxidant, anti-inflammatory, and safety aspects of *P. harmala*, which justifies its use in folk medicine.

Keywords: *Peganum harmala*; anti-inflammatory activity; antioxidant; toxicity; LC-ESI-MS/MS; traditional medicine

1. Introduction

Free radicals are highly reactive species that can cause DNA damage owing to their unstable nature, and are a documented cause of different ailments including inflammation, cancer, ageing, bone diseases, cataracts, and even neurodegenerative disorders [1]. On the other hand, inflammation is the immune system response to injury, infection, or destruction characterized by heat, pain, swelling, redness, and disturbed physiological functions [2]. In contrast, chronic inflammation is usually related to pain, and involves some other events such as membrane disruptions, denaturation of protein, and an increase in vascular permeability [3]. Nonsteroidal anti-inflammatory drugs (NSAIDS) and corticosteroids are the most commonly used to reduce inflammation and relieve pain induced by inflammatory conditions [4]. However, extended exposure to these drugs may cause severe gastric lesions, digestive system upset, and liver and kidney diseases [5].

The genus *Peganum* is amongst the most significant medicinal plants, having two varieties of six species. *Peganum harmala* L., usually known as "Harmal", is a glabrous perennial plant, which belongs to the family *Zygophyllaceae* [6,7]. *P. harmala* is a perennial herb growing in arid and semiarid regions around the globe including Mexico, southern regions of America, Africa, Pakistan, India, and various other countries [8]. *P. harmala* fruit and seeds have been used by the local herbal practitioners in Pakistan, especially in Bajaur Agency (tribal area of Pakistan) and Lakki Marwat, for the treatment of various diseases. Similarly, fruits are employed to relieve heart pain whereas seeds mixed with honey are used to treat fever, colic pain, and as virmifuge [9,10]. Indigenous communities of Wana, a district in South Waziristan Agency, Pakistan, used a decoction of *P. harmala* seeds to reduce irritation in the larynx, to fight against jaundice, as an abortifacient, to increase the flow of milk, and as a stimulant. In addition, seeds of *P. harmala* are known for their antiperiodic, antiparasitic, antispasmodic, narcotic, and galactagogic effects. In addition, *P. harmala* is used to treat asthma, colic, jaundice, and is found efficient in reducing fever in chronic malaria [11].

Moreover, *P. harmala* is used in the traditional medicinal systems of Morocco [12], Algeria [13], China [14], and Spain [15]. The plant extracts exhibit numerous biological activities such as anti-inflammatory, antibacterial, antifungal, antiviral, antioxidant, analgesic, cardio-protective, antitumor, antidiabetic, histofunctional, cerebral protective, anti-proliferative, and anticancer, among others [16–21]. Recently, *P. Harmala* leaf methanol extract was reported to exhibit antioxidant activities comparable to the standard quercetin and rutoside [22]. Research findings indicate that the alkaloids extract of *P. harmala* causes a significant antinociceptive effect in both phases of the formalin test in mice [23]. Similarly, *P. harmala* ethanol extract was found useful in the management of rheumatoid arthritis complications by boosting the intracellular antioxidant defense mechanism [24]. In addition, research findings demonstrate that *P. harmala* oil extract and seed alkaloid extract are not toxic in acute and subacute studies [25,26], however, the ethanol seeds extract of *P. harmala* shows toxicity in *Caenorhabditis elegans* [27].

In light of the previous discussion, it can be concluded that the *P. harmala* herb as a whole has not been investigated for safety aspects despite its use in different parts of the world. Accordingly, the aim of the present study was to evaluate the in vitro antioxidant and anti-inflammatory properties of the crude liquid–liquid partitioning, and RP-HPLC fractions of *P. harmala* obtained with the aid of bioassay-guided techniques. In addition, the present work highlights the safety aspects of *P. harmala* (the whole herb) through acute and subacute toxicity studies in rat models. To the best of our knowledge, these properties have not been fully investigated. Moreover, these tests were conducted to investigate whether the anti-inflammatory action of this plant in traditional medicine is justified.

2. Materials and Methods

2.1. Plant Material

Plant material was collected from DG Khan, in the district of Punjab, Pakistan and submitted to the Department of Botany, Bahauddin Zakariya University, Multan for taxonomic identification and authentication (Figure 1). A voucher specimen (ID 1132/PH) was deposited at the herbarium located at the pharmacy department, Bahauddin Zakariya University, Multan, Pakistan. The plant material was cleaned, washed, air-dried in the shade at room temperature, and then placed in an oven for 72 h at 37 °C, for complete dryness. It was then crushed, powdered by means of an electric grinder, and subjected to extraction processes.

Figure 1. *Peganum harmala.*

2.2. Preparation of Extracts

The dried powder (2 kg) was extracted with *n*-hexane to remove fatty substances, and then sequential extraction was performed with 100% dichloromethane, 100% methanol, and 70% aqueous methanol using a temperature-controlled orbital shaker (Figure 2). The extract was filtered, and the filtrate was concentrated and dried by means of rotary evaporation (Heidolph, Hei-Vap, Schwabach, Germany) under reduced pressure (800 millibar) at 40 °C to afford semisolid crude extracts. Dried extracts thus obtained were stored at −18 °C in an upright ultralow freezer (Sanyo, MDF-U32V, Osaka, Japan) for future experiments.

2.3. Solvents and Reagents

Chemicals reagents and HPLC columns used throughout this investigation, including antioxidant reference/standards such as ascorbic acid, quercetin, ferrous sulfate, phosphate buffer, and dimethyl sulfoxide (DMSO); analytical grade solvents such as *n*-hexane, dichloromethane (DCM), chloroform, methanol, and water; HPLC grade solvents such as water, methanol, and trifluoroacetic acid (TFA); and analytical and preparative HPLC columns (Zorbax-SB-C-18, Agilent, Santa Clara, CA, USA) were purchased from Sigma-Aldrich, St. Louis, MO, USA, and used as received.

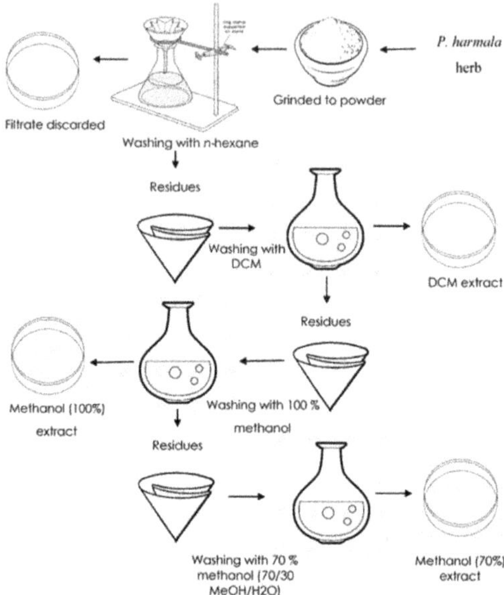

Figure 2. Sequential extraction of *P. harmala* powered using various solvents.

2.4. Animals

In this investigation, we used 28 to 35 day old Wister albino mice (25–30 g) and Wistar albino rats (200–300 g). These animals were procured from the University of Lahore, Pakistan. Animals were housed in cages under standard conditions of 12:12 h light/dark cycle and 25 ± 2 °C. Animals were given free access to water and a standard diet (ad libitum) for 14 h prior to experiments, and kept under standard conditions mentioned in the Animals By-Laws N° 425–2008. All in vivo trials were performed according to the ethical codes set by the Institute of Laboratory Animal Resources Commission on Life Sciences, National Research Council (NRC, 1996), Washington, DC, USA. In addition, the animal care committee at Bahauddin Zakariya University, Multan, Pakistan, under identification number ACC-06-17, approved experimental assays.

2.5. Determination of Total Phenolic Content

We determined the total phenolic contents of the extracts according to the colorimetric Folin–Ciocalteu assay, which was also adopted by Hossain and Shah [28], using gallic acid as standard. Briefly, 0.5 mL of test sample was mixed with 1.5 mL Folin–Ciocalteu reagent and 1.2 mL of 7.5% sodium carbonate (Na_2CO_3) aqueous solution in test tubes. After an incubation period of 30 min in the dark, the absorbance of each test sample was measured at 765 nm using a spectrophotometer (UV-Vis 3000, Dresden, Germany). The phenolic content of each extract was expressed as mg gallic acid (GA) equivalent (E) per gram of dry extract (mg GAE/g); measurements were conducted in triplicate using ethanol as a blank.

2.6. Determination of Total Flavonoid Content

The total flavonoid contents of the extracts were determined using the aluminum chloride ($AlCl_3$) assay as described by Oriakhi et al. [29] with slight modifications. Briefly, 0.5 mL of test sample was added to 0.5 mL distilled water, 0.15 mL of sodium nitrite ($NaNO_2$) solution, and 0.15 mL of $AlCl_3$ solution (2%). After 30 min incubation of the mixture in the dark, absorbance was determined at 510 nm with the aid of a spectrophotometer (UV-Vis 3000, ORI, Germany). Results were expressed as mg quercetin equivalents per gram (mg QE/g) of dry extract; determinations were conducted in triplicate using ethanol as a blank.

2.7. Determination of Antioxidant Activity

2.7.1. DPPH Free Radical Scavenging Assay

The free radical scavenging potential of *P. harmala* crude extracts was determined according to the procedure outlined by Alara et al. [30] with slight modifications. Rather than using Soxhlet extraction and drying at 60 °C we used an orbital shaker for extraction and extracts were dried in an oven at 40 °C. In brief, 1 mL of test sample was mixed with 3 mL of 0.004% methanolic DPPH solution, and allowed to stand in the dark for 30 min. Then, absorbance of the reaction mixture was determined at 517 nm using a spectrophotometer (UV-Vis 3000, ORI, Germany). Quercetin and ascorbic acid were employed as standards, and methanol as the negative control. The percentage inhibition was evaluated using the following equation, and results were given as IC_{50}:

$$\% \text{ inhibition} = 100 \times (A_C - A_S)/A_C,$$

where A_C = absorption of the control sample and A_S = absorption of the test sample.

2.7.2. Ferric Reducing Antioxidant Power (FRAP)

The ferric reducing antioxidant power of *Peganum harmala* crude extracts was evaluated using the method of Zahin et al. [31] with some modifications. According to this method, 100 µL of extract was added to 300 uL of FRAP working solution (containing 300 mmol/L acetate buffer (pH 3.6), 10 mmol/L 2,4,6-tripyridyl-s-triazine (TPTZ) in 40 mmol/L HCl, and 20 mmol/L $FeCl_3$ in a ratio of 10:1:1). After a 15 min incubation, absorbance of the reaction mixture was measured spectrophotometrically at 593 nm (UV-Vis 3000, ORI, Germany); ferrous sulfate was used for standard calibration. Results are expressed as mmol/g, and compared with ascorbic acid and quercetin.

2.7.3. Hydrogen Peroxide (H_2O_2) Scavenging Activity

The hydrogen peroxide (H_2O_2) scavenging ability of *P. harmala* crude extracts was assessed as per the method of Ruch et al. [32] with slight modifications. H_2O_2 solution (40 mM) was added to 50 mM phosphate buffer (7.4 pH). Experimental extracts were then mixed with 0.6 mL H_2O_2 and incubated for 15 min. Absorbance of each mixture was recorded spectrophotometrically (UV-Vis 3000, ORI, Germany) at 230 nm. Ascorbic acid and quercetin were used as the positive control, and a phosphate buffer as the negative control. Percent inhibition of H_2O_2 was calculated using the following formula:

$$H_2O_2 \text{ scavenging activity } (\%) = 100 \times (A_C - A_S)/A_C$$

where A_C = absorption of the control sample and A_S = absorption of the test sample.

2.8. In Vitro Anti-Inflammatory Activity

2.8.1. Membrane Stabilization Assay (Heat Induced Hemolysis)

In vitro anti-inflammatory experiments were performed by collecting blood samples from the cubital veins of healthy human subjects from Karachi, Sindh, Pakistan. All subjects voluntarily gave the blood samples after signing the consent performa. Blood samples were collected according to the guidelines of the International Federation of Blood Donor Organizations (IFBDO) with standard operating procedures, and were approved by the Bioethical Committee, Bahauddin Zakariya University, Multan Reg. no. 06–18. This work was also conducted in accordance with the Declaration of Helsinki. Blood samples were washed with normal saline after centrifugation (3000 rpm, 5 min) and then reconstituted as 10% *v/v* suspension with isotonic buffer solution (10 mM sodium phosphate buffer, pH 7.4) [33,34]. Finally, 1 mL of experimental samples of different concentrations (100, 200, and 300 µg/mL) were added to 1 mL (10%) of red blood cell suspension to make a reaction mixture of 2 mL, which was incubated for 25 min at 50 °C and then cooled to room temperature. After another centrifugation (2500 rpm; 5 min), absorbance of the reaction mixture was measured at 560 nm using a spectrophotometer; diclofenac sodium was employed as a standard drug and phosphate buffer as the control [35]. Doses were selected following the recently published research [36]. The percent inhibition was calculated using the following equation:

$$\% \text{ inhibition of denaturation} = 100 \times (A_C - A_S)/A_C$$

where A_C = absorption of the control sample, and A_S = absorption of the test sample.

2.8.2. Egg Albumin Denaturation Assay

We conducted the egg albumin denaturation assay according to the procedure outlined by Mizushima and Kobayashi [37] with slight modification. In this method, 2 mL of a test sample with a concentration of 100–400 µg/mL was mixed with 0.2 mL of egg albumin and 2.8 mL of phosphate buffer (pH = 6.5) saline. The reaction mixture was incubated for 20 min at 37 °C followed by heating at 70 °C for 5 min. After cooling to room temperature, absorbance was measured at 660 nm with a spectrophotometer. We used diclofenac sodium as the standard drug and phosphate buffer as the control, and calculated the percentage inhibition according to the following equation:

$$\% \text{ inhibition of denaturation} = 100 \times (A_C - A_S)/A_C$$

where A_C = absorption of the control sample, and A_S = absorption of the test sample.

2.8.3. Bovine Serum Albumin Denaturation Assay

The anti-inflammatory activity of the extracts was evaluated by the effect on bovine serum albumin denaturation; the rest was conducted according to the method described by Sakat et al. [34] with slight modifications. According to this method, a reaction mixture of 0.5 mL was made by mixing 0.05 mL experimental extracts (100–400 µg/mL) with 0.45 mL of bovine serum albumin and then incubated for 25 min at 25 °C. Afterwards, 2.5 mL of phosphate buffer (pH = 6.3) was added to the reaction mixture tubes and incubated in a water bath at 70 °C for 15 min. After cooling the mixture to room temperature, absorbance was measured at 660 nm with the aid of a spectrophotometer; diclofenac sodium was employed as a standard drug and phosphate buffer as the control. The percentage inhibition of protein denaturation was calculated using the following equation:

$$\% \text{ inhibition of denaturation} = 100 \times (A_C - A_S)/A_C$$

where A_C = absorption of the control sample, and A_S = absorption of the test sample.

2.9. In Vivo Anti-Inflammatory Activity

2.9.1. Inhibition of Carrageenan-Induced Paw Edema in Wistar Rats

We conducted the carrageenan-intoxicated paw edema study in accordance with the ethical codes set by the Institute of Laboratory Animal Resources, Commission on Life Sciences, National Research Council (NRC, 1996), Washington, DC, USA. Moreover, the animal use protocol was approved by the Institutional Animal Care and Use Committee at Bahauddin Zakariya University, Multan, Pakistan, under the protocol number AEC-06-18 and title: "In vivo anti-inflammatory activity of southern Punjab medicinal plants". We employed the carrageenan-intoxicated inflammation model to evaluate the activity of *P. harmala* sequential crude extracts against inflammation following the method of Morris [38] with slight changes. In this method, *Wistar rats* were randomly divided into eight groups of five animals each (n = 5). Group 1 (control): rats were fed with normal saline. Group 2 (positive control): rats received standard indomethacin (100 mg/kg, b.w). On the other hand, rats in groups 3 and 4 were fed with 100 mg/kg and 200 mg/kg of DCM extract, respectively, whereas rats in groups 5 and 6 were fed with 100 mg/kg and 200 mg/kg of 100% methanol extract, respectively. Finally, rats in groups 7 and 8 were fed with 100 mg/kg and 200 mg/kg of 70% methanol extracts, respectively. In all these groups, the initial value of normal paw volume was measured. Doses were selected as per previous studies reported on *P. harmala* [8,39]. After 30 min of intraperitoneal administration of experimental extracts, freshly prepared 0.1 mL carrageenan in 0.9% normal saline was injected into the plantar aponeurosis surface of the right hind paw of each animal. Then, paw linear circumference was measured after 0, 1, 2, and 3 h of carrageenan injection by means of a plethysmometer (UGO-BASILE 7140, Comerio, Italy). The increase in paw circumference was considered as a way to measure inflammation.

2.9.2. Inhibition of Formaldehyde-Induced Hind Paw Edema in Albino Mice

We have used the formaldehyde-induced hind paw edema in albino mice to evaluate the pain alleviating action of different extracts from *P. harmala* following Brownlee's guidelines with slight modification [40]. In this method, albino mice were randomly divided into eight groups with five mice each (n = 5), and each group was treated according to the previous study. After 30 min of treatment with experimental extracts, 100 μL (4%) formaldehyde was infused into the plantar aponeurosis surface of the right paw of each mouse. Changes in the linear paw circumference were measured after 0, 3, 6, 12, and 24 h of formaldehyde infusion.

2.10. Acute and Subacute Toxicity Assessment
2.10.1. Acute Toxicological Study

The study was performed by following the Organization for Economic Cooperation and Development (OECD) guidelines 407 and 423 for acute oral toxicity tests [41,42]. In this study, rats were given free access to clean drinking water and a standard diet (ad libitum) for 24 h before and after the experiment. *P. harmala* methanol extract doses of 1500 and 3000 mg/kg (per oral) were prepared as stock solutions for 14 days. After 1 h of extract administration, behavioral changes of each animal were noted periodically after 4, 8, 12, and 24 h. In addition, the body weight of all animals was recorded every 7 days, and samples of blood were collected for analysis for hematological and biochemical parameters. Afterwards, animals were sacrificed and organs (heart, liver, and kidney) were isolated for histopathological studies.

2.10.2. Chronic Toxicological Study

All rats involved in this study were starved for 2 h before administration of 400 and 800 mg/kg (per oral) doses of *P. harmala* methanol extract for 28 days. After 1 h of extract administration, behavioral changes of each animal were recorded after 4, 8, 12, and 24 h. Furthermore, body weights of all animals were measured every weekend, and samples of blood were collected for analysis of hematological and biochemical parameters. At the completion of the study, all animals were sacrificed, and organs (liver and kidney) were isolated for histopathology.

2.10.3. Histopathological Examination

Fresh organ portions from the heart, liver, and kidney, collected from normal and treated animals, were cut and fixed in 10% formalin solution. Fixed samples were dehydrated with alcohol dilutions series (60–100%) and embedded in paraffin. The paraffin fixed blocks of organs were cut to 4 μm thickness. These sections were stained with hematoxylin and eosin (H&E), and examined under light microscope for histopathological changes and photomicrographs were taken.

2.11. Liquid–Liquid Partitioning of Active Crude Extract

The 100% methanol extract was further separated using solvent–solvent partitioning by dissolving the crude extract with water, then chloroform, and finally with ethyl acetate. This extract was mixed with 15 mL of distilled water in a beaker and vigorously shaken with an identical volume of chloroform (15 mL). The two layers were separated: aqueous (top) and chloroform (bottom), and placed in different containers. The aqueous layer was again partitioned with chloroform and the process was repeated three times. The combined chloroform extracts were combined and evaporated under reduced pressure with the aid of a rotary evaporator to afford a semisolid thick paste and stored at −18 °C for future use. Using the same procedure, the aqueous layer was extracted with ethyl acetate. Both layers were recovered and evaporated using rotary evaporator. Importantly, the chloroform layer was named as fraction A, ethyl acetate fraction B, and the water layer as fraction C, and all fractions were evaluated for in vitro antioxidant and anti-inflammatory activities.

2.12. Method Optimization for Fractionation Using RP-HPLC

Fraction B, which showed significant antioxidant and anti-inflammatory activities, was further subjected to reversed phase column chromatography by dissolving solidified fractions into methanol as described by Cock [43]. Samples were prepared as 10 mg/mL and filtered using 0.45 mm syringe filter. The sample injection limit and flow rate were adjusted to 70 µL and 0.5 mL/min, respectively, using the Agilent LC technology and an analytical column (4.6 × 150 mm, 5 µm, Agilent, Waldbronn, Germany). Maximum number of peaks was observed with acidified (0.1% formic acid) water (A) and acidified (TFA) acetonitrile (B) at 254 nm.

Various combinations of mobile phases such as methanol: water, acidified methanol (0.1% TFA): acidified water (0.1% TFA), acidified methanol (0.1% FA): acidified water (0.1% FA), acetonitrile: water, acidified acetonitrile (0.1% TFA): acidified water (0.1% TFA), acidified acetonitrile (0.1% FA): acidified water (0.1% FA), and different wavelengths such as 210 nm, 230 nm, 254 nm, 280 nm, 300 nm, and 330 nm were used. A mixture of water (A) and acetonitrile (B), both containing 0.1% formic acid, was selected as the mobile phase for Fraction B of the *P. harmala* methanol extract. Then 10 µL sample was injected into the HPLC system and the linear eluting gradient was as follows: 10% B in 0–5 min, 10–40% B in 5–12 min, 40–60% B in 12–20 min, 60–80% B in 20–25 min, and 100% B in 25–30 min. Maximum number of peaks was observed with acidified (0.1% FA) water (A) and acidified (FA) acetonitrile (B) at 280 nm.

2.13. RP-HPLC Fractionation (Reversed Phase Chromatography)

Reversed phase chromatography was performed through a semipreparative column (C-18, 25 × 250 mm, 5 µm particle size, Agilent, Waldbronn, Germany). Samples were prepared as 50 mg/mL and filtered using a 0.45 mm syringe filter. The sample injection limit and flow rate were adjusted to 1 mL and 10 mL/min; this led to 5 subfractions: PHMF1, PHMF2, PHMF3, PHMF4, and PHMF5 from fraction B of the *P. harmala* methanol extract.

2.14. LC-ESI-MS/MS Analysis

We performed mass spectral analysis on RP-HPLC subfractions that have exhibited significant antioxidant and anti-inflammatory activity using LC-ESI-MS/MS (LTQ XL, Thermo Electron Corporation, Walthan, MA, USA) for tentative identifications of bioactive components according to the protocols suggested by Steinmann and Ganzera [44]. An online software was used to obtain the structures of bioactive compounds identified in the present study and to compare them with previously published data (www.chemspider.com, accessed on 6 October 2021).

2.15. Quantification of Compounds Using HPLC

P. harmala methanol fraction 2 (PHMF2) was dissolved in 1 mL of methanol in order to quantify the identified compounds using standard calibration curves. The mixture was centrifuged at 14,000 rpm for 10 min to collect the supernatant, whereas the filtration of supernatant was achieved using a syringe filter. Finally, we injected a 100-µL sample into the HPLC system for analysis.

2.16. Statistical Analysis

All determinations were conducted in triplicate and data were subjected to one-way analysis of variance (ANOVA). Results are expressed as the mean $\pm$ standard deviation (SD), and values are given as geometric mean with 95% confidence intervals (CI). Statistical analysis was performed using Dunnett's test for significance at 95% confidence intervals (CI) with the aid of GraphPad Prism-6 (GraphPad Software, San Diego, CA, USA, http://www.graphpad.com, accessed on 6 October 2021); differences were considered significant at (* $p < 0.05$, ** $p < 0.01$, *** $p < 0.001$, **** $p < 0.0001$).

3. Results and Discussion

3.1. Phytochemical Constituents and Antioxidant Activity

Phenolic compounds are plant secondary metabolites known to possess radical scavenging properties owing to their redox potentials. Results from this investigation show that the methanol extract has the highest amount of total phenolic (371.4 mg GAE/g) and flavonoid contents (1.3 mg QE/g) followed by DCM and hydro-alcoholic extracts as shown in Table 1. In addition, our findings reveal that the 100% methanol extract of *P. harmala* exhibits the highest antioxidant potential in the DPPH (IC_{50} 49 $\pm$ 3.1 µg/mL), FRAP (39 $\pm$ 0.9 mmol/g), and H_2O_2 (66% inhibition) assays as compared to other crude extracts. These results show a direct link between the quantity of phenolic compounds and antioxidant activity. The antioxidant activity of phenolics is mainly due to their redox properties, which make them act as reducing agents, hydrogen donors, and singlet oxygen quenchers. They also have a metallic chelating potential [45].

Table 1. Total phenolics and flavonoids content, and antioxidant potential of DCM, MeOH, and 70% MeOH extracts of *Peganum harmala* crude extracts and subsequent fractions.

Parameter	DCM Extracts	Methanol Extracts	70% Methanol Extracts	Fraction B (Methanolic Extract)	PHMF2 (Fraction B of Methanolic Extract)	Ascorbic Acid	Quercetin
Total phenolic contents (mg GAE/g)	106.2 $\pm$ 0.31	371.4 $\pm$ 0.2	142.3 $\pm$ 0.1	-	-	-	-
Total flavonoid contents (mg QE/g)	0.31 $\pm$ 0.5	1.3 $\pm$ 0.3	0.81 $\pm$ 0.02	-	-	-	-
FRAP (mmol/g)	9.2 $\pm$ 0.6	39 $\pm$ 0.9	19.2 $\pm$ 0.2	42.9 $\pm$ 0.1	45.3 $\pm$ 0.2	51 $\pm$ 0.02	62 $\pm$ 0.02
DPPH (IC_{50} µg/mL)	146 $\pm$ 2.0	49 $\pm$ 3.1	69 $\pm$ 1.4	44.6 $\pm$ 3.0	35.4 $\pm$ 1.1	29.1 $\pm$ 0.02	25.4 $\pm$ 0.01
H_2O_2 (%)	25 $\pm$ 0.6	66 $\pm$ 0.9	43 $\pm$ 2.10	71 $\pm$ 2.0	75 $\pm$ 0.1	79 $\pm$ 0.02	84 $\pm$ 0.05

Values are means $\pm$ SD. DCM extracts = 100% dichloromethane extracts. MeOH = 100% methanol extracts. 70% MeOH = Methanol: water (70:50 *v/v*).

Published research indicates that the methanol seeds extract of *P. harmala* exhibits a higher free radical scavenging ability in DPPH assay (92% inhibition) than *n*-hexane, benzene, DCM, and chloroform extracts due to its higher phenolic contents of 30.9 mg GAE/g [46]; these results agree with findings of the present study. Recently, *P. harmala* leaf methanolic extract was reported to exert antioxidant activities in DPPH (IC_{50} 21.5 µg/mL) and FRAP (IC_{50} 32.4 mM TEAC/g) assays parallel to the standard quercetin outlined IC_{50} of 21.5 µg/mL (DPPH) and 32.6 mM TEAC/g (FRAP)) was also identified and quantified with a reasonable amount of phenolic compounds [22].

In addition, results displayed in Table 1 indicate that Fraction B exhibits stronger antioxidant potential as compared to others, such as fractions A and C, in all in vitro antioxidant trials. Similarly, PHMF2 of Fraction B (methanol extract) showed significant antioxidant activity, and was comparable to the antioxidant activity of the standard ascorbic acid and quercetin. This is in agreement with results obtained by Abderrahim and colleagues [21] using the DPPH assay. These researchers concluded that the antioxidant activity of *P. harmala* extract is primarily associated with their appreciable amount of flavonoids and polyphenols, which were calculated as 220.94 ± 1.1 and 650.38 ± 30.3 mg GAE/g.

3.2. In Vitro Anti-Inflammatory Activity

3.2.1. Heat Induced Hemolysis (Membrane Stabilization)

Agents that are capable of stabilizing the human red blood cell membrane in response to hypotonicity-induced lysis are recognized as anti-inflammatory drugs [47]. In this investigation, the membrane stabilization potential of *P. harmala* crude methanol extracts, liquid–liquid partitioned fractions, and HPLC fractions at different concentrations (100, 200, 300, and 400 µg/mL) was evaluated. Results reveal that PHMF2 exhibits the greatest inhibition (62.7%, $p < 0.01$) followed by Fraction B (55.3%, $p < 0.01$), and crude methanol extract (48.2%, $p < 0.05$), whereas the standard diclofenac sodium displayed potent inhibition ($p < 0.0001$) of 89.3% against heat-induced hemolysis in a concentration-dependent manner at 400 µg/mL as compared to the control (phosphate buffer), as shown in Table 2. Moreover, drugs that stabilize the red blood cell membrane may also protect the lysosomal membrane due to similar composition [48]. The in vitro anti-inflammatory activity exhibited by the methanol extract is consistent with its phytochemical and antioxidant potential.

Table 2. In vitro anti-inflammatory activity of *Peganum harmala* crude extracts and subsequent fractions.

Treatment	Dose (µg/mL)	Membrane Stabilization		Egg Albumin Denaturation		Serum Albumin Denaturation	
		Absorbance	% Inhibition	Absorbance	% Inhibition	Absorbance	% Inhibition
Control	-	0.94 ± 0.1	-	0.91 ± 0.2	-	0.92 ± 0.1	-
MeOH extract (crude extract)	100	0.75 ± 0.2 [ns]	20.2	0.69 ± 1.3 [ns]	24.1	0.63 ± 0.3 *	31.5
	200	0.69 ± 0.4 [ns]	26.5	0.65 ± 1.1 [ns]	28.5	0.58 ± 0.2 **	36.9
	300	0.58 ± 0.5 *	38.2	0.56 ± 0.2 *	38.4	0.51 ± 0.2 **	44.5
	400	0.43 ± 0.1 **	48.2	0.39 ± 0.2 **	57.1	0.34 ± 0.1 **	63.0
Fraction B (liquid–liquid partitioned fraction)	100	0.66 ± 0.2 *	29.7	0.60 ± 0.2 *	34.0	0.58 ± 0.1 *	36.0
	200	0.65 ± 0.3 *	30.8	0.58 ± 0.2 *	36.2	0.54 ± 0.2 **	41.3
	300	0.55 ± 0.2 **	41.7	0.50 ± 0.2 **	45.0	0.48 ± 0.4 **	47.8
	400	0.42 ± 1.1 **	55.3	0.37 ± 0.2 **	59.3	0.26 ± 0.1 ***	71.7
PHMF2 (RP-HPLC sub-fraction)	100	0.63 ± 0.1 *	32.9	0.59 ± 0.1 *	35.1	0.56 ± 0.2 *	39.1
	200	0.57 ± 0.3 *	39.3	0.54 ± 0.5 *	40.6	0.51 ± 0.2 *	44.5
	300	0.50 ± 0.2 **	46.8	0.49 ± 0.6 **	46.1	0.45 ± 0.1 **	51.0
	400	0.35 ± 0.2 **	62.7	0.29 ± 0.2 ***	68.1	0.25 ± 0.4 ***	72.9
70% MeOH extract (crude extract)	50	0.80 ± 0.4 [ns]	14.8	0.76 ± 0.1 [ns]	16.4	0.70 ± 0.3 [ns]	23.9
	100	0.76 ± 0.5 [ns]	19.1	0.72 ± 0.1 [ns]	20.8	0.69 ± 0.2 *	25.0
	200	0.70 ± 0.4 *	25.5	0.65 ± 1.2 *	28.5	0.63 ± 0.2 *	31.5
	300	0.59 ± 0.1 *	37.3	0.55 ± 1.1 *	39.5	0.52 ± 0.3 **	43.4
DCM (crude extract)	100	0.94 ± 0.1	NA	0.94 ± 0.1	NA	0.93 ± 0.3	NA
	200	0.93 ± 0.1	NA	0.94 ± 0.2	NA	0.94 ± 0.2	NA
	300	0.93 ± 0.2	NA	0.93 ± 0.2	NA	0.94 ± 0.2	NA
	400	0.90 ± 0.2	NA	0.94 ± 0.1	NA	0.93 ± 0.3	NA
Diclofenac sodium (Standard drug)	100	0.19 ± 0.2 ***	79.7	0.16 ± 0.1 ***	82.4	0.05 ± 0.5 ****	94.5
	200	0.14 ± 0.2 ****	85.1	0.11 ± 0.1 ***	87.9	0.03 ± 0.1 ****	96.5
	300	0.11 ± 0.1 ****	88.2	0.09 ± 0.1 ****	90.4	0.02 ± 0.1 ****	97.0
	400	0.10 ± 0.2 ****	89.3	0.09 ± 0.5 ****	91.5	0.02 ± 0.5 ****	97.8

Values are means $\pm$ S.D. PHMF2; *Peganum harmala* methanolic fraction. 70% MeOH extract; methanol: water (70:50 v/v). ns; non significant (* $p < 0.05$, ** $p < 0.01$, *** $p < 0.001$, **** $p < 0.0001$).

Published research indicates that in vitro antioxidant, in vivo anti-inflammatory, and analgesic activities establish a beneficial role of the plant. In these findings, a cream formulation made from the seed oil demonstrated a significant anti-inflammatory effect. A slight peripheral analgesic effect was also observed due to the presence of polyphenols, linoleic acid, and γ-tocopherols, and is primarily due to antioxidant properties [49]. Earlier studies showed that plant extracts with antioxidant activity also display notable membrane stabilization properties in a concentration-dependent manner [50]. Previously, Khlifi et al. [51] studied *Artemisia herba-alba*, *Ruta chalpensis* L., and *P. harmala*, and also showed a direct relationship between phenolic compounds, antioxidant activity, and anti-inflammatory potential.

3.2.2. Inhibition of Protein Denaturation (Serum and Egg Albumin)

Protein denaturation is an apparent cause of inflammation, and reliable literature cited earlier validates the link between inflammatory/arthritic problems and the denaturation of tissue proteins [52,53]. Results displayed in Table 2 show that *P. harmala* crude extracts and subsequent fractions exhibit considerable inhibition against egg albumin denaturation, where PHMF2 was the most potent (68.1%, $p < 0.001$), followed by fraction B (59.3%, $p < 0.01$), and crude methanol extract (57.1%, $p < 0.01$) at 400 µg/mL in a dose-dependent manner when compared to the control. These results are in accord with those obtained by Chandra et al. [2], which showed a concentration-dependent inhibition against denaturation of egg albumin by experimental coffee extracts and standard diclofenac sodium. Other researchers showed that *Enicostemma axillare* methanol extract at a concentration range of 100−500 µg/mL significantly protects the heat-induced protein denaturation [54]. Our findings show that PHMF2 ($p < 0.001$) exhibits notable inhibition against denaturation of serum albumin followed by fraction B ($p < 0.001$), and crude methanol extract ($p < 0.01$) with percentage inhibition of 72.9%, 71.7%, and 63.0%, respectively, at 400 µg/mL as compared to the control. In this respect, diclofenac sodium ($p < 0.0001$) showed significant inhibition against egg albumin and serum albumin denaturation at 400 µg/mL (Table 2).

3.3. In Vivo Anti-Inflammatory Activity of Sequential Crude Extracts
Carrageenan-Induced Paw Edema

Carrageenan-induced paw edema is an important method to investigate the anti-inflammatory potential of experimental extracts. Carrageenan-intoxicated paw swelling is a biphasic response wherein the first phase is associated with the release of pro-inflammatory cytokines such as kinins, histamine, and 5-HT, whereas the release of prostaglandins is associated with the second phase [55]. In the present study, *P. harmala* methanol extract, when dispensed at the rate of 200 mg/kg, causes notable inhibition ($p < 0.001$) of 75.14% after 3 h, which is similar to the 86.3% inhibition ($p < 0.0001$) caused by the standard drug indomethacin at 100 mg/kg when compared to the control (normal saline = 0% inhibition) (Figure 3). In contrast, the hydromethanol and DCM extracts were not as active. In this respect, *P. harmala* methanol extract could inhibit the mediation and release of pro-inflammatory cytokines in both phases of inflammation. The anti-inflammatory activities of the methanol extracts are consistent with their phenolic, flavonoids contents, and antioxidant activity. Our findings agree with those obtained by Edziri and colleagues [56]. These researchers found that both *M. alysson* and *Peganum harmala* methanol extracts exhibit good inhibition against carrageenan-induced paw edema; these extracts displayed good antioxidant activity and have high total phenolic content. In another investigation, the ethyl acetate extract of *P. harmala* caused significant inhibition of 70.3% at 200 mg/kg against carrageenan-induced paw edema. Preliminary phytochemical assessment of the extract indicated the presence of alkaloids and flavonoids in the extract [57]. Recently, *P. harmala* seeds ethanol extract, when dispensed orally at the rate of 100 mg/kg, was reported to reduce complete Freund's Adjuvant intoxicated paw edema by up to 63.09% as compared to the control, and a decrease in synovial/hepatic tissues lipid peroxidation and increase in cellular antioxidants was also observed, supporting the findings of the current investigation [24]. In the present study, *P. harmala* methanol extract, when dispensed at the rate of 200 mg/kg, causes notable inhibition ($p < 0.001$) of 76.32% after 24 h, which is similar to the 89.3% inhibition caused by the standard drug indomethacin ($p < 0.0001$) at 100 mg/kg when compared to the control (normal saline = 0% inhibition) (Figure 4). In contrast, the hydromethanol and DCM extracts did not cause any substantial inhibition against formaldehyde-induced pain.

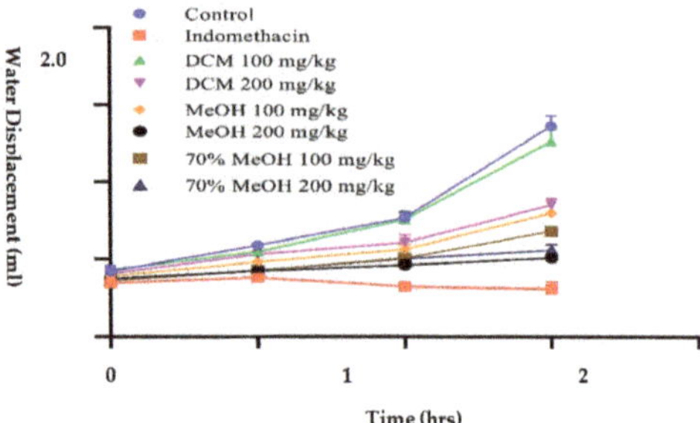

Figure 3. Inhibition against carrageenan-induced paw edema by *Peganum harmala* sequential crude extracts.

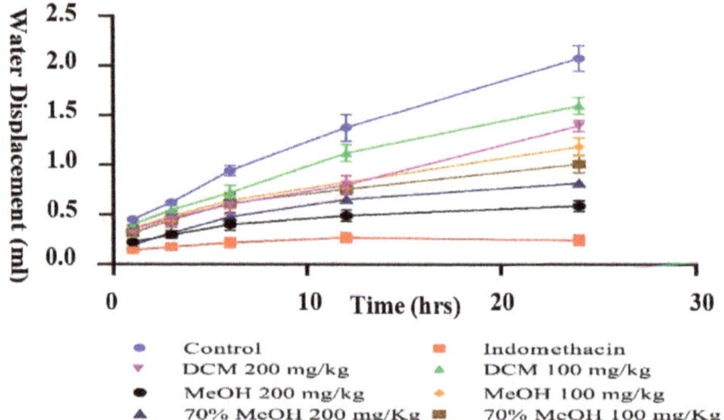

Figure 4. Inhibition against formaldehyde-induced paw edema by *Peganum harmala* sequential crude extracts. Values are mean ± SEM. $p < 0.05$ was considered significant (* $p < 0.05$, ** $p < 0.01$, *** $p < 0.001$, **** $p < 0.0001$).

3.4. Acute and Subacute Toxicity Assessment

Our findings indicate that the *P. harmala* methanol extract exhibits substantial antioxidant and anti-inflammatory activities when compared to other extracts. Consequently, this extract was subjected to toxicity assessment. Results presented in Table 3 reveal no toxicity, body weight changes, behavioral changes, and mortality among animals during a 14 day trial of acute toxicity evaluation when the extract was given at a dose of 3 g/kg. Furthermore, the organ weights of the treated animals were within the normal range. Therefore, it was inferred that the LD_{50} of *P. harmala* methanol extract was above 3 g/kg. These findings are in agreement with those obtained by Selim et al. [25]. These researchers demonstrated no toxicity and no deaths in hamsters after 14 days of treatment with *P. harmala* oil extract at doses of 80, 160, and 320 mg/kg. Furthermore, *P. harmala* methanol extract was subjected to a 28 day trial of subacute toxicity. Results shown in Table 3 reveal no notable changes in the body and organ weights of the treated animals with 500 and 1000 mg/kg as compared to the control group (receiving normal saline). The weights of the heart, kidney, lungs, spleen, and liver of the treated animals were similar to those of the animals in the control group. Along this line, published work indicates that no clinical toxicity signs, changes in body weight, mortality, any gross morphological abnormalities in various organs of the treated mice, and relative weights of the organs were observed when *P. harmala* seed extracts were given to animals at a dose of 18 mg/kg [26].

Table 3. Acute and subacute toxicity assessment of *Peganum harmala* methanol extract.

Parameters	Control Group	Acute Toxicity (14-Day)		Subacute Toxicity (28-Day)	
	Normal Saline	2000 mg/kg PHME	3000 mg/kg PHME	500 mg/kg PHME	1000 mg/kg PHME
Body weight (g)	206.00 ± 4.50	208.00 ± 5.50	205.50 ± 6.33	209.83 ± 5.92	208.50 ± 7.17
Organ Weight					
Heart (g)	0.60 ± 0.15	0.65 ± 0.04	0.80 ± 0.06	0.74 ± 0.05	0.70 ± 0.10
Paired Lungs (g)	2.15 ± 1.15	2.53 ± 0.15	3.36 ± 0.47	3.14 ± 0.31	2.78 ± 0.81
Liver (g)	7.93 ± 1.40	8.10 ± 0.65	7.95 ± 1.33	9.66 ± 0.99	9.33 ± 1.37
Spleen (g)	0.45 ± 0.03	0.50 ± 0.03	0.47 ± 0.03	0.50 ± 0.03	0.67 ± 0.03
Hematological Parameters					
WBCs ($10^5/\mu$L)	3.40 ± 0.02	2.03 ± 0.03	2.73 ± 0.02	3.07 ± 0.02	3.15 ± 0.02
Neutrophils (%)	40.20 ± 3.00	42.95 ± 4.50	43.55 ± 4.00	47.40 ± 4.25	48.40 ± 3.50
Lymphocytes (%)	45.45 ± 4.30	49.05 ± 5.50	49.45 ± 5.20	53.77 ± 5.35	56.57 ± 4.75
Eosinophils (%)	0.94 ± 0.10	0.96 ± 0.02	1.06 ± 0.05	1.09 ± 0.03	1.48 ± 0.07
RBCs ($10^6/\mu$L)	9.05 ± 2.10	10.40 ± 0.50	11.35 ± 1.03	11.30 ± 0.77	12.27 ± 1.57
Hemoglobin (g/dL)	14.55 ± 2.45	16.05 ± 1.00	16.95 ± 1.27	17.12 ± 1.13	17.85 ± 1.86
Hematocrit (%)	45.60 ± 3.70	47.95 ± 3.50	48.75 ± 3.40	50.83 ± 3.45	56.43 ± 3.55
Mean corpuscular volume (MCV (f/L)	57.60 ± 7.50	62.05 ± 5.00	64.70 ± 5.43	66.88 ± 5.22	71.45 ± 6.47
Mean corpuscular hemoglobin (MCH (pg)	17.50 ± 1.65	18.80 ± 1.00	19.75 ± 1.03	19.72 ± 1.02	20.68 ± 1.34
MCHC (%)	29.45 ± 1.50	31.00 ± 2.00	32.25 ± 2.03	32.93 ± 2.02	33.90 ± 1.77
Platelets($10^5/\mu$L)	6.65 ± 1.10	6.95 ± 0.25	7.55 ± 0.53	7.58 ± 0.39	8.05 ± 0.82
Serum Biological Parameters					
Total Protein (g/dL)	4.05 ± 0.65	4.39 ± 0.25	5.10 ± 0.23	4.74 ± 0.24	6.51 ± 0.44
Albumin (g/dL)	1.88 ± 0.35	2.25 ± 0.20	2.45 ± 0.23	2.43 ± 0.22	2.69 ± 0.29
Albumin/Globulin ratio	2.30 ± 0.26	2.67 ± 0.03	2.95 ± 0.02	2.66 ± 0.02	2.75 ± 0.14
Lactate Dehydrogenase (U/L)	1732.00 ± 271.00	2088.50 ± 217.50	2266.50 ± 252.00	2281.00 ± 234.75	2379.00 ± 261.50
Asparate Transaminase (U/L)	111.30 ± 10.10	121.00 ± 14.50	141.10 ± 12.73	137.20 ± 13.62	136.47 ± 11.42
Alanine Transaminase (U/L)	29.85 ± 5.50	36.80 ± 4.50	39.35 ± 4.33	39.67 ± 4.42	43.33 ± 4.92
Alkaline Phosphatase (U/L)	284.00 ± 15.00	298.00 ± 19.00	312.50 ± 18.33	316.50 ± 18.67	320.17 ± 16.67
Total bilirubin (mg/dL)	0.25 ± 0.06	0.29 ± 0.01	0.36 ± 0.04	0.31 ± 0.03	0.29 ± 0.05
Creatinine (mg/dL)	1.43 ± 0.07	1.84 ± 0.45	2.46 ± 0.33	2.00 ± 0.39	1.77 ± 0.20
Uric Acid (mg/dL)	0.96 ± 0.06	1.15 ± 0.01	0.91 ± 0.04	1.04 ± 0.03	1.03 ± 0.05
Total Cholesterol (mg/dL)	53.70 ± 5.00	55.65 ± 3.50	60.20 ± 4.13	60.65 ± 3.82	66.52 ± 4.57
Triglycerides (mg/dL)	116.80 ± 9.40	121.95 ± 11.00	133.95 ± 8.77	133.00 ± 9.88	144.23 ± 9.08
Sodium (mmol/L)	108.00 ± 18.00	124.50 ± 11.00	125.00 ± 13.67	132.83 ± 12.33	139.17 ± 15.83
Chloride (mmol/L)	71.50 ± 20.00	91.50 ± 8.00	92.00 ± 12.33	97.33 ± 10.17	110.00 ± 16.17
Potassium (mmol/L)	3.04 ± 0.06	3.70 ± 0.40	3.07 ± 0.30	3.64 ± 0.35	3.44 ± 0.18

3.4.1. Hematological and Serum Parameters

Results pertaining to hematological and serum parameters of acute and subacute studies are presented in Table 3. Results revealed that in a 14 day trial, *P. harmala* methanol extract treated animals show a slight increase in the hematological and serum parameters, but within normal range. Similarly, *P. harmala* methanol extract revealed a slight increase in the hematological parameters when it was administered at the dose of 2000 and 3000 mg/kg as compared to the control group. In addition, results showed no adverse changes in hematological parameters in a 28 day trial of a subacute toxicity assessment of *P. harmala* methanol extract. However, there was a slight increase in parameters including WBCs, neutrophils, RBCs, hemoglobin, and platelets as compared to the control group (which received normal saline) when *P. harmala* methanol extract was given at the rate of 500 and 1000 mg/kg. Published work by Guergour and coworkers showed that treatment of female mice with alkaloids seeds of *P. harmala* did not cause significant changes in levels of urea and creatinine [26].

The serum profile in the acute and subacute studies did not show significant differences between normal and treated rats, which indicates that *P. harmala* methanol extract has no adverse effects on the liver and kidneys, and on biochemical parameters including serum total protein, albumin, lactate dehydrogenase, aspartate transaminase, total bilirubin, creatinine, and uric acid when given at the doses of 2000 and 3000 mg/kg as compared to the control group. Hence, consecutive oral administration of *P. harmala* methanol extract for 28 days has no toxic impact on the liver and kidneys (Table 3). In addition, research findings revealed that *P. harmala* does not cause significant changes in the hematological profile of female mice when compared with the control [26]; this agrees with findings obtained from this investigation that *P. harmala* is not harmful to the blood system.

3.4.2. Histopathological Analysis

Displayed in Figure 5 are the histopathological sections of the heart, liver, and kidneys of the acute and subacute toxicity assessment. The microscopic observations revealed no substantial histopathological changes in *P. harmala* methanol extract treated rats as compared to those in the normal group. However, some lymphocytes with shrinkage of cardiomyocytes were observed in the heart muscles, which may indicate the presence of inflammation in subacute toxicity experiments. In kidney microscopy, interstitial edema and vacuolar degenerations were seen in subacute toxicity experiments. This indicates that renal epithelial cells inflammation was reversed by *P. harmala* methanol extract. Similarly, the architecture of the liver in the treated animals was similar to normal animals. Published work demonstrated that *P. harmala* in a subacute toxicity trial did not cause destruction to kidney architecture. Moreover, liver histology showed ground glass appearance of hepatocytes in acute toxicity [26].

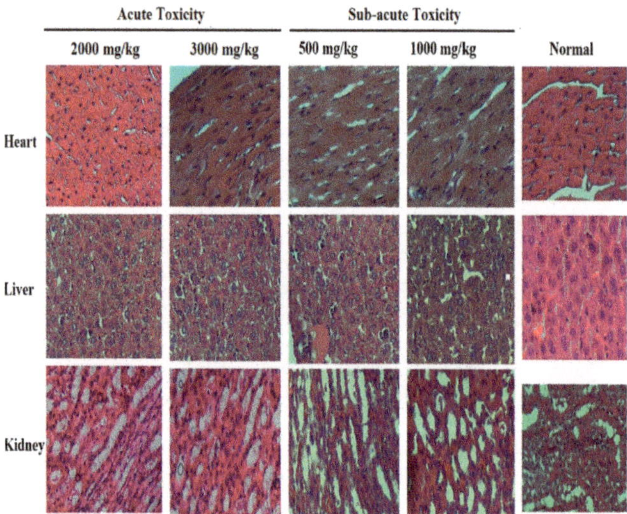

Figure 5. Histopathological investigation.

3.5. ESI-MS/MS Analysis

Since PHMF2 demonstrated significant antioxidant potential as compared to the standards ascorbic acid and quercetin, and in vitro anti-inflammatory capacity as compared to the control, this fraction was subjected to ESI-MS-MS analysis for identification of bioactive compounds. Results revealed the presence of quinic acid, harmaline, harmol, harmine, and pegamine in the fraction as shown in Table 4 and Figure 6. To the best of our knowledge, quinic acid was reported for the first time in the *Peganum harmala* methanol fraction, whereas harmine and harmaline have been reported earlier.

Table 4. LC-ESI-MS/MS identification of bioactive compounds from PHMF2 of *Peganum harmala*.

Fractions	Average Mass	ESI-MS/MS (Ions)	Compound	Chemical Formula	References
PHMF2	191	191, 173.1	Quinic acid	$C_7H_{12}O_6$	
	198	198.08, 181, 171.08	Harmol	$C_{12}H_{10}N_2O$	
	213	213.17, 198.08	Harmine	$C_{13}H_{12}N_2O$	[58]
	214	215, 200.17, 197.17, 171	Harmaline	$C_{13}H_{14}N_2O$	
	204	205, 187, 161	Pegamine	$C_{11}H_{12}N_2O_2$	[59]

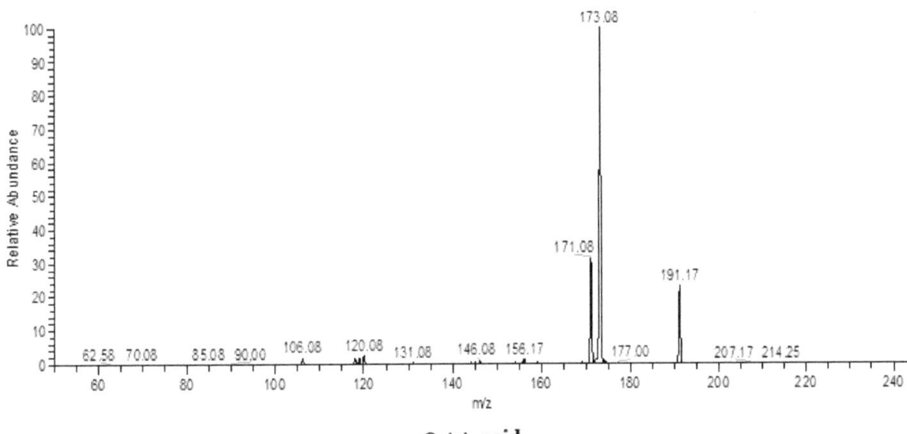

Quinic acid

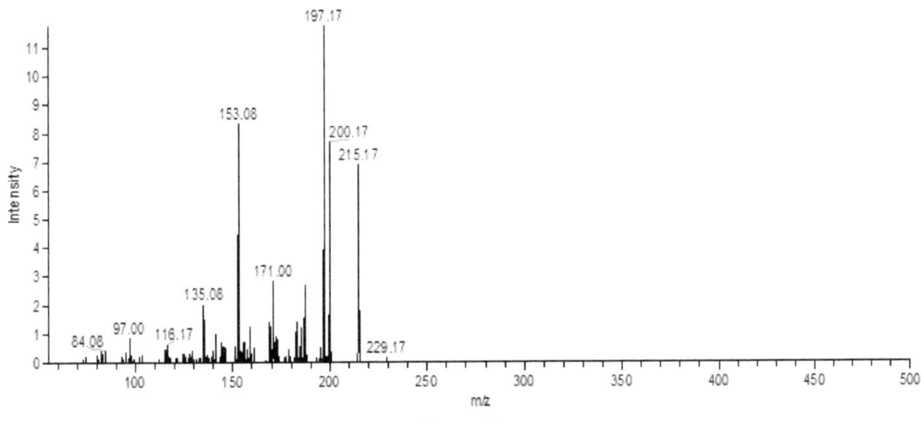

Harmaline

Figure 6. *Cont.*

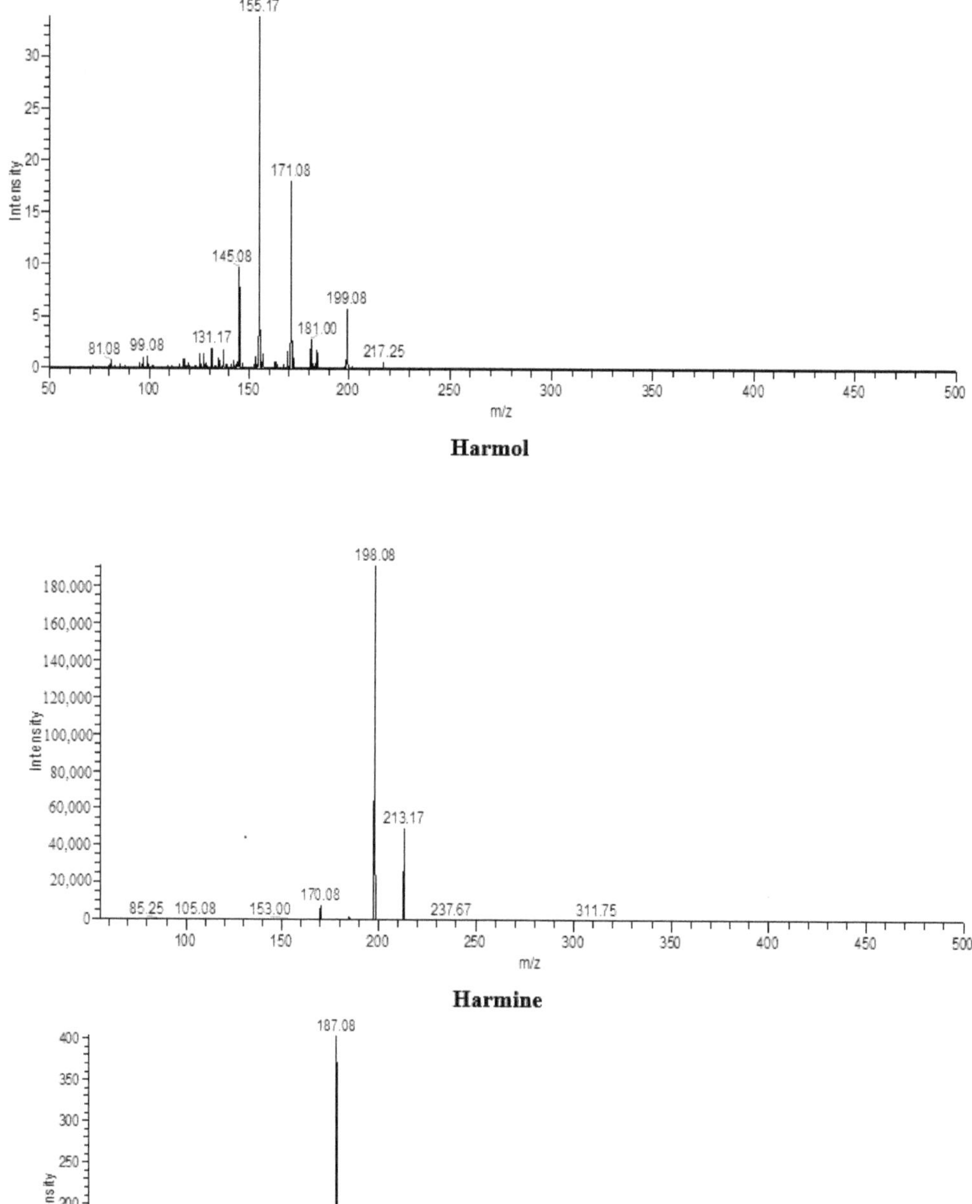

Figure 6. Mass spectra of different bioactive compounds found in **PHMF2** of *Peganum harmala*.

3.6. Quantification of Bioactive Compounds Using External Standards on HPLC

RP-HPLC subfraction PHMF2 was subjected to chromatographic investigation to further confirm and quantify the compounds identified based on mass spectroscopy using external standards (Table 5). External standards of quinic acid, peganine, harmol, harmaline, and harmine were well separated with our methodology, and eluted at retention times of 9.1, 10.1, 14, 21.8, 22.8, and 23.3 min at 280 nm as shown in Figure 7. Likewise, all these compounds were well separated from PHMF2 at the same retention times and wavelength, and quantified as 6.34, 19.2, 1.3, 3.9, and 53.9 µg/mg, respectively (Figure 8). Therefore, identities of these compounds were confirmed by comparing their retention times and mass spectral data with those of commercially available authentic samples as evident from HPLC overlay chromatograms (Figure 9). In this respect, Herraiz and coworkers reported a detectable amount of harmol, harmaline, and harmine as 0.03, 56, and 43.2 mg/g, respectively, from the *P. harmala* seed methanol extract [60]. On the other hand, the *P. harmala* methanol root extract also showed the presence of considerable amounts of harmol (14.1 mg/g) and harmine (20.6 mg/g), whereas harmaline was not detected. Other research groups indicated that harmol, harmaline, and harmine were present in the amounts 874, 488, and, 380 µg/mL, respectively, in the *P. harmala* methanol seeds extract [8]. Moreover, peganine was also identified and quantified earlier in the amounts of 9.6, 5.4, 21.2, 19.5, and 10.9 mg/g, respectively, from dry seeds, green fruits, immature fruits, flowers, and leaves of *P. harmala* [61]. In light of this, more work is required to identify and quantify more biologically active compounds from this traditionally important medicinal plant.

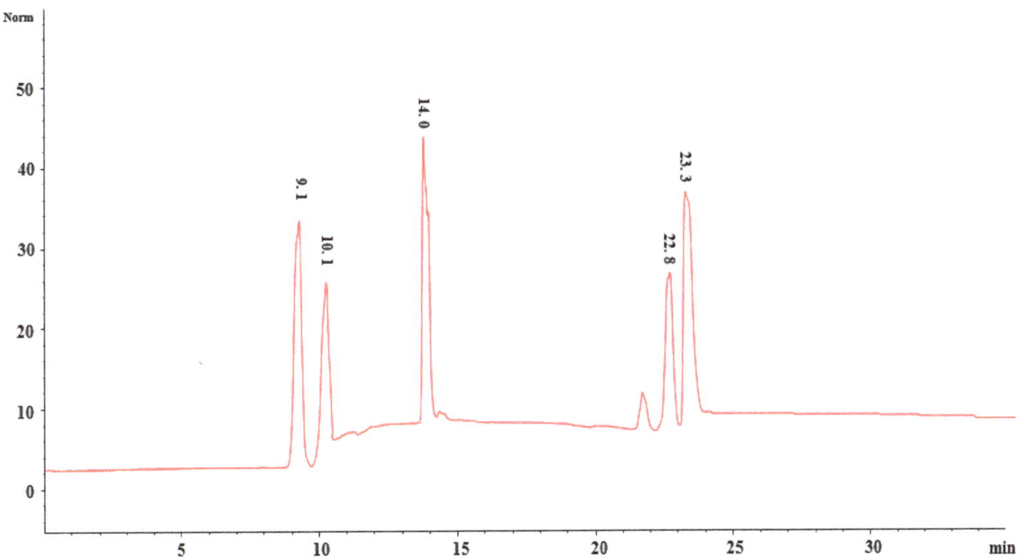

Figure 7. HPLC chromatogram of a mixture of external standards at 280 nm. (1) quinic acid, (2) peganine, (3) harmol, (4) harmaline, and (5) harmine.

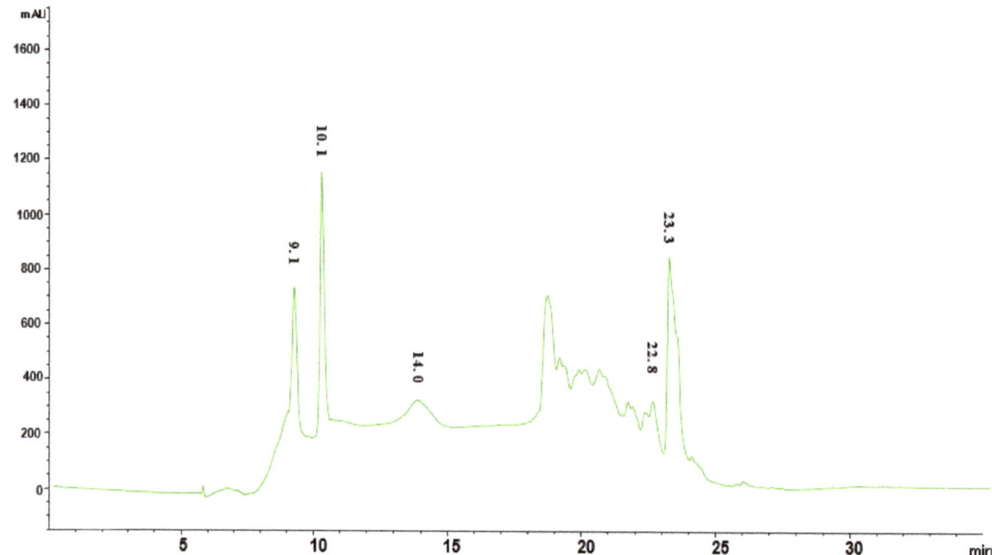

Figure 8. HPLC chromatogram of the PHMF2 subfraction at 280 nm. (1) quinic acid, (2) peganine, (3) harmol, (4) harmaline, and (5) harmine.

Table 5. Quantification of quinic acid, peganine, harmol, harmaline, and harmine of PHMF2 obtained from *Peganum harmala* RP-HPLC subfraction.

Fractions	Compound Name	Wavelength	LOD	LOQ	r^2	R_t min	Concentration (µg/mg)
	Quinic acid		1.1	3.2	0.9999	9.1	6.34
	Peganine		3.2	9.4	0.9989	10.1	19.2
PHMF2 (Methanol extract)	Harmol	280	0.2	0.5	0.9998	14.0	1.3
	Harmaline		0.4	0.9	0.9999	22.8	3.9
	Harmine		2.9	8.1	0.9997	23.3	53.9

Limit of quantification (LOQ), Limit of detection (LOD), Coefficient of regression (r^2), Retention time (Rt), *Peganum harmala* methanol fraction 2 (PHMF2).

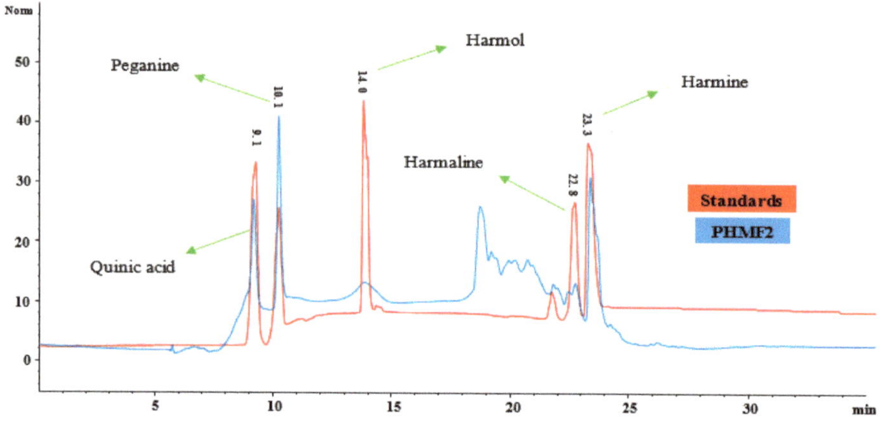

Figure 9. HPLC overlay chromatogram of the mixture of external standards and PHMF2 at 280 nm.

4. Conclusions and Future Prospects

In summary, the findings from this investigation reveal that extracts of *P. harmala* exhibit significant antioxidant and anti-inflammatory activities in both in vitro and in vivo models, which justifies the use of this plant in the traditional medicine of different countries. In addition, LD_{50} of PHME above 3000 mg/kg reflects the practical safety of the traditional herb with no changes in behavior. Most importantly, no mortality was observed during the acute (14 day) and subacute (28 day) toxicity studies. In addition, the presence of quinic acid, harmol, harmine, harmaline, and peganine may be responsible for the bioactivity of the extracts. However, more detailed studies are required to elucidate the possible mechanisms of action and pathways responsible for the antioxidant and anti-inflammatory capacities of the extracts. In conclusion, these results suggest that *P. harmala* possesses anti-inflammatory and antioxidant activities, and can be a promising new source of natural chemotherapeutic agents with no adverse effects.

Author Contributions: Conceptualization, M.H., P.W. and M.S.M.; investigation and resources, M.W.A., M.H., M.Q., S.A., Z.S. and M.S.M.; writing—original draft preparation, M.W.A., M.H., M.Q., S.A. and Z.S.; writing—review and editing, M.H., M.W.A. and M.S.M.; visualization and supervision, M.H., M.W.A. and M.S.M.; project administration, M.H., M.W.A. and M.S.M. All authors have read and agreed to the published version of the manuscript.

Funding: This research received no external funding.

Institutional Review Board Statement: Not applicable.

Informed Consent Statement: Not applicable.

Data Availability Statement: Available data are presented in the manuscript.

Acknowledgments: This manuscript is generated from the doctoral research of Malik Waseem Abbas under the supervision of Mazhar Hussain, Institute of Chemical Sciences, Bahauddin Zakariya University, Multan, Pakistan.

Conflicts of Interest: The authors declare no conflict of interest.

Sample Availability: Samples of the compounds are not available from the authors.

References

1. Battu, G.R.; Ethadi, S.R.; Priya, G.V.; Priya, K.S.; Chandrika, K.; Rao, A.V.; Reddy, S.O. Evaluation of antioxidant and anti-inflammatory activity of *Euphorbia heyneana* Spreng. *Asian Pac. J. Trop. Biomed.* **2011**, *1*, 191–194. [CrossRef]
2. Chandra, S.; Chatterjee, P.; Dey, P.; Bhattacharya, S. Evaluation of in vitro anti-inflammatory activity of coffee against the denaturation of protein. *Asian Pac. J. Trop. Biomed.* **2012**, *1*, 178–180. [CrossRef]
3. Umapathy, E.; Ndebia, E.J.; Meeme, A.; Adam, B.; Menziwa, P.; Nkeh-Chungag, B.N.; Iputo, J.E. An experimental evaluation of *Albuca setosa* aqueous extract on membrane stabilization, protein denaturation and white blood cell migration during acute inflammation. *J. Med. Plants Res.* **2010**, *4*, 789–795.
4. Rodriguez, V.L.; Davoudian, T. Clinical Measurement of Pain, Opioid Addiction, and Functional Status. In *Treating Comorbid Opioid Use Disorder in Chronic Pain*; Springer: Cham, Switzerland, 2016; pp. 47–56.
5. Kazemi, S.; Shirzad, H.; Rafieian-Kopaei, M. Recent findings in molecular basis of inflammation and anti-inflammatory plants. *Curr. Pharm. Des.* **2018**, *24*, 1551–1562. [CrossRef] [PubMed]
6. Mikaili, P.; Sharifi, M.; SHayegh, J.; Sarahroodi, S.H. Etymological review on chemical and pharmaceutical substances of the oriental origin. *Int. J. Anim. Vet. Adv.* **2012**, *4*, 40–44.
7. Li, S.G.; Wang, K.B.; Gong, C.; Bao, Y.; Qin, N.B.; Li, D.H.; Hua, H.M. Cytotoxic quinazoline alkaloids from the seeds of *Peganum harmala*. *Bioorgan. Med. Chem. Lett.* **2018**, *28*, 103–106. [CrossRef]
8. Kartal, M.; Altun, M.L.; Kurucu, S. HPLC method for the analysis of harmol, harmalol, harmine and harmaline in the seeds of *Peganum harmala* L. *J. Pharmaceut. Biomed. Anal.* **2003**, *3*, 263–269. [CrossRef]
9. Aziz, M.A.; Khan, A.H.; Adnan, M.; Izatullah, I. Traditional uses of medicinal plants reported by the indigenous communities and local herbal practitioners of Bajaur Agency, Federally Administrated Tribal Areas. *Pak. J. Ethnopharmacol.* **2017**, *198*, 268–281. [CrossRef]
10. Ullah, S.; Khan, M.R.; Shah, N.A.; Shah, S.A.; Majid, M.; Farooq, M.A. Ethnomedicinal plant use value in the Lakki Marwat District of Pakistan. *J. Ethnopharmacol.* **2014**, *158*, 412–422. [CrossRef]

11. Ullah, M.; Khan, M.U.; Mahmood, A.; Malik, R.N.; Hussain, M.; Wazir, S.M.; Daud, M.; Shinwari, Z.K. An ethnobotanical survey of indigenous medicinal plants in Wana district south Waziristan agency, Pakistan. *J. Ethnopharmacol.* **2003**, *150*, 918–924. [CrossRef]

12. Bellakhdar, J. *La Pharmacop'ee Marocaine Traditionnelle. Medecine Arabe Ancienne et Savoirs Populaires*; Ibis Press: Paris, France, 1997; pp. 529–530.

13. Baba Aissa, F. *Encyclopédie des Plantes Utiles, Flore d'Algérie et du Maghreb, Substances Végétales d'Afrique, d'Orient et d'Occident*; Ed Librairie Modern: Rouiba, Algeria, 2000; Volume 46, p. 46.

14. Liu, L.; Zhao, T.; Cheng, X.M.; Wang, C.H.; Wang, Z.T. Characterization and determination of trace alkaloids in seeds extracts from *Peganum harmala* linn. using LC-ESI-MS and HPLC. *Acta Chromatogr.* **2013**, *25*, 21–240. [CrossRef]

15. Bremner, P.; Rivera, D.; Calzado, M.A.; Obón, C.; Inocencio, C.; Fiebich, B.L.; Muñoz, E.; Heinrich, M. Assessing medicinal plants from South-Eastern Spain for potential anti-inflammatory effects targeting nuclear factor-Kappa Band otherpro-inflammatory mediators. *J. Ethnopharmacol.* **2009**, *124*, 295–305. [CrossRef] [PubMed]

16. Oodi, A.; Norouzi, H.; Amirizadeh, N.; Nikougoftar, M.; Vafaie, Z. Harmine, a novel DNA methyltransferase 1 inhibitor in the leukemia cell line. Indian. *J. Hematol. Blood. Transfus.* **2017**, *33*, 509–515. [CrossRef] [PubMed]

17. Liu, W.; Zhu, Y.; Wang, Y.; Qi, S.; Wang, Y.; Ma, C.; Xuan, Z. Anti-amnesic effect of extract and alkaloid fraction from aerial parts of *Peganum harmala* on scopolamine-induced memory deficits in mice. *J. Ethnopharmacol.* **2017**, *204*, 95–106. [CrossRef] [PubMed]

18. Bensalem, S.; Soubhye, J.; Aldib, I.; Bournine, L.; Nguyen, A.T.; Vanhaeverbeek, M.; Nève, J. Inhibition of myeloperoxidase activity by the alkaloids of *Peganum harmala* L. (Zygophyllaceae). *J. Ethnopharmacol.* **2014**, *154*, 361–369. [CrossRef]

19. Rezaee, M.; Hajighasemi, F. Sensitivity of hematopoietic malignant cells to *Peganum harmala* seed extract in vitro. *J. Basic. Clin. Pathophysiol.* **2019**, *7*, 21–26.

20. Shaheen, H.A.; Issa, M.Y. In vitro and in vivo activity of *Peganum harmala* L. alkaloids against phytopathogenic bacteria. *Sci. Hortic.* **2020**, *264*, 108940. [CrossRef]

21. Abderrahim, L.A.; Taïbi, K.; Abderrahim, C.A. Assessment of the antimicrobial and antioxidant activities of *Ziziphus lotus* and *Peganum harmala. Iran. J. Sci. Technol. A Sci.* **2019**, *43*, 409–414. [CrossRef]

22. Elansary, H.O.; Szopa, A.; Kubica, P.; Ekiert, H.; Al-Mana, F.A.; El-Shafei, A.A. Polyphenols of *Frangula alnus* and *Peganum harmala* leaves and associated biological activities. *Plants* **2020**, *9*, 1086. [CrossRef]

23. Monsef, H.R.; Ghobadi, V.; Iranshahi, M.; Abdollahi, M. Antinociceptive effects of *Peganum harmala* L. alkaloid extract on mouse formalin test. *J. Pharm. Pharm. Sci.* **2004**, *7*, 65–69.

24. Singhai, A.; Patil, U.K. Amelioration of oxidative and inflammatory changes by *Peganum harmala* seeds in experimental arthritis. *Clin. Phytosci.* **2021**, *7*, 13. [CrossRef]

25. Selim, S.A.; Aziz, M.H.A.; Mashait, M.S.; Warrad, M.F. Antibacterial activities, chemical constitutes and acute toxicity of Egyptian *Origanum majorana* L., *Peganum harmala* L. and *Salvia officinalis* L. essential oils. *Afr. J. Pharm. Pharmacol.* **2013**, *7*, 725–735.

26. Guergour, H.; Allouni, R.; Mahdeb, N.; Bouzidi, A. Acute and subacute toxicity evaluation of alkaloids of *Peganum harmala* L. in experimental mice. *Int. J. Pharmacogn. Phytochem. Res.* **2017**, *9*, 1182–1189. [CrossRef]

27. Miao, X.; Zhang, X.; Yuan, Y.; Zhang, Y.; Gao, J.; Kang, N.; Tan, P. The toxicity assessment of extract of *Peganum harmala* L. seeds in Caenorhabditis elegans. *BMC Complement. Med. Ther.* **2020**, *20*, 256. [CrossRef]

28. Hossain, M.A.; Shah, M.D. A study on the total phenols content and antioxidant activity of essential oil and different solvent extracts of endemic plant *Merremia borneensis. Arab. J. Chem.* **2015**, *8*, 66–71. [CrossRef]

29. Oriakhi, K.; Oikeh, E.I.; Ezeugwu, N.; Anoliefo, O.; Aguebor, O.; Omoregie, E.S. Comparative antioxidant activities of extracts of *Vernonia amygdalina* and *Ocimum gratissimum* leaves. *J. Agric. Sci.* **2014**, *6*, 13–20. [CrossRef]

30. Alara, O.R.; Abdurahman, N.H.; Mudalip, S.A.; Olalere, O.A. Effect of drying methods on the free radicals scavenging activity of *Vernonia amygdalina* growing in Malaysia. *J. King Saud Univ. Sci.* **2019**, *31*, 495–499. [CrossRef]

31. Zahin, M.; Aqil, F.; Ahmad, I. Broad spectrum antimutagenic activity of antioxidant active fraction of *Punica granatum* L. peel extracts. *Mutat. Res. Genet. Toxicol. Environ. Mutagen.* **2010**, *703*, 99–107. [CrossRef] [PubMed]

32. Ruch, R.J.; Cheng, S.J.; Klaunig, J.E. Prevention of cytotoxicity and inhibition of intercellular communication by antioxidant catechins isolated from Chinese green tea. *Carcinogenesis* **1989**, *10*, 1003–1008. [CrossRef] [PubMed]

33. Sadique, J.; Al-Rqobahs, W.A.; Bughaith, E.I.; Gindi, A.R. The bioactivity of certain medicinal plants on the stabilization of RBC membrane system. *Fitoterapia* **1989**, *60*, 525–532.

34. Sakat, S.; Juvekar, A.R.; Gambhire, M.N. In vitro antioxidant and anti-inflammatory activity of methanol extract of *Oxalis corniculata* Linn. *Int. J. Pharm. Pharm. Sci.* **2010**, *2*, 146–155.

35. Shinde, U.A.; Phadke, A.S.; Nair, A.M.; Mungantiwar, A.A.; Dikshit, V.J.; Saraf, M.N. Membrane stabilizing activity—A possible mechanism of action for the anti-inflammatory activity of *Cedrus deodara* wood oil. *Fitoterapia* **1999**, *70*, 251–257. [CrossRef]

36. Qamar, M.; Akhtar, S.; Ismail, T.; Yuan, Y.; Ahmad, N.; Tawab, A.; Ziora, Z.M. *Syzygium cumini* (L.), skeels fruit extracts: In vitro and in vivo anti-inflammatory properties. *J. Ethnopharmacol.* **2021**, *271*, 113805. [CrossRef] [PubMed]

37. Mizushima, Y.; Kobayashi, M. Interaction of anti-inflammatory drugs with serum proteins, especially with some biologically active proteins. *J. Pharm. Pharmacol.* **1968**, *20*, 169–173. [CrossRef]

38. Morris, C.J. Carrageenan-induced paw edema in the rat and mouse. *Inflammat. Protocol.* **2003**, *225*, 115–121.

39. Ramadhan, U.H. Study the effect of *Peganum harmala* L. Alkaloids extract in-vivo as anti-inflammatory agent. *Univ. Thi Qar J. Sci.* **2013**, *3*, 58–64.

40. Brownlee, G. Effect of deoxycortone and ascorbic acid on formaldehyde-induced arthritis in normal and adrenalectomised rats. *Lancet* **1950**, *268*, 157–159. [CrossRef]

41. OECD. *OECD Guideline for Testing of Chemicals. Repeated Dose 28-Day Oral Toxicity in Rodents, Test. No. 407*; OECD: Paris, France, 2008.

42. OECD. *OECD Guidelines for Testing of Chemicals: Acute Oral Toxicity—Acute Toxic Class. Method. Test. No. 423, Adopted 22 March 1996, and Revised Method Adopted 17 December 2001*; OECD: Paris, France, 2001.

43. Cock, I.E. Problems of reproducibility and efficacy of bioassays using crude extracts, with reference to *Aloe vera*. *Pharmacogn. Commun.* **2011**, *1*, 52–62. [CrossRef]

44. Steinmann, D.; Ganzera, M. Recent advances on HPLC/MS in medicinal plant, analysis. *J. Pharm. Biomed. Anal.* **2011**, *55*, 744–757. [CrossRef] [PubMed]

45. Rice-evans, C.A.; Miller, N.J.; Bolwell, P.G.; Bramley, P.M.; Pridham, J.B. The relative antioxidant activities of plant-derived polyphenolic flavonoids. *Free Radic. Res.* **1995**, *22*, 375–383. [CrossRef] [PubMed]

46. Iqbal, Z.; Javed, M.; Rafique, G.; Saleem, T. A comparative study of total phenolic contents and antioxidant potential of seeds of *Peganum harmala*. *Int. J. Biosci.* **2019**, *14*, 121–127.

47. Kumari, C.S.; Yasmin, N.; Hussain, M.R.; Babuselvam, M. In vitro anti-inflammatory and anti-arthritic property of *Rhizopora mucronata* leaves. *Intern. J. Pharma. Sci. Res.* **2015**, *6*, 482–485.

48. Scanlon, V.V.; Sanders, T. *Essentials of Anatomy and Physiology*, 6th ed.; FA Davis: Philadelphia, PA, USA, 2010; p. 287.

49. Khadhr, M.; Bousta, D.; El Mansouri, L.; Boukhira, S.; Lachkar, M.; Jamoussi, B.; Boukhchina, S. HPLC and GC–MS analysis of Tunisian *Peganum harmala* seeds oil and evaluation of some biological activities. *Am. J. Ther.* **2017**, *24*, 706–712. [CrossRef] [PubMed]

50. Moussaid, M.; Elamrani, A.E.; Bourhim, N.; Benaissa, M. In vivo anti-inflammatory and in vitro antioxidant activities of Moroccan medicinal plants, *Nat. Prod. Commun.* **2011**, *6*, 1441–1443.

51. Khlifi, D.; Sghaier, R.M.; Amouri, S.; Laouini, D.; Hamdi, M.; Bouajila, J. Composition and anti-oxidant, anti-cancer and anti-inflammatory activities of *Artemisia herba-alba*, *Ruta chalpensis* L. and *Peganum harmala* L. *Food Chem. Toxicol.* **2013**, *55*, 202–208. [CrossRef]

52. Opie, E.L. On the relation of necrosis and inflammation to denaturation of proteins. *J. Exp. Med.* **1962**, *115*, 597–608. [CrossRef] [PubMed]

53. Williams, L.A.D.; O'Connar, A.; Latore, L.; Dennis, O.; Ringer, S.; Whittaker, J.A.; Kraus, W. The in vitro anti-denaturation effects induced by natural products and non-steroidal compounds in heat treated (immunogenic) bovine serum albumin is proposed as a screening assay for the detection of anti-inflammatory compounds, without the use of animals, in the early stages of the drug discovery process. *West. Indian Med. J.* **2008**, *57*, 327–331. [PubMed]

54. Leelaprakash, G.; Dass, S.M. In vitro anti-inflammatory activity of methanol extract of *Enicostemma axillare*. *Int. J. Drug Dev. Res.* **2011**, *3*, 189–196.

55. Wills, A.L. Release of Histamin, Kinin and Prostaglandins during Carrageenin Induced Inflammation of the Rats. *Prostaglandins Pept. Amins* **1969**, 31–48. Available online: https://ci.nii.ac.jp/naid/10006158109/ (accessed on 6 October 2021).

56. Edziri, H.; Marzouk, B.; Mabrouk, H.; Garreb, M.; Douki, W.; Mahjoub, A.; Verschaeve, L.; Najjar, F.; Mastouri, M. Phytochemical screening, butyrylcholinesterase inhibitory activity and anti-inflammatory effect of some Tunisian medicinal plants. *S. Afr. J. Bot.* **2018**, *114*, 84–88. [CrossRef]

57. Kumar, M.P.; Joshi, S.D.; Kulkarni, V.H.; Savant, C. Phytochemical screening and evaluation of analgesic, anti-inflammatory activities of *Peganum harmala* Linn., seeds in rodents. *Appl. Pharm. Sci.* **2015**, *5*, 52–55. [CrossRef]

58. Riaz, M.; Rasool, N.; Iqbal, M.; Tawab, A.; E.-Hbib, F.; Khan, A.; Farhan, M. Liquid chromatography-electrospray ionization-tandem mass spectrometry (LC-ESI-MS/MS) analysis of *Russelia equisetiformis* extract. *Bulg. Chem. Commun.* **2017**, *49*, 354–359.

59. Wang, Z.; Kang, D.; Jia, X.; Zhang, H.; Guo, J.; Liu, C.; Liu, W. Analysis of alkaloids from *Peganum harmala* L. sequential extracts by liquid chromatography coupled to ion mobility spectrometry. *J. Chromatogr. B* **2018**, *1096*, 73–79. [CrossRef] [PubMed]

60. Herraiz, T.; González, D.; Ancín-Azpilicueta, C.; Arán, H.; Guillén, V.J. β-Carboline alkaloids in *Peganum harmala* and inhibition of human monoamine oxidase (MAO). *Food Chem. Toxicol.* **2010**, *48*, 839–845. [CrossRef]

61. Herraiz, T.; Guillén, H.; Arán, V.J.; Salgado, A. Identification, occurrence and activity of quinazoline alkaloids in *Peganum harmala*. *Food Chem. Toxicol.* **2017**, *103*, 261–269. [CrossRef]

Article

Sage, *Salvia officinalis* L., Constituents, Hepatoprotective Activity, and Cytotoxicity Evaluations of the Essential Oils Obtained from Fresh and Differently Timed Dried Herbs: A Comparative Analysis

Hamdoon A. Mohammed [1,2,*], Hussein M. Eldeeb [3,4,*], Riaz A. Khan [1,*], Mohsen S. Al-Omar [1,5], Salman A. A. Mohammed [3], Mohammed S. M. Sajid [3], Mohamed S. A. Aly [6], Adel M. Ahmad [7], Ahmed A. H. Abdellatif [8], Safaa Yehia Eid [9] and Mahmoud Zaki El-Readi [9,10]

[1] Department of Medicinal Chemist and Pharmacognosy, College of Pharmacy, Qassim University, Qassim 51452, Saudi Arabia; m.omar@qu.edu.sa

[2] Department of Pharmacognosy, Faculty of Pharmacy, Al-Azhar University, Cairo 11371, Egypt

[3] Department of Pharmacology and Toxicology, College of Pharmacy, Qassim University, Qassim 51452, Saudi Arabia; m.azmi@qu.edu.sa (S.A.A.M.); su.mohammed@qu.edu.sa (M.S.M.S.)

[4] Department of Biochemistry, Faculty of Medicine, Al-Azhar University, Assiut 71524, Egypt

[5] Department of Medicinal Chemistry and Pharmacognosy, Faculty of Pharmacy, JUST, Irbid 22110, Jordan

[6] Hospital of Police Academy, Nasr city, Cairo 11765, Egypt; mohamedshawky1974@yahoo.com

[7] Department of Pharmaceutical Analytical Chemistry, Faculty of Pharmacy, South Valley University, Qena 83523, Egypt; Adelpharma2004@svu.edu.eg

[8] Department of Pharmaceutics, College of Pharmacy, Qassim University, Qassim 51452, Saudi Arabia; A.Abdellatif@qu.edu.sa

[9] Department of Clinical Biochemistry, Faculty of Medicine, Umm Al-Qura University, Abdia, Makkah 21955, Saudi Arabia; syeid@uqu.edu.sa (S.Y.E.); mzreadi@uqu.edu.sa (M.Z.E.-R.)

[10] Department of Biochemistry, Faculty of Pharmacy, Al-Azhar University, Assiut 71524, Egypt; zakielreadi@azhar.edu.sa

* Correspondence: ham.mohammed@qu.edu.sa (H.A.M.); hu.ali@qu.edu.sa (H.M.E.); ri.khan@qu.edu.sa (R.A.K.); Tel.: +00966566176074 (H.A.M.)

Citation: Mohammed, H.A.; Eldeeb, H.M.; Khan, R.A.; Al-Omar, M.S.; Mohammed, S.A.A.; Sajid, M.S.M.; Aly, M.S.A.; Ahmad, A.M.; Abdellatif, A.A.H.; Eid, S.Y.; et al. Sage, *Salvia officinalis* L., Constituents, Hepatoprotective Activity, and Cytotoxicity Evaluations of the Essential Oils Obtained from Fresh and Differently Timed Dried Herbs: A Comparative Analysis. *Molecules* **2021**, *26*, 5757. https://doi.org/10.3390/molecules26195757

Academic Editor: Raffaele Pezzani

Received: 21 August 2021
Accepted: 16 September 2021
Published: 23 September 2021

Publisher's Note: MDPI stays neutral with regard to jurisdictional claims in published maps and institutional affiliations.

Abstract: Sage, *Salvia officinalis* L., is used worldwide as an aromatic herb for culinary purposes as well as a traditional medicinal agent for various ailments. Current investigations exhibited the effects of extended dryings of the herb on the yields, composition, oil quality, and hepatoprotective as well as anti-cancer biological activities of the hydrodistillation-obtained essential oils from the aerial parts of the plant. The essential oils' yields, compositions, and biological activities levels of the fresh and differently timed and room-temperature dried herbs differed significantly. The lowest yields of the essential oil were obtained from the fresh herbs (FH, 631 mg, 0.16%), while the highest yield was obtained from the two-week dried herbs (2WDH, 1102 mg, 0.28%). A notable decrease in monoterpenes, with increment in the sesquiterpene constituents, was observed for the FH-based essential oil as compared to all the other batches of the essential oils obtained from the different-timed dried herbs. Additionally, characteristic chemotypic constituents of sage, i.e., α-pinene, camphene, β-pinene, myrcene, 1, 8-cineole, α-thujone, and camphor, were present in significantly higher proportions in all the dried herbs' essential oils as compared to the FH-based essential oil. The in vivo hepatoprotective activity demonstrated significant reductions in the levels of AST, ALT, and ALP, as well as a significant increase in the total protein ($p < 0.05$) contents level, as compared to the acetaminophen (AAP) administered experimental group of rats. A significant reduction ($p < 0.05$) in the ALT level was demonstrated by the 4WDH-based essential oil in comparison to the FH-based essential oil. The levels of creatinine, cholesterol, and triglycerides were reduced ($p < 0.05$) in the pre-treated rats by the essential oil batches, with non-significant differences found among them as a result of the herbs dryings based oils. A notable increase in the viability of the cells, and total antioxidant capacity (TAOxC) levels, together with the reduction in malondialdehyde (MDA) levels were observed by the essential oils obtained from all the batches as compared with the AAP-treated cell-lines, HepG-2, HeLa, and MCF-7, that indicated the in vitro hepatoprotective effects of the sage essential oils. However, significant improvements in the in vivo and in vitro hepatoprotective activities with the 4WDH-based oil, as compared to all other essential oil-batches and silymarin standard demonstrated the beneficial effects of the drying protocol for the herb for its medicinal purposes.

Keywords: Sage; *Salvia officinalis*; cytotoxicity; hepatoprotection; MDA; TAOxC; MCF-7; HeLA cells; HepG-2 cells

1. Introduction

Sage, salvia, or *Salvia officinalis* L., the plant belonging to the family Lamiaceae, is a perennial aromatic herb cultivated worldwide for its medicinal, culinary, and flavor properties. The sage, as a culinary and medicinal herb, is popular worldwide and considered an economically important commodity for use in the apothecary and spices business. Herbal shops regularly trade sage as well as other aromatic herbs wherein the herbs are dried at room temperature under shade to retain their aromatic contents. The herb is worldwide available and is used in food preparations and also as an essential part of the tea mixture of the Mediterranean countries [1]. The folk-medicinal and traditional usage of sage is global in nature, and the Latin name of the plant "Salvia" means curative, which is indicative of the health-giving properties, and importance of this plant [2]. The aerial parts of the plant are used for the management of gout, rheumatism, diarrhea, and diabetes in the countries of Northern America, and Asia [3]. Sage is also used for the treatment of gastrointestinal disorders, such as dyspepsia and ulcers, as well as heartburn, and upper respiratory complaints, including sore throat, and inflammations [1]. Sage also has noticeable beneficial effects in age-related cognitive disorders [4]. It is strongly reputed to induce calmness and improve memory in Alzheimer's patients [5]. The plant leaves are frequently used in the food industry due to their flavor and antioxidant, and antimicrobial properties [6]. The plant is also part of the perfumery and cosmetic industries [2].

Chemical analysis of the sage essential oil from different locations has been carried out, and detailed reports are available from different quarters. The major volatile constituents of the plant were determined as being α-thujone, 1,8-cineole, and camphor, which are considered the chemotypic constituents of the *S. officinalis* essential oil [7]. The effects of seasonal variations [8], geographical locations differences in oil quality and yields [8–10], essential oils' extraction procedures [11–13], and the essential oils obtained through different drying methods [14,15] have been reported as parts of the parameters affecting the plant essential oils' yields and constituents. Significant variations among sage essential oil constituents in response to the aforementioned parameters have been reported. The major chemotypic products of the sage essential oil, α-thujone, 1,8-cineole, and camphor also significantly varied in the essential oils obtained from the plant batches dried in an oven at 45 °C and 65 °C, and plant materials dried in a microwave oven (500 W) [14]. Likewise, the α-thujone and camphor's componental ratios fluctuated in the essential oil batches obtained from the herbs collected in different seasons, and from different geographic locations [8].

Nonetheless, the essential oil compositions and yields obtained from *S. officinalis* have been extensively studied [7,16–19], and the effects of the extraction procedures, drying methods, and harvesting frequency on the plant's essential oil constituents have been demonstrated in previous reports [14,15,20,21]. The current work instead, in contrast to the previous reports on the essential oils of sage from different locations, seasons, and multiple drying methods, dealt with the periodic effects of a single drying procedure that is frequently adopted by the apothecary and herbal dealers, i.e., room temperature and natural atmospheric pressure conditions for drying the herbs. To the best of information available, the effects of extended, i.e., 1- to 4-weeks, dryings on the sage essential oils yields and variations in their constituents have not been investigated. This is the first report disclosing the substantial effects of the plants' drying effects on its oils' yields, oils' composition, and effects on the chemotypic constituents.

Hepatoprotective effects of herbal natural products have been reported, and several secondary metabolic origins compounds responsible for potential activity in treating liver dysfunctions are known [22]. The testing protocols, including animal models, in assessing the hepatoprotective activity are one of the important aspects in confirming the in vivo potential of the test material [23,24]. Moreover, for the cell lines–based assays for liver functional assessments, the HepG-2 cell lines are the prime choice, owing to their desirable biochemical and morphological characteristics imitating the normal hepatocytes; hence, they are used as a representative model for the in vitro hepatoprotective assays. In addition, the HepG-2 cell lines have a certain advantage over the normal hepatocytes in a way that they have high survival and maintenance rates in large quantities without changes in their drug-metabolizing enzyme activities, and this occurs in the primary cultures of the human hepatocytes [25]. Hence, the HepG-2 cell lines are widely used as an in vitro model for the assessment of various liver functions, their metabolic activity, and the evaluation of the drug's toxicity [26]. In the context of hepatic disorders and hepatoprotective actions, acetaminophen (AAP, paracetamol) is widely used as an antipyretic, analgesic, and anti-inflammatory standard drug/agent. However, the AAP produces hepatotoxicity upon larger dose administrations, and the mechanisms responsible for in vivo liver toxicity of the AAP are complex [27], as it (AAP) undergoes metabolic activation in a cytochrome-P450–dependent step to produce a highly reactive metabolite, N-acetyl-*p*-benzoquinone imine (NAPQI), as well as free radicals, which can initiate lipid peroxidation. The in vivo toxicity induced by the AAP and the toxicity in cultured hepatocytes involves stimulations of lipid peroxidation, which is detected as an increase in the levels of malondialdehyde (MDA) formations [28].

The hepatoprotective effects of sage essential oil have been investigated by several research groups since the plant is claimed to be curative in nature with context to hepatoprotection, and is still used as part of the traditional medicament for cure and protection of liver-related disorders in several quarters [29–32]. Previous reports describing the cytotoxic activities of sage essential oils are also available [33–35].

The current study specifics were performed to evaluate the impacts of extended dryings of the sage aerial parts on their essential oil yields and constituents' variations, as well as effects on biological activities, in comparison to the essential oil obtained from the fresh herb. Specifically, the current study investigated the variations in the yields and chemical compositions of the essential oil batches, and their impact on cytotoxicity in MCF-7, and HeLa cell lines, together with the hepatoprotective activity against AAP-induced liver damages in the in vivo models in Wistar male rats, and in an in vitro model in HepG-2 cell lines.

2. Materials and Methods

2.1. Plant Materials

The aerial parts of the sage herb were collected in February 2019 from the gardens maintained by the Ministry of Agriculture, Qassim province, KSA. The institutional botanist identified the plants as *Salvia officinalis* L. The fresh plants' aerial parts were divided into 5 equal groups of 400 g each. One group of the fresh aerial parts of the plants were subjected to hydrodistillation, while other groups were dried in shade at room temperature ($25 \pm 2\ °C$) for a pre-defined time (one-week intervals) as defined in Section 2.2.

2.2. Dryings of the Plants and Hydrodistillation for Obtaining Essential Oil Batches

The five fresh herb groups of 400 g each were individually subjected to the procedure as follows. Batch one (fresh aerial parts herb group, 400 g) of the plant was mixed with 700 mL of distilled water in a 2 L round-bottom flask, and instantly subjected to hydrodistillation for 5 h using a Clevenger apparatus. The subsequent batches, two, three, four, and five (400 g each) were allowed to dry at room temperature ($25 \pm 2\ °C$) in shade for one, two, three, and four weeks, respectively, before their hydrodistillation to procure the essential oils in the same conditions as used for the fresh plants batch. The obtained essential oils were dried over anhydrous sodium sulfate, and the percent yields of the essential oils for each batch was calculated according to the following formula:

$$Oil\ Yields\ \% = \left(\frac{A}{400}\right) \times 100 \tag{1}$$

where A is the amount of essential oil in gram obtained from 400 g of the fresh aerial parts of the plant. The batches' weight reductions were calculated after the defined drying periods for each batch.

2.3. Gas Chromatography-Flame Ionization Detector (GC-FID) Analyses

Hydrodistilled essential oil batches of the herb sage were subjected to GC-FID analyses on a Perkin Elmer Auto System XL gas chromatography instrument equipped with a flame ionization detector (FID). The chromatographic separations of the essential oil samples were achieved on a fused silica capillary column ZB5 (60 m × 0.32 mm i.d. × 0.25 μm film thickness). The oven temperature was maintained initially at 50 °C and programmed from 50 °C to 240 °C at a rate of 3 °C/min. Helium gas was used as a carrier at a flow rate of 1.1 mL/min. The injector and detector temperatures were 220 °C and 250 °C, respectively.

2.4. Identification of the Essential oil Constituents

The essential oils obtained from the fresh sage and the dried batches were identified based on the experimental retention index (*RI*) calculated with references to standard *n*-alkenes series (C_8–C_{40}), and the retention indices reported in the literature under similar GC experimental conditions. Additionally, the identification of the compounds was carried out based on their retention time and comparisons with the mass spectral libraries (National Institute of Standards and Technology (NIST-11) and Wiley Database). The relative percentages of the constituents were calculated from the area under the peak obtained from the GC-FID chromatogram.

2.5. In Vivo Hepatoprotective Assay

2.5.1. Experimental Animals

The in vivo hepatoprotective activity of the sage essential oil batches was performed using Wistar male rats weighing about 200–250 g, which were kindly provided by the animal house facility of the College of Pharmacy, Qassim University. The rats were housed in suitable humidity and temperature (25 ± 2 °C) and given a standard diet and water *ad libitum*. The animals were kept in a pathogen-controlled, air-conditioned room in the animal house. All the experiments were performed, according to the guidelines for animal studies that were approved by the Ethical Committee of College of Pharmacy, Qassim University, KSA.

2.5.2. Acute Toxicity Studies

Briefly, ten-weeks-old male Wistar rats ($n = 15$), weighing 200–250 g, being overnight fasted, were weighed, and a single dose of 50, 100, and 200 mg/kg ($n = 5$/group) of *Salvia officinalis* essential oil was administered, using the oral route. The animals were observed for abnormality in behavior and movements for the first three days and mortality for up to two weeks. Based on the results, 20 mg/kg, 10% of the maximum administered dose according to the Hedge and Sterner scale, was selected for evaluation of the hepatoprotective activity [36].

2.5.3. Animal Groups

A total of 40 ten-week-old Wistar male rats, weighing 200–250 g, were divided randomly into eight equal groups ($n = 5$); the first group was considered the control and received oral supplementation of saline, using an orogastric cannula. The second group of animals (negative control) received oral administration of AAP (in a dose of 500 mg/kg) once daily starting from the 11th day of the experiment for 5 consecutive days to induce liver injury. The third, fourth, fifth, sixth, and seventh groups of animals received 20 mg/kg BW (bodyweight), once daily, for 15 days the essential oils obtained from the fresh herb (FH), one-week (1WDH), two-week (2WDH), three-week (3WDH), and four-week dried herb (4WDH) of the *Salvia officinalis,* respectively. The animal groups (third to seventh groups) received AAP to induce liver injury (in a dose of 500 mg/kg) starting from the 11th day of the experiment for 5 days. Standard liver support was administrated to group number eight, which was pretreated with silymarin (oral dose: 100 mg/kg, 15 days), and AAP for the last 5 days. At the end of the experiment, blood samples were collected by retro-orbital puncture, serum was separated from all groups' collected blood for the determination of liver functions (ALP, AST, ALT, and total protein) as well as kidneys functions (urea and creatinine) and the lipid profile (triglycerides and total cholesterol) analyses.

2.5.4. Determination of Liver, Kidneys Functions, and Lipid Profile

The plasma levels of ALT, AST, ALP, and creatinine were determined, using an optimized UV-test, according to the international federation of clinical chemistry (IFCC). The plasma levels of total protein, urea, cholesterol, and triglycerides were measured, using the colorimetric methods. All the reagents were provided by the Crescent Diagnostics Company, KSA.

2.6. Cell-proliferation Assays

Human breast adenocarcinoma cell lines (MCF-7), cervical carcinoma cell lines (HeLa), and hepatocellular carcinoma cell lines (HepG-2) were used to evaluate the cytotoxic activity of the different batches of the sage essential oils in comparison with the normal fibroblast lung cell lines, MRC-5. The cell lines were grown in RPMI1640 or DMEM media (Gibco) supplemental with 10% fetal bovine serum (FBS) (Gibco), penicillin-streptomycin 1%, and L-glutamine 1%. Cells were grown under optimum growth conditions at 37 °C in a humidified atmosphere of 5% CO_2. The proliferation of tested cells after treatment with the sage essential oils from different batches and doxorubicin as a positive cytotoxic drug was determined by the colorimetric 3-(4,5- dimethylthiazol-2-yl)-2,5-diphenyl tetrazolium bromide (MTT) assays. Cells (1×10^4 cells/well) were incubated in 96-wells plates for 24 h. After that, the cells were further incubated for 24 h with and without essential oils and doxorubicin. MTT solution (Sigma-Aldrich, Germany) (10 µL of 5 mg/mL in PBS) was added to each well and incubated for a further 4 h at 37 °C in a CO_2 incubator. The formazan crystals were dissolved in 100 µL of DMSO, and the color density was measured at 570 nm using SpectraMax M II microplate reader (Molecular Devices, LLC. San Jose, CA, USA).

Viability percentage and Selectivity Index:

To calculate the viability percentage, the following formula was used:

$$\text{Viability \%} = (\text{OD}_{\text{treatment}} / \text{OD}_{\text{control}}) \times 100 \qquad (2)$$

IC_{50} values (i.e., the sample concentration that exerts 50% inhibition concerning untreated cells) were determined for all the cell lines. The selectivity index (SI), which indicates the cytotoxic selectivity (i.e., drug safety) for tested samples against cancer cells versus normal cells, was calculated from the following formula:

$$\text{SI} = IC_{50} \text{ calculated for normal cells} / IC_{50} \text{ calculated for cancer cells} \qquad (3)$$

The SI values higher than 2 were considered to be high selectivity [37].

2.7. Acetaminophen (AAP) Induced Hepatotoxicity in Hepg2 Cells

To evaluate the hepatoprotective effects of the sage essential oils on HepG-2 cell lines, the non-toxic concentration of the sage essential oil, 100 μg/mL, was selected to conduct the hepatoprotective activity assessment using the MTT assay as described earlier (Section 2.6). The HepG-2 cell lines were treated with AAP (4 mM) for 24 h, and the HepG-2 cells in the medium were considered as a negative control. Further, the hepatoprotective assays were performed using standard methods as described in each assay in the following sections as detailed below. The HepG-2 cells were pretreated with the essential oils for 12 h and incubated with AAP (4 mM) for 24 h, with and without the essential oils; the HepG-2 cells in medium only were considered as a negative control.

2.8. Measurement of Total Antioxidant Capacity (TAOxC)

The HepG-2 cell lines were treated with AAP in the presence and absence of the sage essential oils obtained from different batches. After treatment, the cells were lysed and suspended by sonication on ice in 0.9% sodium chloride solution containing 0.1% glucose and 5 mM potassium phosphate buffer (pH 7.4). The supernatant of the lysed cells was used to measure TAOxC, using an antioxidant assay kit obtained from Cayman Chemical Company (Ann Arbor, MI, USA). The assay was dependent on the ability of the antioxidants in the sample to inhibit the oxidation of 2,2'-azino-bis-3-ethylbenzothiazoline (ABTS) to $ABTS^+$ by metmyoglobin absorbance in the wells, which were measured after 5 min at a wavelength of 405 nm on a microplate reader, SpectraMax M II (Molecular Devices, LLC. San Jose, CA, USA). The results were expressed as millimoles of the antioxidants utilized [38].

2.9. Measurement of MDA for Lipid Peroxidation

Malondialdehyde (MDA), an end product of the lipid peroxidation, was used as an oxidative stress marker, and its concentration was measured using a thiobarbituric acid reactive substance (TBARS) assay kit obtained from the Cayman Chemical Company. The HepG-2 cells were treated with AAP in the presence and absence of sage essential oils, the supernatant of cells lysate or the standard sodium dodecyl sulfate, and the color reagent was added, heated to 100 °C for 1 h, and immediately cooled in an ice bath and centrifuged. The absorbance of the product was measured at a wavelength of 540 nm on a microplate reader, SpectraMax M II (Molecular Devices, LLC. San Jose, CA, USA). The extent of lipid peroxidation was quantified by estimating the MDA concentration. The results are expressed as micromoles of MDA equivalents formed per liter.

2.10. Statistical Analysis

The results were analyzed using GraphPad Prism V6 (GraphPad Software, San Diego, CA, USA). Data were expressed as mean ± SD. of 3–4 independent experiments performed at least in triplicate. One-way analysis of variance (ANOVA) followed by Tukey's test was used to detect any significant differences among the different mean values. A *p*-value less than 0.05 was considered a significant difference.

3. Results and Discussion

3.1. Sage Essential Oil Obtained from the Fresh Aerial Parts of the Plants and the Extended-Dried Plant Batches

The current study was designed to evaluate the effects of extended dryings on the sage essential oil yields, compositions, and biological activities, wherein the herbs' aerial parts were utilized to obtain the essential oils by the hydrodistillation process. The factors of drying temperatures (25 ± 2 °C), pressure (atmospheric pressure), and the amount of the fresh herbs (400 g) in each batch were constants; however, the variable parameter was the drying period and the weight loss of the dried herbs. From the viewpoint of essential oils production, the overall results in Table 1 show higher essential oil yields through the hydrodistillation method from the dried aerial parts of the herbs batches than that obtained from the fresh herb.

Table 1. Reduction in sage herbs' weights and essential oils obtained by hydrodistillation in response to extended dryings.

Periods of Drying	Fresh Weight	Weight after Drying	Essential Oil (mg)	% Yields *
Fresh Herb (FH)		400 g	631 ± 8.05	0.16
1WDH	400 g	131 g	923 ± 6.34	0.23
2WDH		111 g	1102 ± 15.58	0.28
3WDH		107 g	944 ± 5.73	0.24
4WDH		107 g	702 ± 9.10	0.18

* Yield percentages were calculated from the equation: weight of the essential oil obtained in gram/ 400 × 100.

The results showed a noticeable change in the plant weight after one week of drying from 400 g to 131 g (−67.25%) and a significant increase in the essential oil yields obtained from the one-week dried herb (923 ± 6.34 mg) as compared to the fresh herb material (631 ± 8.05 mg). The herbs dried for two weeks showed a drastic change in their weight; a loss of the weight from 400 g to 131 g, as compared to the fresh sample, was observed. At the same time, the essential oil yields percentage was ~75% higher for the two weeks dried herbs batch than the essential oil yields obtained from the fresh herbs sample. The results in Table 1 displayed that further dryings of the sage for three and four weeks were accompanied by a reduction in the yields of the essential oils in comparison to the two weeks of dried herbs, which showed maximum yields of the essential oil. The weight reduction in response to the extended dryings was a logical culmination of the drying process; nonetheless, the improvements in the essential oil obtained from the herbs after two weeks of drying as compared to the fresh plant-based oil yields need careful consideration. The reduction in the water to essential oil ratio in the herbs during the drying process is a plausible explanation for the higher yields of essential oil recovery from the dried herbs rather than the fresh herbs batch. The drying period extension over two weeks might be accompanied by higher emission of the essential oil from the herbal materials, which possibly explains the significant reduction in the essential oil yields obtained from the dried herbs batches of three and four weeks as compared to the two weeks dried herbs. Notably, similar results in parallel to the current findings for weight loss and essential oil yields were recently reported for rosemary, *Rosmarinus officinalis* [39]. Nevertheless, the essential oil yields percentages from the fresh and extended dried sage materials ranged from 1.5 to 2.8 mL/Kg, and it did not confirm the recommendations of European Pharmacopoeial (EP) standards for *Salvia officinalis* oil yields (10 mL/Kg) [19]. However, the essential oil yields were in alignment with the reported values for the essential oil yields from the aerial parts of the plant, sage, grown in the Middle Eastern regions, which varied between 0.8% to 2.5% [9,40–42]. Additionally, the combined results of the Table 1 confirmed the economic importance of the herbs-drying process for better essential oils production. The results also are in agreement with the normal practice conducted by the herbalist, and home drying of the herbs utilized for different culinary purposes.

3.2. Componental Analysis of the Essential Oil Obtained from Different Batches

The essential oils obtained from all the batches of sage (fresh and dried aerial parts) were analyzed for the presence, in percentages, of their constituents (GC-FID analysis, supplementary file, Table S1–S3). The results exhibited in Table 2 demonstrated distinct differences between the constituents and their percentages in all five batches of sage oil. Among all the identified constituents, five components were identified only in the fresh herbs–based essential oil, i.e., butyl acetate (0.64%), α-phellandrene (0.3%), neral (0.17%), α-cadinene (0.17%), and viridiflorol (0.59%). The differently timed dried herbs exhibited an absence as well as an increment and/or reduction in percent ratios of different components in the essential oils (Table 2). The results showed major differences in the percentages of the chemotypic constituents of the essential oils obtained from the dried herbs–based essential oils as compared to the fresh herb-based essential oil. The non-oxygenated monoterpenes, i.e., α-pinene, camphene, β-pinene, myrcene, and γ-terpinene, were significantly increased ($p < 0.05$) in the dried herbs–based essential oils batches, as compared to the essential oil obtained from the fresh herb. Here, the α-pinene concentration was significantly increased ($p < 0.05$) from 0.07% of the fresh herbs–based essential oil to 1.54, 1.57, 1.46, and 0.73% in the one-, two-, three-, and four weeks dried herbs–based essential oils. Some of the oxygenated monoterpenes, e.g., 1,8-cineole, α-thujone, and camphor, were also significantly increased ($p < 0.05$) in the essential oils batches obtained from dried sage as compared to the essential oil batch obtained from the fresh herbs (Table 2). In contrast, the sesquiterpenes β-caryophyllene, α-humulene, d-cadinene, humulene epoxide III, 14-hydroxy-(Z)-caryophyllene, and α-bisabolol showed drastic reductions ($p < 0.05$) in their contents ratio in the oil yields of the dried herbs–based essential oil batches in comparison to the fresh herbs–based essential oil. For instance, the α-bisabolol percentage was reduced from 3.43% in the fresh herbs–based essential oil to 0.27% in the essential oil batch of the one-week dried herbs, which equals to a 92.12% reduction in the percentage of the α-bisabolol. However, *p*-cymene-8-ol and dihydrocarveol acetate monoterpenes were found in higher concentrations in the fresh herbs–based essential oil.

Table 2. Chemical constituents of the *Salvia officinalis* essential oils batches distilled from the fresh aerial parts of the herbs and the aerial parts after different drying periods.

No.	Components	KI$^{exp.}$	KI$^{rep.}$	FH	1WDH	2WDH	3WDH	4WDH
1	Butyl acetate	815	817	0.64 ± 0.09				
2	α-Thujene	930	932		0.16 ± 0.02	0.20 ± 0.01	0.15 ± 0.01	
3	α-Pinene	938	939	0.07 ± 0.02 [A]	1.54 ± 0.14 [B]	1.57 ± 0.06 [B]	1.46 ± 0.05 [B]	0.73 ± 0.08 [B]
4	Camphene	954	950	0.09 ± 0.02 [A]	1.01 ± 0.05 [B]	1.40 ± 0.08 [B]	1.49 ± 0.08 [B]	0.69 ± 0.05 [B]
5	Sabinene	977	976		0.17 ± 0.02	0.16 ± 0.01	0.14 ± 0.01	
6	β-Pinene	983	980	0.71 ± 0.08 [A]	5.81 ± 0.56 [B]	5.82 ± 0.24 [B]	5.08 ± 0.22 [B]	3.33 ± 0.13 [B]
7	Myrcene	992	992	0.64 ± 0.08 [A]	3.10 ± 0.23 [B]	3.64 ± 0.17 [B]	2.17 ± 0.05 [B]	2.05 ± 0.05 [B]
8	α-Phellandrene	1001	1008	0.30 ± 0.04				
9	α-terpinene	1021	1018		0.18 ± 0.03	0.24 ± 0.02	0.19 ± 0.00	0.16 ± 0.01
10	Limonene	1034	1033	0.75 ± 0.07 [A]	0.41 ± 0.03 [B]	0.44 ± 0.03 [B]	0.32 ± 0.01 [B]	0.24 ± 0.02 [B]
11	1,8-Cineole	1040	1039	16.70 ± 1.62 [A]	35.70 ± 1.49 [B]	38.70 ± 0.49 [B]	37.90 ± 0.92 [B]	33.21 ± 0.15 [B]
12	γ-Terpinene	1063	1064	0.46 ± 0.04 [A]	1.01 ± 0.04 [B]	1.02 ± 0.06 [B]	0.62 ± 0.02 [A]	0.54 ± 0.03 [A]
13	cis-Sabinene hydrate	1074	1076	0.20 ± 0.02 [A]	0.18 ± 0.01 [A]	0.17 ± 0.01 [A]	0.18 ± 0.01 [A]	0.18 ± 0.01 [A]
14	Terpinolene	1094	1089		0.22 ± 0.00	0.27 ± 0.02	0.20 ± 0.01	0.14 ± 0.02
15	Linalool	1104	1104	0.85 ± 0.06 [A]	0.68 ± 0.03 [A]	0.73 ± 0.11 [A]	0.87 ± 0.03 [A]	1.03 ± 0.06 [A]
16	α-Thujone	1112	1117	0.69 ± 0.05 [A]	2.20 ± 0.04 [B]	1.32 ± 0.03 [B]	1.83 ± 0.05 [B]	2.12 ± 0.03 [B]
17	β-Thujone	1124	1127	0.99 ± 0.06 [A]	1.51 ± 0.03 [B]	1.21 ± 0.02 [A]	1.56 ± 0.03 [B]	1.54 ± 0.05 [B]
18	Camphor	1155	1150	8.32 ± 0.53 [A]	7.56 ± 0.08 [A]	10.71 ± 0.15 [B]	11.50 ± 0.24 [B]	12.09 ± 0.06 [B]
19	Borneol	1170	1170	0.66 ± 0.04 [A]	0.60 ± 0.01 [A]	0.90 ± 0.02 [A]	0.22 ± 0.00 [A]	0.85 ± 0.02 [A]
20	erpinen-4-ol	1177	1178	3.50 ± 0.13 [A]	3.13 ± 0.06 [A]	2.71 ± 0.05 [A]	3.66 ± 0.05 [A]	3.42 ± 0.07 [A]
21	*p*-Cymene-8-ol	1185	1183	1.02 ± 0.05 [A]	0.31 ± 0.01 [B]	0.52 ± 0.01 [B]	0.19 ± 0.01 [B]	0.65 ± 0.01 [B]
22	Myrtenol	1188	1194	2.13 ± 0.09 [A]	1.88 ± 0.05 [B]	1.65 ± 0.01 [B]	1.62 ± 0.02 [B]	2.13 ± 0.04 [A]
23	α-Terpineol	1204	1199	5.53 ± 0.15 [A]	7.64 ± 0.22 [A]	4.67 ± 0.02 [A]	5.96 ± 0.07 [A]	6.03 ± 0.07 [A]
24	Neral	1236	1238	0.17 ± 0.01				
25	Carvone	1257	1258	0.13 ± 0.01 [A]	0.06 ± 0.02 [A]	0.16 ± 0.01 [A]	0.15 ± 0.00 [A]	

Table 2. *Cont.*

No.	Components	KI$^{exp.}$	KI$^{rep.}$	FH	1WDH	2WDH	3WDH	4WDH
26	Linalyl acetate	1261	1259	0.22 ± 0.03 [A]	0.10 ± 0.02 [A]			
27	Bornyl acetate	1293	1288	0.24 ± 0.04 [A]	0.17 ± 0.02 [A]	0.16 ± 0.02 [A]	0.29 ± 0.00 [A]	
28	Thymol	1305	1293		0.11 ± 0.01			
29	Carvacrol	1324	1309	0.23 ± 0.01 [A]	0.11 ± 0.00 [A]	0.18 ± 0.00 [A]	0.15 ± 0.04 [A]	
30	Dihydrocarveol acetate	1357	1347	4.42 ± 0.17 [A]	1.82 ± 0.17 [B]	3.12 ± 0.05 [B]	3.21 ± 0.02 [B]	2.67 ± 0.06 [B]
31	Eugenol	1378	1359		0.23 ± 0.01			
32	β-Bourbonene	1385	1384	0.15 ± 0.02 [A]	0.11 ± 0.03 [A]	0.12 ± 0.01 [A]	0.16 ± 0.03 [A]	
33	α-Cadinene	1424		0.17 ± 0.03				
34	β-Caryophyllene	1439	1426	13.88 ± 0.63 [A]	10.63 ± 0.54 [B]	8.73 ± 0.17 [B]	7.02 ± 0.04 [B]	10.14 ± 0.10 [B]
35	Aromadendrene	1444	1440		0.11 ± 0.01			
36	5-Oxobornyl acetate	1452	1484			0.10 ± 0.00		
37	α-Humulene ne	1456	1456	0.75 ± 0.04 [A]	0.77 ± 0.08 [A]	0.86 ± 0.02 [A]	0.74 ± 0.02 [A]	0.81 ± 0.02 [A]
38	*Allo*-Aromadendrene	1471	1462	4.58 ± 0.21 [A]	1.80 ± 0.09 [B]	2.04 ± 0.05 [B]	1.52 ± 0.01 [B]	2.99 ± 0.03 [B]
39	γ-Muurolene	1478	1477		0.11 ± 0.01			
40	(z)-β-Guaiene	1489	1490	0.17 ± 0.04 [A]	0.15 [A]			
41	Viridiflorene	1497	1494	0.19 ± 0.02 [A]	0.07 ± 0.01 [B]			
42	γ-Cadinene	1513	1513	1.89 ± 0.10 [A]	0.84 ± 0.05 [B]	0.68 ± 0.02 [B]	0.57 ± 0.01 [B]	0.64 ± 0.02 [B]
43	(Z)-Calamenene	1529	1526	0.30 ± 0.04 [A]	0.08 ± 0.00 [B]	0.13 ± 0.03 [B]	0.13 ± 0.01 [B]	0.15 ± 0.01 [B]
44	δ-Cadinene	1537	1531	2.74 ± 0.24 [A]	0.36 ± 0.02 [B]	0.45 ± 0.02 [B]	0.47 ± 0.02 [B]	0.72 ± 0.03 [B]
45	UD	1543		1.00 ± 0.14				
46	Viridiflorol	1595	1590	0.59 ± 0.19				
47	Humulene epoxide I	1599	1596	1.43 ± 0.07 [A]	0.63 ± 0.03 [B]	0.56 ± 0.02 [B]	0.60 ± 0.02 [B]	0.66 ± 0.01 [B]
48	Humulene epoxide II	1605	1600	2.96 ± 0.18 [A]	1.28 ± 0.07 [B]	1.01 ± 0.04 [B]	1.71 ± 0.03 [B]	1.59 ± 0.06 [B]
49	Humulene epoxide III	1616	1615	7.62 ± 0.25 [A]	2.54 ± 0.14 [B]	1.94 ± 0.06 [B]	3.70 ± 0.11 [B]	3.46 ± 0.08 [B]
50	muurola-4,10(14)-dien-1-b-ol	1632	1625	0.83 ± 0.11 [A]	0.21 ± 0.01 [B]	0.23 ± 0.02 [B]	0.30 ± 0.01 [B]	0.55 ± 0.02 [B]
51	UD	1656		0.35 ± 0.04				
52	14-hydroxy-(Z)-caryophyllene	1660	1667	2.10 ± 0.16 [A]	0.34 ± 0.12 [B]	0.24 ± 0.02 [B]	0.37 ± 0.10 [B]	0.60 ± 0.03 [B]
53	UD	1667		0.79 ± 0.08				
54	α-Bisabolol	1675	1683	3.43 ± 0.36 [A]	0.27 ± 0.01 [B]	0.22 ± 0.02 [B]	0.24 ± 0.02 [B]	0.43 ± 0.01 [B]
55	UD	1693		2.09 ± 0.06 [A]	0.53 ± 0.03 [B]	0.37 ± 0.01 [B]	0.61 ± 0.04 [B]	0.62 ± 0.03 [B]
	Identified components			46	46	40	39	33
	Total Yields %			98.3 ± 2.39	98.43 ± 1.82	99.37 ± 0.18	99.26 ± 0.67	97.14 ± 0.35
	Non-oxygenated monoterpenes			3.22	13.79	14.93	12.0	8.06
	Non-oxygenated sesquiterpenes			24.82	15.03	13.01	10.61	15.45
	Oxygenated monoterpenes			45.57	63.36	66.66	68.96	65.74
	Oxygenated sesquiterpenes			18.96	5.27	4.2	6.92	7.29
	Phenolics constituents			0.23	0.45	0.18	0.15	0

KI$^{exp.}$: Experimental Kovats retention index; KI$^{rep.}$: Reported Kovats retention index; FH: Fresh herb; 1WDH: One-week dried herb; 2WDH: Two-week dried herb; 3WDH: Three-week dried herb; 4WDH: Four-week dried herb; UD: Unidentified. The mean of 3 independent runs was used to calculate areas under the peaks. Tukey's multiple comparisons were performed for the statistically significant constituents, using a one-way ANOVA. Different superscript letters (A–B) within the corresponding column show significant differences in the constituents' percentage ($p < 0.05$) in the sage essential oil obtained from dried herb compared to the fresh herb.

The overall follow-up of the decreasing and increasing percentages of the sage essential oils' constituents among the fresh and dried herbs reflected that most of the monoterpene constituents, including the oxygenated and non-oxygenated, were steadily increasing for up to three weeks of drying as compared to the fresh herbs–based essential oil (Table 2, Figure 1). Interestingly, the percentages of the sesquiterpenes were also decreased in the dried herbs batches of the sage essential oils, as compared to the fresh herbs–based essential oil batch. Wherein, the fresh herbs–based essential oil analysis showed 24.82% and 18.96% as compared to 15.03% and 5.27% of the non-oxygenated and oxygenated sesquiterpenes, respectively, which was obtained after only one week of drying as shown in the heat-map comparison which is based on the abundance of major components found in the essential oil batches (Figure 1). The number of identified compounds in every essential oil batch was also one of the distinct comparable points among them. The highest numbers of constituents in the essential oil obtained from the fresh, and one-week dried herbs based essential oil batches of the plant (46 compounds representing 98.3%, and 98.43% of the essential oil, respectively), as compared to the 40 and 39 identified constituents in the two- and three-weeks dried herbs–based essential oil batches, which were represented at 99.37% and 99.26% of the essential oil weight ratios according to the peak area analyses, respectively. Nonetheless, the drastic reductions in the identified compounds were shown in the four-week dried herbs–based essential oil, which showed 33 constituents and represented 97.14% of the essential oil componential weight. Moreover, all the common sage constituents were identified in the fresh herbs-based essential oil batch and the batches obtained after dryings of the herbs. A number of 46 constituents were identified in the fresh herbs and one week dried herbs based essential oils, while the two and three weeks dried herbs based oil contained 40 and 39 constituents, respectively, which formed their total yields of the oils, and their identified and the unidentified constituents weight ratios. The four weeks dried herb-based essential oil contained only 33 constituents in it (Table 2). The GC-FID analysis also revealed that the chemotypic constituents of the sage essential oils were represented in all the essential oil batches as major constituents with different proportions. For instance, 1,8-cineole was found as a major component; however, its proportions were significantly different, and vacillating in different essential oil batches. The economical outcome that could be implied from the results demonstrated in Tables 1 and 2, for the essential oil yields percentage and componential representation of the essential oil constituents in each essential oil batch, showed the importance of sage drying for two weeks in shade at room temperature before the procurement of its essential oils. However, the biological activities of these essential oil batches were another factor mostly centered on the biological activity of the constituents rather than the amount of the obtained essential oil yields from each batch.

3.3. Hepatoprotective Effect of the Essential Oil Batches

The quality of sage essential oils obtained from fresh and different-timed dried batches, i.e., 1WDH, 2WDH, 3WDH, and 4WDH, of the herbs were evaluated for their effectiveness to restore the liver and kidneys functions, together with the lipid profile in comparison to the normal behavior, which was impaired by AAP-induced liver toxicity in the experimental animal models, rats. The ability of sage's essential oils from different batches to protect the HepG-2 cells toxicity induced by AAP was also evaluated as an in vitro hepatoprotective assay. The sage decoction was reported to exaggerate the liver toxicity by the CCl_4 in the mice models [43], which is an unexpected result of the sage infusion. However, the hepatoprotective activity of sage infusion against the liver injury inducer, azathioprine, was reported in rats [44]. Sage's essential oil and herbal extract were also reported to protect against liver injury induced by AAP, attributed to the antioxidant potentials of the sage constituents [29,45].

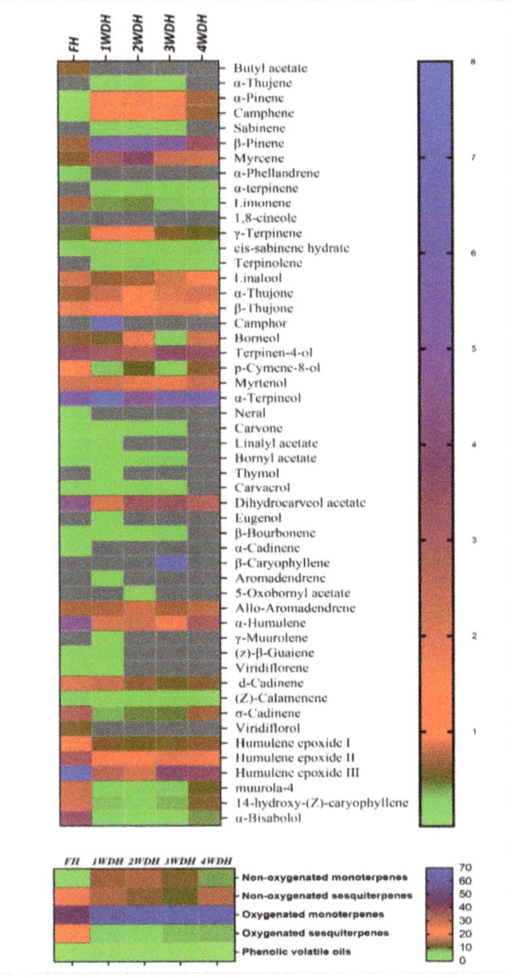

Figure 1. A heat-map comparison based on the abundance of an individual component of sage's essential oils from different batches; gray color indicates un-detected from the identified component.

3.3.1. In vivo Hepatoprotective Effect

The current study showed a significant increase in the serum levels of AST, ALT, and ALP in a concomitant manner with a significant decrease in the total protein contents in the AAP-induced liver injury in several groups of animals, as compared to the control group. These disorders occurred as a result of AAP-induced liver toxicity, which led to hepatic cell damage and necrosis [46] culminating from reactive oxygen species accumulation, lipid peroxidation [47], and calcium release [48], which represented a plausible biomechanism of the AAP-induced liver injury. The pre-treated rats by sage herbs' essential oils showed a significant decrease in AST, ALT, ALP, and a significant increase in total protein contents ($p < 0.001$) in comparison to the AAP-induced liver injury in groups of animals under study. Similar results were obtained from the silymarin-treated rats, which is considered a standard liver support drug which restores hepatic cells functions in liver injury conditions induced by the drug toxicity, and the interplay of free radicals [49]. These findings are in complete alignment with previous reports [29,50], wherein the sage essential oil showed hepatoprotective effects, due to its antioxidant potential. Recently, another study reported that sage essential oil reduces oxidative stress and the toxic effects in liver injury which was induced by vanadium metal in rat liver [31]. Similarly, another study observed that sage protected the liver against isoniazid-induced hepatic toxicity [51]. In this regard, the *Salvia officinalis* leaves' methanolic extract also protected the liver injury induced by aflatoxins [52]. The possible hepatoprotective mechanism of sage may be attributed to its high contents of the oxygenated mono- and sesquiterpenes, e.g., 1,8-cineole, camphor, and humulene epoxide II, that were detected in high concentrations in the current study. These compounds are reported to have antioxidant effects as well as free radical scavenging properties and are in abundance in higher quantities in the hepatoprotective essential oils from other plants also [53,54]. In contrast, a previous study also reported that *Salvia officinalis* aggregated the CCl_4-induced liver toxicity in mice, and the bio-mechanism for it can possibly be attributed to herbs–drug interactions [43]. There were no significant differences detected in the liver functions of all the essential oils (1WDH–4WDH batches) administered animal groups, except that the 4WDH-dried herb-based essential oil showed a significant decrease in the ALT enzymatic activity in comparison to the fresh herb-based essential oil ($p < 0.05$). These effects may be attributed to the presence of 1,8-cineole, and camphor, accumulated as a result of the herbs drying. Further investigations are needed to clarify the issue (Table 3).

Table 3. The effects of sage's essential oils on liver functions in the AAP-induced liver toxicity in rats as compared with the control and silymarin groups.

Test	AST IU/L	ALT IU/L	ALP IU/L	Total Protein gm/dL
Control group	93.32 ± 41.39	28.82 ± 2.731	111.3 ± 11.11	7.903 ± 0.28
AAP group	202.6 ± 36.60 [a]	59.56 ± 21.55 [a]	260.5 ± 40.72 [a]	4.133 ± 0.195 [a]
AAP + FH	132.2 ± 14.95 [b]	46.55 ± 17.25 [b]	103.0 ± 4.29 [b]	8.597 ± 0.22 [b]
AAP + 1WDH	123.9 ± 6.671 [b]	25.89 ± 18.27 [b]	138.2 ± 15.32 [b]	8.670 ± 0.81 [b]
AAP + 2WDH	122.4 ± 15.13 [b]	23.75 ± 3.064 [b]	144.6 ± 19.29 [b]	8.810 ± 0.61 [b]
AAP + 3WDH	139.2 ± 31.64 [b]	25.89 ± 3.397 [b]	102.0 ± 1.08 [b]	9.210 ± 0.20 [b,c]
AAP + 4WDH	141.9 ± 18.35 [b]	20.40 ± 5.143 [b,d]	114.4 ± 8.46 [b]	9.035 ± 0.25 [b,c]
AAP + silymarin	132.4 ± 23.30 [b]	36.10 ± 9.336 [b]	118.7 ± 9.90 [b]	7.910 ± 0.25 [b]
p-value	0.001 **	0.001 **	0.001 **	0.001 **

AAP, acetaminophen (paracetamol); FH, essential oil obtained from a fresh sample of sage; 1WDH, 2WDH, 3WDH, and 4WDH referred to the essential oil obtained from sage herb sample dried for one week, two weeks, three weeks, and four weeks, respectively. The results are expressed as mean ± SD; a = Significant difference, compared to controls, b = Significant difference compared to AAP group, c = Significant difference compared to silymarin group, d = Significant difference compared to fresh extract of sage, *p*-value is significant if it ≤ 0.05. ** is indicated for *p*-value < 0.001.

The present work showed significant increase in the serum levels of cholesterol, triglycerides, urea, and creatinine on the AAP-administered groups, as compared to the control group ($p = 0.001$***; $p < 0.001$***; $p < 0.01$*** and $p < 0.001$***), respectively. The observations are in tune with the outcomes of AAP-induced toxicity also documented previously [46–48]. The current study showed a significant decrease in the cholesterol, triglycerides, and creatinine levels in the sage's essential oil pre-treated and silymarin-treated groups, with a *p*-value (0.001 ***), as compared to the AAP-administered groups. Additionally, no significant differences were detected in the kidneys' functions, cholesterol, and triglycerides levels among all the animal groups when compared for the use of oil obtained from herbs drying (1WDH–4WDH). These observations are commensurate with the study which reported that *Salvia officinalis* use normalizes the lipid disturbance, and lipoprotein metabolism in rat liver injury induced by vanadium metal [31]. Sage also decreased cholesterol, triglycerides, and creatinine levels in streptozotocin-induced diabetic rats [55] (Table 4).

Table 4. The prophylactic effect of sage essential oil on blood cholesterol, triglycerides, urea, and creatinine in acetaminophen-induced liver toxicity in rats as compared with the control and silymarin groups.

Test (Unit/L)	Cholesterol mg/dL	Triglycerides mg/dL	Urea mg/dL	Creatinine mg/dL
Control group	86.87 ± 10.37	137.4 ± 2.117	34.28 ± 8.15	0.62 ± 0.33
AAP group	119.3 ± 37.25 [a]	160.1 ± 18.81 [a]	52.28 ± 6.61 [a]	1.43 ± 0.39 [a]
AAP + FH	73.20 ± 8.39 [b]	130.3 ± 2.20 [b]	23.23 ± 17.57 [b]	0.92 ± 0.06 [b]
AAP + 1WDH	84.66 ± 6.99 [b]	141.8 ± 7.91 [b]	42.66 ± 20.69	0.76 ± 0.21 [b]
AAP + 2WDH	85.89 ± 10.45 [b]	162.0 ± 1.69	47.75 ± 5.383	0.54 ± 0.37 [b]
AAP + 3WDH	84.10 ± 14.47 [b]	137.1 ± 7.12 [b]	41.29 ± 12.38	0.86 ± 0.16 [b]
AAP + 4WDH	79.13 ± 11.53 [b]	138.3 ± 7.42 [b]	39.46 ± 11.00	0.91 ± 0.08 [b]
AAP + silymarin	75.45 ± 6.54 [b]	152.0 ± 19.56 [b]	46.13 ± 19.91	0.89 ± 0.09 [b]
p-value	0.001 ***	0.001***	0.01 **	0.001 ***

AAP, acetaminophen (paracetamol); FH, essential oil obtained from a fresh sample of sage; 1WDH, 2WDH, 3WDH, and 4WDH referred to the essential oil obtained from sage herb sample dried for one week, two weeks, three weeks, and four weeks, respectively. The results are expressed as mean ± SD, a = Significant difference compared to control group, b = Significant difference compared to AAP group. ** and *** are indicated for *p*-value < 0.01 and *p*-value < 0.001, respectively.

3.3.2. In vitro Hepatoprotective Effects

The current study used an in vitro cell culture model (HepG-2 cells) to evaluate the hepatoprotective activity of the fresh and differently timed dried sage herbs–based essential oils obtained by the hydrodistillation procedure against liver damages induced by the AAP. The strategy was used to evaluate the hepatoprotective effects of the sage's essential oil and to support the findings obtained from in vivo studies. The cytotoxic effects of AAP were determined in the presence, and absence of the essential oils obtained from the fresh and other differently timed dried herbs–based essential oils as well as with the standard hepatic support, silymarin (Figure 2A).

The cytotoxic activity results of the current study demonstrated that the selected doses of sage essential oils were non-toxic at 100 μg/mL concentrations. It was also found that the sage's essential oil significantly improved the viability of the cells of AAP-treated HepG-2 from 40% to 56% by FH, to 65% by 1WDH, to 80% by 2WDH, to 71% by 3WDH, and 83% by 4WDH as compared to the 78% viability of the silymarin-treated animals group (Figure 2A).

The hepatoprotective effects of the sage essential oils on HepG-2 cells that were pretreated with a hepatoprotective agent, and subsequently exposed to APP to induce damage are shown in Figure 2. The pretreated HepG-2 cells with FH, 1WDH, 2WDH, 3WDH, and 4WDH essential oils significantly decreased the MDA levels of the AAP treated cells from 3.1 μM to 1.1, 1.4, 1.1, 1 and 1.2 μM, respectively. In addition, a significant increase in the TAOxC levels of the AAP-treated cells from 0.2 mM to 0.4, 0.3, 0.5, 0.45, and 0.6 mM, respectively, was observed. Furthermore, the pretreatment with silymarin significantly decreased the MDA levels to 1.1 μM as well as an increase in TAOxC levels to 0.4 mM of the AAP-treated HepG-2 cells.

The exposure of HepG-2 cells to AAP demonstrated a significant reduction in the viability of the cells as indicated by their inability to metabolize the tetrazolium salt. A significant decrease in TAOxC, as well as a significant increase in the levels of MDA (Figure 2B,C), was detected. The underlying mechanisms of the in vitro liver damage caused by the AAP may be attributed to the AAP concentration and the exposure time [38].

The HepG-2 cells were exposed to the toxic dose of AAP that led to the generation of reactive oxygen species (ROS) interacting with the macromolecules inside of the cells [56]. This interaction results in DNA damage, lipid peroxidation of the lipids bilayers of the cell membrane, as well as denaturation of many essential proteins of the cells, and finally, exhibits cells death as observed in the loss of 40% of the viability of the cells by treatment with 4 mM of AAP. The exposure of hepatic cell lines to a high concentration of AAP causes cells injury and reduces viability as also reported previously [57]. The balancing between the oxidant and antioxidant capacities inside of the cells is important for the cells' survival. Therefore, two parameters, MDA and TAOxC, including the cell viability, were evaluated to assess the hepatoprotective effects of all the essential oils batches obtained from sage. MDA is a biomarker of ROS effects, especially lipo-peroxidation, and TAOxC is an indicator marker for the general antioxidant status of cells.

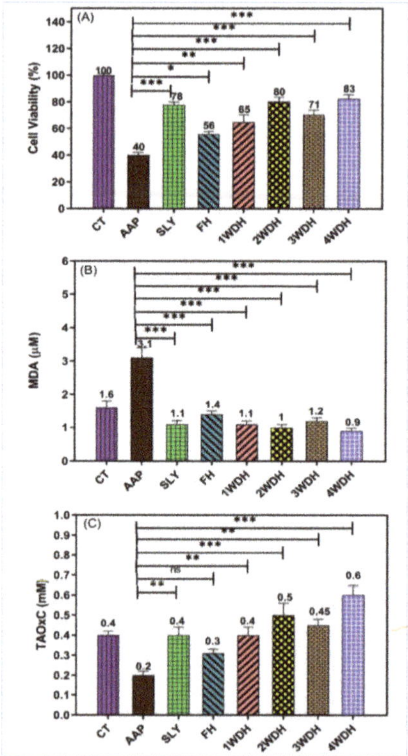

Figure 2. Hepatoprotective effects of sage essential oils against damage induced by 4 mM acetaminophen (AAP) in HepG-2 cells for 24 h in comparison to silymarin. The cytotoxicity of AAP with and without selected dose (100 µg/mL < IC_{50} values) of sage essential oils and silymarin (SLY) on hepatic cell lines (HepG-2) (**A**) for hepatoprotective activity tests MDA levels (µM) (**B**), and TAOxC levels (mM) (**C**) in HepG-2 cells after exposure to 4 mM AAP and pretreated with sage essential oils or silymarin. Controls: supplemented media (CT); AAP 4 mM (AAP), silymarin (100 µg/mL) (SLY). Values are the mean ± SD of three independent experiments performed in triplicate. * For $p < 0.05$, ** for $p < 0.01$, and *** for $p < 0.001$.

Oxidative stress plays a major role in AAP-induced toxicity as observed by decreases in the TAOxC, and an increase in the MDA levels after treatment of HepG-2 cells with AAP. Several studies have suggested that the oxidative stress that leads to apoptosis is the cause of cell death in the HepG-2 cell lines. It was found that the pre-treated

HepG-2 cells with different essential oils (100 µg/mL) obtained in the current study showed significant improvements in the cell viability. It also showed an increase in the TAOxC and a reduction in the MDA levels (Figure 1). These results suggest that the sage essential oil exerts hepatoprotective effects in AAP-induced damages in the HepG-2 cell lines. It is presumed that the hepatoprotective effects of the sage essential oil are mainly owing to their antioxidant contents, i.e., 1, 8-cineole, β-pinene, camphor, β-caryophyllene, and α-pinene. The significant improvements in the HepG-2 protective effects demonstrated by the essential oils obtained from differently-timed dried herbs, especially the 4WDH, as compared to the FH-based essential oil of the sage herbs. This can be attributed to the significant increase in the 1,8-cineole, β-pinene, camphor, and α-pinene presence in the dried essential oil batches as compared to the FH-based essential oil. Notably, the results also confirmed the in vivo observations, wherein the 4WDH-based sage essential oil significantly decreased the ALT enzymatic activity compared to the essential oil obtained by the FH ($p < 0.05$). It was also revealed that the 4WDH-based essential oil-induced significant elevation of TAOxC as compared to the standard hepatoprotective drug, silymarin. These effects seemed attributed to the cumulative effects of the major essential oil constituents in the 2WDH- and 4WDH-based essential oils that possessed comparatively strong antioxidant activity, owing to the higher contents of the constituents, e.g., 1, 8-cineole, and camphor. All the dried herb-based essential oil batches significantly increased the TAOxC. However, the 1WDH and 3WDH essential oils showed comparable results to the silymarin-treated cells. Similar results were also obtained for the levels of MDA, which were significantly reduced in the cells treated by the silymarin and the dried herbs–based essential oil batches, compared to the fresh sage essential oil. The fresh sage essential oil also showed a significant reduction in the MDA levels as compared to the AAP-treated cells.

3.4. Anticancer Effects of Essential Oils Obtained from Different-Timed Drying Herbs Batches

The effects of the sage essential oil obtained from the fresh herbs, and dried herbs were evaluated by the MTT assay for the cell viability of cancer and normal cell lines. The results showed that all the essential oil batches from sage showed moderate cytotoxicity against cancer cell lines; hepatocellular carcinoma HepG-2, cervical carcinoma HeLa, and the breast carcinoma cell lines, MCF-7 (Table 5). Figure 3 represents the dose-response curve of the sage essential oil batches on the viability of all the cell lines used in this study. As shown in the curves, sage's essential oils inhibited all the cancer cells, HepG-2, HeLa, and MCF-7, in a dose-dependent manner (1–500 µg/mL), while the curves showing an upgrade pattern indicated the low cytotoxicity of the tested samples on the normal cells (MRC-5).

Table 5. IC_{50} and selective index values obtained for fresh and dried essential oil batches of sage against different cancer and normal fibroblast lung (MRC-5) cell lines.

Groups	MCF-7		HepG-2		HeLa		MRC-5
	$IC_{50} \pm SD$	SI	$IC_{50} \pm SD$	SI	$IC_{50} \pm SD$	SI	$IC_{50} \pm SD$
DOX	1.4 ± 0.1	1.1	1.04 ± 0.1	1.5	1.02 ± 0.1	1.6	1.6 ± 0.1
FH	181.3 ± 18.3	3.3	161.7 ± 15.3	3.7	142.1 ± 12.4	4.2	596.5 ± 20
1WDH	189.3 ± 17.45	2.2	179.12 ± 16.5	2.4	164.1 ± 15.5	2,6	425.6 ± 30.2
2WDH	194.1 ± 20.1	2.6	182.4 ± 14.2	2.7	174.7 ± 13.6	2.9	499.4 ± 32
3WDH	176.3 ± 16.5	2.3	174.9 ± 13.2	2.3	127.5 ± 11.4	3.2	405.2 ± 18.3
4WDH	215.7 ± 18.4	1.9	195.4 ± 20	2.1	176.2 ± 18	2.4	414.7 ± 27.1

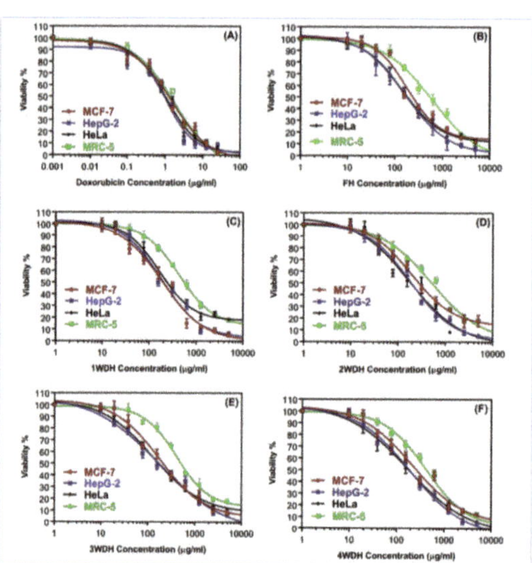

Figure 3. Dose-response curve of cytotoxicity of doxorubicin as a positive control (**A**), sage essential oils (1–500 µg/mL) of FH (**B**), 1WDH (**C**), 2WDH (**D**), 3WDH (**E**), and 4WDH (**F**) on cancer cell lines breast (MCF-7), hepatic (HepG-2), and cervical (HeLa) in comparison to normal cells (MRC-5).

Figure 4 shows the summarized comparisons between the IC_{50} values of the tested samples and the positive control, doxorubicin, on the cancer cell lines, and the normal fibroblast cells. The 4WDH essential oil showed the lowest cytotoxicity on all the tested cancer cell lines, i.e., HepG-2, HeLa, and MCF-7, with IC_{50} values ranging from 176.2 µg/mL for HeLa cells to 215.7 µg/mL for the MCF-7 cell lines. The FH-based essential oil seemed to be having the highest cytotoxic effects on the cancer cells, and the lowest cytotoxic effects on the normal cells with the highest SI: HeLa (IC_{50} = 142.1 µg/mL, SI = 4.2), HepG-2 (IC_{50} = 161.7 µg/mL, SI = 3.7), and MCF-7 cells (IC_{50} = 181.3, SI = 3.3) (Table 5). A previous study also demonstrated the ability of the *S. officinalis* polar solvents extracted material to inhibit MCF-7 cancer cell lines without significant cytotoxic effects against normal human umbilical vein endothelial cells with SI > 3, and this is in agreement with our current findings [58].

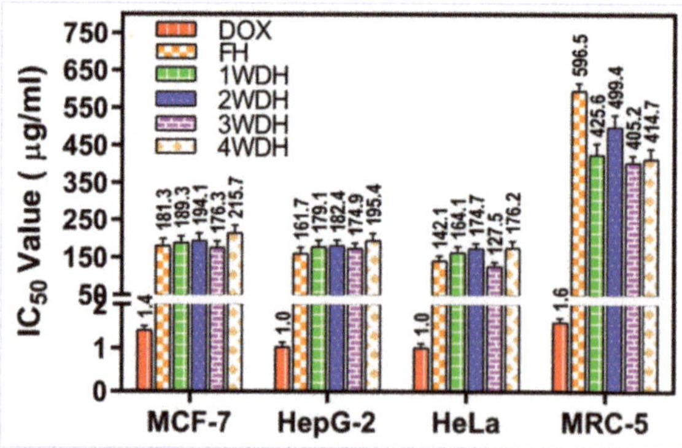

Figure 4. The cytotoxicity IC_{50} values of the sage essential oils (µg/mL) on cancer cell lines breast (MCF-7), hepatic (HepG-2), and cervical (HeLa) in comparison to the normal cells (MRC-5). Doxorubicin was used as a cytotoxic drug (positive control).

4. Conclusions

This report comprehensively demonstrated the extended drying effect of aerial parts of *S. officinalis* on oil yields, its quality, and its effects on biological activities. Sage yielded the highest yields of essential oil from the two-week shade and room-temperature dried plant materials, with 0.28% of essential oil procurement containing over 99.82 $\pm$ 0.18% of the sage oil's reported componential constituents. The typical chemotypic components were present in abundance, and presumably are part of the active constituents of the essential oil against induced liver damages and active cancer cell lines, i.e., MCF-7, HepG-2, and HeLA. The herb and its essential oil exhibited beneficial properties and are recommended for use in liver disorders. Their prophylactic efficacy in the liver damage caused by acetaminophen lends credence to their widespread use in culinary purposes worldwide. The essential oil yields modulation study from the differently timed drying of fresh herbs established the preferable drying period of two weeks for maximum essential oil yields from the herbs. Nonetheless, the essential oil obtained from two-week dried sage herbs also exhibited better liver protection, as well as anti-cancer activities in the in vitro conditions. The potential for bulk scale essential oil procurement from the herbs and its household use for perceived health benefits were thus verified.

Supplementary Materials: The following are available online at, Excel sheet of the GC-FID analysis raw data.

Author Contributions: Conceptualization, H.A.M., M.S.A.-O., and R.A.K.; methodology, H.A.M., H.M.E., M.S.M.S., S.A.A.M., M.S.A.A., R.A.K., S.Y.E., A.M.A., and M.Z.E.-R.; software, H.A.M., H.M.E., M.S.M.S., and M.Z.E.-R.; validation, H.A.M., H.M.E., M.S.M.S., M.S.A.-O., R.A.K., and M.Z.E.-R.; formal analysis, H.A.M., M.S.A.-O., S.A.A.M., H.M.E., and R.A.K.; investigation, H.A.M., and M.Z.E.-R.; resources, H.A.M.; data curation, H.A.M., A.A.H.A., M.S.A.-O., S.A.A.M., H.M.E., and R.A.K.; writing—original draft preparation, H.A.M., H.M.E., and M.Z.E.-R.; writing—review and editing, H.A.M., H.M.E., S.A.A.M., R.A.K., and M.Z.E.-R.; visualization, H.A.M.; supervision, H.A.M.; project administration, H.A.M.; funding acquisition, H.A.M. All authors have read and agreed to the published version of the manuscript.

Funding: This research was funded by the Deputyship for Research & Innovation, Ministry of Education in Saudi Arabia under the project number (QU-IF-1-2-2).

Institutional Review Board Statement: The animal-work protocol was approved by the Ethics Committee for Animal Care and Use in the College of Pharmacy-Qassim University (Approval ID 2019-CP-11).

Informed Consent Statement: Not applicable.

Data Availability Statement: Data are provided in the main text and supplementary file.

Acknowledgments: The authors extend their appreciation to the Deputyship for Research & Innovation, Ministry of Education and, Saudi Arabia for funding this research work through the project number (QU-IF-1-2-2). The authors also thank the technical support of Qassim University.

Conflicts of Interest: The authors declare no conflict of interest.

References

1. Sharifi-Rad, M.; Ozcelik, B.; Altın, G.; Daşkaya-Dikmen, C.; Martorell, M.; Ramírez-Alarcón, K.; Alarcón-Zapata, P.; Morais-Braga, M.F.B.; Carneiro, J.N.P.; Leal, A.L.A.B. *Salvia* spp. plants-from farm to food applications and phytopharmacotherapy. *Trends Food Sci. Technol.* **2018**, *80*, 242–263.
2. Khedher, M.R.B.; Khedher, S.B.; Chaieb, I.; Tounsi, S.; Hammami, M. Chemical composition and biological activities of *Salvia officinalis* essential oil from Tunisia. *EXCLI J.* **2017**, *16*, 160. [PubMed]
3. Ghorbani, A.; Esmaeilizadeh, M. Pharmacological properties of *Salvia officinalis* and its components. *J. Tradit. Complement. Med.* **2017**, *7*, 433–440.
4. Adams, M.; Gmünder, F.; Hamburger, M. Plants traditionally used in age-related brain disorders—A survey of ethnobotanical literature. *J. Ethnopharmacol.* **2007**, *113*, 363–381. [PubMed]
5. Perry, E.K.; Pickering, A.T.; Wang, W.W.; Houghton, P.J.; Perry, N.S.L. Medicinal plants and Alzheimer's disease: From ethnobotany to phytotherapy. *J. Pharm. Pharmacol.* **1999**, *51*, 527–534. [PubMed]

6.   Eidi, M.; Eidi, A.; Zamanizadeh, H. Effect of *Salvia officinalis* L. leaves on serum glucose and insulin in healthy and streptozotocin-induced diabetic rats. *J. Ethnopharmacol.* **2005**, *100*, 310–313.

7.   Craft, J.D.; Satyal, P.; Setzer, W.N. The chemotaxonomy of common Sage (*Salvia officinalis*) based on the volatile constituents. *Medicines* **2017**, *4*, 47.

8.   Perry, N.B.; Anderson, R.E.; Brennan, N.J.; Douglas, M.H.; Heaney, A.J.; McGimpsey, J.A.; Smallfield, B.M. Essential oils from Dalmatian Sage (*Salvia officinalis* L.): Variations among individuals, plant parts, seasons, and sites. *J. Agric. Food Chem.* **1999**, *47*, 2048–2054.

9.   Russo, A.; Formisano, C.; Rigano, D.; Senatore, F.; Delfine, S.; Cardile, V.; Rosselli, S.; Bruno, M. Chemical composition and anticancer activity of essential oils of Mediterranean Sage (*Salvia officinalis* L.) grown in different environmental conditions. *Food Chem. Toxicol.* **2013**, *55*, 42–47.

10.   Bouajaj, S.; Benyamna, A.; Bouamama, H.; Romane, A.; Falconieri, D.; Piras, A.; Marongiu, B. Antibacterial, allelopathic and antioxidant activities of essential oil of *Salvia officinalis* L. growing wild in the Atlas Mountains of Morocco. *Nat. Prod. Res.* **2013**, *27*, 1673–1676. [PubMed]

11.   Reverchon, E.; Taddeo, R.; Porta, G.D. Extraction of Sage oil by supercritical C02: Influence of some process parameters. *J. Supercrit. Fluids* **1995**, *8*, 302–309.

12.   Glisic, S.; Ivanovic, J.; Ristic, M.; Skala, D. Extraction of Sage (*Salvia officinalis* L.) by supercritical CO_2: Kinetic data, chemical composition and selectivity of diterpenes. *J. Supercrit. Fluids* **2010**, *52*, 62–70.

13.   Durling, N.E.; Catchpole, O.J.; Grey, J.B.; Webby, R.F.; Mitchell, K.A.; Foo, L.Y.; Perry, N.B. Extraction of phenolics and essential oil from dried Sage (*Salvia officinalis*) using ethanol-water mixtures. *Food Chem.* **2007**, *101*, 1417–1424.

14.   Sellami, I.H.; Rebey, I.B.; Sriti, J.; Rahali, F.Z.; Limam, F.; Marzouk, B. Drying Sage (*Salvia officinalis* L.) plants and its effects on content, chemical composition, and radical scavenging activity of the essential oil. *Food Bioprocess Technol.* **2012**, *5*, 2978–2989.

15.   Venskutonis, P.R. Effect of drying on the volatile constituents of thyme (*Thymus vulgaris* L.) and Sage (*Salvia officinalis* L.). *Food Chem.* **1997**, *59*, 219–227.

16.   Ibraliu, A.; Doko, A.; Hajdari, A.; Gruda, N.; Šatović, Z.; Karanfilova, I.C.; Stefkov, G. Essential oils chemical variability of seven populations of *Salvia officinalis* L. in North of Albania. *Maced. J. Chem. Chem. Eng.* **2020**, *39*, 31–39.

17.   Cvetkovikj, I.; Stefkov, G.; Karapandzova, M.; Kulevanova, S.; Satović, Z. Essential oils and chemical diversity of southeast European populations of *Salvia officinalis* L. *Chem. Biodivers.* **2015**, *12*, 1025–1039.

18.   Jug-Dujaković, M.; Ristić, M.; Pljevljakušić, D.; Dajić-Stevanović, Z.; Liber, Z.; Hančević, K.; Radić, T.; Šatović, Z. High diversity of indigenous populations of dalmatian Sage (*Salvia officinalis* L.) in essential-oil composition. *Chem. Biodivers.* **2012**, *9*, 2309–2323.

19.   Raal, A.; Orav, A.; Arak, E. Composition of the essential oil of *Salvia officinalis* L. from various European countries. *Nat. Prod. Res.* **2007**, *21*, 406–411.

20.   Boutebouhart, H.; Didaoui, L.; Tata, S.; Sabaou, N. Effect of extraction and drying method on chemical composition, and evaluation of antioxidant and antimicrobial activities of essential oils from *Salvia officinalis* L. *J. Essent. Oil Bear. Plants* **2019**, *22*, 717–727.

21.   Putievsky, E.; Ravid, U.; Dudai, N. The influence of season and harvest frequency on essential oil and herbal yields from a pure clone of Sage (*Salvia officinalis*) grown under cultivated conditions. *J. Nat. Prod.* **1986**, *49*, 326–329.

22.   Madrigal-Santillán, E.; Madrigal-Bujaidar, E.; Álvarez-González, I.; Sumaya-Martínez, M.T.; Gutiérrez-Salinas, J.; Bautista, M.; Morales-González, Á.; y González-Rubio, M.G.-L.; Aguilar-Faisal, J.L.; Morales-González, J.A. Review of natural products with hepatoprotective effects. *World J. Gastroenterol. WJG* **2014**, *20*, 14787.

23.   Mohammed, S.A.A.; Khan, R.A.; El-Readi, M.Z.; Emwas, A.-H.; Sioud, S.; Poulson, B.G.; Jaremko, M.; Eldeeb, H.M.; Al-Omar, M.S.; Mohammed, H.A. Suaeda vermiculata Aqueous-Ethanolic Extract-Based Mitigation of CCl_4-Induced Hepatotoxicity in Rats, and HepG-2 and HepG-2/ADR Cell-Lines-Based Cytotoxicity Evaluations. *Plants* **2020**, *9*, 1291. [CrossRef]

24.   Eldeeb, H.M.; Mohammed, H.A.; Sajid, M.S.M.; Eltom, S.E.M.; Al-Omar, M.S.; Mobark, M.A.; Ahmed, A.S. Effect of Roasted Date Palm Rich Oil Extracts in Liver Protection and Antioxidant Restoration in CCl_4-induced Hepato Toxicity in Rats. *Int. J. Pharmacol.* **2020**, *16*, 367. [CrossRef]

25.   Duthie, S.J.; Melvin, W.T.; Burke, M.D. Bromobenzene detoxification in the human liver-derived HepG-2 cell line. *Xenobiotica* **1994**, *24*, 265–279. [PubMed]

26.   Sassa, S.; Sugita, O.; Galbraith, R.A.; Kappas, A. Drug metabolism by the human hepatoma cell, Hep G2. *Biochem. Biophys. Res. Commun.* **1987**, *143*, 52–57.

27.   Bajt, M.L.; Knight, T.R.; Lemasters, J.J.; Jaeschke, H. Acetaminophen-induced oxidant stress and cell injury in cultured mouse hepatocytes: Protection by N-acetyl cysteine. *Toxicol. Sci.* **2004**, *80*, 343–349.

28.   Ingawale, D.K.; Mandlik, S.K.; Naik, S.R. Models of hepatotoxicity and the underlying cellular, biochemical and immunological mechanism (s): A critical discussion. *Environ. Toxicol. Pharmacol.* **2014**, *37*, 118–133. [PubMed]

29.   Banna, H.; Soliman, M.; Wabel, N. Hepatoprotective effects of *Thymus* and *Salvia* essential oils on paracetamol-induced toxicity in rats. *J. Physiol. Pharmacol. Adv.* **2013**, *3*, 41.

30.   Fahmy, M.A.; Diab, K.A.; Abdel-Samie, N.S.; Omara, E.A.; Hassan, Z.M. Carbon tetrachloride-induced hepato/renal toxicity in experimental mice: Antioxidant potential of Egyptian *Salvia officinalis* L essential oil. *Environ. Sci. Pollut. Res.* **2018**, *25*, 27858–27876.

31. Koubaa, F.G.; Chaâbane, M.; Turki, M.; Ayadi, F.M.; El Feki, A. Antioxidant and hepatoprotective effects of *Salvia officinalis* essential oil against vanadium-induced oxidative stress and histological changes in the rat liver. *Environ. Sci. Pollut. Res.* **2021**, *28*, 11001–11015.

32. Lima, C.F.; Carvalho, F.; Fernandes, E.; Bastos, M.D.L.; Santos-Gomes, P.C.; Fernandes-Ferreira, M.; Pereira-Wilson, C. Evaluation of toxic/protective effects of the essential oil of *Salvia officinalis* on freshly isolated rat hepatocytes. *Toxicol. Vitr.* **2004**, *18*, 457–465.

33. Loizzo, M.R.; Tundis, R.; Menichini, F.; Saab, A.M.; Statti, G.A.; Menichini, F. Cytotoxic activity of essential oils from Labiatae and Lauraceae families against in vitro human tumor models. *Anticancer Res.* **2007**, *27*, 3293–3299. [PubMed]

34. Foray, L.; Bertrand, C.; Pinguet, F.; Soulier, M.; Astre, C.; Marion, C.; Pélissier, Y.; Bessière, J.-M. In vitro cytotoxic activity of three essential oils from *Salvia* species. *J. Essent. Oil Res.* **1999**, *11*, 522–526.

35. Sertel, S.; Eichhorn, T.; Plinkert, P.K.; Efferth, T. Anticancer activity of *Salvia officinalis* essential oil against HNSCC cell line (UMSCC1). *HNO* **2011**, *59*, 1203–1208. [PubMed]

36. OECD. *Test No. 425: Acute Oral Toxicity: Up-and-Down Procedure*; OECD Publishing: Paris, France, 2008; pp. 1–27.

37. Machana, S.; Weerapreeyakul, N.; Barusrux, S.; Nonpunya, A.; Sripanidkulchai, B.; Thitimetharoch, T. Cytotoxic and apoptotic effects of six herbal plants against the human hepatocarcinoma (HepG-2) cell line. *Chin. Med.* **2011**, *6*, 39. [CrossRef]

38. González, L.T.; Minsky, N.W.; Espinosa, L.E.M.; Aranda, R.S.; Meseguer, J.P.; Pérez, P.C. In vitro assessment of hepatoprotective agents against damage induced by acetaminophen and CCl_4. *BMC Complement. Altern. Med.* **2017**, *17*, 39. [CrossRef]

39. Mohammed, H.A.; Al-Omar, M.S.; Mohammed, S.A.; Aly, M.S.; Alsuqub, A.N.; Khan, R.A. Drying Induced Impact on Composition and Oil Quality of Rosemary Herb, *Rosmarinus Officinalis* Linn. *Molecules* **2020**, *25*, 2830. [CrossRef]

40. Abu-Darwish, M.S.; Cabral, C.; Ferreira, I.V.; Gonçalves, M.J.; Cavaleiro, C.; Cruz, M.T.; Al-Bdour, T.H.; Salgueiro, L. Essential oil of common Sage (*Salvia officinalis* L.) from Jordan: Assessment of safety in mammalian cells and its antifungal and anti-inflammatory potential. *Biomed Res. Int.* **2013**, *2013*, 538940.

41. Abu-Darwish, M.S.; Al-Fraihat, A.H.; Al-Dalain, S.Y.A.; Afifi, F.M.U.; Al-Tabbal, J.A. Determination of Essential Oils and Heavy Metals Accumulation in *Salvia officinalis* Cultivated in three Intra-raw Spacing in Ash-Shoubak, Jordan. *Int. J. Agric. Biol.* **2011**, *13*, 981–985.

42. Edris, A.E.; Jirovetz, L.; Buchbauer, G.; Denkova, Z.; Stoyanova, A.; Slavchev, A. Chemical composition, antimicrobial activities, and olfactive evaluation of a *Salvia officinalis* L.(Sage) essential oil from Egypt. *J. Essent. Oil Res.* **2007**, *19*, 186–189.

43. Lima, C.F.; Fernandes-Ferreira, M.; Pereira-Wilson, C. Drinking of *Salvia officinalis* tea increases CCl_4-induced hepatotoxicity in mice. *Food Chem. Toxicol.* **2007**, *45*, 456–464. [PubMed]

44. Amin, A.; Hamza, A.A. Hepatoprotective effects of *Hibiscus*, *Rosmarinus* and *Salvia* on azathioprine-induced toxicity in rats. *Life Sci.* **2005**, *77*, 266–278. [PubMed]

45. Foruozandeh, H.; Vosughi Niri, M.; Kalantar, M.; Azadi, M.; Samadani, M. Protective Effect of Hydroalcoholic Extract of *Salvia officinalis* L. against Acute Liver Toxicity of Acetaminophen in Mice. *Horiz. Med. Sci.* **2016**, *22*, 185–191.

46. Ray, S.D.; Mumaw, V.R.; Raje, R.R.; Fariss, M.W. Protection of acetaminophen-induced hepatocellular apoptosis and necrosis by cholesteryl hemisuccinate. *J. Pharm. Exptl. Ther.* **1996**, *279*, 1470–1483.

47. Kamiyama, T.; Sato, C.; Liu, J.; Tajiri, K.; Miyakawa, H.; Marumo, F. Role of lipid peroxidation in acetaminophen-induced hepatotoxicity: Comparison with carbon tetrachloride. *Toxicol. Lett.* **1993**, *66*, 7–12.

48. Salas, V.M.; Corcoran, G.B. Calcium-dependent DNA damage and adenosine 3′, 5′-cyclic monophosphate-independent glycogen phosphorylase activation in an in vitro model of acetaminophen-induced liver injury. *Hepatology* **1997**, *25*, 1432–1438.

49. Freitag, A.F.; Cardia, G.F.E.; Da Rocha, B.A.; Aguiar, R.P.; Silva-Comar, F.M.D.S.; Spironello, R.A.; Grespan, R.; Caparroz-Assef, S.M.; Bersani-Amado, C.A.; Cuman, R.K.N. Hepatoprotective effect of silymarin (*Silybum marianum*) on hepatotoxicity induced by acetaminophen in spontaneously hypertensive rats. *Evid.-Based Complement. Altern. Med.* **2015**, *2015*. [CrossRef]

50. El-Hosseiny, L.S.; Alqurashy, N.N.; Sheweita, S. Oxidative stress alleviation by Sage essential oil in co-amoxiclav induced hepatotoxicity in rats. *Int. J. Biomed. Sci. IJBS* **2016**, *12*, 71.

51. Shahrzad, K.; Mahya, N.; Fatemeh, T.B.; Maryam, K.; Mohammadreza, F.B.; Jahromy, M.H. Hepatoprotective and antioxidant effects of *Salvia officinalis* L. hydroalcoholic extract in male rats. *Chin. Med.* **2014**, *2014*, 47465.

52. Parsai, A.; Eidi, M.; Sadeghipour, A. Hepatoprotective effect of Sage (*Salvia officinalis* L.) Leaves hydro-methanolic extract against *Aspergillus parasiticus* aflatoxin-induced liver damage in male rats. *Bull. Pharm. Res* **2014**, *4*, 129–132.

53. Elshibani, F.; Alshalmani, S.; Mohammed, H.A. *Pituranthos tortuosus* Essential Oil from Libya: Season Effect on the Composition and Antioxidant Activity. *J. Essent. Oil Bear. Plants* **2020**, *23*, 1095–1104.

54. Rašković, A.; Milanović, I.; Pavlović, N.; Ćebović, T.; Vukmirović, S.; Mikov, M. Antioxidant activity of rosemary (*Rosmarinus officinalis* L.) essential oil and its hepatoprotective potential. *BMC Complement. Altern. Med.* **2014**, *14*, 225.

55. Eidi, A.; Eidi, M. Antidiabetic effects of Sage (*Salvia officinalis* L.) leaves in normal and streptozotocin-induced diabetic rats. *Diabetes Metab. Syndr. Clin. Res. Rev.* **2009**, *3*, 40–44.

56. Cover, C.; Mansouri, A.; Knight, T.R.; Bajt, M.L.; Lemasters, J.J.; Pessayre, D.; Jaeschke, H. Peroxynitrite-induced mitochondrial and endonuclease-mediated nuclear DNA damage in acetaminophen hepatotoxicity. *J. Pharmacol. Exp. Ther.* **2005**, *315*, 879–887. [PubMed]

57. Lin, J.; Schyschka, L.; Mühl-Benninghaus, R.; Neumann, J.; Hao, L.; Nussler, N.; Dooley, S.; Liu, L.; Stöckle, U.; Nussler, A.K. Comparative analysis of phase I and II enzyme activities in 5 hepatic cell lines identifies Huh-7 and HCC-T cells with the highest potential to study drug metabolism. *Arch. Toxicol.* **2012**, *86*, 87–95. [PubMed]
58. Shahneh, F.Z.; Baradaran, B.; Orangi, M.; Zamani, F. In vitro cytotoxic activity of four plants used in Persian traditional medicine. *Adv. Pharm. Bull.* **2013**, *3*, 453.

Article

Antihypertensive Effect of Galegine from *Biebersteinia heterostemon* in Rats

Weien Wang [1,2,3,*] and Xiaofeng Zhang [2]

[1] School of Chemistry and Chemical Engineering, Qinghai Normal University, Xining 810008, China
[2] Northwest Institute of Plateau Biology, Chinese Academy of Sciences, Xining 810007, China; xfzhang@126.com
[3] School of Chemistry, University of Chinese Academy of Sciences, Beijing 100049, China
* Correspondence: weienwang@126.com or 2010044@qhnu.edu.cn; Fax: +86-971-6307635

Abstract: The aerial part of *Biebersteinia heterostemon* Maxim. (Geraniaceae Biebersteiniaceae) known as *ming jian na bao* in Chinese, has been traditionally used in Tibetan folk medicine for treatment of diabetes and hypertension. The aim of the present study was to evaluate the effects of galegine obtained from an ethanol extract of the entire *Biebersteinia heterostemon* plant on the rat's cardiovascular system in order to characterize its contributions as an antihypertensive agent. The antihypertensive effect of galegine was investigated in pentobarbital-anesthetized hypertensive rats at three dose levels based on the LD_{50} of galegine. Meanwhile a positive control group received dimaprit with the same procedure. Dimaprit infusion induced a significant hypotension which declined by an average margin of 20%. Simultaneously, single administration of galegine at the doses of 2.5, 5, and 10 mg/kg by intraperitoneal injection induced an immediate and dose-dependent decrease in mean arterial blood pressure (MABP) by an average margin of 40% with a rapid increase in heart rate (HR). We demonstrated that galegine is effective in reducing blood pressure in anesthetized hypertensive rats with rapid onset and a dose-related duration of the effects. The results indicate that galegine was the bioactive compound which can be used as a pharmacophore to design new hypertensive agents.

Keywords: *Biebersteinia heterostemon*; galegine; hypotensive; toxicity

Citation: Wang, W.; Zhang, X. Antihypertensive Effect of Galegine from *Biebersteinia heterostemon* in Rats. *Molecules* **2021**, *26*, 4830. https://doi.org/10.3390/molecules26164830

Academic Editors: Raffaele Pezzani and Sara Vitalini

Received: 30 May 2021
Accepted: 5 August 2021
Published: 11 August 2021

Publisher's Note: MDPI stays neutral with regard to jurisdictional claims in published maps and institutional affiliations.

1. Introduction

Hypertension is a major risk factor for stroke, myocardial infarction, heart failure, and kidney failure. Worldwide, hypertension is estimated to cause 9.4 million premature deaths and counts for 4.5% of disease [1,2]. Treatment of hypertension is associated with a reduction in the risk of stroke of approximately 40% and a reduction in the risk of myocardial infarction of approximately 15% [3]. Consequently, guidelines on clinical practice have identified lowering of blood pressure (BP) as a priority in hypertension treatment [4]. However, hypertension can be managed in a suboptimal manner in many countries.

Natural products have made many unique and vital contributions to drug discovery. Several hypertensive agents have been derived from pharmacophores (i.e., a part of a molecular structure responsible for a particular biologic/pharmacologic interaction that it undergoes) from natural products. The treatment of hypertension with plant-derived products is well known, such as (+)-Dicentrine, Rhynchophylline, Stevioside, ACE inhibitory peptides and so on [5]. The potential value of herbal medicines for hypertension treatment has been rediscovered [6]. Therefore, pharmacologic validation of medicinal plants or ethnomedical treatment methods could benefit development of new drugs.

The aerial part of *Biebersteinia heterostemon* Maxim. (Geraniaceae), known as *ming jian na bao* in Chinese, has been used in Tibetan folk medicine for treatment of diarrhea, edema, apoplexy, stomach pain, anthrax, erysipelas, and malaria. In Qinghai (Tibet, China), the entire plant is used by traditional healers for treatment of diabetes mellitus and hypertension [7,8]. However, preparation of *B. heterostemon* as an antihypertensive agent has not been described in detail. Only phytochemical studies carried out with the entire *B. heterostemon* plant, which have led to the isolation of guanidine alkaloids comprising mainly galegine and 4-hydroxygalegine [9], have been mentioned. Galegine has been reported to cause hyperglycemia if isolated as an alkaloid from *Verbesina encelioides* and *Galega officinalis* (which contains 0.1–0.3% galegine). Study of the hypoglycemic properties of galegine led to the discovery of metformin [10,11]. Studies have also shown that galegine can reduce weight indirectly by inhibiting the synthesis and stimulating the oxidation of fatty acids [12]. However, *B. heterostemon* has not been studied specifically for its cardiovascular effects or mechanism of action of its antihypertensive effects. Therefore, the aim of the present study was to evaluate the effects of galegine obtained from an ethanol extract of the entire *B. heterostemon* plant on the cardiovascular system of rats so that its contributions as an antihypertensive agent could be characterized.

2. Materials and Methods

2.1. Isolation of Plant Material

The aerial parts of *B. heterostemon* in the blooming phase were collected in Tongren County (Qinghai province, Tibet, China) in August 2014. These aerial parts were identified by Professor Xuefeng Lu (Department of Botany, Northwest Institute of Plateau Biology, Chinese Academy of Sciences, Qinghai Sheng, China). A voucher specimen (number 98,018) was deposited at the Herbarium of Tibetan Medicinal Plants (0028, holotypus) at the Northwest Institute of Plateau Biology.

Air-dried and finely ground aerial parts (20 kg) of *B. heterostemon* were extracted three times, once every 5 days, with 20 L 90% EtOH at room temperature. The concentrated syrup was suspended in H_2O then partitioned successively with petroleum ether, AcOEt, and n-BuOH, with a residue yield of 5 g, 138 g, and 460 g, respectively, after solvent removal. As part of a search for antihypertensive principles by bioassay-directed separation, the AcOEt extract was named GAP, which was selected for study because previous phytochemical studies have revealed that GAP contains galegine.

Half of the GAP (50 g) was chromatographed over a normal silica-gel column (40–63 μm, 5 × 120 cm) eluted with solvents of increasing polarity in the order petrol-AcOEt (10:1–2:1), petrol-acetone (10:1–10:3), $CHCl_3$-acetone (10:1–1:1), and acetone. Chromatography was monitored by thin-layer chromatography (petrol-AcOEt, 1:1; $CHCl_3$-acetone, 4:1). Fractions eluted with $CHCl_3$-acetone 10:3 (following elution with 3 L of this solvent, and fractions of 250 mL were collected, and were collectively named PCF according to the elution order) gave compound **4** (990 mg), which was purified by column chromatography and recrystallization.

The content of galegine in the GAP extract is 1.98% and the participation of galegine is a pure compound in the pharmacological effect.

2.2. Animals

The experimental protocol for animal studies was approved by local Animal Ethics Committees in accordance with the guidelines for the care and use of laboratory animals set by the Faculty of Medicine of Qinghai University (Xining, China), and incompliance with national (GB/T 35892-2018) and international rules on care and use of laboratory animals (NIH Publication No. 85-23, rewised by 1985). All tests were performed during the light phase.

Male Sprague Dawley rats (4 weeks, 90 ± 6 g) and Kunming rats of both sexes (10–12 weeks, 25–30 g) were purchased from the Institute of Local Disease (Xining, China). Rats were kept in a room under automatically controlled conditions of 22 ± 1 °C and a 12-h light–dark cycle. Rats were fed standard laboratory diet provided by the Institute of Local Disease and were allowed to acclimatize to their surroundings for ≥1 week before experimentation.

2.3. Acute Toxicity

Kunming rats of both sexes (10–12 weeks, 25–30 g) were used for acute toxicity studies. Rats were divided into five groups with graded randomization to make the mean weight and sex distribution as equal as possible. Each group comprised 5 males and 5 females. An acute study for calculation of the median lethal dose (LD_{50}) was carried out using Karber's method as modified by Sun and colleagues [13]. In this method, the galegine dose was determined through pre-testing. Rats were fasted overnight before conducting the experiment but had free access to water. The diluted drug was injected by intraperitoneal injection (i.p.) taking 1.25 as the geometric proportion between groups to administrate the dosage volume of galegine by base dosage (80 mg/kg. body weight) in each group.

Detailed clinical observations were made for all rats throughout the study. Body weights were recorded on the day of treatment and on test days 4, 7, 10, 13, and 16. Every 24 h, the dose for each group and the number of dead rats were recorded. Necropsies were carried out as soon as possible after death on all rats that died during the study. At the end of the study, all surviving rats were sacrificed.

For acute toxicity, we used death rates in the group with a minimum dose of 0% and the group with a maximum dose of 100%. LD_{50} values and 95% confidence limits (A) were calculated as follows:

$$LD_{50} = lg^{-1}\left[X_m - i\left(\sum p - 0.5\right)\right] \tag{1}$$

$$A = lg^{-1}\left(lgLD_{50} \pm 1.96 \times i \times \sqrt{\frac{\sum p(1-p)}{n}}\right) \tag{2}$$

where X_m is the logarithm of maximum dose; i represents the logarithm difference between two adjacent doses; $\sum p$ is the sum of the mortality of animals; n is number of rats per group.

Galegine was administered to five groups in doses of 32.8, 40.96, 51.5, 64, and 80 mg/kg body weight, and the death rate in each group was 0, 10%, 40%, 70%, and 100%, respectively.

2.4. Feeding of a Refined-Sugar Preparation to Rats

Studies were carried out in male Sprague Dawley (4 weeks, 90 ± 6 g). Rats were separated into two groups: I (control rats given tap water) and II (rats given 30% of commercially refined sugar in drinking water). These rats were given the respective drinks for 16 weeks. Sugar treatment ended immediately after the mean BP of group-II rats increased significantly from 90–110 mmHg to 120–150 mmHg. This model is characterized by hyperinsulinemia, loss of tissue sensitivity towards insulin, hypertriglyceridemia, arterial hypertension, and an increase in oxidative stress [14]. Systolic arterial pressure (SAP) measurements were taken every month. Rats were maintained at 32 °C in a LE 5650/6 heater and scanner heating unit (Letica, Rochester Hills, MI, USA). A pulse transducer and pressure cuff (LE 5160/R, pulse transducer and pressure cuff for rats, Harvard Appratus, Holliston, MA, USA) were placed around the tail of each rat and connected to an automatic blood-pressure system (LE 5007, RCA connector, Ningbo Hysound Electronic Co., Ltd., Ningbo, China). After 16 weeks, we selected rats whose SAP had increased $\leq 30\%$ to be the hypertensive rats from group II.

2.5. Experimental Protocol

Six male hypertensive Sprague Dawley rats (20 weeks, 280–320 g) were anesthetized successively (pentobarbital sodium, 40 mg/kg body weight, i.p.). The common carotid artery was cannulated and BP monitored using a pressure transducer (BL-410 biological experimental system; Sichuan Tai Meng Technology, Chengdu, China), which was triggered by the pressure pulse and recorded on a separate polygraph channel. Various concentrations of galegine were injected (i.p.) into rats, and the effects on arterial blood pressure (ABP) and heart rate (HR) recorded. Each rat was tested with only one concentration. ABP was expressed as the mean arterial blood pressure (MABP) according to the following equation:

$$\text{MABP} = P_\mathrm{d} + \frac{1}{3}(P_\mathrm{s} - P_\mathrm{d}) \, \text{mmHg} \tag{3}$$

where P_s denote systolic BP and P_d denotes diastolic BP.

The common carotid artery was excised rapidly. Then, the distal part of the heart was ligated with thread tightly. The proximal part of the heart was clamped by artery forceps. A V-shaped incision was made near the ligation point. An arterial cannula filled with 0.1% heparinized saline was inserted. The arterial cannula was connected to the force–displacement transducer linked to the physiologic-pressure detector. After the pressure detector had been adjusted, the artery forceps were opened and normal BP recorded. After a stabilization period of 15 min, single administration of galegine (2.5, 5 or 10 mg/kg) or saline solution (0.2 mL/100 g) was injected (i.p.) in six experimental rats. An additional positive control group received dimaprit (5 mg/kg body weight, 99% purity, Sigma–Aldrich, St. Louis, MO, USA) under the same procedure. In this series, recordings of MABP and HR were taken immediately over 60 min and their values were registered every 5 min. In sum, each group contained 6 rats and their MABP and HR were recorded 12 times over 60 min (before and every 5 min). We obtained six values for each of those 12 recordings, from which we present a median. The male Sprague Dawley rats were sacrificed after the experiment.

2.6. Statistical Analyses

Data are the mean $\pm$ SEM. Student's *t*-test, one-way analysis of variance (ANOVA), and post hoc least-significant difference tests were used to determine significant differences between groups. $p < 0.05$ was considered significant.

3. Results

The structure was demonstrated to be galegine (compound **4**, Figure 1) by two-dimensional nuclear magnetic resonance (NMR) and high-resolution mass spectrometry.

GAP was subjected to reversed-phase column chromatography (C_{18}, 5 µm, 250 mm $\times$ 4.6 mm i.d.; Thermo Fisher Scientific, Waltham, MA, USA) to high-performance liquid chromatography with diode-array detection (HPLC–DAD) analyses. Spectrometric analyses were carried out with a HPLC system (Waters, Milford, MA, USA) comprising a 1525 binary pump and 2996 photodiode array detector.

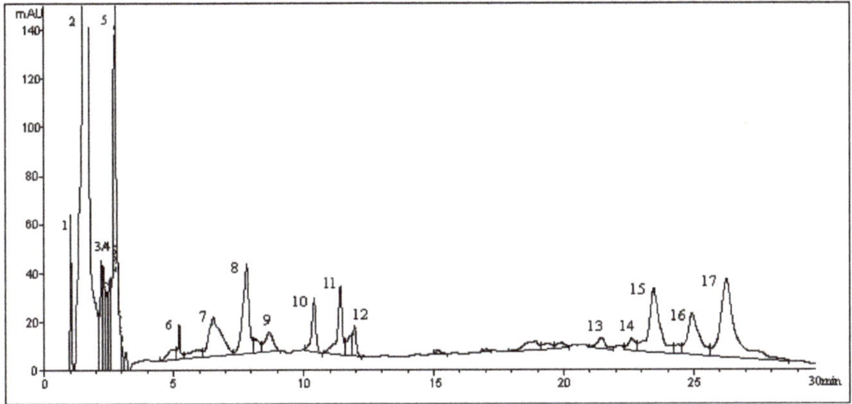

Figure 1. Scheme of the structure of galegine.

PCF (20 mg) was dissolved in 1 mL of H_2O–MeOH (80:20) and injected into the C_{18} cartridge. Then, 2 mL of H_2O–MeOH 80:20 (v/v) was applied to the cartridge for rinsing. The achieved retained sample was eluted with a mixture of 2 mL H_2O–MeOH 50:50 (v/v). This mixture displaced GAP and showed an intense narrow ring proceeding downwards, which was monitored by the naked eye. Parameters for this process were: flow rate = 1 mL/min; injection volume = 20 μL; concentration of galegine sample = 10 mg/mL in H_2O–MeOH 50:50 (v/v); DAD conditions = 205 nm. The HPLC profile of GAP and galegine (peak 1) is shown in Figure 2.

Figure 2. HPLC profile recorded at 255 nm. *Biebersteinia heterostemon* bark antihypertensive fraction (GAP).

Compound **4** was obtained as colorless, needle-like crystals. m.p. 104~105 °C. A positive Sakaguchi reaction suggested that this compound could be a guanidine alkaloid. High-resolution electrospray ionization mass spectrometry exhibited a molecular ion peak $[M + H]^+$ at m/z 128, which corresponded to the molecular formula $C_6H_{13}N_3$. Infrared absorption bands at 3405, 3201, and 1676 cm^{-1} suggested the presence of primary amines and secondary amines. According to ^1H and ^{13}C NMR data (Table 1), the structure of compound **4** was determined to be galegine. These data were identical to those of galegine [9].

The galegine (purity = 99.2%; 990 mg) used in the experiments was prepared fresh by dissolving in distilled water.

Table 1. ^{1}H and ^{13}C NMR spectral data of galegine.

	δH [a]	δC [b]
1(-CH2)	3.69 (d, 6.4)	40.1
2(-CH)	5.17 (*br t*, 6.0)	118.9
3		136.3
4(-CH$_3$)	1.69 *br s*	25.2
3(-CH$_3$)	1.63 *br s*	17.7
C=NH	7.50 *br s*	156.5

Measured in DMSO-d$_6$. [a] 400 MH$_Z$. [b] 150 MH$_Z$. Data are expressed as mean $\pm$ S.E.M. (n = 6). Significantly different between before and after treatment.

3.1. Lethality and Clinical Signs

Galegine was administered by intraperitoneal injection (IP) to five groups of fifty mice at the doses of 32.8, 40.96, 51.5, 64, and 80 mg kg^{-1}, and we recorded the clinical signs and calculated the toxin median lethal dose (LD$_{50}$), based on 24 h lethality data.

At the dose of 41 mg kg^{-1} and above, galegine administration provoked an onset of clinical signs (prostration, tremors, followed by abdominal breathing, paralysis of the hindlimbs, and cyanosis), which led to the death of mice within less than 18 h. In particular, the lowest lethal dose (41 mg kg^{-1}) provoked the death of 1/10 mice, while 80 mg kg^{-1} was lethal for 10/10 mice (Table 2). These results are presented in Figure 3 as percentage of mice mortality versus the administered toxin doses. Based on lethality data, the oral LD$_{50}$ of galegine was calculated at 54.75 mg kg^{-1} (95% confidence limits: 49.15–61.51 mg kg^{-1}).

Table 2. Lethality and signs of toxicity of mice after intraperitoneal injection administration of galegine.

Group of Treatment	Dose (mg·kg^{-1})	Lethality	Survival Times (hour)	Signs of Toxicity
Controls	-	0/10	-	
Galgine	32.8	0/10	-	-
	41	1/10	18	Debility, abdominal breathing, paralysis of the hindlimbs, cyanosis
	51.2	4/10	9.5–10–10.5–10.5	Prostration, tremors, abdominal breathing, paralysis of the hindlimbs, cyanosis
	64	7/10	9–9–9.5–9.5–9.6–9.7–9.7	Prostration, tremors, abdominal breathing, paralysis of the hindlimbs, cyanosis
	80	10/10	6–6–7-7–7–7.5–7.6–7.6–8–8	Prostration, tremors, abdominal breathing, jumping, paralysis of the hindlimbs, cyanosis

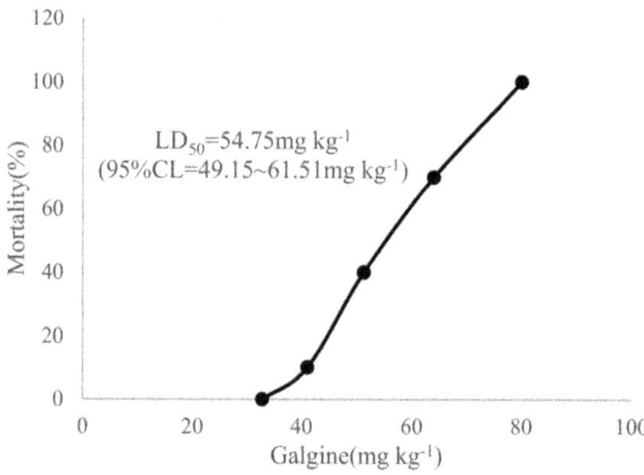

Figure 3. Dose-response mortality curve of galegine after i.p. injections administration in mice. Percentage lethality is plotted against the administered doses of galegine.

3.2. Effect of Galegine on Arterial BP

To evaluate the acute hypotensive effects of dimaprit and galegine in vivo, arterial blood pressure (ABP) was continuously measured through a pressure transducer inserted in the carotid artery of anesthetized rats. Intraperitoneal injection (i.p.) of dimaprit and galegine significantly decreased ABP and MABP, as expected (Figure 4A,B). To assess the intoxication protocol, we designed a negative control experiment (control group) where rats received an injection of distilled water: no significant change in ABP was observed (Figure 5, Table 3). Intraperitoneal injection (i.p.) of dimaprit was designed as THE positive control group.

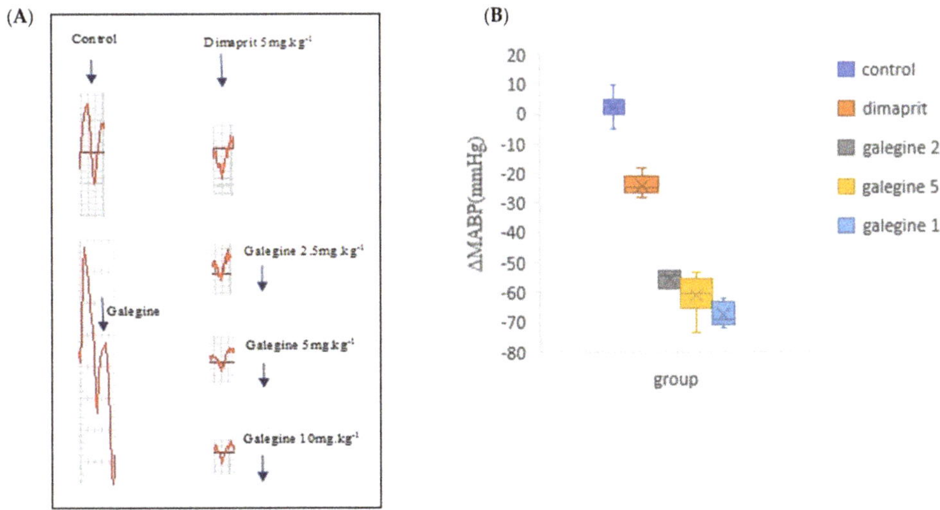

Figure 4. Influence of dimaprit and galegine on pentobarbital-anesthetized hypertensive rats MABP. (**A**) Representative raw traces of rat arterial blood pressure (ABP) following dimaprit or galegine infusion (blue arrow with doses used) compared to the control group. As expected, protocol assessment molecules dimaprit (5 mg·kg^{-1}) and galegine (2.5 mg·kg^{-1}, 5 mg·kg^{-1}, 10 mg·kg^{-1}) induced ABP, respectively. (**B**) Box and whisker plots of ΔMABP data in the control, dimaprit and galegine group following i.p. injections, 5 mg·kg^{-1} for dimaprit and 2.5 mg·kg^{-1}, 5 mg·kg^{-1}, 10 mg·kg^{-1} for galegine, respectively.

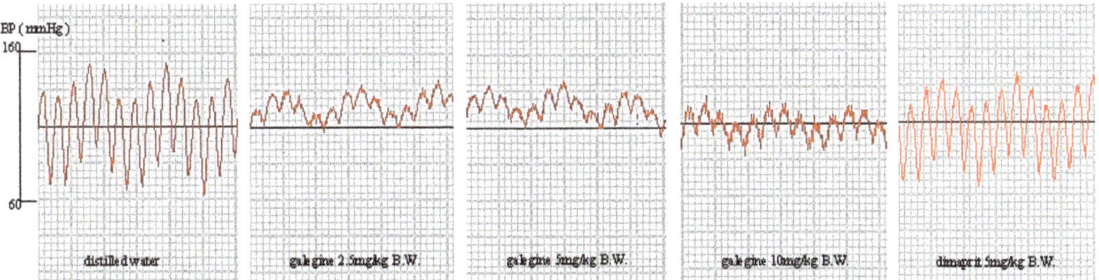

Figure 5. Effect of galegine on blood pressure of pentobarbital-anesthetized hypertensive rats. *n* = 6.

Table 3. Effect of galegine on mean arterial blood pressure (MABP) of pentobarbital-anesthetized hypertensive rats.

Group	MABP (mmHg)		Decrease of MABP (%)	Heart Rate (beats/min)	Duration of Hypotension Period (min)
	Before Treatment	After Treatment			
Control	98.25 ± 3.22	100.65 ± 4.32	0.66 ± 0.07	427 ± 20	0.15 ± 0.16
Dimaprit 5 mg/kg	153.29 ± 5.21	122.62 ± 9.65	19.99 ± 0.02	457 ± 11	15.60 ± 0.20
Galegine 2.5 mg/kg	152.83 ± 7.03	96.83 ± 10.38	36.64 ± 0.05	602 ± 15	28.30 ± 0.20
Galegine 5 mg/kg	159.07 ± 4.92	93.08 ± 13.91	40.42 ± 0.08	850 ± 12	32.20 ± 0.16
Galegine 10 mg/kg	153.58 ± 9.13	86.60 ± 4.40	43.56 ± 0.02	1150 ± 20	45.50 ± 0.15

A dose of 5 mg of dimaprit infusion induced significant hypotension (MABP declined by an average of 20%) and was associated with the slight increase of HR. The blood pressure decreased and was maintained for 15 min. Simultaneously, injection of galegine (2.5, 5, and 10 mg·kg^{-1}, i.p.) induced an immediate and dose-dependent decrease in MABP by an average of 36%, 40%, and 43% respectively (Figure 4, Table 3) with a rapid increase in HR (Table 3). The blood pressure decreased and was maintained for 28, 32, and 45 min, respectively. In contrast, at the same dose, the hypotensive effect of galegine was better and lasted longer. However, after this peak in hypotension, the MABP increased progressively and reached the initial basal value in approximately 10–15 min depending on galegine dose.

4. Discussions

The guanidine functional group has been reported to possess hypotensive properties. However, few investigations have been done on the hypotensive effects of galegine, an archetypal guanidine alkaloid. Here, we demonstrated that galegine is effective in reducing ABP in anesthetized hypertensive rats with a rapid onset and dose-related duration of effects using a pressure transducer to record ABP. The safe dose of galegine was determined to be one-tenth of the LD$_{50}$ value after acute toxicity experiments. Galegine showed physiological effects such as vasodilatation and hypotension [15]. A dose of 2.5 mg of galegine appears to be effective in managing hypertensivity, and the hypotensive effect of galegine was better and lasted longer. This indicated that galegine could be effective in managing hypertensive urgency and controlling blood pressure [5]. Although the antihypertensive mechanisms were not clear, it has been reported that the hypotensive mechanism of galegine is related to the H$_2$-receptor agonist [16]. Thus, a highly selective agonist of the histamine H$_2$ receptor, dimaprit, could mimic the excitatory effect of histamine on rubral cells, and the histaminergic nervous system may have a modulatory role in motor control through its excitatory effects on almost all subcortical motor structures [17–19]. Several ionic channels and intracellular signaling pathways have also been suggested to mediate the excitatory response of central neurons to histamine stimulation [20]. Thus, further studies, such as drug receptor-specific interaction tests or chronic toxicity tests, should be carried out to confirm the long-term safety of galegine for BP control.

Author Contributions: Experiments were performed by W.W. Analysis was performed by W.W., Manuscript preparation and revision was done by W.W. Resources was provided by X.Z. Both authors have read and agreed to the published version of the manuscript.

Funding: This research received no external funding.

Institutional Review Board Statement: All animal experiments were carried out in compliance with the EU Directive 2010/63/EU on the care and use of laboratory animals. The study was conducted according to the guidelines of the Declaration of Helsinki, and approved by the Ethics Committee of the Second Affiliated Hospital of Qinghai Medical University, Qinghai, China. Protocol codes is SYXK 2016-0001 for rats, and date of approval is 8 January 2016.

Data Availability Statement: All the data are contained within the article.

Acknowledgments: This research was supported in part by Northwest Institute of Plateau Biology for Basic Research of Tibetan Medicine. We thank Shen Minghua for kind assistance in animal experiment.

Conflicts of Interest: We declare all authors have no conflict of interest.

Sample Availability: Samples of the compounds are not available from the authors.

References

1. World Health Organization. A Global Brief on Hypertension. Silent killer, Global Public Health Crisis. Available online: http://www.who.int/cardiovascular_disease/publications/global_brief_hypertension/en (accessed on 1 May 2016).
2. Rapsomaniki, E.; Timmis, A.; George, J.; Pujades-Rodriguezet, M.; Shah, A.D.; Denaxas, S.; White, I.R.; Caulfield, M.J.; Deanfield, J.E.; Smeeth, L.; et al. Blood pressure and incidence of twelve cardiovascular disease: Lifetime risks, healthy life-years lost, and age-specific associations in 1.25 million people. *Lancet* **2014**, *383*, 1899–1911. [CrossRef]
3. Writing Group Members; Roger, V.L.; Go, A.S.; Lloyd-Jones, N.M.; Benjamin, E.J.; Berry, J.D.; Borden, W.B.; Bravata, D.M.; Dai, S.; Ford, E.S.; et al. Heart Disease and Stroke Statistics—2012 Update. *Circulation* **2012**, *125*. [CrossRef]
4. Lonn, E. The clinical relevance of pharmacological blood pressure lowering mechanisms. *Can. J. Cardiol.* **2004**, *20*, 83–88.
5. Bai, R.R.; Wu, X.M.; Xu, J.Y. Current natural products with antihypertensive activity. *Chin. J. Nat. Med.* **2015**, *13*, 721–729. [CrossRef]
6. Rates, S.M. Plants as source of drugs. *Toxicon* **2001**, *39*, 603–613. [CrossRef]
7. Tibetan Medicine Administration. *Tibetan Medicine*; Tibetan People Press: Tibet, China, 2003; pp. 65–66.
8. Yang, Y.C. *Tibetan Medicine Glossary*; Qinghai People Press: Xining, China, 1991; pp. 262–2633.
9. Wang, W.E.; Zhang, X.F.; Shen, J.W.; Lou, D.J. Chemical constituents in aerial parts of *Biebersteinia heterostemon* Maxim. *China Nat. Prod. Res. Dev.* **2009**, *21*, 199–202.
10. Bailey, C.J.; Day, C. Metformin: Its botanical background. *Pract. Diabetes Int.* **2004**, *21*, 115–117. [CrossRef]
11. Eichholzer, J.V.; Lewis, I.; Macleod, J.; Oelrichs, P.; Vallelyl, P. Galegine and a new dihydroxyalkylacetamide from *Verbesina enceloiodes*. *Phytochemistry* **1982**, *21*, 97–99. [CrossRef]
12. Mooney, M.H.; Fogarty, S.; Stevenson, C.; Gallagher, A.M.; Palit, P.; Hawley, S.A.; Hardie, D.G.; Coxon, G.D.; Waigh, R.D.; Tate, R.J.; et al. Mechanisms underlying the metabolic actions of galegine that contribute to weight loss in mice. *Br. J. Pharmacol.* **2008**, *135*, 1669–1677. [CrossRef] [PubMed]
13. Yuan, B.J. *Drug Toxicological Method and Technic*; Chemical Industry Press: Beijing, China, 2007; p. 142.
14. Hafidi, M.E.; Cuellar, A.; Ramirez, J.; Banos, G. Effect of sucrose addition to drinking water, that induces hypertension in the rats, on liver microsomal Δ9 and Δ5-desaturase activities. *J. Nutri. Biochem.* **2001**, *12*, 396–403. [CrossRef]
15. Monachea, G.D.; Volpea, A.R.; Monachea, F.D.; Vitalia, A.; Botta, B.; Espinalc, R.; De Bonnevaux, S.C.; De Luca, C.; Botta, M.; Corellie, F.; et al. Further hypotensive metabolites from *Verbesina caracasana*. *Bioorg. Med. Chem. Lett.* **1999**, *9*, 3249–3254. [CrossRef]
16. Camille, G. *Wermuth the Practiceof Medicinal Chemistry*; Academic Press: London, UK, 2003; p. 85.
17. Shen, B.; Li, H.Z.; Wang, J.J. Excitatory effects of histamine on cerebellar interpositus nuclear cells of rats through H$_2$ receptors in vitro. *Brain Res.* **2002**, *948*, 64–71. [CrossRef]
18. Sittig, N.; Davidowa, H. Histamine reduces firing and bursting of anterior and intralaminar thalamic neurons and activates striatal cells in anesthetized rats. *Behav. Brain Res.* **2001**, *124*, 137–143. [CrossRef]
19. Tian, L.; Wen, Y.Q.; Li, H.-Z.; Zuo, C.-C.; Wang, J.-J. Histamine excites rat cerebellar Purkinje cells via H$_2$ receptors In Vitro. *Neurosci. Res.* **2000**, *30*, 61–66. [CrossRef]
20. Brown, R.E.; Stevens, D.R.; Haas, H.L. The physiology of brain histamine. *Prog. Neurobiol.* **2001**, *63*, 637–672. [CrossRef]

Article

Pharmacoinformatics and UPLC-QTOF/ESI-MS-Based Phytochemical Screening of *Combretum indicum* against Oxidative Stress and Alloxan-Induced Diabetes in Long–Evans Rats

Md. Shaekh Forid [1], Md. Atiar Rahman [2,*], Mohd Fadhlizil Fasihi Mohd Aluwi [3], Md. Nazim Uddin [4], Tapashi Ghosh Roy [5], Milon Chandra Mohanta [6], AKM Moyeenul Huq [7,*] and Zainul Amiruddin Zakaria [8,*]

[1] Department of Pharmacy, School of Science and Engineering, Southeast University, Dhaka 1213, Bangladesh; foridpharmacy91@gmail.com
[2] Department of Biochemistry & Molecular Biology, University of Chittagong, Chittagong 4331, Bangladesh
[3] Faculty of Industrial Science and Technology, Universiti Malaysia Pahang, Kuantan 26300, Pahang, Malaysia; fasihi@ump.edu.my
[4] Institute of Food Science and Technology, Bangladesh Council of Scientific and Industrial Research, Dhaka 1205, Bangladesh; nazimbio@yahoo.com
[5] Department of Chemistry, University of Chittagong, Chittagong 4331, Bangladesh; tapashir57@cu.ac.bd
[6] Department of Chemistry, School of Science and Engineering, Tulane University, New Orleans, LA 70118, USA; milonmohanta@yahoo.com
[7] Department of Pharmacy, School of Medicine, University of Asia Pacific, 74/A, Green Road, Dhaka 1205, Bangladesh
[8] Department of Biomedical Science, Faculty of Medicine and Health Sciences, Universiti Putra Malaysia, UPM Serdang, Serdang 43400, Selangor, Malaysia
* Correspondence: atiar@cu.ac.bd (M.A.R.); moyeenul.rph@uap-bd.edu (A.M.H.); zaz@upm.edu.my (Z.A.Z.); Tel.: +880-3126-060011-4 (M.A.R.); +880-1906-790224 (A.M.H.); +60-1-9211-7090 (Z.A.Z.)

Citation: Forid, M.S.; Rahman, M..A.; Aluwi, M.F.F.M.; Uddin, M..N.; Roy, T.G.; Mohanta, M.C.; Huq, A.M.; Amiruddin Zakaria, Z. Pharmacoinformatics and UPLC-QTOF/ESI-MS-Based Phytochemical Screening of *Combretum indicum* against Oxidative Stress and Alloxan-Induced Diabetes in Long–Evans Rats. *Molecules* **2021**, *26*, 4634. https://doi.org/10.3390/molecules26154634

Academic Editors: Raffaele Pezzani and Sara Vitalini

Received: 8 June 2021
Accepted: 25 July 2021
Published: 30 July 2021

Publisher's Note: MDPI stays neutral with regard to jurisdictional claims in published maps and institutional affiliations.

Abstract: This research investigated a UPLC-QTOF/ESI-MS-based phytochemical profiling of *Combretum indicum* leaf extract (CILEx), and explored its in vitro antioxidant and in vivo antidiabetic effects in a Long–Evans rat model. After a one-week intervention, the animals' blood glucose, lipid profile, and pancreatic architectures were evaluated. UPLC-QTOF/ESI-MS fragmentation of CILEx and its eight docking-guided compounds were further dissected to evaluate their roles using bioinformatics-based network pharmacological tools. Results showed a very promising antioxidative effect of CILEx. Both doses of CILEx were found to significantly ($p < 0.05$) reduce blood glucose, low-density lipoprotein (LDL), and total cholesterol (TC), and increase high-density lipoprotein (HDL). Pancreatic tissue architectures were much improved compared to the diabetic control group. A computational approach revealed that schizonepetoside E, melianol, leucodelphinidin, and arbutin were highly suitable for further therapeutic assessment. Arbutin, in a Gene Ontology and PPI network study, evolved as the most prospective constituent for 203 target proteins of 48 KEGG pathways regulating immune modulation and insulin secretion to control diabetes. The fragmentation mechanisms of the compounds are consistent with the obtained effects for CILEx. Results show that the natural compounds from CILEx could exert potential antidiabetic effects through in vivo and computational study.

Keywords: *Combretum indicum* L.; antidiabetic activity; histopathology; UPLC-QTOF/ESI-MS; network pharmacology

1. Introduction

Diabetes mellitus (DM) is a major public health concern associated with many debilitating health conditions. It is a complex and chronic metabolic disease in which the body's ability to produce or respond to insulin is impaired. It was reported that about 451 million people aged 18–99 years were afflicted by diabetes in 2017, with 5 million deaths recorded worldwide. The prevalence is expected to rise to 693 million by 2045 [1]. The long-term symptoms of diabetes include vascular disorders such as retinopathy, nephropathy, neuropathy, cardiomyopathy, liver dysfunction, and dystrophy in skeletal muscle and adipose tissue [2,3]. The commonly available therapeutic agents, including sulfonylurea and biguanides, are not devoid of side effects, such as vomiting, anorexia, skin rashes, heartburn, and gastrointestinal discomfort [4,5]. Moreover, the cost of modern antidiabetic agents is also a big concern for patients, especially in underdeveloped and developing countries. Traditional medicines—predominantly herbal therapies—are attracting more interest worldwide to treat a wide range of ailments, including diabetes. Many plants have already been reported to be useful in the management of diabetes.

Combretum indicum L. (Combretaceae family), commonly known as Rangoon creeper, is an impressive tropical vine for outdoor gardens. It is distributed all over the world, especially in India, Sri Lanka, Nepal, Bangladesh, South China, Myanmar, and along the East Asian countries [6,7]. Traditionally, the plant is used for cough relief [8]. Its fruits and seeds are used as a vermifuge and for rickets in the Philippines, Thailand, and the Indochina region [9]. In Indonesia, the flowers are used in salads to add color [10]. In Bangladesh, its seeds are used for diarrhea, fever, boil, ulcers, and helminthiasis. Four diphenylpropanoids can be isolated from the stem bark [11], and the presence of linoleic, oleic, palmitic, and stearic acids is also reported [12].

A number of pharmacological studies—such as immunomodulatory, antimicrobial, antioxidant, antipyretic, anthelmintic, antirheumatic, antiviral, antifungal, antiseptic, antidiarrheal, and anti-hyperlipidemic studies—have also been reported [11,13–23]. In addition, the flower showed a hypoglycemic effect in rats [24]. However, there is no study available for its effect on diabetes-associated abnormal lipid profiles. Therefore, the current investigation was carried out to evaluate the antidiabetic and anti-hyperlipidemic effects of ethanol extracts of *C. indica* leaves (CILEx) on an alloxan-induced diabetic Long–Evans rat model.

2. Results

2.1. Determination of Total Phenolic and Total Flavonoid Contents

The total phenolic and total flavonoid contents of CILEx are presented in Table 1. Total phenolic content was expressed by the GAE equivalent per gram of extract, which was determined as 155 ± 7.35 mg/g dry weight. The regression equation of the calibration curve ($y = 0.005x + 0.022$; $r2 = 0.985$) is shown in Figure 1A. The total flavonoid content of CILEx was 164.33 ± 2.71 mg/g dry weight expressed by quercetin (QE) equivalent per gram of extract. The regression equation of the calibration curve ($y = 0.003x + 0.022$; $r2 = 0.990$) is shown in Figure 1B.

Table 1. Antioxidative status of CILEx.

Indices	Values (Unit)	Reference/Standard
Total phenolic content	155 ± 7.35 mg/g as GAE	
Total flavonoid content	164.33 ± 2.71 mg/g as QE	80.65 ± 2.8 μg/mL
Inhibition concentration (IC_{50})	165.6 ± 3.1 μg/mL	

Data presented as mean $\pm$ SEM.

2.2. DPPH Scavenging Activity

The results of DPPH free radical scavenging activity are presented in Figure 1C. The mean percentage scavenging activity of standard ascorbic acid ($78.1 \pm 0.63\%$) was significantly different from CILEx ($72.9 \pm 0.68\%$). The half-maximal inhibitory concentration (IC_{50}) of the extract was 162.6 ± 3.10 μg/mL, which is significantly different ($p < 0.05$) from that of ascorbic acid (80.65 ± 2.8 μg/mL) (Figure 1D). As the cutoff value of radical scavenging activity is 1000 μg/mL, it is clear that the extract possess a high antioxidant potential. An IC_{50} value of any substance for this activity higher than the cutoff limit is considered to be ineffective.

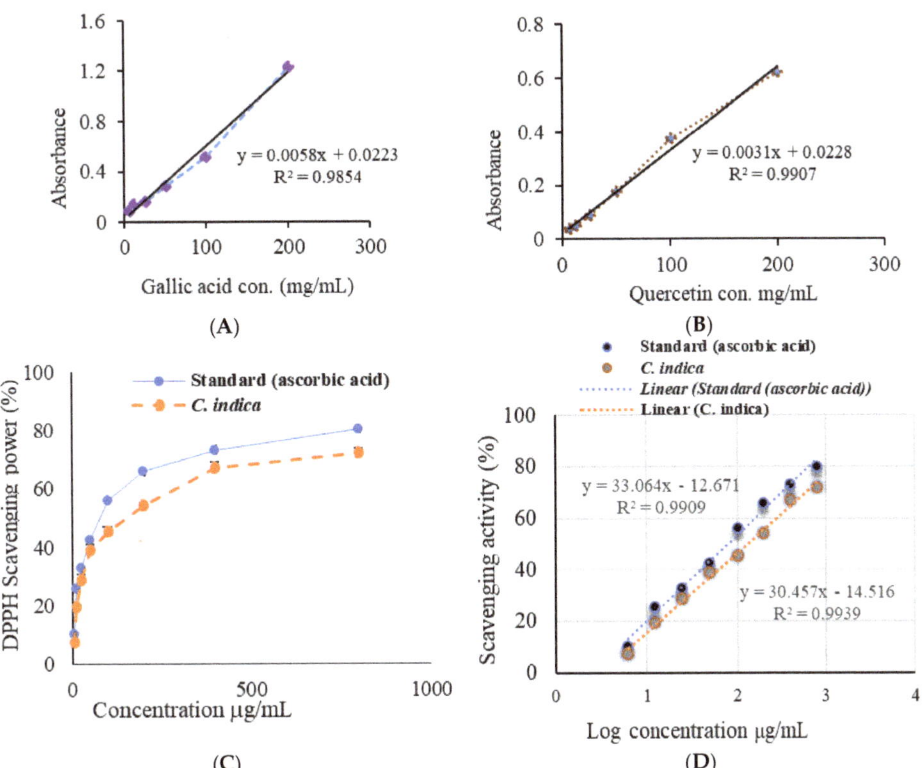

Figure 1. Antioxidative potential of CILEx: (**A**) total phenolic content, and (**B**) total flavonoid content calculated from the standard curve extrapolated against gallic acid and quercetin, respectively, used as standards; (**C**) 1,1-diphenyl-2-picrylhydrazyl (DPPH) free radical scavenging effect measured using ascorbic acid as a standard; (**D**) extrapolation of inhibition concentration (IC_{50}) using the linear curve as a standard, and sampling by *t*-test.

2.3. Acute Oral Toxicity Study and Selection of Dose

In an acute toxicity study, the oral administration of CILEx in rats did not show any change in their behavioral patterns. No toxic effects were observed at the higher dose of 1000 mg/kg body weight. Therefore, CILEx at a dose of 1000 mg/kg was considered to be safer for administration in biological systems.

2.4. Effect of CILEx on Blood Glucose Levels

The changes in blood sugar levels over one week are shown in Figure 2. The introduction of alloxan drastically increased the blood glucose level, which was highly significant ($p < 0.001$) compared to the normal control group, and remained constant throughout the study period. The administration of CILEx was able to reduce the elevated blood glucose level significantly ($p < 0.001$), compared to the diabetic control group, after every dosing with 250 and 500 mg/kg body weight A gradual blood glucose lowering effect was observed as the intervention continued for 7 days.

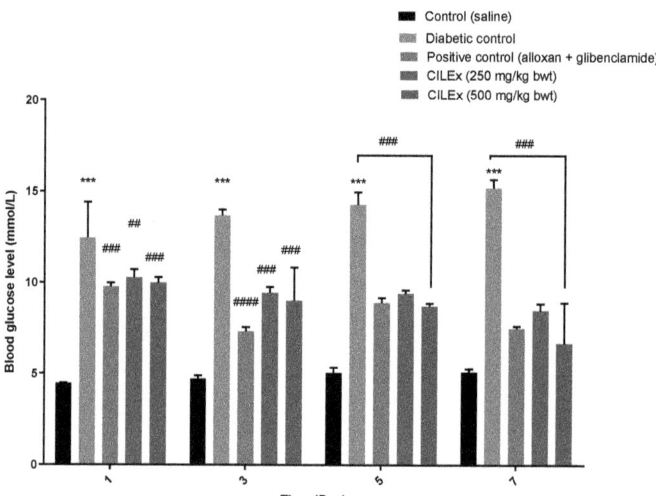

Figure 2. Effect of CILEx on blood glucose levels after a 7-day treatment. Glibenclamide was used as a positive control. The values are represented as mean $\pm$ SEM; n = 4, significant difference ($p < 0.05$) was calculated as compared to the untreated diabetic control group. In the figure, (***) over the bar denotes significant different from control, and (##, ###, ####) indicates significant difference from diabetic control.

2.5. Effect of CILEx on Lipid Profiles in Animal Intervention

The serum lipid profiles of all of the groups were measured, as shown in Figure 3. Both of the doses of CILEx (250 and 500 mg/kg body weight) significantly decreased the LDL levels, to 5.8 $\pm$ 0.2 and 3.1 $\pm$ 0.2 mmol/L, respectively ($p < 0.001$), while HDL was simultaneously increased compared to the diabetic control group (8.1 $\pm$ 1.6 and 8.3 $\pm$ 1.0 mmol/L, respectively, $p < 0.05$). Reduced total blood cholesterol levels were found to be achieved, with 8.6 $\pm$ 0.7 mmol/L ($p < 0.05$) by 250 mg/kg, and 7.5 $\pm$ 0.7 mmol/L ($p < 0.001$) by 500 mg/kg. Triglyceride levels were also found to be lowered, but the change was statistically insignificant.

2.6. Effect of CILEx on Animals' Tissue Architecture

The morphology of pancreatic tissues of different groups was examined through the hematoxylin and eosin staining method. The histopathological slides of pancreatic islets are shown in Figure 4A–E. It can be seen from the slides that group I, with normal rats, showed normal histological characteristics and islet structures. The group II alloxan-induced (150 mg/kg) rats were found to have a reduced number and size of the islet cells, thus causing shrinkage. The animals in groups IV and V—with diabetes, and treated with CILEx (250 mg/kg and 500 mg/kg body weight, respectively)—were able to restore and improve the morphology of pancreatic islets, as did glibenclamide at a 5 mg/kg dose.

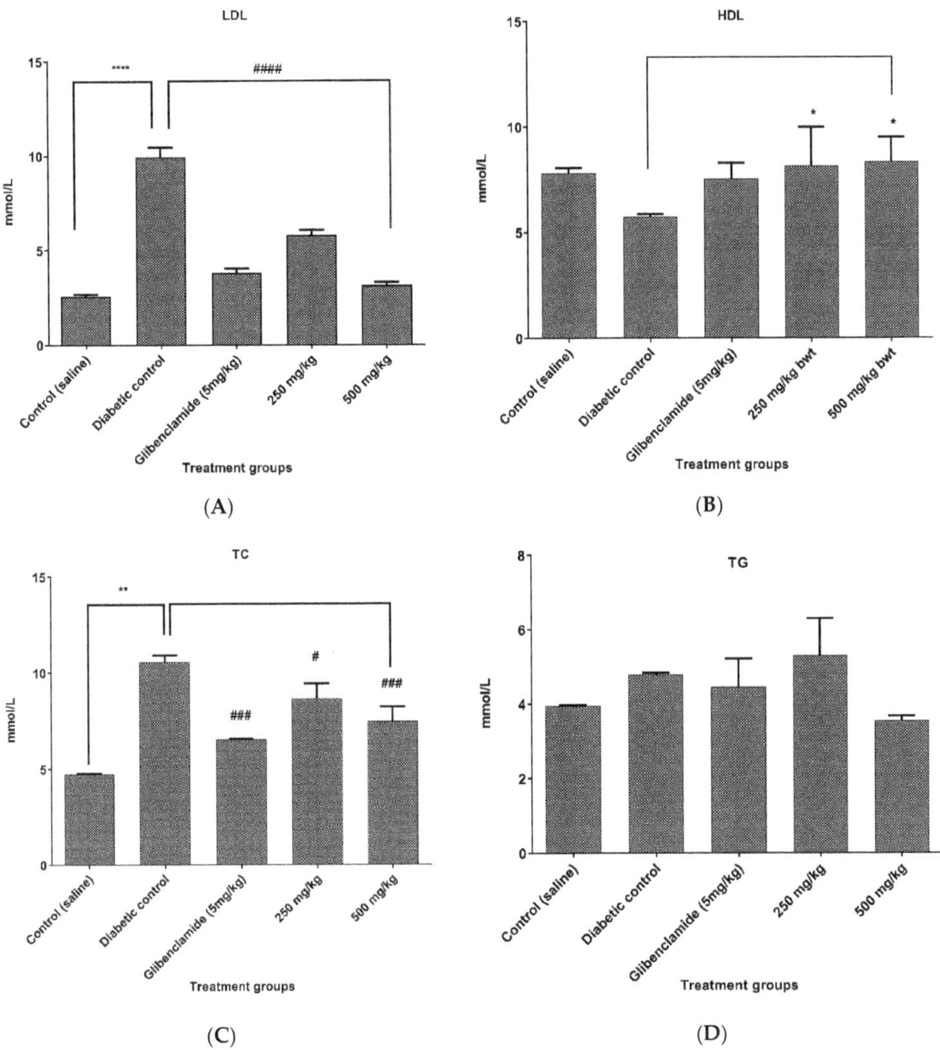

Figure 3. Effect of CILEx on serum lipid profiles: (**A**) low-density lipoprotein (LDL), (**B**) high-density lipoprotein (HDL), (**C**) total cholesterol (TC), and (**D**) triglyceride levels of different treatment groups. Glibenclamide (5 mg/kg body weight) was used as a positive control. Data were represented as mean ± SD of four animals, and were analyzed via one-way ANOVA using Tukey's multiple range post hoc test. (*, **, ****) over the bar denotes significant different from control, and (#, ###, ####) indicates significant difference from diabetic control $p < 0.05$ was considered to be significant.

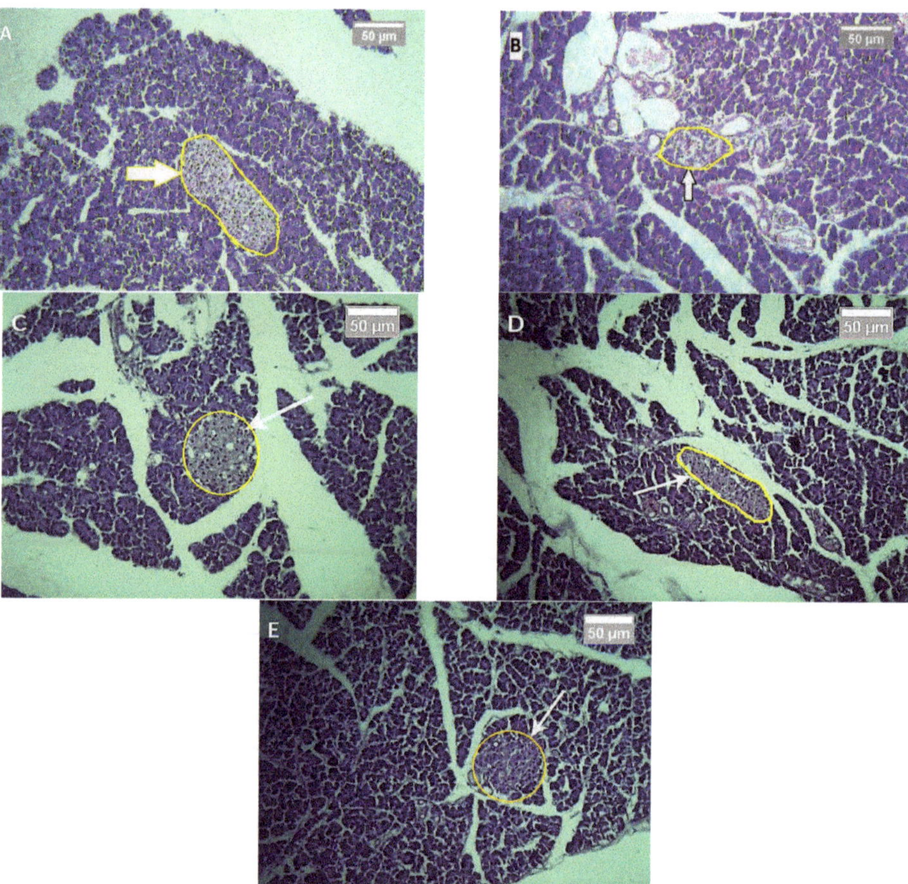

Figure 4. Histopathological slides of rat pancreas cells after 7 days of treatment: (**A**) normal control (presence of normal islet cells), (**B**) diabetic control (shrinkage of the islet cells), (**C**) diabetic + glibenclamide 150 mg/kg (increased number and size of islet cells), (**D**) diabetic + CILEx 250 mg/kg (improved size and shape of islet cells), and (**E**) diabetic + CILEx 500 mg/kg (restored number and size of islets).

2.7. UPLC-QTOF/ESI-MS Characterization of CILEx

Ultra-performance liquid chromatography coupled with time-of-flight mass spectrometry (UPLC-QTOF/ESI-MS) is one of the most promising tools for the analysis of the phytochemical constituents in extracts. It can effectively separate and analyze the compounds by giving the inclusive mass of different ions and accurate chemical formulae [25]. In the present study, flavonoids, flavonoid glycosides, saponins, and terpenoids were identified by using either positive mode ((+) ESI-MS) or negative mode ((−) ESI-MS). The presence of compounds was determined on the basis of the pattern of mass fragments, low mass error (±5 mDa), and ion response. Automatic elucidation of fragment ions by mass fragments eases the process of verification. The identified compounds were classified as a good match with ±5 mDa error, or a poor match with ±10 mDa error, by the UNIFY software. A list of identified flavonoids, flavonoid glycosides, saponins, and terpenoids is shown in detail in Table 2, while the full liquid chromatogram and the mass spectra of high-intensity compounds identified are presented in Figure 5A. Chromatograms for the most promising individual compounds are presented in Figure 5B–E.

Table 2. Tentatively identified compounds of CILEx.

SL No.	RT (min)	Molecular Formula	Observed Mass (m/z)/Neutral mass	$[M - H]^-$ (m/z)	Main Fragments	Compound
A.	17.92	$C_{40}H_{56}O_2$	568.4282	567	593.27, 585.43,55 9.27, 353.16, 635.28, 636.28,683.33, 834.24	Zeaxanthin
B.	15.75	$C_{15}H_{14}O_8$	345.0603	344	345.06, 313.03, 285.03, 229.04 346.06, 536.06, 728.00	Leucodelphinidin
C.	17.02	$C_{32}H_{42}O_{10}$	609.2711	608	609.27, 591.26,531.23, 251.14, 173.09, 625.26, 667.27, 854.58	Azedarachin C
D.	7.29	$C_{27}H_{32}O_5$	437.2363	436	437.23, 415.20 133.08, 438.23 459.21	3-o-Benzoyl-2-o-deoxyingenol
E.	17.44	$C_{30}H_{46}O_{13}$	621.29	620	621.30, 607.29, 594.28, 653.29 675.27, 676.28, 977.72	Picrasinoside E
F.	5.89	$C_{16}H_{28}O_8$	349.1834	348	349.18, 350.18	Schizonepetoside E
G.	12.03	$C_{22}H_{36}O_{10}$	461.2358	460	461.23, 375.19, 462.23, 553.30 863.48	1β,3β,6α-Trihydroxy-4α(15)-dihydrocostic acid methyl ester-1-o-β-D-glucopyranoside
H.	9.40	$C_{35}H_{48}O_9$	613.3410	612	613.34, 608.38, 133.08, 615.34	Melianol
I.	6.27	$C_{12}H_{16}O_7$	273.0944	272	205.01, 265.14, 273.09	Arbutin

Note: The compounds are arranged according to their fragmentation details in the text.

The mass spectrum of zeaxanthin displays a molecular ion (M+) peak at 568.4235, which undergoes a fragmentation to give a peak at m/z 559.27. corresponding to fragment ion [M–H–4H$_2$] in Figure 6A (Path 1), which was further cleaved to make a daughter ion [M–H–4H$_2$–C$_{14}$H$_{22}$O] at m/z 353.16 by removal of a fragment $C_{14}H_{22}O$ (Path 2). The compound leucodelphinidin did not show any molecular ion peak corresponding to m/z 322.07. However, the molecular ion produced an ion [M–H–2H$_2$O] in Figure 6B (Path 1) at m/z 285.03 upon the expulsion of one H radical and successive removal of two H_2O molecules. Furthermore, the molecular ion resulted in a daughter ion [M–C$_2$H$_2$O$_2$–H-2OH] (Path 2) at m/z 229.02 upon the removal of a $C_2H_2O_2$ fragment by homolytic fission, followed by successive expulsion of one H radical and two OH radicals. Azedarachin C did not give any peak at m/z corresponding to molecular ion M+. However, it produced a fragment [M–CH$_3$–2OH–3H$_2$] (Figure 6C) (Path 1) at m/z 531 via the removal of one CH_3 radical and two OH- (hydroxyl) radicals, followed by the expulsion of three H_2 molecules. The other fragment [M–C$_{16}$H$_{17}$O$_4$–OH–3CH$_3$·] in Figure 6C (Path 2) resulted from the cleavage to remove a large fragment $C_{16}H_{17}O_4$, and the successive removal of one OH radical and three CH_3 radicals at m/z 251. On the other hand, the molecular ion underwent cleavage to afford a daughter ion [M–C$_{16}$H$_{24}$O$_6$–C$_4$H$_4$O–H–OH–CH$_3$] in Figure 6C (path 3) at m/z 173 by removing two large fragments $C_{16}H_{24}O_6$ and C_4H_4O, followed by the expulsion of one H radical, one OH radical, and one CH_3 radical. The compound 3-o-benzoyl-2-o-deoxyingenol shows (M + 1) and (M + 2) peaks at m/z 437.23 and 438.23, respectively. A fragment [M–OH–2H$_2$] 6D (Path 1) at m/z 415.20 was observed due to the successive removal of one OH radical and two molecules of H_2. Further double cleavage (Path 2) of the compound (Figure 6D) produces a fragment [M–C$_{19}$H$_{25}$O$_3$–H$_2$] at m/z 133.08 via the removal of a large $C_{19}H_{25}O_3$ fragment, followed by the expulsion of one H_2 molecule. Picrasinoside E in its mass spectrum did not reveal any peak at the m/z value corresponding to its molecular ion M+. However, one fragment ion [M–H–3H$_2$] at m/z 607.29 in Figure 6E (Path 1) may be the result of the successive expulsion of three H_2 molecules and one H radical. The ion [M–H–2H$_2$–CH$_3$]$^+$ in Figure 6E (Path 2) at m/z

594.28 can be attributed to the fragment resulting from the removal of one H radical, two H_2 molecules, and a CH_3 radical. In Figure 6F, schizonepetoside E afforded (M + 1) and (M + 2) peaks corresponding to *m/z* 349.18 and 350.18, respectively, instead of a molecular ion peak at *m/z* 348.18. It is surprising to note that this compound did not undergo any fragmentation. The compound 1β,3β,6α-trihydroxy-4α(15)-dihydrocostic acid methyl ester-1-O-β-D glucopyranoside revealed (M + 1) and (M + 2) peaks at *m/z* 461.23 and 462.23, respectively. Furthermore, the molecular ion cleaved to create a fragment ion [M–C₄H₅O₂] at *m/z* 375.19 in Figure 6G, via the removal of a $C_4H_5O_2$ fragment. Figure 6H shows the fragmentation pattern of melianol, which did not reveal any molecular ion peak at *m/z* 472.70, even when a fragment for this compound was identified. Importantly, the LCMS produced molecular weight does not befit with the molecular formula which makes the current fragmentation proposal ambiguous for future research. Finally, arbutin displays an (M + 1) peak at *m/z* 273.09, which upon successive removal of 4OH radicals can result in a daughter ion [M + 1 − 4OH] (Figure 6I (path 1)) at *m/z* 205.01. On the other hand, upon the removal of one H radical and three H_2 molecules, the molecular ion produced an ion [M–H−3H₂] (Figure 6I (Path 2)) at *m/z* 265.

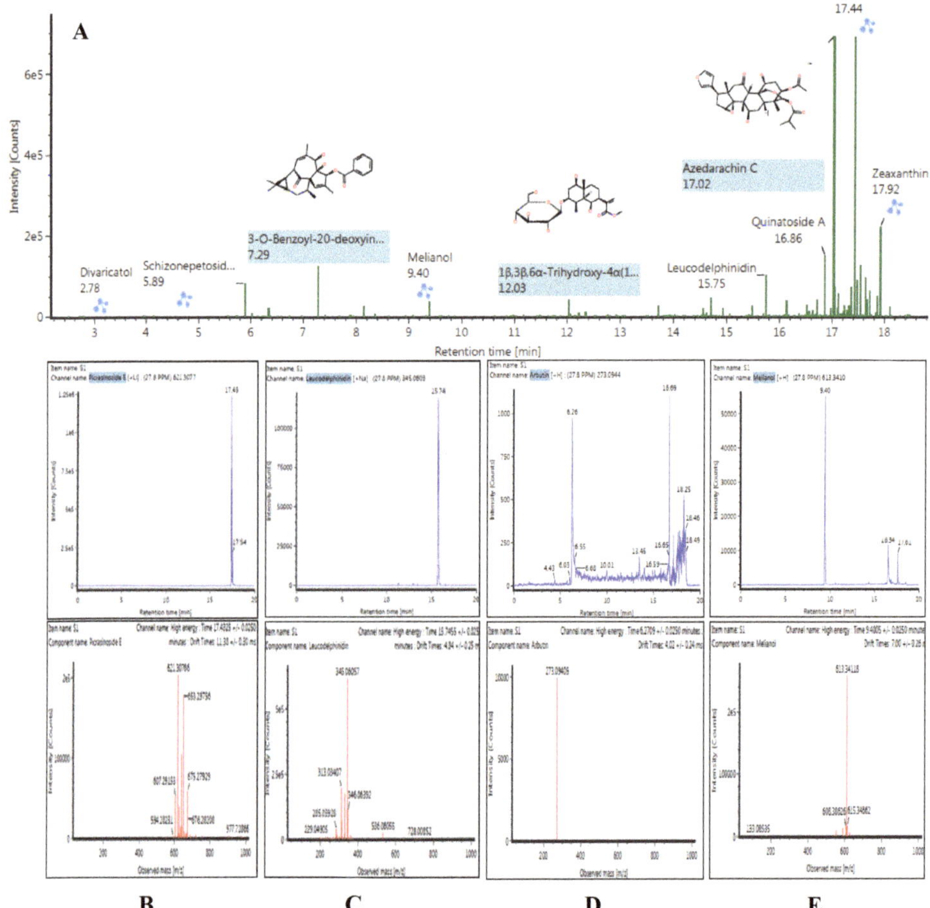

Figure 5. (**A**) Full chromatogram of CILEx in negative ion mode. LC chromatograms and mass spectra of CILEx for high-intensity compounds identified and vertically arranged as (**B**) picrasinoside E, (**C**) schizonepetoside E, (**D**) melianol, and (**E**) arbutin.

2.8. Molecular Docking

The phytochemicals picrasinoside E, azedarachin C, arbutin, 3-O-benzoyl-20-deoxyingenol, leucodelphinidin, melianol, schizonepetoside E, zeaxanthin, reference drugs metformin and gliclazide, and 1XU9, 1XU7, 6R4F, and 3A5J were used as the receptors for molecular interaction. The predicted active sites for all of the target proteins are shown in Table 3. All of the ligands were bound to the receptor and produced scores, except for zeaxanthin. The proteins yielded multiple binding sites, and we used the sites with the best scores for each protein. Glide-docking resulted in scores for different parameters, including docking score, glide emodel, and glide energy, the three of which were used to evaluate the docking study presented in Table 4. The two-dimensional binding of the highest affinity compounds based on the site map interactions is displayed in Figure 7.

Table 3. Active site prediction of enlisted proteins using SiteMap. Only the top ranked scores are mentioned on the table for each protein.

PDB ID	SiteScore	Size	Dscore	Volume	Exposure	Enclosure	Contact	Phobic	Philic	Balance	Don/acc
1XU9	1.069	666	0.969	1496.166	0.481	0.801	1.079	0.264	1.384	0.191	0.714
1XU7	1.05	182	0.904	336.14	0.504	0.772	1.06	0.199	1.532	0.13	0.33
2BEL	1.093	122	1.117	471.282	0.556	0.812	0.904	0.606	0.951	0.637	0.674
6R4F	1.155	308	1.12	766.262	0.274	0.929	1.195	0.762	1.157	0.659	0.628
3A5J	0.807	53	0.635	214.032	0.662	0.707	0.859	0.129	1.453	0.089	0.443

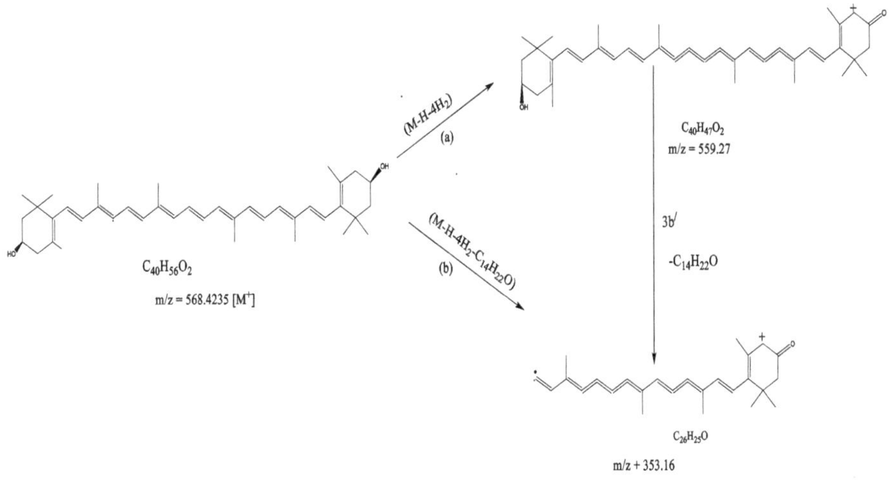

(A)

Figure 6. *Cont.*

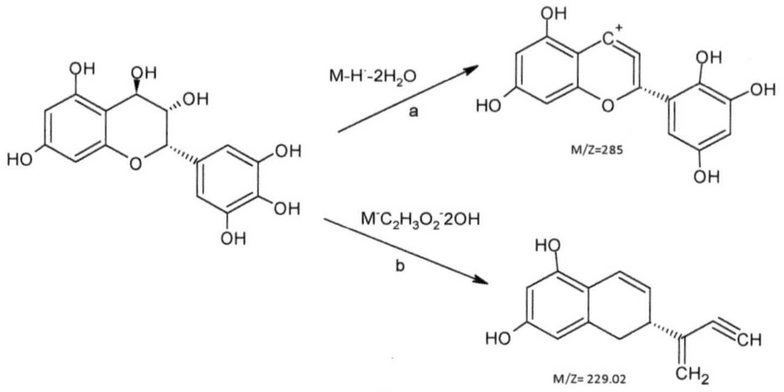

$M-H^--2H_2O$
a

M/Z=285

$M^-C_2H_3O_2^--2OH$
b

M/Z= 229.02

(B)

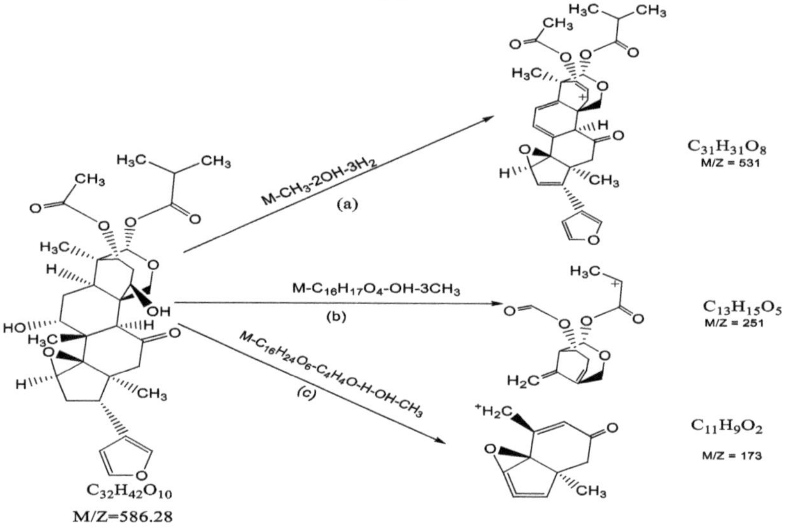

$C_{32}H_{42}O_{10}$
M/Z=586.28

$M-CH_3-2OH-3H_2$
(a)

$C_{31}H_{31}O_8$
M/Z = 531

$M-C_{16}H_{17}O_4-OH-3CH_3$
(b)

$C_{13}H_{15}O_5$
M/Z = 251

$M-C_{16}H_{24}O_6-C_4H_4O-H-OH-CH_3$
(c)

$C_{11}H_9O_2$
M/Z = 173

(C)

$C_{27}H_{32}O_5$
M/Z=436.23

$M-OH-2H_2$
(a)

$C_{27}H_{27}O_4$
M/Z = 415.20

$M-C_{19}H_{25}O_3-H_2$
(b)

$C_8H_5O_2$
M/Z = 133.08

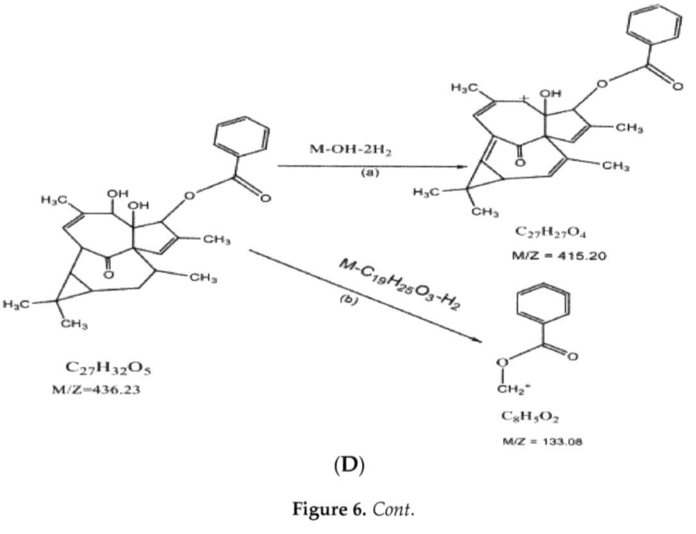

(D)

Figure 6. *Cont.*

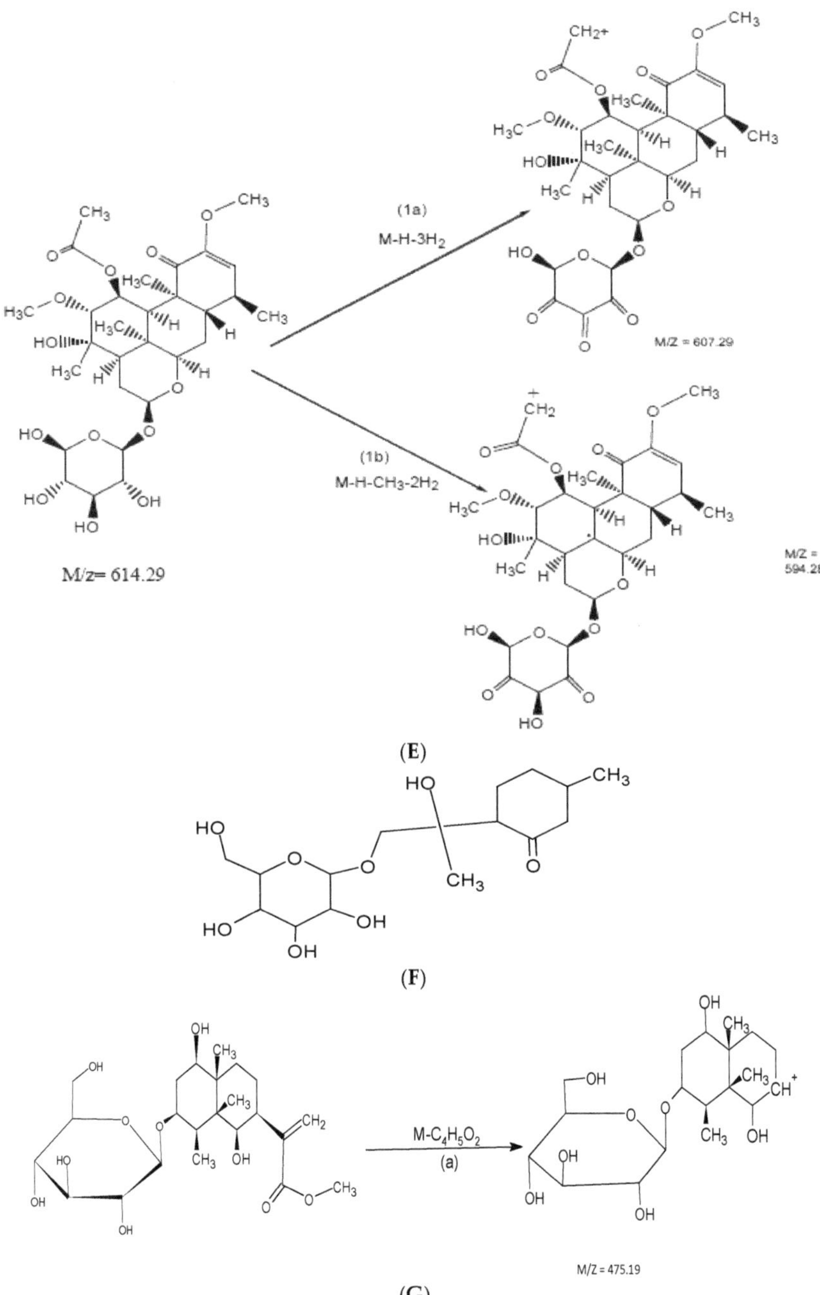

Figure 6. *Cont.*

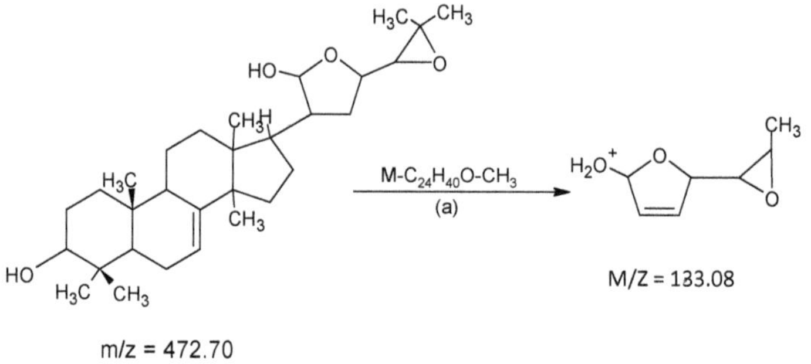

Figure 6. Mass fragmentation mechanisms of (**A**) zeaxanthin, (**B**) leucodelphinidin, (**C**) azedarachin C, (**D**) 3-O-benzyl-2-O-deoxyingenol, (**E**) picrasinoside E, (**F**) schizonepetoside E, (**G**) 1β,3β,6α-trihydroxy-4α(15)-dihydrocostic acid methyl ester-1-O β-D glucopyranoside, (**H**) melianol, and (**I**) arbutin, as identified in the CILEx.

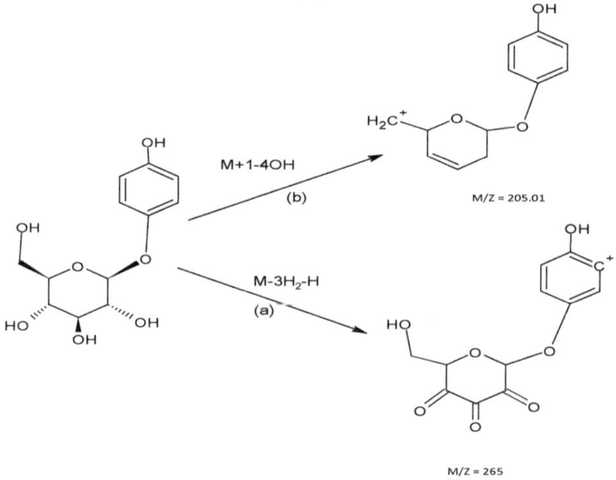

Table 4. Docking scores of 8 PubChem-available compounds out of 10 identified compounds from LCMS data for CILEx.

Compounds	Docking Score	Glide Model	Glide Energy
	1XU9		
Metformin	−2.947	−25.206	−18.318
Gliclazide	−6.115	−49.527	−37.472
Picrasinoside E	−2.343	−17.099	−20.358
Azedarachin C	−2.575	−32.511	−28.533
Arbutin	−6.172	−46.152	−33.944
3−O−Benzoyl−20−deoxyingenol	−6.008	−44.674	−23.01
Leucodelphinidin	−6.468	−53.684	−39.835
Melianol	−8.363	−49.99	−36.079
Schizonepetoside E	−8.145	−59.072	−47.635
	1XU7		
Metformin	−4.163	−29.164	−23.565
Gliclazide	−4.913	−34.925	−26.387
Picrasinoside E	−4.196	−39.489	−36.295
Azedarachin C	−5.662	−31.181	−34.502
Arbutin	−5.163	−37.436	−29.202

Table 4. *Cont.*

Compounds	Docking Score	Glide Model	Glide Energy
1XU7			
3−o−Benzoyl−20−deoxyingenol	−6.049	−35.397	−31.4
Leucodelphinidin	−6.166	−46.923	−35.177
Melianol	−8.475	−66.188	−47.383
Schizonepetoside E	−6.641	−47.271	−35.517
2BEL			
Metformin	−2.99	−22.089	−17.491
Gliclazide	−5.885	−53.992	−39.404
Picrasinoside_E	−6.358	−57.254	−45.617
Azedarachin_C	−7.246	−53.762	−42.76
Arbutin	−5.957	−49.719	−37.458
3−o−Benzoyl−20−deoxyingenol	−5.948	−54.747	−41.134
Leucodelphinidin	−6.744	−64.075	−45.816
Melianol	−8.995	−77.557	−52.946
Schizonepetoside_E	−7.685	−60.926	−44.759
6R4F			
Metformin	−4.031	−29.186	−19.423
Arbutin	−7.492	−74.393	−57.298
Leucodelphinidin	−5.279	−32.25	−23.607
Schizonepetoside_E	−6.859	−64.027	−53.418
3A5J			
Metformin	−2.71	−18.439	−16.472
Gliclazide	−4.252	−50.298	−36.799
Picrasinoside E	−4.055	−50.347	−42.864
Azedarachin C	−2.39	−29.319	−30.194
Arbutin	−4.962	−45.683	−35.8
3−o−Benzoyl−20−deoxyingenol	−3.532	−48.425	−40.094
Leucodelphinidin	−6.123	−58.309	−43.265
Melianol	−4.453	−52.271	−42.912
Schizonepetoside E	−5.716	−50.39	−44.064

2.9. Analysis of Interactions between Active Ingredients and Target Proteins

According to the previous analysis in this study, the active ingredients of CILEx show good pharmacological lipolytic and antidiabetic effects in a synergistic way. Arbutin, in this study, was found to interact significantly (PPI enrichment p-value $<1.0 \times 10^{-16}$) with 205 target proteins (Supplementary Table S1). Target proteins with the highest confidence scores for arbutin are displayed in Figure 8A. Cytoscape 3.6.1 was used to analyze the interaction between arbutin and the top 20 target proteins. Based on the arbutin–target protein relationships, it is now clear that arbutin acts on the target proteins. Recently, it was stated that arbutin alleviates diabetic symptoms by attenuating oxidative stress in mice through inhibiting the increasing blood glucose [26]. These results indicate that CILEx performs substantial biological and physiological activities via arbutin's multitarget interactions.

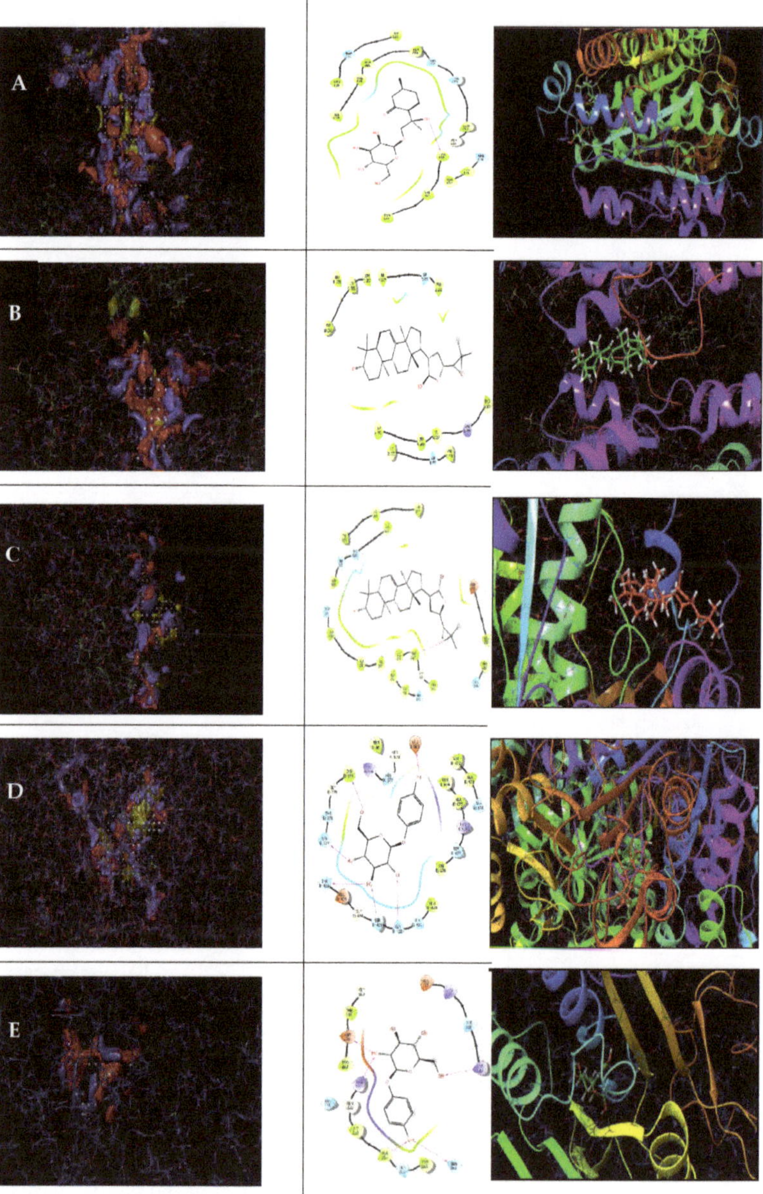

Figure 7. Based on site maps (**left**), interactions of five receptor proteins with seven ligands (**right**). Highest docking scorers are shown: (**A**) schizonepetoside showed the highest docking score (−8.145) with IXU9; (**B**) melianol (highest docking score, −8.475) with receptor IXU7; (**C**) melianol (highest docking score −8.995) with receptor 2BEL; (**D**) arbutin (highest docking score, −7.492) with receptor GR4F, and (**E**) 3-O-benzoyl-20-deoxyingenol (highest docking score, −6.123) with receptor 3A5J. In each case, gliclazide and metformin were used as reference antidiabetic drugs (Supplementary Materials).

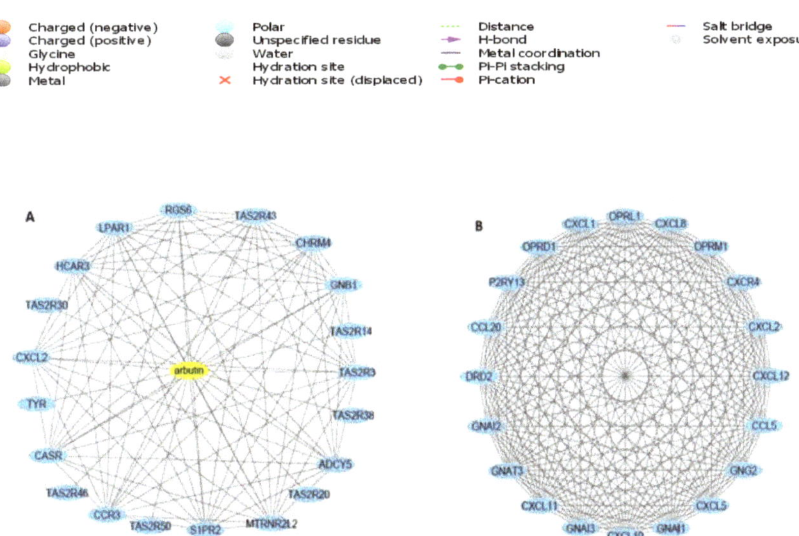

Figure 8. (**A**) CILEx arbutin–target interaction network—the yellow node represents the arbutin and the blue node represents the top 20 proteins—and (**B**) protein–protein interaction (PPI) network of top 20 ranked target proteins.

2.10. Construction and Analysis of Target Proteins' PPI Network

PPI networks play substantial roles in molecular processes, and abnormal PPI is the basis of many pathological conditions [27]. Using the STRING42 database and Network Analyst software [28], all target proteins (205) were mapped into the PPI network. Interestingly, we found that 203 target proteins are involved in the PPI network, with 17,310 edges, and an average node degree of 169, while the PPI enrichment p-value was less than 1.0×10^{-16}. In this PPI network, the larger the node degree, the stronger the relationship between the proteins corresponding to the node in this network, which indicates that the target proteins play a key role in the whole interaction network, highlighting their importance. Only the HELZ and MTRNR2L proteins were not included in the PPI network. We delineated two subnetworks in the PPI network: subnetwork 1 included 200 target proteins (listed with degree of interaction in Supplementary Table S2), while subnetwork 2 included only 3 target proteins (ERCC1, ERCC4, and CHEK2). The 20 top ranked target proteins, along with their greatest degree of interactions with other proteins, are illustrated in Figure 8B. Cytoscape 3.6.1 was used to analyze the interaction among the top 20 target proteins. GNAI1, the top hub target, is one of the crucial genes for type 2 diabetes [29]. Figure 8B shows that most of the immunological target proteins—including CCL4, CXCR4, CXCL12, CXCL8, CXCL10, CXCL1, CXCL11, CXCL5, CCL20, and CXCL2—are centrally located in the PPI networks, with top degrees of interaction, indicating that this PPI network is associated with immunological activities. It has been stated that human chemokines are associated with or implicated in the pathogenesis of type 1 diabetes [30].

2.11. Gene Ontology (GO) Analysis of Interacted Target Proteins

GO enrichment analysis of interacted target proteins (total 203) that act with arbutin was performed using DAVID (https://david.ncifcrf.gov/, accessed on 3 August 2020). The top 10 significantly enriched terms in the biological process (BP), molecular function (MF), and cellular component (CC) categories were selected, according to Benjamini–Hochberg corrected p-values < 0.05. A total of 61 significant BPs are listed in Supplementary Table S3, and the top 10 BPs are represented in Figure 9A. In BP analysis, the target proteins are mainly involved in inflammatory response, immune response, and some other metabolic processes. We also found 31 significant molecular functions (Supplementary Table S4), the top 10 of which are illustrated in Figure 9B, including G-protein-coupled receptor activity and many chemokine-mediated immune responses. In addition, we also identified 10 significant CCs, as shown in Figure 9C and Supplementary Table S5, which mainly included integral components of the plasma membrane, heterotrimeric G-protein complex, extracellular space, exterior of the plasma membrane, and cell. Inflammatory response— the most significant biological process—is documented to be associated with diabetes [31]. G-protein-coupled receptor signaling pathways are related with the crosstalk with insulin signaling [32], while chemokines have been found to be associated with or implicated in the pathogenesis of type 1 diabetes [30]. Type 2 diabetes has broad impact on immune responses [33]. The GO analysis indicates that the target proteins may bind with the plasma membranes of cells to mediate the process of immunological activities, so as to exert the anti-inflammatory and antidiabetic potential of arbutin in CCs.

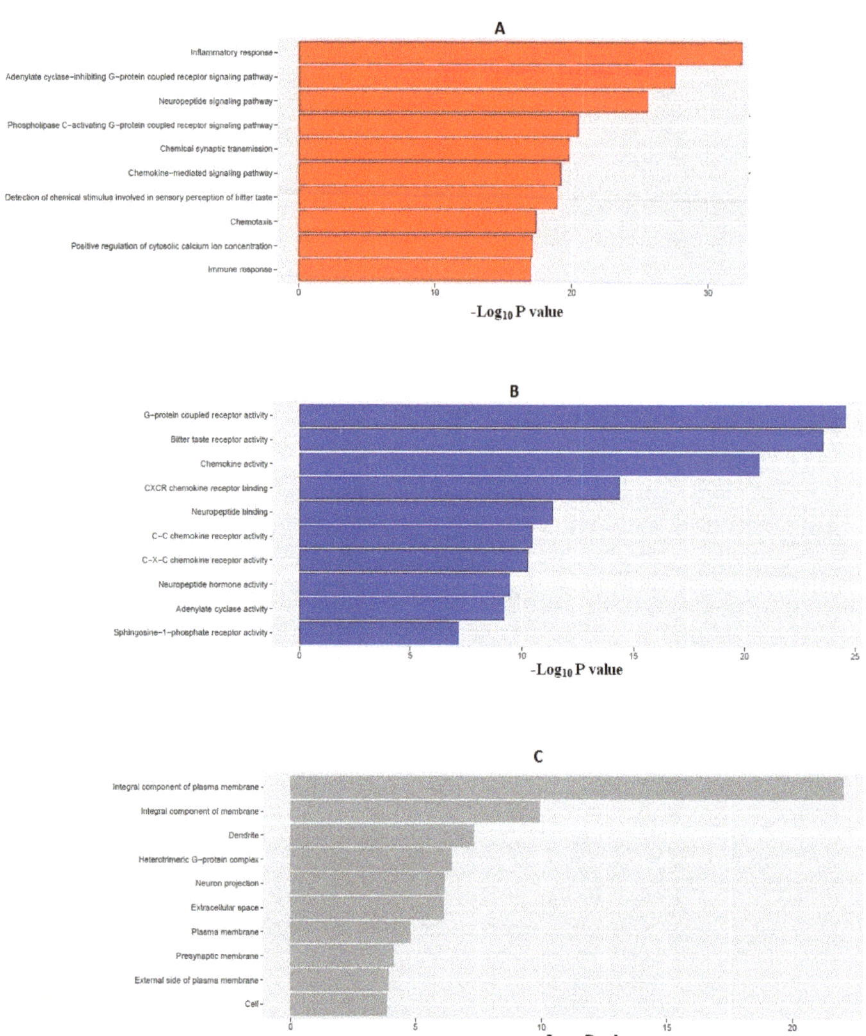

Figure 9. Gene Ontology (GO) enrichment analysis of the interacted target proteins: (**A**) top 10 biological processes (red); (**B**) top 10 molecular functions (blue); and (**C**) top 10 cellular components (gray).

2.12. Target Proteins Set Enrichment Analysis of KEGG Pathways

To further elucidate the relationship between the target proteins and the pathways, we identified 48 KEGG pathways that were significantly associated with the target proteins (Figure 10 and Supplementary Table S6). These pathways were mainly involved in immune regulation (chemokine signaling pathway, cytokine–cytokine receptor interaction, platelet activation, inflammatory mediator regulation of TRP channels, complement and coagulation cascades, and intestinal immune network for IgA production), secretion (gastric acid secretion, bile secretion, salivary secretion, aldosterone synthesis and secretion, insulin secretion, pancreatic secretion, and renin secretion), neurological regulation (neuroactive ligand–receptor interaction, taste transduction, glutamatergic synapse, morphine addiction, circadian entrainment, cholinergic synapse, GABAergic synapse, serotonergic synapse, cocaine addiction, and dopaminergic synapse), metabolism (regulation of lipolysis in adipocytes, thyroid hormone synthesis), cellular development (gap junction, progesterone-mediated oocyte maturation, oocyte meiosis, and vascular smooth muscle contraction), and cellular signaling (cAMP and cGMP-PKG signaling pathways, retrograde endocannabinoid signaling, Rap1 signaling pathway, estrogen signaling pathway, sphingolipid signaling pathway, adrenergic signaling in cardiomyocytes, oxytocin signaling pathway, GnRH signaling pathway, and calcium signaling pathway).

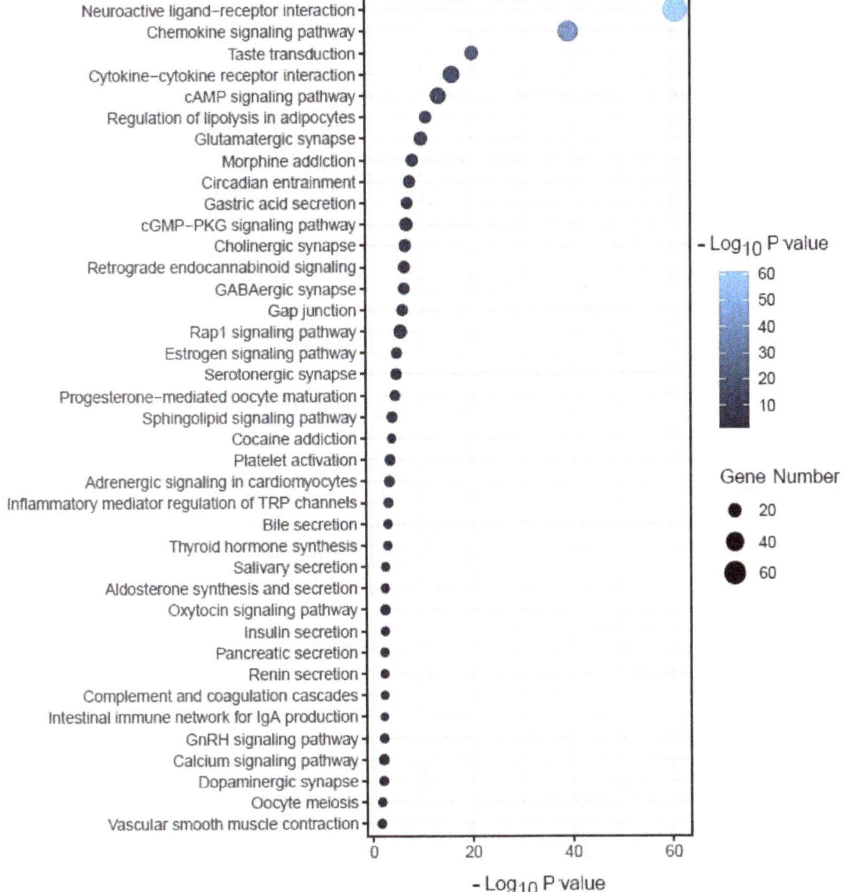

Figure 10. Some of the enriched KEGG pathways significantly associated with target proteins.

It was found that the neuroactive ligand–receptor interaction was predicted to be a major modulated pathway in an antidiabetic study [34]. Chemokines have been associated with or implicated in the pathogenesis of type 1 diabetes [30], while cytokines are crucial immunotherapeutic targets in diabetes [35].

2.13. Target Proteins Involved in Regulating the Diabetes-Associated Pathways

We found various pathways associated with diabetes (Figure 10 and Supplementary Table S6). For example, target proteins ADCY1, ADCY3, ADCY4, ADCY5, ADCY6, ADCY7, ADCY8, and ADCY9 are associated with insulin secretion (Figure 11A). Similarly, these target proteins are also associated with pancreatic secretion, bile secretion, and gastric acid secretion (Supplementary Table S6). In addition, ADCY3, ADCY4, ADCY1, PTGER3, GNAI3, GNAI2, ADCY7, GNAI1, ADCY8, ADCY5, ADCY6, NPY1R, ADORA1, NPY, and ADCY9 are linked with the regulation of lipolysis in adipocytes (Figure 11B). Gastric acid secretion is correlated with diabetic pathophysiology [36]. Fat cell lipolysis (i.e., fat cell triacylglycerol breakdown into fatty acids and glycerol in the absence of stimulatory factors) is elevated during obesity, and is correlated with insulin resistance [37].

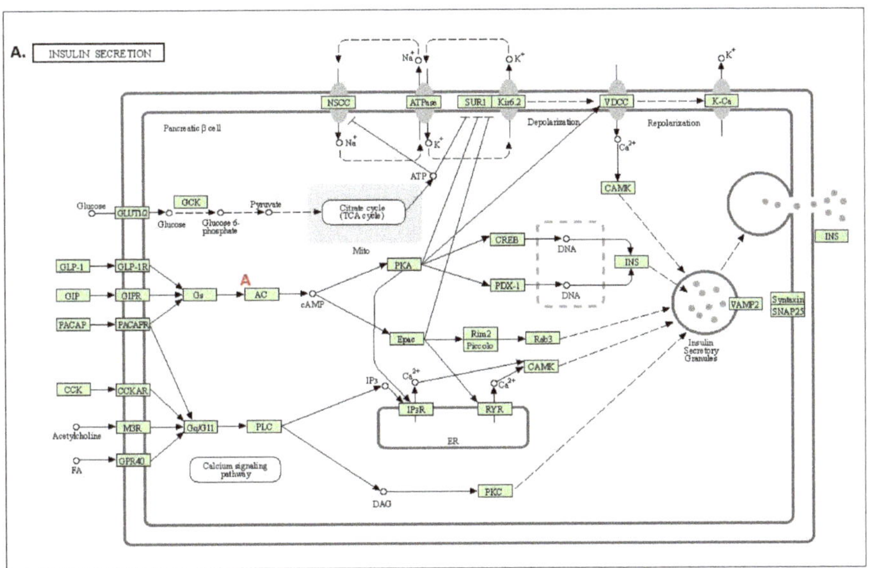

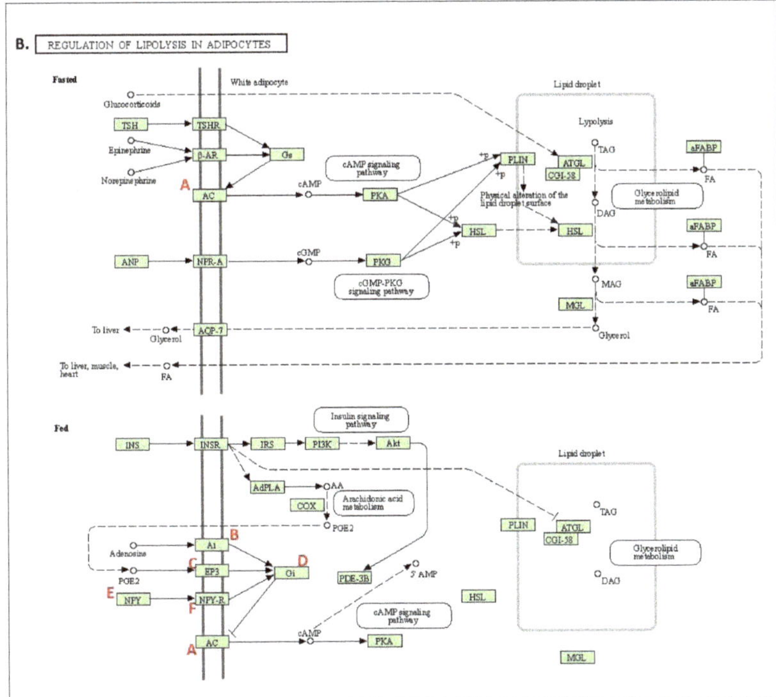

Figure 11. Involvement of target proteins in (**A**) insulin secretion and (**B**) regulation of lipolysis in adipocytes. A = ADCY1, ADCY3, ADCY4, ADCY5, ADCY6, ADCY7, ADCY8, and ADCY9; B = ADORA1; C = PTGER3; D = GNAI1, GNAI2, and GNAI3; E = NPY; F = NPY1R.

3. Discussion

An effective in vivo model for antidiabetic testing ensures reliable results to evaluate the antidiabetic activity of plant extracts and compounds. Diabetes in rats was caused by alloxan, which is a popular and long-established agent for the induction of type 1 diabetes mellitus. The evidence from experiments and clinical studies reveals that the reactive oxygen species (ROS) level is higher in both type 1 and type 2 diabetes, playing prominent role in the development and progression of diabetic complications. The administration of alloxan causes sudden and drastic insulin secretion in the presence or absence of glucose. As a consequence, total suppression of beta cells occurs by inhibiting the glucose-sensing glucokinase enzyme. Thus, beta cells fail to recognize the glucose levels in the blood. Alloxan also produces ROS and superoxide free radicals, leading to the rapid damage of the islets of beta cells [38]. Elevation of ROS occurs via glucose autoxidation, and glucose–protein reaction increases glycation [39]. Anabolic enzymatic cofactors such as nitric oxide synthase (NOS), nicotinamide adenine diphosphate (NADP) oxidase, and xanthine oxidase are the sources of reactive species in diabetics, leading to various associated complications. Hence, the use of antioxidants would be one of the effective measures to reduce oxidative damage in diabetes.

Studying with a plant extract, plant-based product, or any formulation thereof requires the evaluation of their safety profiles; this is particularly important for unknown plant extracts or compounds. The acute oral toxicity test in animal models helps select the safe dose that could be extrapolated in human clinical trials. However, different pharmacokinetic behavior could still be observed in humans than in animals. It has been reported that different pathophysiological disorders related to the gastrointestinal tract and blood in animals have similarities with those of humans [40]. Thus, the acute oral toxicity study helps measure various toxic effects of the extract when given in a single dose, which is also useful for the researchers to adjust the doses in an experiment. Our observation shows that CILEx leaf ethanol extract has no toxic effects when given at a higher dose (1000 mg/kg body weight).

A significantly reduced fasting blood sugar of the treatment groups in this study was displayed throughout the study period due to the administration of CILEx (250 and 500 mg/kg body wt.), which potentially improved blood sugar levels in alloxan-induced diabetic rats. The increased oxidative stress combined with increased blood glucose levels and free fatty acids badly affect insulin secretion and function [41]. The extract might have shown a beneficial effect on the islet of beta cells, as well as antioxidant activity through the high total phenolic and flavonoid contents of CILEx. Additionally, plant polyphenols exhibit antioxidant properties by donating hydrogen from their hydroxyl groups [42] and, thus, participate in the antioxidant activity. Previous reports revealed a positive correlation between the phenolic contents and antioxidant activity of *C. sericeum* and *C. acutum* [43]. Other studies have also reported the antioxidant activity of certain *Combretum* species [44]. Therefore, the use of antioxidants in the treatment of diabetes-associated complications has been an effective strategy of choice [39]. Some natural antioxidants—such as vitamin C, vitamin E, and α-lipoic acid treatments—were found to reduce the oxidative stress in animals, as well as in humans [39].

In diabetes mellitus, an abnormal lipid profile is a common manifestation. Literature suggests that hyperlipidemic conditions are among the most common consequences of alloxan induction in experimental rats [45]. In our study, elevation of total cholesterol levels was observed in diabetic rats, which was probably due to elevated free fatty acid levels having harmful effects in the body through free radical accumulation and stimulation of protein kinase C [46]. Hyperlipidemia causes decreased glucose transport to the cells, making lipids available in the form of LDL fat, which deposits in the blood vessels and is transported to the liver by HDL for elimination. Thus, the increased HDL and decreased LDL levels are expected to be beneficial for the therapeutic application of CILEx, which was achieved in this study.

Any abnormality of the architecture of the pancreatic tissue may alter the secretion, sensitivity, and function of insulin from islets of beta cells. Cellular atrophy and degeneration of pancreatic beta cells are marked as damage to the pancreas [47]. In the diabetic control group of this study, the number of beta cells was reduced and shrinkage was seen, while the normal control group retained regular cellular integrity. In the treatment groups, the beta cells were well recovered compared to in the normal rats.

The phytochemical composition of CILEx was characterized using the UPLC-QTOF/ESI-MS technique. This helps in the profiling and subsequent standardization of phytochemicals in the extract. UPLC-QTOF/ESI-MS analysis shows that the ethanol leaf extract contains a complex mixture of flavonoid, glycosides, saponins, and terpenoids of different classes. Among the 71 identified compounds, 6 were previously reported for their hypoglycemic and lipid-lowering effects. It has been reported that arbutin significantly inhibits α-amylase and α-glucosidase activity in vitro [28]. Arbutin-rich *Pyrus boissieriana* Buhse leaves also reduced glucose and lipid levels in blood, with an increased antioxidant state in alloxan-induced hyperglycemic rats [48]. Geetha et al. revealed that leucodelphinidin was found in the bark of *Ficus benghalensis*, and exerted hypoglycemic effects on either normal or diabetes-induced rats [49]. Saponins are known for their ability to lower plasma cholesterol levels and the risk of many chronic diseases in humans [50]. Ginsenoside Rh2—a glycosylated triterpene—when administered in fructose-rich chow-fed rats, causes plasma glucose to fall with enhanced insulin sensitivity and secretion [51]. Early literature also reveals that loganin, lycopene, and zeaxanthin have significant hypoglycemic effects in diabetic rats, and decrease fasting blood glucose levels in diabetes mellitus mice. The supplementation of lycopene significantly reduces diabetic plasma glucose levels [52].

Arbutin (ARB) has been associated with protecting HK-2 cells against high-glucose-induced apoptosis and autophagy in diabetic nephropathy (DN) through regulating the miR-27a/JNK/mTOR axis [53]. In CC analysis, we found that target proteins are associated with membranes, while in MF analysis, we found that many of the target proteins were linked with G-protein-coupled receptors. It has been stated that islet G-protein-coupled receptors are potential therapeutic targets for diabetes [54]. Thus, ARB targeted proteins may be associated with antidiabetic activities through membrane-receptor-mediated cellular signaling. Pathway analysis further reveals that ARB is associated with insulin secretion and pancreatic secretion. Altogether, our analyses indicate that CILEx is a potential antidiabetic agent [55]. These results indicate that CILEx exerts its antidiabetic and lipid-lowering activities through potentially regulating insulin secretion, pancreatic secretion, and lipolysis regulation in adipocytes.

4. Materials and Methods

4.1. Chemicals and Reagents

Analytical grade chemicals and reagents were used in this research, except where specified otherwise. Ethanol, alloxan, gallic acid, potassium acetate and quercetin were obtained from Sigma-Aldrich Chemicals, CA, USA. Glibenclamide (Chadwell Heath Essex, England), DPPH (1,1-diphenyl-2-picrylhydrazyl), ascorbic acid, Folin–Ciocalteu reagent, and sodium carbonate were purchased from Merck, India, while aluminum chloride was procured from Fine Chemicals, Delhi, India.

4.2. Collection and Identification of Plant Material

The fresh leaves of *Combretum indicum* (CILEx) were collected in April 2018 from Tangail District. The plant was identified by Professor Dr. Sheikh Bokhtear Uddin, a taxonomist at the Department of Botany, University of Chittagong, Bangladesh. A voucher specimen (accession No. 47044) has been deposited to the Bangladesh National Herbarium, Mirpur, Dhaka for future reference.

4.3. Preparation of Crude Extract

The fresh leaves of CILEx were washed with distilled water and shade-dried for 7 days at room temperature. The dried leaves were ground to powder (500 g) using a mechanical grinder (Miyako, Model No: DL-718 Jiaxing China) and stored in an airtight container. Then, 500 g of dried powder was soaked in 2500 mL of 96% ethanol at room temperature (25 $\pm$ 1 °C), with occasional stirring. After 14 days, the extract was filtered and concentrated in vacuum using a rotary evaporator (Barloworld, Berkshire, UK). The concentrated extract was then allowed to air dry for complete evaporation of the solvent. Finally, a blackish-green semisolid extract was preserved at 4 °C until further use.

4.4. Experimental Animals and Their Maintenance

Long–Evans rats (26 rats; age: 5–6 weeks; average body weight: 92 $\pm$ 9 g) of both sexes were obtained from the International Centre for Diarrheal Disease Research, Bangladesh (ICDDR, B), Mohakhali, Dhaka, Bangladesh. During the experimental period, the rats were kept in a well-ventilated animal house at room temperature and were supplied with a standard commercial rat pellet diet from ICDDR, B, and fresh drinking water. The animals were housed in plastic cages, and soft wood shavings were used as bedding. Animals were maintained under standard environmental conditions (temperature: 25 $\pm$ 1 °C; relative humidity: 55–65%; and a 12 h/12 h day/night cycle) in a properly ventilated room. Animals were handled and maintained according to the local animal ethical guidelines approved by the institutional animal ethics committee of Southeast University, Dhaka, Bangladesh (Approval No.: SEU/Pharm/CECR/102/2019).

4.5. Determination of Total Phenolic and Flavonoid Contents

A slightly modified Folin–Ciocalteu method was used to determine the total phenolic content (TPC) [56]. Briefly, a standard gallic acid (6.25–200 µg/mL) calibration curve was prepared, and leaf extract was prepared at a concentration of 200 µg/mL. Next, 1 mL of the extract solution or standard gallic acid solution was taken in a screw cap tube, and 5 mL of Folin–Ciocalteu reagent (previously prepared as 10% v/v dilution in distilled water) was added. Then, 4 mL of anhydrous sodium carbonate (7.5%) was added and incubated for 30 min at 40 °C. A typical blank solution contained the vehicle solvent. Absorbance was taken at 765 nm with a UV–Vis spectrophotometer (Shimadzu, Kyoto, Japan). The total phenolic content was calculated as gallic acid equivalent (GAE) by the following equation:

$$C = (c \times V)/m$$

where C = TPC (mg/g plant extract in GAE), c = the concentration of the sample obtained from the calibration curve (mg/mL), V = the volume of the sample, and m = the sample weight (g).

The total flavonoid content (TFC) of CILEx was determined according to the method of Rahman et al. [57].

4.6. DPPH Radical Scavenging Assay

The free radical scavenging effect of CILEx was evaluated with the stable scavenger DPPH described by Rahman et al. [58]. Briefly, 100 µL of CILEx and standard (ascorbic acid) solution in different concentrations was taken, and 3 mL of DPPH solution (0.004%) was mixed separately. These solutions were kept in the dark for 30 min to read absorbance at 517 nm using a UV–Vis spectrophotometer. Lower absorbance of the reaction mixture indicated higher free radical scavenging activity. Percentage inhibition was determined by the following formula:

Percentage of scavenging activity (%) = [(A − B)/A] × 100, where A is the absorbance of the control (DPPH solution without the sample), and B is the absorbance of the DPPH solution in the presence of the sample (extract/ascorbic acid). Then, % scavenging was plotted against concentration, and IC_{50} was calculated.

4.7. Acute Oral Toxicity Test

Acute oral toxicity testing of CILEx was performed on Long–Evans rats, according to OECD-423 guidelines (acute toxic class method), with slight modifications. The animals were overnight fasted, providing only water. Two groups of three rats each were used for this study. Group I received a single oral dose of CILEx (500 mg/kg body weight), and Group II received a single oral dose of CILEx (1000 mg/kg body weight). After the oral administration of CILEx, animals were observed individually at least once in the first 30 min, and periodically over the first 24 h, with special attention given during the first 4 h, for 10 consecutive days. All observations were systematically recorded for each animal. The animals were observed for gross behavioral, neurological, and autonomic effects. Additional conditions such as tremors, convulsions, salivation, diarrhea, lethargy, sleep, coma, and lethality were also observed. The effective therapeutic dose was calculated as one-tenth of the median lethal dose using the arithmetic method of Karber G in association with the Hodge–Sterner scale (LD50 > 2.0 g/kg) (58).

4.8. Induction of Diabetes and Experimental Design

Sixteen Long–Evans rats (average body weight 92 ± 9 g) of both sexes were used for the induction of diabetes. Diabetes was induced in overnight-fasted rats by a single intraperitoneal (IP) injection of alloxan monohydrate (150 mg/kg). Two days after alloxan injection, fasting blood glucose levels of all of the animals were recorded from tail vein blood using a portable glucometer (Accu-Chek, Japan), and rats with plasma glucose levels of >7.5 mmol/L were confirmed for the study. Treatment with CILEx was started after 48 h of alloxan injection. The plant sample CILEx, standard glibenclamide, and saline were administered with the help of feeding cannulas. Fasting blood glucose estimation was carried out on days 3, 5, and 7 of the study. The animals were grouped as follows:

Normal control (I): Normal rats received saline water only.
Diabetic control (II): Non-treated diabetic rats (alloxan treated; 150 mg/kg; IP).
Positive control (III): Alloxan-treated diabetic rats (150 mg/kg; IP) + glibenclamide (5 mg/kg; PO)
Treatment group (IV): Alloxan (150 mg/kg; IP) + CILEx (250 mg/kg; PO)
Treatment group (V): Alloxan (150 mg/kg; IP) + CILEx (500 mg/kg; PO)

4.9. Collection of Blood and Serum Analysis

After 7 days of treatment, animals were fasted for 12 h and their blood glucose levels were measured. The animals were then anaesthetized using diethyl ether and euthanized by decapitation. Blood was collected in a dry test tube from cardiac vessels using a disposable syringe via the heart puncture method [59], centrifuged (Hitachi, Japan) at 112 g for 15 min, and then the plasma samples were stored at 4 °C until biochemical estimations. Total cholesterol (TC), triglyceride (TG), high-density lipoprotein (HDL), and low-density lipoprotein (LDL) were measured using wet reagent diagnostic kits according to the manufacturer's protocol, using a biochemistry analyzer (BAS 100TS, Spectronics Corporation, LA, U.S.A) [60–63].

4.10. Histopathological Studies

After euthanizing the animals, pancreases of two animals from each group were excised and stored in 10% buffered formalin solution after washing with normal saline water. The pancreas was washed, dehydrated with alcohol, and cleared with xylene, and then paraffin blocks were made. Serial sections of 4–5 μm in thickness were cut using a microtome (semi-automated, Biobase BK-MT390S (BK-2488, Jinan, Shandong, China). Then, the sections were deparaffinized with xylene and hydrated in descending grades of alcohol. The slides were then transferred to hematoxylin for 10 min, followed by rinsing with water. These were examined and later stained with eosin, rinsed with water, dehydrated with ascending grades of alcohol, cleared with xylene, and mounted. Different parameters of pancreatic cellular condition were observed under a compound microscope, and the histopathological images were taken with the help of an Optica DP20 system (Italy).

4.11. UPLC-QTOF/MS Analysis

The phytochemical profiling of the CILEx was determined using UPLC-MS. UPLC-MS analysis was performed using Waters ACQUITY UPLC IClass/Xevo in line with a Waters Xevo G2 Q-TOF mass spectrometer (Milford, MA, USA). Extract samples were prepared by dissolving 100 mg of CILEx in 1 mL of methanol. Separation was conducted on a Zorbax Eclipse plus Acquity UPLC BEH C18 (1.7 μm particle size) 2.1 mm × 50 mm. The UPLC was interfaced with a Q-TOF mass spectrometer integrated with positive and negative electrospray ionization (ESI) sources. Full-scan mode from m/z 50 to 1000 was performed with a source temperature of 120 °C. Solvent A was water with 0.1% formic acid, while solvent B was acetonitrile with 0.1% formic acid. A gradient elution was performed, starting with 99% solvent A and 1% solvent B for the first 15 min, and then 65% solvent A and 35% solvent B for 1 min, followed by a gradual increase in solvent A to 100% over 2 min, and finally a slow increase in solvent B to 99% and solvent A 1% over 2 min. Highly purified nitrogen (N_2) and ultra-high-purity helium (He) were used as a nebulizing gas and collision gas, respectively. In terms of positive electrospray mode, the capillary voltage was set at 2.0 kV. Other instrument conditions implied were: source offset, 100 V; desolvation temperature at 550 °C; 50 L/h cone gas flow with temperature 120 °C; and desolvation gas flow, 800 L/h.

4.12. Computational Molecular Docking Analysis

4.12.1. Preparation of Ligands

LigPrep (ver. 2018, New York, NY, USA) was used in this regard to prepare the ligands, and yielded 3D structures with accurate chiralities [64]. It generated possible states at a target pH of 7.0 ± 2.0 using Epik v4.6.12 [65], and also desalted and produced tautomers. Computationally, specified chiralities were retained and generated at a rate of 32 per ligand at most. Then, output was saved as Maestro on the device. All of the ligands were imported in SDF format from PubChem.

4.12.2. Protein Preparation

Protein Preparation Wizard was used to modify the crystallographic structures of the proteins that were taken from the PDB (Protein Data Bank) [66]. The proteins 1XU9, 1XU7, 2BFL, 6R4F, 3A5J from the PDB were imported to Protein Preparation Wizard. Glide v8.1.12 was used in this regard, to optimize the structures from their raw state. The proteins were preprocessed by assigning bond orders using the CCD database, adding hydrogens, creating zero-order bonds to metals and disulfide bonds, deleting waters beyond 5.00 Å from the het groups, and generating het states using Epik at pH of 7.0 ± 2.0. The H-bond assignment was done by orienting water molecules, (Epik v4.6.12, Schrödinger, LLC, 2018-4, and PROPKA) at a specific pH of 7.0 by Schrödinger Release 2018-4 (SiteMap, Schrödinger, LLC, 2018-4), to determine the states of protonation, and to predict the pKa values of the residues [67,68]. Restrained minimization was done by converging heavy atoms to RMSD 0.30 Å.

4.12.3. SiteMap: Active Site Prediction

The proteins 1XU9, 1XU7, and 6R4F had multiple binding sites, while 3A5J did not have any binding sites for ligand–protein interaction. To look for the possible binding sites, we used SiteMap from Schrödinger, 2018-4 [69], so that the ligands could bind to the receptor tightly [70]. The tool produced the maps based on hydrophobic and hydrophilic (donor, acceptor, and metal-binding portions) maps. For selection of binding sites, SiteScore and druggability score (Dscore)—including site size, volume, exposure, enclosure, contact, hydrophobic and hydrophilic character, balance (phobic/philic ratio), and donor/acceptor of hydrogen bond—were used to evaluate each active site [71].

4.12.4. Receptor Grid Generation and Molecular Docking

The sites visualized by SiteMap were used as the entry, and Glide v8.1.12 (Schrödinger, LLC, 2018-4) was used to discover the suitable interaction between a ligand and a protein [72]. Van der Waals radii of receptor atoms with partial charge (absolute value) were scaled at a scaling factor of 1.0 and partial charge cutoff of 0.25 to soften the potential for the non-polar part. Site constraints, rotatable groups, and excluded volumes were set to default settings, as provided by Maestro 11.8.

In ligand–receptor interaction, Van der Waals radii were fixed at a scaling factor of 0.80, and partial charge cutoff was scaled at 0.15 for the non-polar parts of the ligands. SP (standard precision) was set for ligand screening and sampling. The energy window for ring sampling to generate conformers was 2.5 kcal/mol. Initial poses for docking were kept at 5000 poses per ligand, and the scoring window was 100–400 poses for energy optimization. Post-docking minimization was performed for 5 poses per ligand, the strain-correcting threshold was 4.00 Kcal/mol, and excess strain energy was scaled at 0.25. The parameters used were defaults, as provided by Maestro 11.8 [73–75].

4.12.5. Bioactive Compound–Target Protein Network Construction

On the basis of network pharmacology-based prediction, STITCH 5 (http://stitch. embl.de/, ver. 5.0, accessed on 3 August 2020) was used to identify target proteins related to the bioactive phytochemicals that were identified in CILEx [43]. It calculated a score for each pair of protein–chemical interactions. Chemical names of bioactive compounds (picrasinoside E, azedarachin C, zeaxanthin, quinatoside A, 3-O-Benzoyl-20-deoxyingenol, leucodelphinidin, schizonepetoside E, 1β,3β,6α-Trihydroxy-4α(15)-dihydrocostic acid methyl ester-1-O-β-D-glucopyranoside, melianol, and arbutin) were put into STITCH 5 individually to match their potential targets, with the organism selected as "Homo sapiens" and the medium required interaction score being $\geq$0.4.We predicted 205 target proteins with medium confidence score for arbutin, which was confirmed. The compound targets with no relationship with the compound–protein interactions were not considered for further analysis. The obtained compound–protein interaction data of the top 20 target proteins were imported into Cytoscape 3.6.1 software to construct a compound–protein interaction network.

4.12.6. Construction of Protein–Protein Interaction (PPI) Network of the Predicted Genes

We constructed a PPI network of the predicted genes by using the search tool for the retrieval of interacting genes (STRING) database (https://string-db.org/cgi/input.pl; STRING-DB v11.0, accessed on 3 August 2020) [76]. The rank of the target proteins based on degree of interactions in the PPI network was identified using the node explorer module of NetworkAnalyst software [77]. The obtained protein interaction data of the top 20 target proteins were imported into Cytoscape 3.6.1 software to construct a PPI network.

4.12.7. Gene Ontology (GO) and Kyoto Encyclopedia of Genes and Genomes (KEGG) Pathway Enrichment Analyses of the Target Proteins

To identify the role of target proteins that interact with the active ingredients of CILEx in gene function and signaling pathways, the Database for Annotation, Visualization, and Integrated Discovery (DAVID, https://david.ncifcrf.gov/, accessed on 3 August 2020) v6.8 was employed [78]. The KEGG [79] pathways significantly associated with the predicted genes were identified. We analyzed the Gene Ontology (GO) function and KEGG pathway enrichment of proteins (203 target proteins) involved in the PPI network. The target proteins involved in the cellular components (CCs), molecular functions (MFs), biological processes (BPs), and the KEGG pathways were also described. An adjusted p-value < 0.05, calculated by the Benjamini–Hochberg method, was considered to be significant [80].

4.13. Statistical Analysis

The data on fasting blood sugar and biochemical estimations were expressed as mean ± standard deviation (SD), and statistical comparisons were performed by one-way analysis of variance (ANOVA), followed by Tukey's post hoc test, using GraphPad Prism (version 6 for Windows, GraphPad Software, San Diego, CA, USA, www.graphpad.com (accessed on 3 August 2020)). p-values less than 0.05 were considered to be significant.

5. Conclusions

The current study results indicate that *C. indicum* leaves have potential benefits for the treatment of diabetes and its associated complications, by protecting the pancreases through the normalization of damaged beta islets and improvement of the lipid profile. However, more studies are recommended in order to understand the mechanism and to isolate the bioactive compounds.

Supplementary Materials: Figure S1: LC chromatogram and mass spectra of CILEx for high-intensity compounds identified and vertically arranged as (**A**) Picrasinoside E; (**B**) Azedarachin C; (**C**) Zeaxanthin; (**D**) Quinatoside A; (**E**) 3-O-Benzoyl-20- deoxyingenol; (**F**) Leucodelphinidin; (**G**) Schizonepetoside E; (**H**) 1β,3β,6α-Trihydroxy-4α(15)-dihydrocostic acid methyl ester-1-O-β-Dglucopyranoside; (**I**) Melianol; (**J**) Arbutin., Figure S2: list of chemical structures of the ligand molecules available in PubChem., Figure S3: Interactions based on **A**. site map of seven ligands **D**. Picrasonide E; **E**. Azedarachin C; **F**. 3-O-Benzoyl-20-deoxyingenol; **G**. Leucodelphinidin; **H**. Schizonepetoside E; **I**. Melianol; and **J**. Arbutin with the receptor **IXU9**. Schizonepetosie showed the highest docking score (−8.145) compared with the reference antidiabetic drugs B. Gliclazide and C. Metformin., Figure S4: Interactions based on **A**. site map of seven ligands **D**. Picrasonide_E; **E**. Azedarachin C; **F**. 3-O-Benzoyl-20-deoxyingenol; **G**. Leucodelphinidin; **H**. Schizonepetoside E; **I**. Melianol; and **J**. Arbutin with the receptor **IXU7**. Melianol showed the highest docking score (−8.475) compared with the reference antidiabetic drugs B. Gliclazide and C. Metformin. The 3D protein binding has been presented with respective amino acid grooves for ligand interaction., Figure S5: Interactions based on **A**. site map of seven ligands B. Gliclazide; C. Metformin; **D**. Picrasonide E; **E**. Azedarachin C; **F**. 3-O-Benzoyl-20-deoxyingenol; **G**. Leucodelphinidin; **H**. Schizonepetoside E; **I**. Melianol; and **J**. Arbutin with the receptor **2BEL**. Melianol showed the highest docking score (−8.995) compared with the reference antidiabetic drugs B. Gliclazide and C. Metformin. The 3D protein binding has been presented with respective amino acid grooves for ligand interaction, Figure S6: Interactions based on **A**. site map of three ligands C. Leucodelphinidin; D. Schizonepetoside E; and **E**. Arbutin with the receptor **GR4F**. Arbutin showed the highest docking score (−7.492) compared with the reference antidiabetic drug B. Metformin. The 3D protein binding has been presented with respective amino acid grooves for ligand interaction., Figure S7: Interactions based on **A**. site map of seven ligands B. Gliclazide; C. Metformin; **D**. Picrasonide E; **E**. Azedarachin C; **F**. 3-O-Benzoyl-20-deoxyingenol; **G**. Leucodelphinidin; **H**. Schizonepetoside E; **I**. Melianol; and **J**. Arbutin with the receptor **3A5J**. 3-O-Benzoyl-20-deoxyingenol showed the highest docking score (−6.123) compared with the reference antidiabetic drugs B. Gliclazide and C. Metformin. The 3D protein binding has been presented with respective amino acid grooves for ligand interaction.

Author Contributions: M.S.F.: Experimental, data acquisition, writing–original draft preparation; M.A.R.: conceptualization, planning, manuscript editing, finalization; M.F.F.M.A.: resources, proofreading; M.N.U.: computational study, data acquisition; T.G.R.: UPLC/MS data analysis; M.C.M.: conceptualization; A.M.H.: planning, project administration, supervision, funding acquisition, manuscript editing; Z.A.Z.: funding acquisition. All authors have read and agreed to the published version of the manuscript.

Funding: This work was supported by the Ministry of Education, Malaysia, and awarded the Fundamental Research Grant Scheme (FRGS, Reference Number: 04-01-18-1984FR).

Institutional Review Board Statement: The institutional animal ethics committee of Southeast University, Dhaka, Bangladesh approved the protocol for conducting this study, with Approval No. SEU/Pharm/CECR/102/2019.

Informed Consent Statement: Not applicable.

Data Availability Statement: All data are available upon request to the authors.

Acknowledgments: The authors are thankful to the Department of Pharmacy, Southeast University for providing all of the facilities, and for partial financial support. We also extend our kind appreciation to the EXIM Bank Hospital, Dhaka for their cooperation in histopathological assay.

Conflicts of Interest: The authors have no conflict of interest.

Sample Availability: Samples of the compounds are not available from the authors.

Abbreviations

HDL	High-density lipoprotein
LDL	Low-density lipoprotein
TC	Total cholesterol
TFC	Total flavonoid content
TG	Triglyceride
TPC	Total phenolic content
UPLC-QTOF	Ultra-performance liquid chromatography coupled with time-of-flight mass spectrometry

References

1. Cho, N.; Shaw, J.; Karuranga, S.; Huang, Y.; Fernandes, J.D.R.; Ohlrogge, A.; Malanda, B. IDF Diabetes Atlas: Global Estimates of Diabetes Prevalence for 2017 and Projections for 2045. *Diabetes Res. Clin. Pract.* **2018**, *138*, 271–281. [CrossRef]
2. UK Prospective Diabetes Study (UKPDS). VIII Study Design, Progress and Performance. *Diabetologia* **1991**, *34*, 877–890.
3. Pooya, S.; Jalali, M.D.; Jazayery, A.D.; Saedisomeolia, A.; Eshraghian, M.R.; Toorang, F. The Efficacy of Omega-3 Fatty Acid Supplementation on Plasma Homocysteine and Malondialdehyde Levels of Type 2 Diabetic Patients. *Nutr. Metab. Cardiovasc. Dis.* **2010**, *20*, 326–331. [CrossRef] [PubMed]
4. Mawa, J.; Rahman, A.; Hashem, M.; Hosen, J. Leea Macrophylla Root Extract Upregulates the mRNA Expression for Antioxidative Enzymes and Repairs the Necrosis of Pancreatic β-cell and Kidney Tissues in Fructose-fed Type 2 Diabetic Rats. *Biomed. Pharmacother.* **2019**, *110*, 74–84. [CrossRef]
5. Panneerselvam, A. Drug Management of Type 2 Diabetes Mellitus-clinical Experience at a Diabetes Center in South India. *Int. J. Diab. Dev. Ctries* **2004**, *24*, 40–46.
6. Khare, C.P. *Indian Medicinal Plants: An Illustrated Dictionary*; Springer: New York, NY, USA, 2007; p. 533.
7. Kirtikar, K.R.; Basu, B.D. *Indian Medicinal Plant*, 2nd ed.; Prashant Gahlot at Valley Offset Publishers: New Delhi, India, 2006; p. 1037.
8. Islam, M.Z.; Sarker, M.; Hossen, F.; Mukharjee, S.K.; Akter, M.S.; Hossain, M.T. Phytochemical and Biological Studies of the *Quisqualis indica* Leaves Extracts. *J. Noakhali. Sci. Technol. Univ.* **2017**, *1*, 9–17.
9. DeFilipps, R.A.; Krupnick, G.A. The Medicinal Plants of Myanmar. *PhytoKeys* **2018**, *102*, 1–341. [CrossRef]
10. Gurib Fakim, A. Combretum Indicum (L.) De Filipps. In *Prota 11(2): Medicinal Plants/Plantes Médicinales 2.* PROTA.; Schmelzer, G.H., Gurib-Fakim, A., Eds.; Wageningen: Pays Bas, The Netherlands, 2012.
11. Jahan, F.N.; Rahman, M.S.; Hossain, M.; Rashid, M.A. Antimicrobial Activity and Toxicity of *Quisqualis indica*. *Orient. Pharm. Exp. Med.* **2008**, *8*, 53–58. [CrossRef]
12. Lin, T.-C.; Ma, Y.-T.; Wu, J.; Hsu, F.-L. Tannins and Related Compounds from *Quisqualis Indica*. *J. Chin. Chem. Soc.* **1997**, *44*, 151–155. [CrossRef]
13. Ferris, H.; Zheng, L. Plant Sources of Chinese Herbal Remedies: Effects on *Pratylenchus vulnus* and *Meloidogyne javanica*. *J. Nematol.* **1999**, *31*, 241–263. [PubMed]
14. Yadav, Y.; Mohanty, P.K.; Kasture, S.B. Evaluation of Immunomodulatory Activity of Hydroalcoholic Extract of *Quisqualis indica* Linn. Flower in Wistar Rats. *Int. J. Life Sci. Biotechnol. Pharma Res.* **2011**, *2*, 689–696.
15. Das, A.; Samal, K.C.; Das, A.B.; Rout, G.R. Quantification, Antibacterial Assay and Cytotoxic Effect of Combretastatin, an Anticancer Compound from Three Indian Combretum species. *Int. J. Curr. Microbiol. Appl. Sci.* **2018**, *7*, 687–699. [CrossRef]
16. Kambar, Y.; Asha, M.M.; Chaithra, M.; Prashith, K.T. Antibacterial Activity of Leaf and Flower Extract of *Quisqualis indica* Linn. against Clinical Isolates of *Staphylococcus aureus*. *J. Sci. Technol.* **2014**, *6*, 23–24.
17. Afify, A.E.-M.M.R.; Hassan, H.M.M. Free Radical Scavenging Activity of Three Different Flowers-*Hibiscus rosa-sinensis*, *Quisqualis indica* and *Senna surattensis*. *Asian Pac. J. Trop. Biomed.* **2016**, *6*, 771–777. [CrossRef]
18. Shinozaki, H.; Izumi, S. A New Potent Excitant, Quisqualic Acid: Effects on Crayfish Neuromuscular Junction. *Neuropharmacology* **1974**, *13*, 665–672. [CrossRef]
19. Efferth, T.; Kahl, S.; Paulus, K.; Adams, M.; Rauh, R.; Boechzelt, H.; Hao, X.; Kaina, B.; Bauer, R. Phytochemistry and Pharmacogenomics of Natural Products Derived from Traditional Chinese Medicine and Chinese Materia Medica with Activity Against Tumor Cells. *Mol. Cancer Ther.* **2008**, *7*, 152–161. [CrossRef] [PubMed]

20. Mollik, A.H.; Hassan, A.I.; Paul, T.K.; Sintaha, M.; Khaleque, H.N.; Noor, F.A.; Nahar, A.; Seraj, S.; Jahan, R. A Survey of Medicinal Plant Usage by Folk Medicinal Practitioners in Two Villages by the Rupsha River in Bagerhat District, Bangladesh. *Am. Eurasian J. Sustain. Agric.* **2010**, *4*, 349–356.

21. Surasak, L.; Supayang, P.V. Anti-*Streptococcus pyogenes* Activity of Selected Medicinal Plant Extracts Used in Thai Traditional Medicine. *Trop. J. Pharm. Res.* **2013**, *12*, 535–540.

22. Sahu, J.; Patel, P.K.; Dubey, B. *Quisqualis indica* Linn: A Review of Its Medicinal Properties. *Int. J. Pharm. Phytopharmacol. Res.* **2012**, *1*, 313–321.

23. Singh, N.; Mohan, G.; Sharma, R.K.; Gnaneshwari, D. Evaluation of Antidiarrhoeal Activity of *Quisqualis indica* leaves. *Ind. J. Nat. Prod. Resourc.* **2013**, *4*, 155–160.

24. Bairagi, V.A.; Sadu, N.; Senthilkumar, K.L.; Ahire, Y. Anti-diabetic Potential of *Quisqualis indica* Linn. in Rats. *Int. J. Pharm. Phytopharm. Res.* **2012**, *1*, 166–171.

25. Yousefi, F.; Mahjoub, S.; Pouramir, M.; Khadir, F. Hypoglycemic Activity of *Pyrus biossieriana* Buhse Leaf Extract and Arbutin: Inhibitory Effects on Alpha Amylase and Alpha Glucosidase. *Casp. J. Intern. Med.* **2013**, *4*, 763–767.

26. Shahaboddin, M.E.; Pouramir, M.; Moghadamnia, A.A.; Parsian, H.; Lakzaei, M.; Mir, H. *Pyrus biossieriana* Buhse Leaf Extract: An Antioxidant, Antihyperglycaemic and Antihyperlipidemic Agent. *Food Chem.* **2011**, *126*, 1730–1733. [CrossRef]

27. Geetha, B.S.; Mathew, B.C.; Augusti, K.T. Hypoglycemic Effects of Leucodelphinidin Derivative Isolated from *Ficus bengalensis* (Linn). *Indian J. Physiol. Pharmacol.* **1994**, *38*, 220–222.

28. Søndergaard, C.R.; Olsson, M.H.; Rostkowski, M.; Jensen, J.H. Improved Treatment of Ligands and Coupling Effects in Empirical Calculation and Rationalization of pK_a values. *J. Chem. Theory Comput.* **2011**, *7*, 2284–2295. [CrossRef] [PubMed]

29. Singh, B.; Singh, J.P.; Singh, N.; Kaur, A. Saponins in Pulses and Their Health Promoting Activities: A review. *Food Chem.* **2017**, *233*, 540–549. [CrossRef] [PubMed]

30. Li, H.; Cao, W.; Wei, L.-F.; Xia, J.-Q.; Gu, Y.; Gu, L.-M.; Pan, C.-Y.; Liu, Y.-Q.; Tian, Y.-Z.; Lu, M. Arbutin Alleviates Diabetic Symptoms by Attenuating Oxidative Stress in a Mouse Model of Type 1 Diabetes. *Int. J. Diabetes Dev. Ctries.* **2021**, 1–7. [CrossRef]

31. Lee, W.-K.; Kao, S.-T.; Liu, I.-M.; Cheng, J.-T. Increase of Insulin Secretion by Ginsenoside Rh2 to Lower Plasma Glucose in Wistar Rats. *Clin. Exp. Pharmacol. Physiol.* **2006**, *33*, 27–32. [CrossRef]

32. Zhu, H.; Zhu, X.; Liu, Y.; Jiang, F.; Chen, M.; Cheng, L.; Cheng, X. Gene Expression Profiling of Type 2 Diabetes Mellitus by Bioinformatics Analysis. *Comput. Math. Methods Med.* **2020**, *2020*, 1–10. [CrossRef]

33. Lee, W.-K.; Kao, S.-T.; Liu, I.-M.; Cheng, J.-T. Ginsenoside Rh2 is One of the Active Principles of Panax Ginseng Root to Improve Insulin Sensitivity in Fructose-rich Chow-fed Rats. *Horm. Metab. Res.* **2007**, *39*, 347–354. [CrossRef] [PubMed]

34. Tsalamandris, S.; Antonopoulos, A.; Oikonomou, E.; Papamikroulis, G.-A.; Vogiatzi, G.; Papaioannou, S.; Deftereos, S.; Tousoulis, D. The Role of Inflammation in Diabetes: Current Concepts and Future Perspectives. *Eur. Cardiol. Rev.* **2019**, *14*, 50–59. [CrossRef]

35. Fu, Q.; Shi, Q.; West, T.M.; Xiang, Y.K. Cross-Talk Between Insulin Signaling and G Protein–Coupled Receptors. *J. Cardiovasc. Pharmacol.* **2017**, *70*, 74–86. [CrossRef]

36. Berbudi, A.; Rahmadika, N.; Tjahjadi, A.; Ruslami, R. Type 2 Diabetes and its Impact on the Immune System. *Curr. Diabetes Rev.* **2020**, *16*, 442–449. [CrossRef]

37. Khanal, P.; Patil, B.; Mandar, B.K.; Dey, Y.N.; Duyu, T. Network Pharmacology-based Assessment to Elucidate the Molecular Mechanism of Anti-diabetic Action of *Tinospora cordifolia*. *Clin. Phytoscience* **2019**, *5*, 1–9. [CrossRef]

38. Lu, J.; Liu, J.; Li, L.; Lan, Y.; Liang, Y. Cytokines in Type 1 Diabetes: Mechanisms of Action and Immunotherapeutic Targets. *Clin. Transl. Immunol.* **2020**, *9*, e1122. [CrossRef] [PubMed]

39. Marks, I.N.; Shuman, C.R.; Shay, H. Gastric Acid Secretion in Diabetes Mellitus. *Ann. Intern. Med.* **1959**, *51*, 227–237. [CrossRef]

40. Morigny, P.; Houssier, M.; Mouisel, E.; Langin, D. Adipocyte Lipolysis and Insulin Resistance. *Biochimie* **2016**, *125*, 259–266. [CrossRef]

41. He, K.; Song, S.; Zou, Z.; Feng, M.; Wang, D.; Wang, Y.; Li, X.; Ye, X. The Hypoglycemic and Synergistic Effect of Loganin, Morroniside, and Ursolic Acid Isolated from the Fruits of Cornus officinalis. *Phytother. Res.* **2015**, *30*, 283–291. [CrossRef]

42. Kou, L.; Du, M.; Zhang, C.; Dai, Z.; Li, X.; Zhang, B. The hypoglycemic, Hypolipidemic, and Anti-diabetic Nephritic Activities of Zeaxanthin in di-et-Streptozotocin-Induced Diabetic Sprague Dawley Rats. *Appl. Biochem. Biotech.* **2017**, *182*, 944–955. [CrossRef]

43. Coulidiati, T.H.; Millogo-Kone, H.; Lamien-Meda, A.; Yougbare-Ziebrou, M.; Millogo-Rasolodimby, J.; Nacoulma, O.G. Antioxidant and Antibacterial Activities of Two *Combretum* Species from Burkina Faso. *Rese. J. Med. Plants* **2011**, *5*, 42–53. [CrossRef]

44. Shelley, J.C.; Cholleti, A.; Frye, L.L.; Greenwood, J.R.; Timlin, M.R.; Uchimaya, M. Epik: A Software Program for pK a Prediction and Protonation State Generation for Drug-like Molecules. *J. Comput. Mol. Des.* **2007**, *21*, 681–691. [CrossRef]

45. Sastry, G.M.; Adzhigirey, M.; Day, T.; Annabhimoju, R.; Sherman, W. Protein and Ligand Preparation: Parameters, Protocols, and Influence on Virtual Screening Enrichments. *J. Comput. Mol. Des.* **2013**, *27*, 221–234. [CrossRef]

46. *Release, S. 2: Schrödinger Release 2018-4: Prime, Schrödinger*, LLC: New York, NY, USA, 2018.

47. *Protein Preparation Wizard 2018-4, Epik version 4. 6.12, Impact version 8.1.12, Prime 5.4.12, Schrödinger*, LLC: New York, NY, USA, 2018.

48. *Schrödinger Release 2018-4: SiteMap, Schrödinger*, LLC: New York, NY, USA, 2018.

49. Nayal, M.; Honig, B. On the Nature of Cavities on Protein Surfaces: Application to the Identification of Drug-binding Sites. *Proteins Struct. Funct. Bioinform.* **2006**, *63*, 892–906. [CrossRef]

50. Halgren, T. New Method for Fast and Accurate Binding-site Identification and Analysis. *Chem. Biol. Drug Des.* **2007**, *69*, 146–148. [CrossRef] [PubMed]

51. Halgren, T.A. Identifying and Characterizing Binding Sites and Assessing Drug Ability. *J. Chem. Inf. Model* **2009**, *49*, 377–389. [CrossRef]

52. *Glide, Version 8.1.12, Schrödinger*, LLC: New York, NY, USA, 2018.

53. Friesner, R.A.; Banks, J.L.; Murphy, R.B.; Halgren, T.A.; Klicic, J.J.; Mainz, D.T.; Repasky, M.P.; Knoll, E.H.; Shelley, M.; Perry, J.K.; et al. Glide: A New Approach for Rapid, Accurate Docking and Scoring. 1. Method and Assessment of Docking Accuracy. *J. Med. Chem.* **2004**, *47*, 1739–1749. [CrossRef]

54. Halgren, T.A.; Murphy, R.B.; Friesner, R.A.; Beard, H.S.; Frye, L.L.; Pollard, W.T.; Banks, J.L. Glide: A New Approach for Rapid, Accurate Docking and Scoring. 2. Enrichment Factors in Database Screening. *J. Med. Chem.* **2004**, *47*, 1750–1759. [CrossRef] [PubMed]

55. Friesner, R.A.; Murphy, R.B.; Repasky, M.P.; Frye, L.L.; Greenwood, J.R.; Halgren, T.A.; Sanschagrin, P.C.; Mainz, D.T. Extra precision glide: Docking and Scoring Incorporating a Model of Hydrophobic Enclosure for Protein−ligand Complexes. *J. Med. Chem.* **2006**, *49*, 6177–6196. [CrossRef] [PubMed]

56. Sagbo, I.J.; Afolayan, A.J.; Bradley, G. Antioxidant, Antibacterial and Phytochemical Properties of Two Medicinal Plants Against the Wound Infecting Bacteria. *Asian Pac. J. Trop. Biomed.* **2017**, *7*, 817–825. [CrossRef]

57. Rahman, M.A.; Chowdhury, J.K.H.; Aklima, J.; Azadi, M.A. *Leea macrophylla* Roxb. Leaf extract Potentially Helps Normalize islet of β-cells Damaged in STZ-induced Albino Rats. *Food Sci. Nutr.* **2018**, *6*, 943–952. [CrossRef] [PubMed]

58. Zaoui, A.; Cherrah, Y.; Mahassini, N.; Alaoui, K.; Amarouch, H.; Hassar, M. Acute and Chronic Toxity of *Nigella sativa* fixed oil. *Phytomedicine* **2002**, *9*, 69–74. [CrossRef]

59. Karber, G. Beitrag zur kollecktiven Behandlung Pharmakologischer Reihenversuche. *Arc. Exp. Pathol. Pharmakol.* **1931**, *162*, 480–483. [CrossRef]

60. Hodge, A.; Sterner, B. *Toxicity Classes*; Canadian Centre for Occupational Health Safety: Hamilton, ON, Canada, 2005. Available online: http://www.ccohs.ca/oshanswers/chemicals/id50.htm (accessed on 3 August 2020).

61. OECD/OCDE. *Guideline for the Testing of Chemicals. Revised Draft Guideline 423: Acute Oral Toxicity*; OECD: Paris, France, 2000.

62. Parthasarathy, R.; Ilavarasan, R.; Karrunakaran, C.M. Antidiabetic Activity of *Thespesia populnea* bark and Leaf Extract Against Streptozotocin induced Diabetic Rats. *Int. J. Pharmtech. Res.* **2009**, *1*, 1069–1072.

63. Mueller, P.H.; Schmuelling, R.M.; Liebich, H.M.; Eggstein, M. A Fully Enzymatic Triglyceride Determination. *J. Clin. Chem. Clin. Biochem.* **1977**, *15*, 457.

64. Allain, C.C.; Poon, L.S.; Chan, C.S.; Richmond, W.; Fu, P.C. Enzymatic Determination of Total Serum cholesterol. *Clin. Chem.* **1974**, *20*, 470–475. [CrossRef] [PubMed]

65. Hoff, J. Methods of Blood Collection in Mouse. *Lab Anim. Technique.* **2000**, *29*, 47–53.

66. Roeschlau, P.; Bernt, E.; Gruber, W. Enzymatic Determination of Total Cholesterol in Serum. *Z Klin Chem. Klin Biochem.* **1974**, *12*, 226. [PubMed]

67. Friedewald, W.T.; Levy, R.; Fredrickson, D.S. Estimation of the Concentration of Low-Density Lipoprotein Cholesterol in Plasma, Without Use of the Preparative Ultracentrifuge. *Clin. Chem.* **1972**, *18*, 499–502. [CrossRef]

68. Guan, J.; Lai, C.; Li, S. A Rapid Method for the Simultaneous Determination of 11 saponins in *Panax notoginseng* Using Ultra Performance Liquid Chromatography. *J. Pharm. Biomed. Anal.* **2007**, *44*, 996–1000. [CrossRef]

69. Rohilla, A.; Ali, S. Alloxan Induced Diabetes: Mechanisms and Effects. *Int. J. Res. Pharm. Biomed. Sci.* **2012**, *3*, 819–823.

70. Tripathi, V.; Verma, J. Different Models Used to Induce Diabetes: A Comprehensive Review. *Int. J. Pharm. Pharm. Sci.* **2014**, *6*, 29–32.

71. Johansen, J.S.; Harris, A.K.; Rychly, D.J.; Ergul, A. Oxidative Stress and the use of Antioxidants in Diabetes: Linking Basic Science to Clinical Practice. *Cardiovasc. Diabetol.* **2005**, *4*, 5. [CrossRef]

72. Jebur, A.B.; Mokhamer, M.H.; El-Demerdash, F.M. A Review on Oxidative Stress and Role of Antioxidants in Diabetes Mellitus. *Austin Endocrinol. Diabetes Case Rep.* **2016**, *1*, 1006.

73. Ogbonnia, S.; Mbaka, G.; Igbokwe, N.; Anyika, E.; Alli, P.; Nwakakwa, N. Antimicrobial evaluation, acute and subchronic toxicity studies of Leone Bitters, a Nigerian polyherbal formulation, in rodents. *Agric. Biol. J. N. Am.* **2010**, 366–376. [CrossRef]

74. Rahimi, R.; Nikfar, S.; Larijani, B.; Abdollahi, M. A Review on the role of Antioxidants in the Management of Diabetes and Its Complications. *Biomed. Pharmacother.* **2005**, *59*, 365–373. [CrossRef] [PubMed]

75. Aberoumand, A.; Deokule, S.S. Comparison of Phenolic Compounds of Some Edible Plants of Iran and India. *Pak. J. Nutr.* **2008**, *7*, 582–585. [CrossRef]

76. Masoko, P.; Eloff, J.N. Screening of Twenty-Four South African Combretum and Six *Terminalia* species (Combretaceae) for Anti-oxidant Activities. *Afr. J. Tradit. Complement Altern. Med.* **2007**, *4*, 231–239.

77. Coulidiati, T.H.; Millogo-Koné, H.; Lamien-Méda, A.; Lamien, C.E.; Lompo, M.; Kiendrébéogo, M.; Bakasso, S.; Yougbaré-Ziébrou, M.; Millogo-Rasolodimby, J.; Nacoulma, O.G. Antioxidant and Antibacterial Activities of *Combretum nioroense* Au-brev. Ex keay (Combretaceae). *Pak. J. Biol. Sci.* **2009**, *12*, 264. [CrossRef] [PubMed]

78. Krishnakumar, K.; Augusti, K.T.; Vijayammal, P.L. Anti-peroxidative and Hypoglycaemic Activity of *Salacia oblonga* Extract in Diabetic Rats. *Pharm. Biol.* **2000**, *38*, 101–105. [CrossRef]

79. Inoguchi, T.; Li, P.; Umeda, F.; Yu, H.Y.; Kakimoto, M.; Imamura, M.; Aoki, T.; Etoh, T.; Hashimoto, T.; Naruse, M.; et al. High Glucose Level and Free Fatty Acid Stimulate Reactive Oxygen Species Production Through Protein Kinase C-dependent Activation of NAD(P)H Oxidase in Cultured Vascular Cells. *Diabetes* **2000**, *49*, 1939–1945. [CrossRef]
80. Zhou, J.Y.; Zhou, S.W.; Zhang, K.B.; Tang, J.L.; Guang, L.X.; Ying, Y.; Xu, Y.; Le, Z.; Li, D.D. Chronic Effects of Berberine on Blood, Liver Glucolipid Metabolism and Liver PPARs Ex-pression in Diabetic Hyperlipidemic Rats. *Biol. Pharm. Bull.* **2008**, *31*, 1169–1176. [CrossRef] [PubMed]

 *molecules*

Article

Bioactive Abietane-Type Diterpenoid Glycosides from Leaves of *Clerodendrum infortunatum* (Lamiaceae)

Md. Josim Uddin [1,2], Daniela Russo [3,4], Md. Anwarul Haque [5,6], Serhat Sezai Çiçek [1], Frank D. Sönnichsen [7], Luigi Milella [3] and Christian Zidorn [1,*]

1   Pharmazeutisches Institut, Abteilung Pharmazeutische Biologie, Christian-Albrechts-Universität zu Kiel, Gutenbergstrasse 76, 24118 Kiel, Germany; juddin@pharmazie.uni-kiel.de (M.J.U.); scicek@pharmazie.uni-kiel.de (S.S.Ç.)
2   Department of Pharmacy, International Islamic University Chittagong, Chittagong 4318, Bangladesh
3   Department of Science, University of Basilicata, Viale dell' Ateneo Lucano 10, 85100 Potenza, Italy; daiela.russo@unibas.it (D.R.); luigi.milella@unibas.it (L.M.)
4   Spinoff BioActiPlant s.r.l., Viale dell' Ateneo Lucano 10, 85100 Potenza, Italy
5   Department of Experimental Pathology, Graduate School of Comprehensive Human Sciences, University of Tsukuba, Ibaraki 305-8575, Japan; a.haque5314@gmail.com
6   Department of Pharmacy, University of Rajshahi, Rajshahi 6205, Bangladesh
7   Otto Diels Institute for Organic Chemistry, University of Kiel, Otto-Hahn-Platz 4, 24118 Kiel, Germany; fsoennichsen@oc.uni-kiel.de
*   Correspondence: czidorn@pharmazie.uni-kiel.de; Tel.: +49-431-880-1139

Citation: Uddin, M..J.; Russo, D.; Haque, M..A.; Çiçek, S.S.; Sönnichsen, F.D.; Milella, L.; Zidorn, C. Bioactive Abietane-Type Diterpenoid Glycosides from Leaves of *Clerodendrum infortunatum* (Lamiaceae). *Molecules* 2021, 26, 4121. https://doi.org/10.3390/molecules26144121

Academic Editors: Raffaele Pezzani and Sara Vitalini

Received: 31 May 2021
Accepted: 29 June 2021
Published: 6 July 2021

Publisher's Note: MDPI stays neutral with regard to jurisdictional claims in published maps and institutional affiliations.

Abstract: In this study, two previously undescribed diterpenoids, (5*R*,10*S*,16*R*)-11,16,19-trihydroxy-12-*O*-β-D-glucopyranosyl-(1→2)-β-D-glucopyranosyl-17(15→16),18(4→3)-*diabeo*-3,8,11,13-abietatetraene-7-one (**1**) and (5*R*,10*S*,16*R*)-11,16-dihydroxy-12-*O*-β-D-glucopyranosyl-(1→2)-β-D-glucopyranosyl-17(15→16),18(4→3)-*diabeo*-4-carboxy-3,8,11,13-abietatetraene-7-one (**2**), and one known compound, the C_{13}-nor-isoprenoid glycoside byzantionoside B (**3**), were isolated from the leaves of *Clerodendrum infortunatum* L. (Lamiaceae). Structures were established based on spectroscopic and spectrometric data and by comparison with literature data. The three terpenoids, along with five phenylpropanoids: 6'-*O*-caffeoyl-12-glucopyranosyloxyjasmonic acid (**4**), jionoside C (**5**), jionoside D (**6**), brachynoside (**7**), and incanoside C (**8**), previously isolated from the same source, were tested for their in vitro antidiabetic (α-amylase and α-glucosidase), anticancer (Hs578T and MDA-MB-231), and anticholinesterase activities. In an in vitro test against carbohydrate digestion enzymes, compound **6** showed the most potent effect against mammalian α-amylase (IC$_{50}$ 3.4 ± 0.2 µM) compared to the reference standard acarbose (IC$_{50}$ 5.9 ± 0.1 µM). As yeast α-glucosidase inhibitors, compounds **1**, **2**, **5**, and **6** displayed moderate inhibitory activities, ranging from 24.6 to 96.0 µM, compared to acarbose (IC$_{50}$ 665 ± 42 µM). All of the tested compounds demonstrated negligible anticholinesterase effects. In an anticancer test, compounds **3** and **5** exhibited moderate antiproliferative properties with IC$_{50}$ of 94.7 ± 1.3 and 85.3 ± 2.4 µM, respectively, against Hs578T cell, while the rest of the compounds did not show significant activity (IC$_{50}$ > 100 µM).

Keywords: *Clerodendrum infortunatum*; terpenoids; phenylpropanoids; antidiabetic; breast cancer

1. Introduction

Clerodendrum (Lamiaceae) is a diverse genus with about 580 species [1] of small trees, shrubs, or herbs, mostly distributed throughout tropical and subtropical regions of the world [2]. *Clerodendrum infortunatum* L. (Syn.: *Clerodendrum viscosum* Vent), locally known as Bhat, is a terrestrial shrub with a noxious odor, distributed throughout mixed deciduous and evergreen to semi-evergreen forests of Bangladesh and the Indian state of West Bengal [3]. Due to its easy availability and presumed beneficial activities, various parts of the plant, particularly the leaves and roots, are extensively used in Indian and Bangladeshi traditional medicine for some common ailments. In folk medicine, the leaves and roots are used to cure helminthiasis, tumors, skin diseases, snakebites, and scorpion stings. Infusions of the leaves are also used as a bitter tonic and antiperiodic for the treatment of malaria. The freshly extracted juice of the leaves is considered to be a good laxative and cholagogue [4,5]. Some experimental evidence has proven the traditional claims, showing various biological effects, such as anti-snake venom activity [6], analgesic and anticonvulsant activities [7], nootropic activity [8], antimicrobial activity [9], antioxidative potential [10], and hepatoprotective activity [11]. Earlier, phytochemical studies of *C. infortunatum* leaves revealed the presence of flavonoids, phenolic compounds, terpenoids, steroids, and phenylpropanoids [12].

Due to the increasing life expectancy worldwide, the prevalence of age-associated diseases (including cardiovascular disease, cancer, type 2 diabetes, neurodegenerative disorders, osteoporosis, pancreatitis, and hypertension) is rising exponentially with age. Consequently, the treatment and prevention of these conditions is turning into a priority in medicine, due to the rapid increase of elderly populations, particularly in Western countries [13,14]. Natural products remain a rich source of anticancer, antidiabetic, and anti-neurodegenerative disorder agents; more than half of all drugs used for the treatment of cancer are either natural products or derived from natural products [15].

Terpenoids, abundant in medicinal plants, structurally constitute complex and diverse groups of natural products. Naturally occurring terpenoids have been shown to have significant preventive properties against age-related diseases such as tumors, diabetes, inflammation, cardiovascular diseases, and neurodegenerative disorders [16].

Phenylpropanoids are a group of natural products with widespread distribution in plants, and are also considered to be potential agents against age-related disorders due to possessing anticancer, antidiabetic, neuroprotective, cardioprotective, antimicrobial, antioxidative, and enzyme inhibitory activities at comparatively low concentrations [17,18].

In view of the important role of terpenoids and phenylpropanoids in treating age-related disorders, eight bioactive compounds, including five previously isolated phenyl-propanoids, were investigated for their antiproliferative and antimetastatic effects by two human triple negative breast cancer (TNBC) cell lines (Hs578T and MDA-MB-231). Additionally, their anti-diabetic properties were assessed through α-amylase and α-glucosidase enzyme inhibition, and their cholinesterase inhibitory properties were also assessed.

2. Results and Discussion

2.1. Phytochemical Investigation

In the present study, we analyzed the specialized natural products from *C. infortunatum* leaves, resulting in the isolation and structural characterization of three terpenoids, including two previously undescribed diterpenoids. A butanol fraction of acetone extract of *C. infortunatum* leaves was subjected to open column chromatography using a silica gel and subsequent semi-preparative HPLC with reversed phase column, and the abietanes (**1** and **2**) were obtained as amorphous solids, together with the previously reported C_{13} nor-isoprenoid glycoside (**3**) (Figure 1). In our previous study, five phenylpropanoid glyco-sides: 6′-*O*-caffeoyl-12-glucopyranosyloxyjasmonic acid (**4**), jionoside C (**5**), jionoside D (**6**), brachynoside (**7**), and incanoside C (**8**) were reported from the same source (Figure 1) [12].

Structures (**1–3**) were identified based on the ^{1}H, ^{13}C NMR, and high-resolution mass spectrometry data (Figures S1–S13). Based on their spectra, the isolates were found to be novel abietane glycosides (**1** and **2**) with a sophorose moiety at C-12.

Compound **1** was obtained as a brown powder, and its molecular formula $C_{32}H_{46}O_{15}$ was confirmed by HR-ESI-MS (m/z = 669.2763 [M − H]$^-$). In the ^{1}H NMR spectrum (Table 1), compound **1** showed the presence of one aromatic proton at δ_H 7.42 (1H, s), which was assumed to be located in the para position, and suggested one penta-substituted benzene ring. The ^{1}H NMR also exhibited two anomeric protons at δ_H 4.64 (1H, *d*, *J* = 8.0 Hz), and 4.75 (1H, *d*, *J* = 8.0 Hz); a methine at δ_H 2.83 (1H, m); an oxygenated methine at δ_H 3.97 (1H, *dd*, *J* = 12.5, 6.0 Hz); two oxygenated methylene protons at δ_H 3.85 (1H, *d*, *J* = 12.0 Hz) and δ_H 4.13 (1H, *d*, *J* = 12.0 Hz); four methylene groups at δ_H 1.43 (1H, *td*, *J* = 12.5, 6.5 Hz) and 3.31 (1H, *dd*, *J* = 13.0, 6.5 Hz), 2.00 (1H, *dd*, *J* = 18.0, 6.0 Hz) and 2.20 (1H, *t*, *J* = 10.0 Hz), 2.66 (1H, *dd*, *J* = 13.5, 6.0 Hz) and 2.92 (1H, m), 2.90 (1H, m) and 2.96 (1H, *t*, *J* = 3.0 Hz); three methyl groups at 1.70 (3H, *d*, *J* = 2.0 Hz), 1.18 (3H, s), and 1.00 (3H, *d*, *J* = 6.0 Hz).

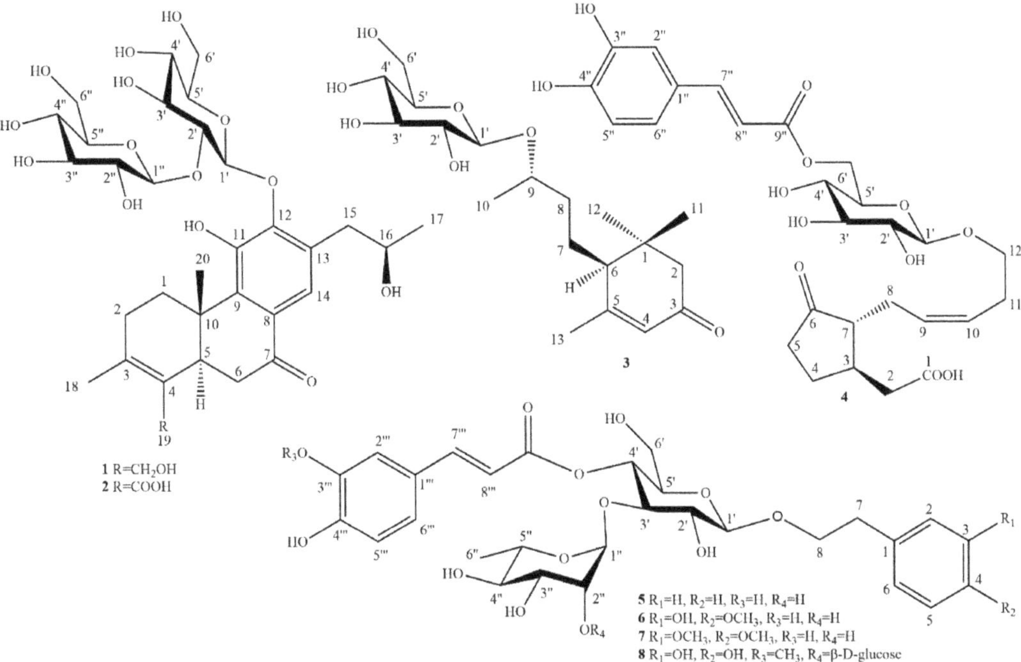

Figure 1. Structures of compounds **1–8** isolated from leaves of *Clerodendrum infortunatum*.

The ¹³C NMR spectrum (Table 2) revealed the presence of a quaternary carbon, indicated by a signal at δ_C 197.8, typical of a ketone; one 8,9,11,12,13-pentasubstituted benzene ring supported by signals at δ_C 128.3, 137.2, 147.8, 148.4, 131.4, respectively; two anomeric carbons displaying the same shifts at δ_C 103.6; two methine carbons at δ_C 42.7 and 65.5; five methylenes at δ_C 31.1, 29.7, 57.5, 36.6, and 39.2. An olefinic moiety was deduced from signals at δ_C 129.9 (C-3) and 129.2 (C-4).

Further, ten oxygenated aliphatic carbons (δ_C 80.8, 76.1, 69.5, 69.9, 61.1, 74.1, 76.2, 77.5, 77.4, and 60.9) together with two anomeric carbons at δ_C 103.6 reflected the presence of two glucose units. The coupling values of both anomeric protons ($J_{1'-2'}$ = 8.0 Hz, $J_{1''-2''}$ = 8.0 Hz) indicated that the sugar chains of **1** were glucopyranosyl-(1→2)-glucopyranosyl, and that both anomeric protons were in β-position. This was confirmed by HMBC data, and thus the linkage of the β-D-sophoroside in position C-12 was also established.

In the HMBC spectrum, correlations between H-1' (δ_H 4.64) and C-12 (δ_C 148.4), and H-15 (δ_H 2.66) and C-13 (δ_C 131.4) were observed (Figure 2), which proved that the glucose unit and the propanol moiety were connected to the benzene ring via C-12 and C-13, respectively.

A majority of the known plant-derived abietane-type diterpenes possess the same carbon skeleton, displaying a trans-fused system of two six-membered rings, a β-oriented methyl at C-10, and α-orientation of the proton at C-5 [19]. In order to identify the absolute configuration of C-16, the chemical shifts at C-15, C-16, and C-17 were compared with that of three known compounds, szemaoenoid A, szemaoenoid C [20], and (5*R*,10*S*,16*R*)-11,16-dihydroxy-12-methoxy-17(15→16)-abeoabieta-8,11,13-trien-3,7-dione [21]. Structures of these three compounds had been established by Mosher esterification and X-ray crystallography (Table 3). In view of the identical NMR data, the absolute configuration of C-16 of **1** was assigned as *R* conformation. Therefore, based on these findings and the supporting correlations along with the identical NMR data and biogenesis, compound **1** was established as (5*R*,10*S*,16*R*)-11,16,19-trihydroxy-12-*O*-β-D-glucopyranosyl-(1→2)-β-D-glucopyranosyl-17(15→16),18(4→3)-*diabeo*-3,8,11,13-abietatetraene-7-one, a previously undescribed natural product.

Table 1. 1D-^{1}H NMR (600 MHz) spectroscopic data for compounds **1** and **2**.

Position	1 [a] δ_H (*J* in Hz)	1 [b] δ_H (*J* in Hz)	2 [a] δ_H (*J* in Hz)	2 [b] δ_H (*J* in Hz)	2 [c] δ_H (*J* in Hz)
1	1.54, *td* (12.5, 6.5)	1.43, *td* (12.5, 6.5)	1.58, *td* (12.5, 6.5)	1.45, *m*	1.40, *m*
	3.46, *m*	3.31, *dd* (13.0, 6.5)	3.48, *m* [d]	3.34, *m*	3.16, *m*
2	2.11, *dd* (18.5, 6.0)	2.00, *dd* (18, 6.0)	2.06, *m*	1.23, *m*	1.98, *m*
	2.32, *m*	2.20, *t* (10.0)	2.29, *m*		2.14, *m*
5	2.94, br *d* (15.5)	2.83, br *d* (15.5)	3.08, *m*	2.53, *m*	2.90, *m*
6	2.59, *dd* (16.5, 15.5)	2.94, *m* [d]			2.39, *m*
	3.01, *dd* (17.0, 3.0)	2.50, *m* [d]			2.50, *m*
14	7.47, *s*	7.42, *s*	7.41, *s*	7.41, *s*	7.35, *s*
15	2.67, *dd* (13.5, 7.0)	2.66, *dd* (13.5, 6.0)	2.67, *m* [d]	2.67, *dd* (13.5, 6.0)	2.62, *dd* (13.5, 6.0)
	3.25, *dd* (13.5, 6.0)	2.92, *m* [d]	3.27, *dd* (13.5, 6.5)	2.93, *dd* (13.5, 6.5)	2.93, *dd* (13.5, 6.5)
16	4.17, *dd* (13.0, 6.5)	3.97, *dd* (12.5, 6.0)	4.16, *dd* (13.0, 6.5)	3.97, *dd* (12.5, 6.0)	3.99, *dd* (12.5, 6.0)
17	1.10, *d* (6.0)	1.00, *d* (6.0)	1.00, *d* (6.0)	1.00, *d* (6.0)	0.97, *d* (6.0)
18	1.78, *s*	1.70, *s*	1.75, *s*	1.70, *s*	1.60, *s*
19	4.08, *d* (12.0)	3.85, *d* (12.0)			
	4.28, *d* (12.0)	4.13, *d* (12.0)			
20	1.28, *s*	1.18, *s*	1.32, *s*	1.21, *s*	1.10, *s*
1′	4.72, *d* (8.0)	4.64, *d* (8.0)	4.87, *d* (8.0)	4.74, *d* (8.0)	4.74, *m* [d]
2′	3.88, *dd* (9.0, 8.0)	3.70, *m* [d]	3.88, *dd* (17.0, 8.0)	3.71, *m* [d]	3.79, *m* [d]
3′	3.41, *m* [d]	3.50, *m* [d]	3.42, *m* [d]	3.51, *m* [d]	3.35, *m* [d]
4′	3.47, *m* [d]	3.23, *m* [d]	3.48, *m* [d]	3.23, *m* [d]	3.39, *m* [d]
5′	3.37, *m* [d]	3.10, *m* [d]	3.39, *m* [d]	3.11, *m* [d]	3.26, *m* [d]
6′	3.67, *m* [d]	3.42, *dd* (12.0, 6.0)	3.68, *dd* (12.0, 6.0)	3.44, *dd* (12.0, 6.0)	3.49, *dd* (12.0, 6.0)
	3.83, *m* [d]	3.66, *m* *d*	3.82, *dd* (6.0, 1.5)	3.65, *m* [d]	3.62, *m* [d]
1″	4.87, *d* (8.0)	4.75, *d* (8.0)	4.73, *d* (8.0)	4.64, *d* (8.0)	4.71, *m* [d]
2″	3.38, *m* [d]	3.12, *m* [d]	3.40, *m* [d]	3.12, *m* [d]	3.25, *m* [d]
3″	3.66, *m* [d]	3.20, *m* [d]	3.67, *m* [d]	3.21, *m* [d]	3.58, *m* [d]
4″	3.30, *m* [d]	3.24, *m* [d]	3.36, *m* [d]	3.23, *m* [d]	3.28, *m* [d]
5″	3.36, *m* [d]	3.18, *m* [d]	3.31, *m* [d]	3.19, *m* [d]	3.21, *m* [d]
6″	3.72, *dd* (12.0, 5.5)	3.47, *m* [d]	3.72, *dd* (12.0, 5.0)	3.48, *m* [d]	3.59, *m* [d]
	3.83, *m* [d]	3.68, *m* [d]	3.84, *dd* (6.0, 2.0)	3.68, *m* [d]	3.64, *m* [d]

[a] Spectra were referenced to solvent residual and solvent signals of CD$_3$OD at 3.31 ppm (^{1}H NMR, 600 MHz). [b] Spectra were referenced to solvent residual and solvent signals of (CD$_3$)$_2$SO at 2.50 ppm (^{1}H NMR, 600 MHz). [c] Spectra were referenced to solvent residual D$_2$O at 4.59 ppm (^{1}H NMR, 600 MHz). [d] Overlapping.

Table 2. 1D-^{13}C NMR (150 MHz) spectroscopic data for compounds **1** and **2**.

Position	**1** [a] δ_C, Type	**1** [b] δ_C, Type	**2** [a] δ_C, Type	**2** [b] δ_C, Type	**2** [c] δ_C, Type
1	32.7, CH_2	31.1, CH_2	32.5, CH_2	30.7, CH_2	32.3, CH_2
2	31.1, CH_2	29.7, CH_2	29.7, CH_2	29.0, CH_2	29.6, CH_2
3	130.2, C	129.2, C	130.5, C	129.4, C	131.2, C
4	132.5, C	129.9, C	132.5, C	129.6, C	133.3, C
5	44.4, CH	42.7, CH	43.2, CH	40.4, CH	42.5, CH
6	38.1, CH_2	36.6, CH_2	38.6, CH_2	36.9, CH_2	38.9, CH_2
7	200.9, C	197.8, C	200.3, C	197.0, C	202.9, C
8	129.5, C	128.3, C	128.4, C	128.4, C	130.4, C
9	139.7, C	137.2, C	139.6, C	136.8, C	140.5, C
10	39.0, C	37.3, C	39.1, C	37.5, C	38.7, C
11	149.4, C	147.8, C	149.3, C	147.8, C	149.1, C
12	150.2, C	148.4, C	150.1, C	148.4, C	150.4, C
13	133.8, C	131.4, C	134.6, C	131.6, C	133.0, C
14	122.5, CH	120.7, CH	122.7, CH	121.0, CH	123.4, CH
15	41.1, CH_2	39.2, CH_2	41.1, CH_2	39.2, CH_2	40.0, CH_2
16	68.1, CH	65.5, CH	68.1, CH	65.5, CH	68.6, CH
17	22.7, CH_3	23.3, CH_3	22.6, CH_3	23.3, CH_3	23.4, CH_3
18	19.0, CH_3	18.8, CH_3	20.5, CH_3	20.1, CH_3	21.2, CH_3
19	59.2, CH_2	57.5, CH_2	—	166.2, COOH	170.7, COOH
20	15.7, CH_3	15.2, CH_3	16.1, CH_3	15.3, CH_3	16.8, CH_3
1′	105.5, CH	103.6, CH	105.3, CH	103.6, CH	104.8, CH
2′	82.7, CH	80.8, CH	82.7, CH	80.7, CH	82.0, CH
3′	77.8, CH	76.1, CH	77.8, CH	76.1, CH	77.1, CH
4′	70.8, CH	69.5, CH	70.8, CH	69.5, CH	70.4, CH
5′	71.4, CH	69.9, CH	71.4, CH	69.9, CH	71.0, CH
6′	62.6, CH_2	61.1, CH_2	62.6, CH_2	61.1, CH_2	62.1, CH_2
1″	105.4, CH	103.6, CH	105.5, CH	103.7, CH	104.9, CH
2″	75.7, CH	74.1, CH	75.7, CH	74.1, CH	75.2, CH
3″	77.9, CH	76.2, CH	77.9, CH	76.2, CH	77.2, CH
4″	78.6, CH	77.5, CH	78.6, CH	77.5, CH	77.9, CH
5″	78.5, CH	77.4, CH	78.6, CH	77.4, CH	77.8, CH
6″	62.3, CH_2	60.9, CH_2	62.2, CH_2	60.8, CH_2	61.7, CH_2

[a] Spectra were referenced to solvent residual and solvent signals of CD_3OD at 49.0 ppm (^{13}C NMR, 150 MHz). [b] Spectra were referenced to solvent residual and solvent signals of $(CD_3)_2SO$ at 39.52 ppm (^{13}C NMR, 150 MHz). [c] Spectra were referenced to solvent residual and solvent signals of D_2O.

Table 3. Comparison of partial NMR data of **1** and **2** with known compounds [a].

Position	Szemaoenoid A δ_C	Szemaoenoid A δ_H	Szemaoenoid C δ_C	Szemaoenoid C δ_H	E δ_C	E δ_H	Compound 1 δ_C	Compound 1 δ_H	Compound 2 δ_C	Compound 2 δ_H
15	40.9	2.71, 3.20	33.6	2.87, 3.17	40.4	2.70, 2.82	41.1	2.67, 3.25	41.1	2.67, 3.27
16	68.3	4.10	68.3	4.15	68.7	4.04	68.1	4.17	68.1	4.16
17	22.8	1.12	22.9	1.12	23.3	1.15	22.7	1.10	22.6	1.00

[a] Spectra of all compounds were measured in CD_3OD. E = (5R,10S,16R)-11,16-dihydroxy-12 methoxy-17(15→16)-abeoabieta-8,11,13-trien-3,7-dione.

Compound **2** was isolated as a colorless amorphous powder. The molecular formula of **2** was deduced as $C_{32}H_{44}O_{16}$ based on the HR-ESI-MS (m/z = 683.2556 [M − H]$^-$), which has one COOH instead of CH_2OH at the same position as that of **1**. An abietane-type diterpenoid derivative was evident based on its UV maxima at 219, 273, 319 nm, and NMR data. Analysis of the ^{1}H and ^{13}C NMR data of **2** (Tables 1 and 2) revealed similar substituent patterns to that of **1**, except a carboxylic group at C-19 rather than the methyleneoxy group. The ^{1}H and ^{13}C NMR data assignments were based on ^{1}H-^{1}H, COSY, HSQC, and HMBC spectra (Figures S7–S11). The ^{13}C NMR spectrum displayed signals for a ketone group at δ_C 202.9, a carboxylic group at δ_C 170.7, two tertiary methyl groups, and four double bonds including an aromatic ring characteristic of an abieta- 3,8,11,13-tetraene.

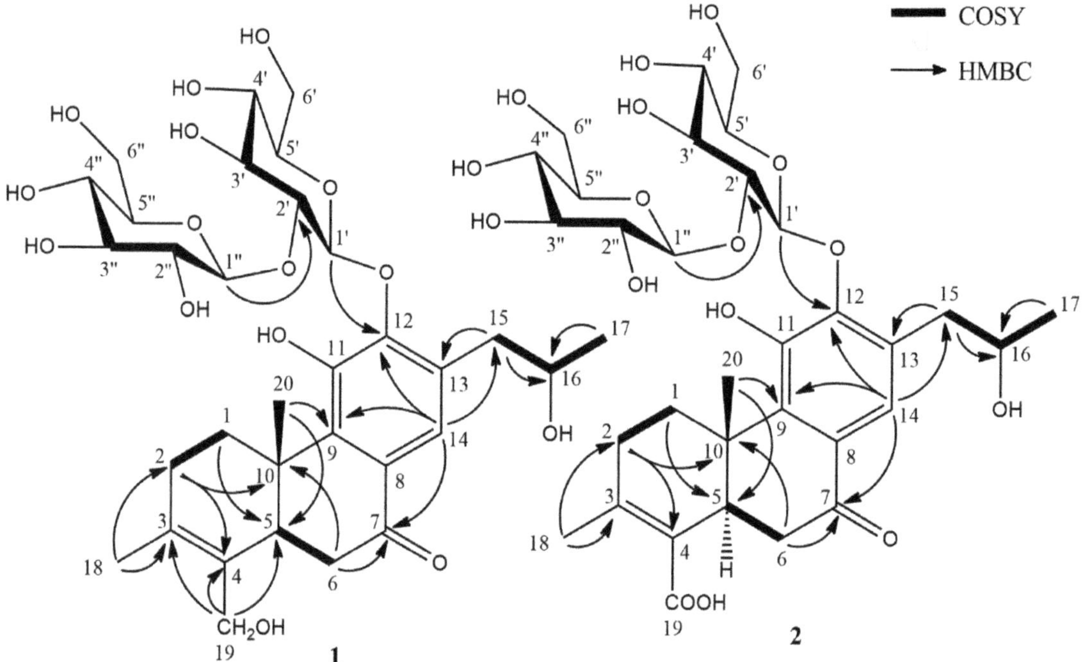

Figure 2. Main ^{1}H-^{1}H COSY and HMBC correlations of compounds **1** and **2**.

Comparing NMR data of **2** with **1** revealed that the aglycon part of both compounds was linked with the same sugar moiety. Based on 1D and 2D spectra, all sugar protons and carbons were assigned β-D-sophorose. Besides NMR spectra, acidic hydrolysis employing GLC-MS/MS analysis of both compounds **1** and **2** also confirms two β-D-glucose units as its sugar component (Section 3.9). From the biogenetic considerations and identical NMR data, **2** was inferred as possessing an identical absolute configuration to **1**. Thus, compound **2** was identified as (5*R*,10*S*,16*R*)-11,16-dihydroxy-12-*O*-β-D-glucopyranosyl-(1→2)-β-D-glucopyranosyl-17(15→16),18(4→3)-*diabeo*-4-carboxy-3,8,11,13-abietatetraene-7-one, a previously undescribed natural product.

The structure of the known C_{13}-nor-isoprenoid glycoside was confirmed as byzantionoside B (**3**) by spectrometric and spectroscopic methods (HR-ESI-MS, and ^{1}H NMR, ^{13}C NMR, COSY, HSQC, HMBC), and by comparison with literature data [22,23].

2.2. α-Amylase and α-Glucosidase Inhibition

α-Amylase (pancreatic enzyme) and α-glucosidase (intestinal enzyme) inhibitors reduce the conversion of carbohydrates into monosaccharides and are considered adjunctive therapeutics for the treatment of diabetes mellitus type 2. Natural products displaying α-amylase and α-glucosidase inhibitory properties could therefore be beneficial for the management of diabetes and obesity by controlling peak blood glucose levels. Compounds **1–3** (terpenoids) and **4–8** (phenylpropanoids) (Figure 1) demonstrated α-amylase and α-glucosidase inhibition, reflecting their previous records (Table 4) [16,18]. One terpenoid and one phenylpropanoid were found to have a significant mammalian α-amylase and yeast α-glucosidase inhibition compared to acarbose, a drug to treat type 2 diabetes mellitus, which was used as a positive control. In the α-amylase inhibition assay, compound **6** showed the most potent activity (IC$_{50}$ 3.4 ± 0.2 µM), and it was found to be almost two times more active than acarbose (IC$_{50}$ 5.9 ± 0.1 µM). Compounds **4**, **1**, **8**, and **5** displayed a slightly lower potency, ranging from IC$_{50}$ 13.0–24.9 µM, which was comparable to that of acarbose, while compounds **3** and **7** were almost inactive.

Table 4. Enzyme inhibition activity of three terpenoids (**1–3**) and five phenylpropanoids (**4–8**) against α-amylase, α-glucosidase, AChE, and BChE in comparison with standard acarbose and galanthamine.

Compound	IC$_{50}$ (µM)			
	α-Amylase	α-Glucosidase	AChE	BChE
1	18.5 ± 0.6 [b]	24.6 ± 0.2 [a]	191 ± 10.2 [a]	>1000
2	64.6 ± 7.1 [c]	78.3 ± 3 [a]	139 ± 7.2 [a]	>1000
3	284 ± 13.2 [d]	>1000	>1000	>1000
4	13.0 ± 1.3 [a,b]	>1000	>1000	>1000
5	24.9 ± 0.4 [b]	96 ± 10.5 [a]	160 ± 12.2 [a]	>1000
6	3.4 ± 0.2 [a]	55.8 ± 0.2 [a]	>1000	>1000
7	221 ± 24.5 [d]	>1000	>1000	>1000
8	19.8 ± 0.50 [b]	>1000	178 ± 11.3 [a]	>1000
acarbose	5.9 ± 0.1	665 ± 42		
galanthamine			2.9 ± 0.4	22.5 ± 1.9

Values are expressed as mean ± SD (n = 3). Different superscript letters correspond to values considered statistically different ($p \leq 0.05$).

Acarbose, a clinically used glycosidase inhibitor, usually demonstrates a weak inhibitory effect against yeast α-glucosidase compared to mammalian glycosidases. Therefore, by using acarbose as a positive control in the yeast α-glucosidase assay (IC$_{50}$ 665 ± 42 µM), compounds **1**, **2**, **5**, and **6** were established as moderate α-glucosidase inhibitors, with IC$_{50}$ values ranging from 24.6 to 96.0 µM (Table 4). The remaining four compounds showed no activity against yeast α-glucosidase at the tested concentrations.

Phenylpropanoid glycosides and abietane diterpenoids have previously been reported as being active against α-glucosidase and α-amylase [24,25]. The number and positions of hydroxy groups on natural compounds are crucial structural features to understand their enzyme inhibition [26]. In our compounds, the presence of one additional hydroxy group at C-4 at compound **6** seems to be involved in the more pronounced α-amylase inhibitory activity in comparison with compound **7**, where the methoxy substituents and sugar moiety could affect negatively on the inhibitory activity [27,28].

2.3. Cholinesterase Inhibitory Properties

Alzheimer's disease (AD) displays low levels of acetylcholine due to neurons degeneration, for this reason accpted therapeutic strategies for a symptomatic treatment of this illness include cholinesterases, acetylcholinesterase (AChE) and butyrylcholinesterase (BChE) inhibitors, as galanthamine. These enzymes are responsible for acetylcholine's hydrolysis, which plays an essential role in the proper functioning of the central cholinergic system, respectively. Due to having antioxidant, antiaging, and neuroprotective properties, terpenoids and phenylpropanoids were tested for their effects in managing AD [16,18]. However, from the tested compounds, **1**, **2**, **5**, and **8** displayed only low AChE inhibitory effects ranging from IC$_{50}$ values of 139–191 µM, while **3**, **4**, **6**, and **7** showed no inhibitory activity on AChE in the tested concentration range. None of the tested compounds demonstrated any activity towards BChE. Galanthamine was used as a positive control for both the AChE (IC$_{50}$ 2.9 ± 0.5 µM) and BChE (IC$_{50}$ 23 ± 2 µM) inhibition assays (Table 4).

The presence of sugar moiety in compounds may interfere with their ChE inhibitory activities, which modify the affinities toward enzymes [28]. Comparing our results with the activity of abietane diterpenoids isolated from *Caryopteris mongolica*, it implies that the presence of sugar moiety in tested compounds might reduce their inhibitory activity [21].

2.4. Antiproliferative and Cytotoxic Activities

Triple-negative breast cancers are a highly aggressive, heterogeneous subtype of breast cancer, with a poor survival rate. These breast cancers are characterized by a lack of expression of estrogen and progesterone receptors as well as a lack of amplification human epidermal growth factor receptor 2 [29]. There are no approved targeted therapies for triple-negative breast cancers, because they do not respond to available targeted therapies. However, patients usually receive chemotherapy with cytotoxic agents such as taxanes [30].

Terpenoids and phenylpropanoids are well-known for their cytotoxic and anticancer activity [31–33]. Compounds **1–8** were therefore tested for their cytotoxicity and effects on the cell migration of Hs578T and MDA-MB-231, which are triple negative breast cancer cell lines. Among the tested compounds, the concentration-related cytotoxic responses were observed with IC_{50} values of 85.3 ± 2.4 and 96.5 ± 1.5 µM against Hs578T and MDA-MB-231, respectively for compound **5**, and 94.7 ± 1.3 µM against Hs578T for compound **3** (Table 5). The rest of the compounds did not show significant activity within the tested range ($IC_{50} > 100$ µM).

Table 5. Cytotoxic activity of the tested compounds at a half maximal inhibitory concentration (IC_{50}) in the TNBC cell lines Hs578T and MDA-MB-231.

Compounds	IC_{50} (µM)	
	Hs578T	**MDA-MB-231**
1	128 ± 2.2	122 ± 1.2
3	94.7 ± 1.3	118 ± 3.3
4	210 ± 5.1	199 ± 3.1
5	85.3 ± 2.4	96.5 ± 1.5

2.4.1. Effects on TNBC Cell Proliferation

To explore the antiproliferative activity of compounds tested in the TNBC cell lines Hs578T and MDA-MB-231, colony formation assays were employed. To fix the effective concentration, the half maximal inhibitory concentration (IC_{50}) of each compound was determined (Table 5), and was used as a working concentration for all experiments. In the cell proliferation assay, the compounds triggered a significant reduction in the number of colony formations compared to that of the control (DMSO-treated) cells (Figure 3A,D). The quantified colonies are represented in bar graphs at Figure 3B,E). Moreover, the cell viability outcome also justified the antiproliferative activity of the tested compounds in the MTS assay (Figure 3C,F). The obtained data revealed that compounds **3** and **5** possess moderate antiproliferative effects: they reduced the number of TNBC cell Hs578T (44 and 42%, respectively) and MDA-MB-231 (48 and 43%, respectively) after three days of treatment at IC_{50}. Chemotherapeutic compounds interrupt the signaling pathways of cancer and control accelerated proliferation to induce cancer cell death [34]. Natural products are considered a key source in the search for new anticancer compounds [15]. This study displayed that tested compounds moderately suppressed breast cancer cell proliferation and viability.

2.4.2. Effects on Cell Migration

The migration of cancer cells are critical determinant steps of tumor metastasis. To evaluate the anti-metastatic effect on breast cancer cells, the inhibition of the cell migration rate is a reliable indicator. Figure 4 shows the inhibition ability of the tested compounds on the migration of the breast cancer cells compared to control cells in DMSO ($p < 0.01$). All tested compounds inhibited the migration of Hs578T cell slightly more than MDA-MB-231. Compound **3** and **5** displayed an interesting activity profile: they were able to inhibit 50%

and 43% of the migration of Hs578T cell, and 40% and 37% for MDA-MB-231, respectively, at IC_{50}. Cancer metastasis, a multistep process, is a major cause of cancer-associated mortalities. During this process, cancer cells escape and travel from the primary tumor site to a distant area through various cascades of events such as cell adhesion, cell motility and invasion, cell movement, and degradation of the cellular matrix [35,36]. The inhibition of cancer cell migration is a novel strategy for the treatment of metastatic cancers. Our results showed that compounds **3** and **5** effectively suppressed breast cancer cell migration.

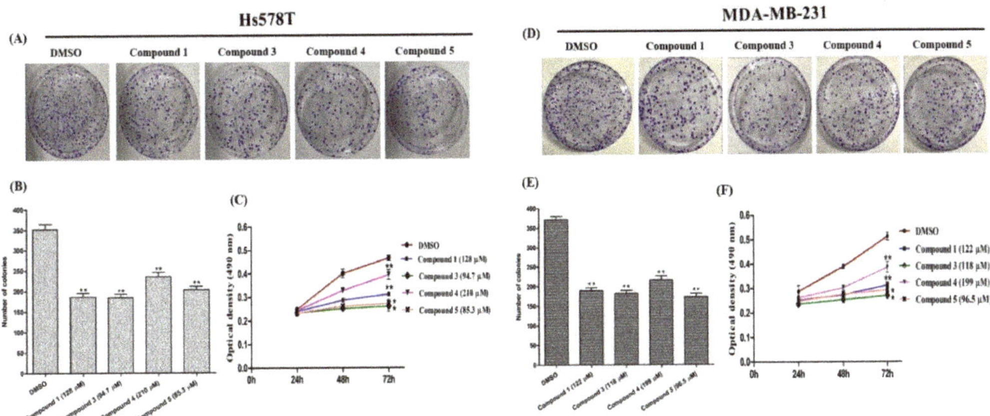

Figure 3. Inhibition of TNBC cell proliferation. (**A,D**): visualized colonies after two weeks of treatment of TNBC cells with tested compounds. (**B,E**): quantified colonies from colony formation assay. (**C,F**): the amount of viable cells after treatment of the cells with compounds for 24, 48, and 72 h, determined using an MTS assay. Data are presented as the mean ± SD of three independent experiments. Bars and lines with asterisks indicate significant differences from the control at $p \leq 0.05$ (*) or $p \leq 0.01$ (**).

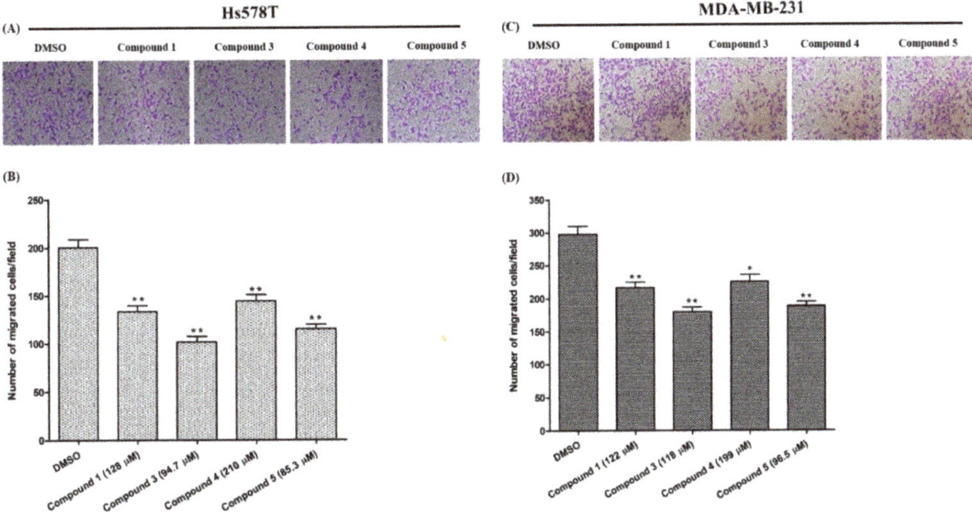

Figure 4. Attenuation of TNBC cell migration. (**A,C**): visualized migrated cells after treating with compounds. (**B,D**): the migrated cells stained with crystal violet, and photographs taken for the inhibition calculation. The results are presented as the mean ± SD of three independent experiments. Bars with asterisks indicate significant differences from the control at $p \leq 0.05$ (*) or $p \leq 0.01$ (**).

2.4.3. Effects on Tumor-Sphere Formation

The capability of compounds to reduce cell size is considered a good indicator in cancer therapy. The in vitro tumor-sphere formation tests demonstrated that compound **5** reduced the cell size (Figure 5A,C) and the cell number significantly (Figure 5B,D) in both cell lines. In vitro tumor-sphere formation is a frequently used new and inexpensive method considered a potential alternative for in vivo screening of anticancer drugs [37]. In the present study, the number and size of tumor spheres were sharply reduced by the compound **5**. However, more intensive research is needed to find out the mechanism of these activities in relation with the respective compound.

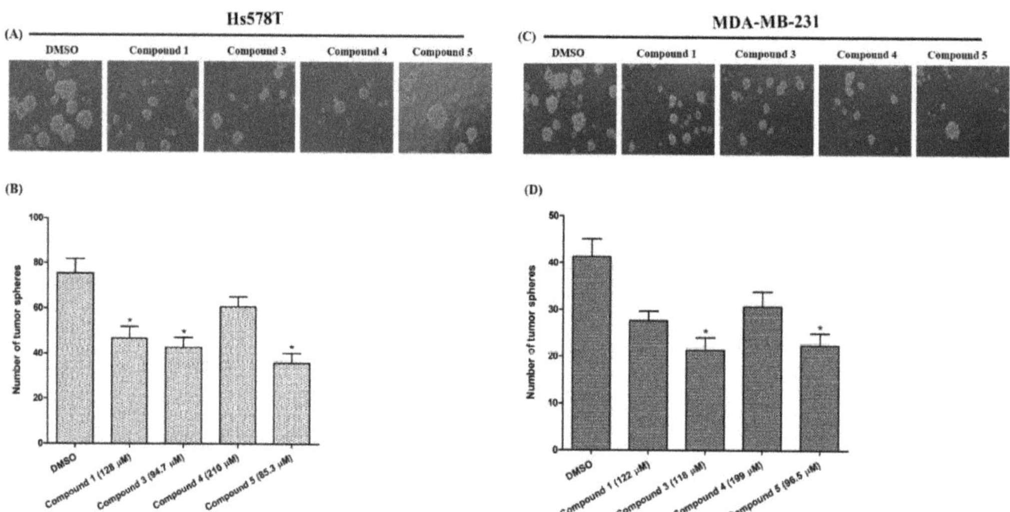

Figure 5. Reduction of the formation of the tumor-sphere. (**A,C**): the captured tumor-spheres after five days of treatment with tested compounds. (**B,D**): the number of tumor-spheres. Data are presented as the mean $\pm$ SD of three independent experiments. The bars with asterisks indicate significant differences from the control at $p \leq 0.05$ (*).

3. Materials and Methods

3.1. General Experimental Procedures

The NMR spectra, 1D (^{1}H, ^{13}C) and 2D (COSY, HMQC, HMBC), were acquired at 600 MHz on a Fourier transform-NMR "Avance III 600" spectrometer equipped with a cryogenically cooled triple resonance Z-gradient probe head operating at 300 K and pH 7.5 (Bruker BioSpin GmbH, Rheinstetten, Germany). TMS was used as the internal reference standard where chemical shifts reported as δ values. ESI-MS data were obtained via Nexera X2 system (Shimadzu, Kyoto, Japan) connected to an autosampler, column heater, PDA, and a Shimadzu LC-MS 8030 Triple Quadrupole Mass Spectrometer. HR-ESI-MS spectra were recorded on a Q-Exactive Plus spectrometer (Thermo Scientific, Bremen, Germany).

UHPLC experiments were performed employing a VWR Hitachi Chromaster UltraRS liquid chromatograph equipped with ELSD 100 and DAD 6430 detectors. A column, Phenomenex Luna Omega C_{18}, 1.6 µm, 100 $\times$ 2.1 mm, was used for all the analyses, with the following settings: mobile phase A: 0.1% formic acid in water; mobile phase B: 100% acetonitrile (linear gradient: 0 min 5% B, 35 min 40% B, 50 min 95% B, 60 min 95% B, 60.1–70 min 5%); flow rate: 0.20 mL/min; injection volume: 2 µL; oven temperature: 30 °C (Figure S14).

Semi-preparative HPLC was performed on a Waters HPLC system (Waters) equipped with Waters Alliance e2695 Separations Module, Alliance 2998 detector, WFC III fraction collector (Waters, Milford, MA, USA), and VP Nucleodur C18 column (250 $\times$ 10 mm, 5 µm particle size, Macherey-Nagel, Düren, Germany). Column chromatography was performed with silica gel (40–63 µm; 230–400 mesh, Carl Roth GmbH, Karlsruhe, Germany) and Sephadex LH-20 (GE Healthcare AB, Uppsala, Sweden). Thin-layer chromatography (TLC) was carried out on precoated TLC plates (Silica gel 60 F_{254}, Merck, Darmstadt, Germany) using ethyl acetate-water-acetic acid-formic acid (15:5:2:2) as the mobile phase, and the spots were visualized by heating after vanillin-sulphuric acid spray.

3.2. Chemicals

Acetylcholinesterase from electric eel (*Electrophorus electricus*, type VI-s, lyophilized powder), acetylthiocholine iodide (ATCI), butyrylcholinesterase from equine serum (lyophilized powder), butyrylthiocholine iodide (BTCI), 5,5'-dithio-bis2-nitrobenzoic acid (DTNB), α-amylase from porcine pancreas, α-glucosidase from *Saccharomyces cerevisiae*, 4-*p*-nitrophenyl-α-D-glucopyranoside, starch, galanthamine, and acarbose were purchased from Sigma-Aldrich (St. Louis, MO, USA). Trizma hydrochloride (Tris-HCl) and bovine serum albumin (BSA) were obtained from Sigma–Aldrich (Steinheim, Germany). Deionized water was produced using a Milli-Q water purification system (Millipore, Bedford, MA, USA). Dulbecco's Modified Eagle Medium (DMEM) was collected from Gibco (Japan), and fetal bovine serum (FBS) was also collected from Gibco (Waltham, MA, USA). Insulin and penicillin G/streptomycin were collected from Wako (Fujifilm Wako Pure Chemical Corporation, Osaka, Japan), the MTS kit was collected from Promega (Madison, WI, USA), and poly-2-hydroxyethyl methacrylate was collected from Sigma-Aldrich (Taufkirchen, Germany).

3.3. Plant Material

The leaves of *C. infortunatum* were collected from Pabna, Bangladesh, in March 2018 at 23 m above mean sea level (coordinates: N 24°03'45.0"; E 89°04'14.0"). Prof. Dr. A.H.M. Mahbubur Rahman, Department of Botany, University of Rajshahi, Bangladesh carried out the botanical identification [38]. A voucher specimen (CV-20180321-04) was deposited in the Department of Botany, University of Rajshahi, Bangladesh.

3.4. Extraction and Isolation

The air-dried fresh leaves of *C. infortunatum* (1.15 kg) were powdered and subjected to cold extraction with acetone (5 L) at room temperature five times, for one day each time. The obtained solution was combined, filtered, and evaporated under reduced pressure at 35 °C, yielding 58.0 g of crude acetone extract. The concentrated extract was solvated in a solution of water:methanol (2:1) and partitioned with ethyl acetate and *n*-butanol, respectively, resulting in the ethyl acetate (35.7 g), *n*-butanol (15.0 g), and water (7.30 g) fractions. The butanol extract (15.0 g) was chromatographed to a silica gel column chromatography (CC) eluted with a gradient of increasing methanol (0–100%) in dichloromethane to attain 14 fractions (CV 1 to CV 14).

Fraction CV 14 was subjected to chromatographic separation by Sephadex LH-20 gel column (3 × 100 cm) eluted with methanol to yield eight subfractions (CV 14 A-H). Subfraction CV 14C was treated by semi-preparative HPLC on an RP-18 column (VP 250 × 10 mm Nucleodur C18, 5 µm, flow rate: 2 mL/min) using 0.025% formic acid in water, developing methanol:water solvent mixtures (40:60, isocratic) that yielded three pure compounds: 2 (6.0 mg, t_R 18 min), 1 (8.5 mg, t_R 30 min), and 3 (3.0 mg, t_R 39 min).

(4a*S*,10a*R*)-6-((((2*S*,3*S*,4*S*,6*S*)-4,5-dihydroxy-6-(hydroxymethyl)-3-(((2*S*,3*S*,4*S*,5*S*,6*R*)-3,4,5-trihydroxy-6-(hydroxymethyl)tetrahydro-2*H*-pyran-2-yl)oxy)tetrahydro-2*H*-pyran-2-yl)oxy)-5-hydroxy-1-(hydroxymethyl)-7-((*R*)-2-hydroxypropyl)-2,4a-dimethyl-4,4a,10,10a-tetrahydrophenanthren-9(3*H*)-one (1). Brown powder; $[\alpha]^{25}_D$ +0.035 (MeOH); For ^{1}H NMR (methanol-*d4*, DMSO-*d$_6$*, 600 MHz) and ^{13}C NMR (methanol-*d4*, DMSO-*d$_6$*, 150 MHz) data, see Tables 1 and 2. HR-ESI-MS *m/z* 669.2763 [M − H]$^-$ (calcd. for $C_{32}H_{46}O_{15}$, 669.2758).

(4a*S*,10a*R*)-6-(((3*S*,4*S*,6*S*)-4,5-dihydroxy-6-(hydroxymethyl)-3-(((2*S*,3*S*,4*S*,5*S*,6*R*)-3,4,5-trihydroxy-6-(hydroxymethyl)tetrahydro-2*H*-pyran-2-yl)oxy)tetrahydro-2*H*-pyran-2-yl)oxy)-

5-hydroxy-7-(2-hydroxypropyl)-2,4a-dimethyl-9-oxo-3,4,4a,9,10,10a hexahydrophenanthrene-1-carboxylic acid (**2**). Pale brown powder; $[\alpha]^{25}_{D}$ +0.030 (MeOH); For ^{1}H NMR (methanol-*d4*, DMSO-d_6, D_2O, 600 MHz) and ^{13}C NMR (methanol-*d4*, DMSO-d_6, D_2O, 150 MHz) data, see Tables 1 and 2. HR-ESI-MS *m/z* 683.2556 $[M - H]^-$ (calcd. for $C_{32}H_{46}O_{15}$, 683.2551).

3.5. α-Amylase Inhibition Assay

α-Amylase inhibitory activity was determined following the starch-iodine method [39] with some modifications. A 1% starch solution was prepared by 1 g of starch in 10 mL of distilled water following gentle boiling and cooling into 100 mL. A reaction mixture, 25 µL sample (0–1 mM) and 50 µL α-amylase (5 U/mL) in phosphate buffer, was incubated at 37 °C for 10 min. Afterwards, the starch (100 µL, 1% *w/v*) solution was added to the mixture and incubated again at 37 °C for 10 min. The enzymatic reaction was suspended by adding HCl (25 µL, 1 N) followed by the incorporation of 50 µL of iodine reagent (2.5 mM I_2 and 2.5 mM KI). After adding the iodine/iodide solution, based on the colour change, the absorbance was monitored at 630 nm for 10 min. Acarbose was used a positive control. The percentage of inhibition was calculated and results were expressed as IC_{50} (µM).

3.6. α-Glucosidase Inhibition Assay

To assess the inhibitory activity of the tested compounds on α-glucosidase, all solutions were prepared according to the previously described method [39]. Different concentrations (0–1 mM) of the sample (50 µL) and α-glucosidase enzyme (40 µL, 0.1 U/mL) dissolved in phosphate buffer were incubated at 37 °C for 10 min. After combining the substrate 4-*p*-nitrophenyl-α-D-glucopyranoside (40 µL, 2.5 mM) to the enzyme mixture, it was incubated again at 37 °C for 10 min. Na_2CO_3 (100 µL, 0.2 M) was used to stop the enzymatic reaction. The release of glucose and *p*-nitrophenol (yellow) was detected spectrophotometrically at 405 nm. Acarbose was used as positive control and the results were expressed as IC_{50} (µM).

3.7. Determination of Cholinesterase Inhibitory Activities

The cholinesterase inhibitory (AChE/BChE) properties were ascertained based on Ellman's method, as previously described [40]. The enzyme activity was detected by spectrophotometric exposure (405 nm), with increasing yellow colour produced from thiocholine, while it reacted with 5,5'-dithio bis-2 nitrobenzoate ions (DTNB). In the AChE inhibitory assay, 25 µL of sample solution (0–1 mM) along with 50 µL of buffer B (50 mM Tris-HCl, pH 8 containing 0.1% BSA), 125 µL of DTNB (3 mM), and 25 µL of 0.05 U/mL AChE were incubated at 37 °C for 10 min. After incubation, 25 µL of acetylthiocholine iodide (5 mM) as AChE substrate was incorporated to the solution. The BChE activity was determined following the same protocol using 25 µL of 5 mM *S*-butyrylthiocholine chloride as BChE substrate and 0.05 U/mL BChE as enzyme. The inhibitory abilities of the compounds (**1–8**) were assessed at different concentrations. Galanthamine (dissolved in 10% DMSO in methanol) was used as a positive control, while 10% DMSO in methanol was used as a negative control for both assays. The percentage of inhibition was calculated and the results were expressed as IC_{50} (µM).

3.8. Determination of Anticancer Activities

3.8.1. Cell Lines and Culture Condition

The human TNBC cell lines Hs578T and MDA-MB-231 were obtained from the American Type Culture Collection and cultured in Dulbecco's Modified Eagle Medium (DMEM) supplemented with fetal bovine serum (FBS) (10% *v/v*), insulin (Hs578T cell only), and penicillin G/streptomycin 1% (*v/v*) at 37 °C under 5% CO_2. The absence of culture contamination by *Mycoplasma* species was confirmed before the experiments.

3.8.2. Colony Formation Assay

The breast cancer cell proliferation activity of the tested compounds was assessed using a colony formation assay [41]. Approximately 400 viable Hs578T and MDA-MB-231 cells were seeded in a 10 cm culture plate containing DMEM medium without or with the tested compounds (at their respective IC_{50} concentration), and incubated for 2 weeks. After incubation, the medium was discarded, and the colonies were washed twice with phosphate-buffered saline (PBS). Subsequently, the colonies were fixed with 4% paraformaldehyde and stained with crystal violet solution. Colonies consisting of more than 20 individual cells were counted by ImageJ software.

3.8.3. Cell Viability Assay

The activity of the selected compounds on breast cancer cell viability were determined employing the MTS assay [42]. Briefly, 5×10^3 cells were seeded in each well of a 96-well plate for 24 h, and growth medium containing different compounds was added and incubated in different periods. A total of 20 µL of the MTS kit was added to each well, and the cells were incubated for 2 h. After incubation, the absorbance was measured at 490 nm with an enzyme-linked immunosorbent assay microplate reader (BioTek Instruments, USA). The absorbance value is directly proportional to the number of living cells.

3.8.4. Transwell Cell Migration Assay

The effects of the selected compounds on breast cancer cell migration were evaluated in transwell chambers according to a published protocol [42]. Briefly, the Hs578T and MDA-MB-231 cells were treated with the tested compounds for 24 h, trypsinized, and washed twice with serum-free medium. Approximately 3×10^4 pretreated cells, suspended in 100 µL of the serum-free medium, were seeded to the upper chamber. The lower chambers were filled with approximately 500 µL of DMEM medium with 10% FBS and incubated for 12 h. After incubation, the cells from the upper surface were wiped off with cotton swabs, while the migrated cells on the opposite side of the transwell were washed twice with PBS, fixed with 4% paraformaldehyde, and finally stained with crystal violet solution. After 2 washes with water (Milli Q), several microscopic fields were taken randomly, and the migrated cells were counted using ImageJ software.

3.8.5. Tumor-Sphere Formation Assay

Three-dimensional or tumor-sphere culture is a recently introduced in vitro technique which maintains a physiological environment which closely resembles that of in vivo conditions [37]. This technique has now been widely used for the screening of anti-cancer moieties [41]. In vitro tumor-sphere formation assay was performed following a reported protocol [43]. Briefly, approximately 3×10^3 cells were resuspended in a poly-2-hydroxyethyl methacrylate coated 6-well plate containing a sphere-forming medium with or without tested compounds and incubated for one week. The number of tumor spheres were counted, and the diameter of each tumor sphere was measured.

3.9. Determination of the Absolute Sugar Configuration

The absolute configuration of the sugar moieties of compound **1** and **2** was determined through GLC-MS/MS analysis of the octylated sugar moiety, after hydrolysis, employing the methods described previously [44,45], with some modifications. For hydrolysis, 0.5 mg of sample and 1 mL 2 M trifluoroacetic acid (TFA) were combined in a glass vial. The mixture was heated to 120 °C for 1 h. Afterwards, the mixture was washed three times, adding 5 mL water each time, by evaporating to dryness under reduced pressure. For octylation, 1 mL of (*R*)-(−)-2-octanol, and one drop of TFA (conc.) were added to the mixture. The sample was kept at 120 °C for 12 h. Subsequently, the sample was transferred to a separation funnel incorporating 5 mL of methanol with a few drops of water and separated three times with 5 mL *n*-hexane each time to remove excess octanol. The methanol fraction was evaporated under reduced pressure. For acetylation, the sample was heated in a vial at 100 °C for 20 min after adding 0.5 mL anhydrous acetic anhydride and 0.5 mL anhydrous pyridine.

After cooling the mixture at room temperature, 10 mL of water, 1 mL 0.1 M H_2SO_4, and 1 mL CH_2Cl_2 were added and the mixture was shaken vigorously. The CH_2Cl_2 layer was used for GLC–MS/MS analysis. For comparison with standard, 0.5 mg of D-glucose and 0.5 mg of L-glucose each were separately treated following the same procedure. GLC-MS analysis was performed using the column TG-5 SILMS (Thermo Scientific, 15 m × 0.25 mm × 0.25 µm) with the following settings: injection volume 1 µL; flow rate 1.2 mL/min; mobile phase helium; split ratio 1:10; ion source temperature 280 °C; injector temperature 290 °C; MS transfer line temperature 280 °C; scanning range for full scan: (m/z) 43–700; ionization mode EI; temperature gradient: 0 min 60 °C, 2 min 60 °C, 4.6 min 180 °C, 5.16 min 180 °C, 39.5 min 280 °C, and 41.13 min 280 °C.

The GLC-MS signals of the (R)-(–)-2-octanyl derivatives of standard D-glucose and the split off sugar moieties from compounds **1** and **2** had the same retention times (t_R = 32.08 and 33.16 min) and nearly identical MS spectra. In contrast, the signals obtained from the derivative of the standard L-glucose had significantly different retention times (t_R = 31.73 and 32.27 min).

3.10. Statistical Analysis

All experiments were repeated in triplicate, and results were expressed as mean ± standard deviation. The analysis of variance (one way ANOVA) was performed to assess statistically significant differences among the tested compounds. Differences in the mean values were assessed by the Tukey test at a significance level of $p < 0.05$ by using GraphPad Prism v. 6.0 (GraphPad Software Inc., San Diego, CA, USA).

4. Conclusions

In this study, three terpenoids, along with five previously isolated phenylpropanoids from *C. infortunatum*, revealed their antidiabetic, anticholinesterase, and anticancer potentials. Among the tested compounds, compound **6** was confirmed to have the best therapeutic potential against mammalian α-amylase compared to the reference standard acarbose. On the other hand, compounds **3** and **5** displayed moderate antiproliferative, antimetastatic, and antitumor properties against TNBC cell lines. In this view, the findings extended the chemical diversity of *C. infortunatum* and hold promise for identifying further potential nor-isoprenoid and phenylpropanoids as lead compounds against diabetes and TNBC.

Supplementary Materials: The following are available online. Supplementary data associated with this article are ^{1}H, ^{13}C, COSY, HSQC, and HMBC NMR spectra of compounds **1** and **2** in CD$_3$OD, DMSO-d_6, and D$_2$O at 600 MHz; HR-ESI-MS in acetone for compounds **1** and **2**; ^{1}H and ^{13}C NMR data in CD$_3$OD, and DMSO-d_6 at 600 MHz of known compound **3**.

Author Contributions: M.J.U.: Methodology, Investigation, Writing—original draft. D.R.: Investigation, Writing—review & editing. M.A.H.: Investigation, Writing—review & editing. S.S.Ç.: Investigation, Writing—review & editing. F.D.S.: Resources. L.M.: Resources, Writing—review & editing. C.Z.: Conceptualization, Supervision, Project administration, Writing—review & editing. All authors have read and agreed to the published version of the manuscript.

Funding: This research received no external funding.

Institutional Review Board Statement: Not Applicable.

Informed Consent Statement: Not Applicable.

Data Availability Statement: Not Applicable.

Acknowledgments: C.Z. and M.J.U. acknowledge support from the German Academic Exchange Service (DAAD-91649878). In addition, C.Z. and M.J.U. wish to thank Gitta Kohlmeyer-Yilmaz and Ulrich Girreser for their expert NMR support. We are indebted to A.H.M. Mahbubur Rahman, Department of Botany, University of Rajshahi, Bangladesh for the identification of plant species.

Conflicts of Interest: The authors declare no conflict of interest.

Sample Availability: Samples are not available from the authors as they were consumed during measurements and bioassays.

References

1. Munir, A.A. A taxonomic revision of the genus *Clerodendrum* L. (Verbenaceae) in Australia. *J. Adel. Bot. Gard.* **1989**, *11*, 101–173.
2. Verdcourt, B. *Flora of Tropical East Africa—Verbenaceae*; Polhill, R.M., Ed.; A.A. Balkema: Rotterdam, The Netherlands, 1992.
3. Nandi, S.; Lyndem, M.L. Clerodendrum viscosum: Traditional uses, pharmacological activities and phytochemical constituents. *Nat. Prod. Res.* **2016**, *30*, 497–506. [CrossRef]
4. Ghani, A. *Medicinal Plants of Bangladesh*, 2nd ed.; The Asiatic Society of Bangladesh: Dhaka, Bangladesh, 2003; pp. 1–398.
5. Uddin, M.S. Traditional Knowledge of Medicinal Plants in Bangladesh. Nature Info. Electronic Database. Available online: https://www.natureinfo.com.bd/category/flora/medicinal-palnts (accessed on 12 January 2021).
6. Lobo, R.; Punitha, I.S.R.; Rajendran, K.; Shirwaikar, A. Prelimenary study on the antisnake venom activity of alcoholic root extract of *Clerodendrun viscosum* in *Naja naja* venom. *J. Nat. Prod. Sci.* **2006**, *129*, 153–156.
7. Pal, D.K.; Sannigrahi, S.; Mazumder, U.K. Analgesic and anticonvulsant effect of saponin isolated from the leaves of *Clerodendrum infortunatum* Linn in mice. *Ind. J. Exp. Biol.* **2009**, *47*, 743–747.
8. Gupta, R.; Singh, H.K. Nootropic potential of *Alternanthera sessilis* and *Clerodendrum infortunatum* leaves on mice. *Asian Pac. J. Trop. Dis.* **2012**, *2*, 465–470. [CrossRef]
9. Ghosh, G.; Sahoo, S.; Das, D.; Dubey, D.; Padhy, R.N. Antibacterial and antioxidant activities of methanol extract and fractions of *Clerodendrum viscosum* Vent. leaves. *Ind. J. Nat. Prod. Resour.* **2014**, *5*, 134–142.
10. Swargiary, A.; Daimari, A.; Daimari, M.; Basumatary, N.; Narzary, E. Phytochemicals, antioxidant, and anthelmintic activity of selected traditional wild edible plants of lower Assam. *Indian J. Pharmacol.* **2016**, *48*, 418–423. [CrossRef]
11. Shendge, A.K.; Basu, T.; Panja, S.; Chaudhuri, D.; Mandal, N. An ellagic acid isolated from *Clerodendrum viscosum* leaves ameliorates iron-overload induced hepatotoxicity in Swiss albino mice through inhibition of oxidative stress and the apoptotic pathway. *Biomed. Pharmacother.* **2018**, *106*, 454–465. [CrossRef] [PubMed]
12. Uddin, M.J.; Çiçek, S.S.; Willer, J.; Shulha, O.; Abdalla, M.A.; Sönnichsen, F.; Girreser, U.; Zidorn, C. Phenylpropanoid and flavonoid glycosides from the leaves of *Clerodendrum infortunatum* (Lamiaceae). *Biochem. Syst. Ecol.* **2020**, *92*, 104131. [CrossRef]
13. Buford, T.W. Hypertension and aging. *Ageing Res. Rev.* **2016**, *26*, 96–111. [CrossRef]
14. Belikov, A.V. Age-related diseases as vicious cycles. *Ageing Res. Rev.* **2019**, *49*, 11–26. [CrossRef]
15. Newman, D.J.; Cragg, G.M. Natural products as sources of new drugs over the nearly four decades from 01/1981 to 09/2019. *J. Nat. Prod.* **2020**, *83*, 770–803. [CrossRef]
16. Yang, W.; Chen, X.; Li, Y.; Guo, S.; Wang, Z.; Yu, X. Advances in pharmacological activities of terpenoids. *Nat. Prod. Commun.* **2020**, *15*, 1934578X20903555. [CrossRef]
17. Panda, P.; Appalashetti, M.; Judeh, Z.M.A. Phenylpropanoid sucrose esters: Plant derived natural products as potential leads for new therapeutics. *Curr. Med. Chem.* **2011**, *18*, 3234–3251. [CrossRef]
18. Neelam; Khatkar, A.; Sharma, K.K. Phenylpropanoids and its derivatives: Biological activities and its role in food, pharmaceutical and cosmetic industries. *Crit. Rev. Food Sci. Nutr.* **2020**, *60*, 2655–2675. [CrossRef]
19. González, M.A. Aromatic abietane diterpenoids: Their biological activity and synthesis. *Nat. Prod. Rep.* **2015**, *32*, 684–704. [CrossRef] [PubMed]
20. Pu, D.B.; Wang, T.; Zhang, X.J.; Gao, J.B.; Zhang, R.H.; Li, X.N.; Wang, Y.M.; Li, X.L.; Wang, H.Y.; Xiao, W.L. Isolation, identification and bioactivities of abietane diterpenoids from *Premna szemaoensis*. *RSC Adv.* **2018**, *8*, 6425–6435. [CrossRef]
21. Murata, T.; Ishikawa, Y.; Saruul, E.; Selenge, E.; Sasaki, K.; Umehara, K.; Yoshizaki, F.; Batkhuu, J. Abietane-type diterpenoids from the roots of *Caryopteris mongolica* and their cholinesterase inhibitory activities. *Phytochemistry* **2016**, *130*, 152–158. [CrossRef] [PubMed]
22. Matsunami, K.; Otsuka, H.; Takeda, Y. Structural revisions of blumenol C glucoside and byzantionoside B. *Chem. Pharm. Bull.* **2010**, *58*, 438–441. [CrossRef] [PubMed]
23. Kang, U.; Ryu, S.M.; Lee, D.; Seo, E.K. Chemical constituents of the leaves of *Brassica oleracea* var. acephala. *Chem. Nat. Compd.* **2018**, *54*, 1023–1026. [CrossRef]
24. Devaraj, S.; Yip, Y.M.; Panda, P.; Ong, L.L.; Wong, P.W.K.; Zhang, D.; Judeh, Z. Cinnamoyl sucrose esters as alpha glucosidase inhibitors for the treatment of diabetes. *Molecules* **2021**, *26*, 469. [CrossRef] [PubMed]
25. Etsassala, N.G.; Cupido, C.N.; Iwuoha, E.I.; Hussein, A.A. Abietane diterpenes as potential candidates for the management of type 2 diabetes. *Curr. Pharm. Des.* **2020**, *26*, 2885–2891. [CrossRef] [PubMed]
26. Güvenalp, Z.; Özbek, H.; Dursunoğlu, B.; Yuca, H.; Gözcü, S.; Çil, Y.M.; Demirezer, Ö.L. α-Amylase and α-glucosidase inhibitory activities of the herbs of *Artemisia dracunculus* L. and its active constituents. *Med. Chem. Res.* **2017**, *26*, 3209–3215. [CrossRef]
27. Santoro, V.; Parisi, V.; D'Ambola, M.; Sinisgalli, C.; Monné, M.; Milella, L.; Tommasi, N.D. Chemical profiling of *Astragalus membranaceus* roots (Fish.) bunge herbal preparation and evaluation of its bioactivity. *Nat. Prod. Commun.* **2020**, *15*, 1934578X20924152. [CrossRef]
28. Pinho, B.R.; Ferreres, F.; Valentão, P.; Andrade, P.B. Nature as a source of metabolites with cholinesterase-inhibitory activity: An approach to Alzheimer's disease treatment. *J. Pharm. Pharmacol.* **2013**, *65*, 1681–1700. [CrossRef]

29. Robles, A.J.; Du, L.; Cichewicz, R.H.; Mooberry, S.L. Maximiscin induces DNA damage, activates DNA damage response pathways, and has selective cytotoxic activity against a subtype of triple-negative breast cancer. *J. Nat. Prod.* **2016**, *79*, 1822–1827. [CrossRef] [PubMed]

30. Hirshfield, K.M.; Ganesan, S. Triple-negative breast cancer: Molecular subtypes and targeted therapy. *Curr. Opin. Obstet. Gynecol.* **2014**, *26*, 34–40. [CrossRef]

31. Chudzik, M.; Korzonek-Szlacheta, I.; Król, W. Triterpenes as potentially cytotoxic compounds. *Molecules* **2015**, *20*, 1610–1625. [CrossRef]

32. Huang, M.; Lu, J.J.; Huang, M.Q.; Bao, J.L.; Chen, X.P.; Wang, Y.T. Terpenoids: Natural products for cancer therapy. *Expert Opin. Invest. Drugs* **2012**, *21*, 1801–1818. [CrossRef]

33. Dinkova-Kostova, A.T. Protection against cancer by plant phenylpropenoids: Induction of mammalian anticarcinogenic enzymes. *Mini-Rev. Med. Chem.* **2002**, *2*, 595–610. [CrossRef]

34. Schwartz, G.K.; Shah, M.A. Targeting the cell cycle: A new approach to cancer therapy. *J. Clin. Oncol.* **2005**, *23*, 9408–9421. [CrossRef]

35. Geiger, T.R.; Peeper, D.S. Metastasis mechanisms. *Biochim. Biophys. Acta* **2009**, *1796*, 293–308. [CrossRef]

36. Aznavoorian, S.; Murphy, A.N.; Stetler-Stevenson, W.G.; Liotta, L.A. Molecular aspects of tumor cell invasion and metastasis. *Cancer* **1993**, *71*, 1368–1383. [CrossRef]

37. Zhu, Z.W.; Chen, L.; Liu, J.X.; Huang, J.W.; Wu, G.; Zheng, Y.F.; Yao, K.T. A novel three-dimensional tumorsphere culture system for the efficient and low-cost enrichment of cancer stem cells with natural polymers. *Exp. Ther. Med.* **2018**, *15*, 85–92. [CrossRef]

38. Hooker, J.D. *The Flora of British India*; L. Reeve and Company: London, UK, 1954.

39. Faraone, I.; Rai, D.K.; Russo, D.; Chiummiento, L.; Fernandez, E.; Choudhary, A.; Milella, L. Antioxidant, antidiabetic, and anticholinesterase activities and phytochemical profile of *Azorella glabra* Wedd. *Plants* **2019**, *8*, 265. [CrossRef]

40. Lelario, F.; De Maria, S.; Rivelli, A.R.; Russo, D.; Milella, L.; Bufo, S.A.; Scrano, L. A complete survey of glycoalkaloids using LC-FTICR-MS and IRMPD in a commercial variety and a local landrace of eggplant (*Solanum melongena* L.) and their anticholinesterase and antioxidant activities. *Toxins* **2019**, *11*, 230. [CrossRef]

41. Haque, M.A.; Islam, M.A.U. *Pleurotus highking* mushroom induces apoptosis by altering the balance of proapoptotic and antiapoptotic genes in breast cancer cells and inhibits tumor sphere formation. *Medicina* **2019**, *55*, 716. [CrossRef] [PubMed]

42. Haque, M.A.; Reza, A.S.M.A.; Nasrin, M.S.; Rahman, M.A. *Pleurotus highking* mushrooms potentiate antiproliferative and antimigratory activity against triple-negative breast cancer cells by suppressing Akt signaling. *Integr. Cancer Ther.* **2020**, *19*, 1–9. [CrossRef] [PubMed]

43. Lee, C.H.; Yu, C.C.; Wang, B.Y.; Chang, W.W. Tumorsphere as an effective in vitro platform for screening anti-cancer stem cell drugs. *Oncotarget* **2016**, *7*, 1215–1226. [CrossRef] [PubMed]

44. Blakeney, A.B.; Harris, P.J.; Henry, R.J.; Stone, B.A. A simple and rapid preparation of alditol acetates for monosaccharide analysis. *Carbohydr. Res.* **1983**, *113*, 291–299. [CrossRef]

45. Leontein, K.; Lindberg, B.; Lönngren, J. Assignment of absolute configuration of sugars by GLC of their acetylated glycosides formed from chiral alcohols. *Carbohydr. Res.* **1978**, *62*, 359–362. [CrossRef]

 *molecules*

Article

Glochidion littorale Leaf Extract Exhibits Neuroprotective Effects in *Caenorhabditis elegans* via DAF-16 Activation

Abdel Fawaz Bagoudou [1], Yifeng Zheng [2], Masahiro Nakabayashi [2], Saroat Rawdkuen [3], Hyun-Young Park [4], Dhiraj A. Vattem [4,5], Kenji Sato [6], Soichiro Nakamura [1] and Shigeru Katayama [1,2,*]

1   Graduate School of Medicine, Science and Technology, Shinshu University, 8304 Minamiminowa, Kamiina, Nagano 399-4598, Japan; 18hs552b@shinshu-u.ac.jp (A.F.B.); snakamu@shinshu-u.ac.jp (S.N.)
2   Institute for Biomedical Sciences, Shinshu University, 8304 Minamiminowa, Kamiina, Nagano 399-4598, Japan; zhengyf@shinshu-u.ac.jp (Y.Z.); 15aa408a@shinshu-u.ac.jp (M.N.)
3   School of Agro-Industry, Mae Fah Luang University, 333 Moo 1, Thasud, Muang, Chiang Rai 57100, Thailand; saroat@mfu.ac.th
4   Edison Biotechnology Institute, Konneker Research Laboratories, Ohio University, Athens, OH 45701, USA; parkh4@ohio.edu (H.-Y.P.); vattem@ohio.edu (D.A.V.)
5   College of Health Sciences & Professions, Ohio University, Athens, OH 45701, USA
6   Graduate School of Agriculture, Kyoto University, Kyoto 606-8502, Japan; kensato@kais.kyoto-u.ac.jp
*   Correspondence: skata@shinshu-u.ac.jp; Tel.: +81-265-77-1603

Abstract: A number of plants used in folk medicine in Thailand and Eastern Asia are attracting interest due to the high bioactivities of their extracts. The aim of this study was to screen the edible leaf extracts of 20 plants found in Thailand and investigate the potential neuroprotective effects of the most bioactive sample. The total phenol and flavonoid content and 2,2-diphenyl-1-picrylhydrazyl radical-scavenging activity were determined for all 20 leaf extracts. Based on these assays, *Glochidion littorale* leaf extract (GLE), which showed a high value in all tested parameters, was used in further experiments to evaluate its effects on neurodegeneration in *Caenorhabditis elegans*. GLE treatment ameliorated H_2O_2-induced oxidative stress by attenuating the accumulation of reactive oxygen species and protected the worms against 1-methyl-4-phenylpyridinium-induced neurodegeneration. The neuroprotective effects observed may be associated with the activation of the transcription factor DAF-16. The characterization of this extract by LC-MS identified several phenolic compounds, including myricetin, coumestrin, chlorogenic acid, and hesperidin, which may play a key role in neuroprotection. This study reports the novel neuroprotective activity of GLE, which may be used to develop treatments for neurodegenerative diseases such as Parkinson's syndrome.

Keywords: *Caenorhabditis elegans*; leaf extract; neuroprotection; antioxidant activity; DAF-16

Citation: Bagoudou, A.F.; Zheng, Y.; Nakabayashi, M.; Rawdkuen, S.; Park, H.-Y.; Vattem, D.A.; Sato, K.; Nakamura, S.; Katayama, S. *Glochidion littorale* Leaf Extract Exhibits Neuroprotective Effects in *Caenorhabditis elegans* via DAF-16 Activation. *Molecules* **2021**, *26*, 3958. https://doi.org/10.3390/molecules26133958

Academic Editors: Raffaele Pezzani and Sara Vitalini

Received: 31 May 2021
Accepted: 24 June 2021
Published: 28 June 2021

Publisher's Note: MDPI stays neutral with regard to jurisdictional claims in published maps and institutional affiliations.

1. Introduction

Neurodegenerative disorders including Alzheimer's disease and Parkinson's disease (PD) pose major health and financial concerns to global health care organizations [1]. Although the human lifespan has increased in the last few decades in industrialized countries, the prevalence of age-related diseases has also increased. The incidence of late-onset disorders such as neurological disruptions is expected to increase rapidly over the next few decades. Therefore, it is crucial to encourage studies and perform clinical trials on compounds that may have the potential to cure, prevent, or at least delay the onset of neurodegenerative diseases [2]. One of the characteristic features of PD is the progressive loss of dopaminergic (DA) neurons in the substantia nigra [3]. In PD pathogenesis, increased production of reactive oxygen species (ROS) plays a key role in the loss of DA cells [4]. Therefore, the reduction in oxidative stress is considered a promising therapeutic approach in PD treatment [5]. The 1-methyl-4-phenylpyridinium (MPP^+), which inhibits mitochondrial complex I activity, can induce PD-like symptoms in humans and animal models [6].

The use of *Caenorhabditis elegans* as an in vivo model provides certain advantages in the study of PD [7]. The nematode is simple, inexpensive, and has a short life cycle. It supports studies involving large-scale analyses. Moreover, the neuronal network of *C. elegans* has been mapped completely. It contains 8 DA neurons and PD-related homologous genes [8]. Neurodegeneration, which mimics parkinsonian symptoms, can be induced in *C. elegans* via treatment with neurotoxins such as MPP^+ [9].

Natural antioxidant compounds represent attractive sources for developing drugs to treat neurodegenerative diseases due to their neuroprotective effects in animal models and low toxicity [3]. Polyphenols are known to be among the most abundant antioxidants in the human diet [10]. It has also been established that oxidative processes are involved in many pathologies, including neurodegeneration, cancer, diabetes, cardiovascular and anti-inflammatory diseases. Hence, finding polyphenols exhibiting antioxidant properties from natural sources could contribute toward preventing or treating those pathologies. This study focused on extracts from the edible leaves of plants found in Thailand. Most varieties cultivated widely in northern and southern Thailand have been used as folk medicine against general injuries and diseases; however, there are few reports concerning their neuroprotective effects.

In this study, we first screened the extracts of edible leaves from 20 plants cultivated in Thailand and assessed their phenolic and flavonoid contents and their 2,2-diphenyl-1-picrylhydrazyl (DPPH) radical-scavenging activity. The effects of *Glochidion littorale* leaf extract (GLE), which showed a high value in all tested parameters, were evaluated on *C. elegans* with neurodegeneration. Furthermore, the potential pathways involved in the neuroprotective effect of GLE were examined, along with the identification of the main components in GLE.

2. Results

2.1. Screening of Thai Plant Leaves

Crude extracts of edible leaves from plants cultivated in Thailand were prepared by ultrasonication. The leaf extracts of 20 plants were screened for their phenolic and flavonoid contents and antioxidant activity by DPPH radical-scavenging assay. Few of the tested samples, such as *Glochidion sphaerogynum* and *Mentha piperita*, were found to possess high radical-scavenging activity with low phenolic and flavonoid content, whereas certain samples, such as *Clinacanthus nutans* and *Ocimum* × *citriodorum*, exhibited the opposite trend (Table 1). The leaf extract of *G. littorale* showed high DPPH radical-scavenging activity as well as high phenolic and flavonoid content. Therefore, the bioactivities associated with *G. littorale* were further investigated.

2.2. GLE Enhanced Resistance against Oxidative Stress via DAF-16 in C. Elegans

The effect of GLE on the survival of N2 worms under oxidative stress was investigated. Treatment with H_2O_2 (5 mM) induced 75% death in the control group, whereas co-treatment with 50 μg/mL and higher concentrations of GLE was associated with a high survival rate (Figure 1A). Among the tested concentrations of GLE, 100 μg/mL and 200 μg/mL were associated with the highest survival rates (82.0% and 88.2%, respectively). Therefore, these two concentrations were used in subsequent experiments. To evaluate the antioxidant effect of GLE in vivo, the intracellular ROS levels were measured in wild-type nematodes using 2′,7′-dichlorodihydrofluorescein diacetate (H_2DCF-DA), a well-known fluorescence probe for detecting intracellular ROS production. Significant decreases in the fluorescence intensities in the GLE-treated groups were observed compared to that in the untreated group (Figure 1B), confirming the antioxidant property of GLE.

Table 1. Properties of the plants investigated in this study.

No.	Scientific Name	Phenolic Content (mg GAE [1]/g)	Flavonoid Content (mg quercetin/g)	DPPH [2] Radical-Scavenging Activity (%)
1	*Clinacanthus nutans*	3.306	1.799	41.058
2	*Gymnema inodorum*	1.183	0.959	33.890
3	*Glochidion sphaerogynum*	0.709	0.781	51.705
4	*Anethum graveolens*	0.825	0.387	29.367
5	*Spilanthes acmella*	0.552	0.243	18.580
6	*Acacia pennata*	5.031	1.563	51.914
7	*Mentha piperita*	1.107	0.761	59.151
8	*Glochidion littorale*	20.104	4.527	78.984
9	*Ocimum sanctum* Linn.	0.131	0.076	45.442
10	*Ocimum basilicum* Linn.	1.342	0.771	51.635
11	*Ocimum* × *citriodorum*	2.446	1.612	39.666
12	*Azadirachta indica*	13.744	2.725	79.819
13	*Morus Alba*	6.190	4.019	54.488
14	*Moringa oleifera*	1.696	5.696	18.928
15	*Psidium guajava* Linn.	3.414	2.937	61.865
16	*Melientha suavis* Pierre	2.263	2.433	40.362
17	*Pandanus amaryllifolius*	1.409	1.050	36.395
18	*Zanthoxylum limonella*	3.128	1.469	46.555
19	*Piper sarmentosum*	0.584	0.487	24.217
20	*Citrus maxima*	11.690	2.461	79.193

[1] GAE, gallic acid equivalent; [2] DPPH, 2,2-diphenyl-1-picrylhydrazyl.

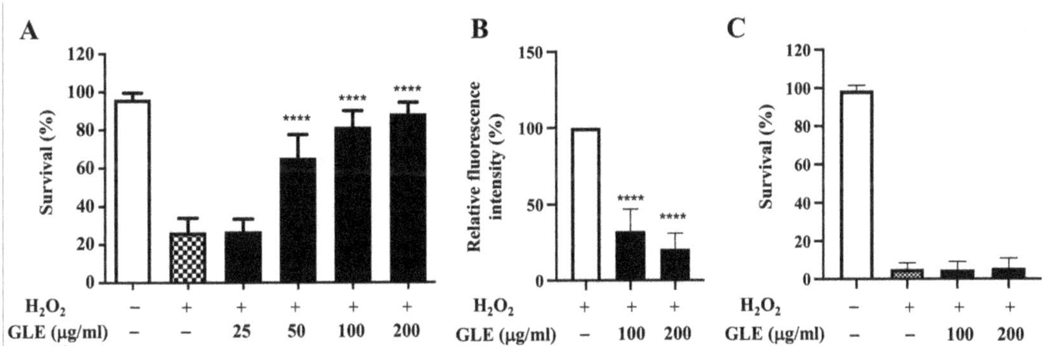

Figure 1. Effect of *Glochidion litorale* leaf extract (GLE) on stress resistance in wild-type and *daf-16* mutant *Caenorhabditis elegans*. (**A**) Effect of GLE against H_2O_2-induced toxicity in wild-type worms. (**B**) Intracellular reactive oxygen species (ROS) contents in wild-type worms. (**C**) Effect of GLE against H_2O_2-induced toxicity in *daf-16* mutant worms. Experiments were performed in triplicate. Data are presented as mean ± standard deviation (SD). **** $p < 0.0001$ compared to H_2O_2-treated worms.

As the transcription factor DAF-16 is known to play a key role in regulating oxidative stress [11], it was hypothesized that GLE may target DAF-16. The *C. elegans* strain CF1038, which is a DAF-16 loss-of-function mutant strain, was used to determine the survival rate of worms treated with and without GLE. In H_2O_2-induced oxidative stress, GLE treatment did not increase the survival rate of transgenic worms (Figure 1C).

2.3. GLE Treatment Reduced the Lethality of MPP$^+$-Induced DA Neurotoxicity via DAF-16 in C. Elegans

C. elegans possesses 8 DA neurons [8]. Selective degeneration of these DA neurons was evaluated after exposure to MPP$^+$. The treatment of wild-type N2 worms with 0.75 mM MPP$^+$ resulted in a remarkable decrease in survival (Figure 2). However, co-treatment with GLE significantly increased the survival of the worms. The effect of GLE treatment on *daf-16* mutant worms was investigated. As shown in Figure 3 and Table 2, GLE treatment did not increase the survival of these worms after exposure to MPP$^+$ compared to that in the control group. These results suggest that DAF-16 may be required for mediating the neuroprotective effect of GLE in *C. elegans*. Next, a DAF-2 loss-of-function mutant strain, CB1370, was used to determine whether DAF-2 was involved in the observed neuroprotective effects. As shown in Figure 4 and Table 3, the median and maximum survival significantly increased in *daf-2* mutant worms treated with GLE.

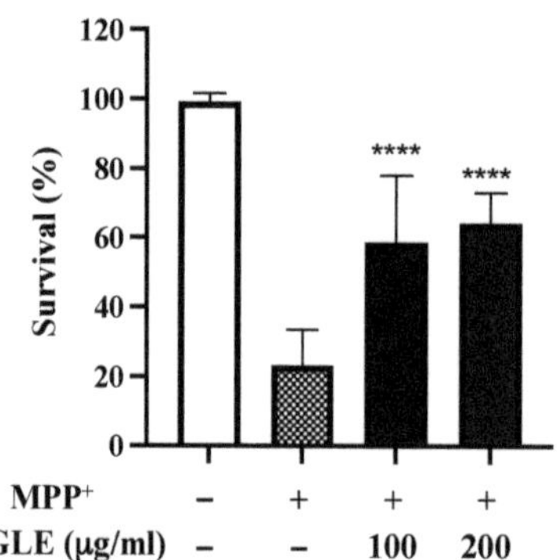

Figure 2. Effect of GLE on 1-methyl-4-phenylpyridinium ion (MPP$^+$)-induced neurotoxicity in N2 *C. elegans*. The effects of GLE (100 and 200 µg/mL) on MPP$^+$-induced toxicity were evaluated. Experiments were performed in triplicate. Data are presented as mean $\pm$ SD. **** $p < 0.0001$ compared to MPP$^+$-treated worms.

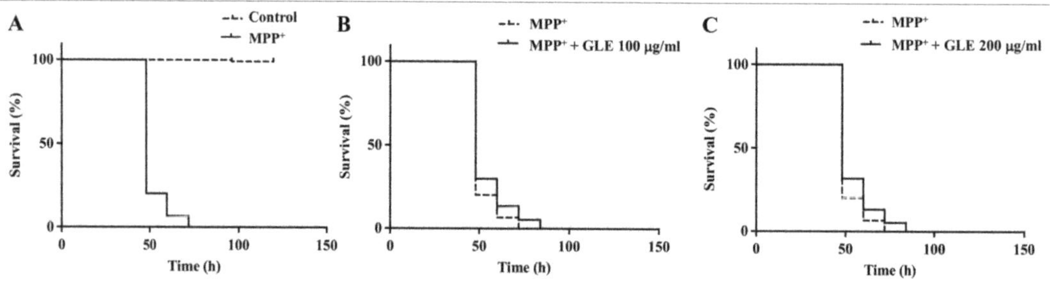

Figure 3. Effect of GLE on MPP$^+$-induced neurotoxicity in *daf-16* mutant *C. elegans*. (**A**) Lifespan curve of worms in the presence or absence of MPP$^+$. (**B**) Lifespan curve of worms with MPP$^+$-induced toxicity treated with 100 µg/mL GLE. (**C**) Lifespan curve of worms with MPP$^+$-induced toxicity treated with 200 µg/mL GLE. Each experiment was repeated independently at least thrice, and one of the representative data is shown.

Table 2. Survival of *daf-16* mutant *C. elegans* treated with MPP$^+$.

Survival Time	MPP$^+$ [1]	MPP$^+$ + GLE (100 µg/mL)	MPP$^+$ + GLE (200 µg/mL)
Median (h)	48.0 ± 1.2	48.0 ± 1.7	48.0 ± 1.6
Maximum (h)	72.0 ± 1.5	72.0 ± 2.1	72.0 ± 1.8

[1] MPP$^+$, 1-methyl-4-phenylpyridinium.

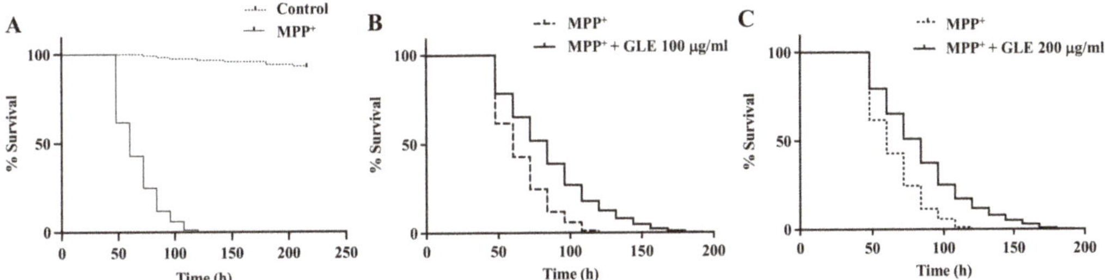

Figure 4. Effect of GLE on MPP⁺-induced neurotoxicity in *daf-2* mutant *C. elegans*. (**A**) Lifespan curve of worms in the presence or absence of MPP⁺. (**B**) Lifespan curve of worms with MPP⁺-induced toxicity treated with 100 μg/mL GLE. (**C**) Lifespan curve of worms with MPP⁺-induced toxicity treated with 200 μg/mL GLE. Each experiment was repeated independently at least thrice, and one of the representative data is shown.

Table 3. Survival of *daf-2* mutant *C. elegans* treated with MPP⁺.

Survival Time	MPP⁺ [1]	MPP⁺ + GLE (100 μg/mL)	MPP⁺ + GLE (200 μg/mL)
Median (h)	60.0 ± 2.6	84.0 ± 3.9 [***]	84.0 ± 4.0 [***]
Maximum (h)	108.0 ± 5.3	180.0 ± 5.5 [****]	192.0 ± 9.6 [****]

[***] $p < 0.001$, [****] $p < 0.0001$ vs. [1] MPP⁺-treated worms.

2.4. Effects of GLE on DAF-16 Localization

It has been demonstrated that DAF-16 activation is regulated by its nuclear accumulation [12]. Subsequently, we investigated whether GLE could induce the nuclear accumulation of DAF-16 in a transgenic strain TJ356 that expresses a DAF-16::GFP fusion protein. Results showed that after 48 h of incubation with 100 μg/mL GLE, the green fluorescence intensity of DAF-16 in the nucleus increased significantly compared to that in the untreated group (Figure 5).

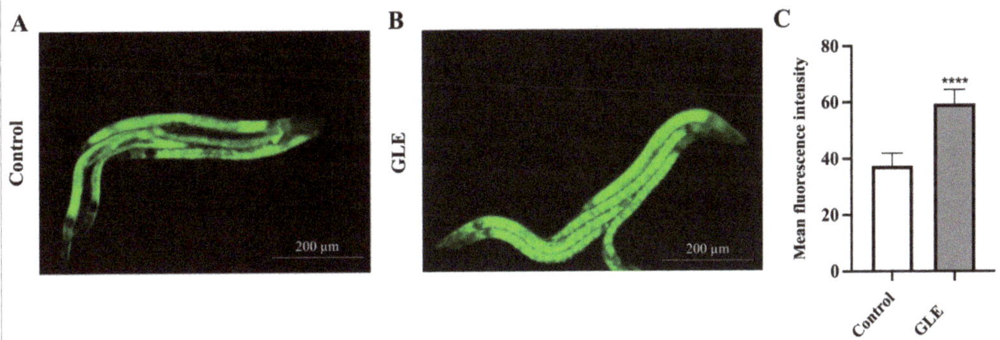

Figure 5. Effect of GLE on DAF-16 localization. (**A**) Untreated worms. (**B**) Worms treated with 100 μg/mL GLE. (**C**) Quantification of DAF-16::GFP nuclear accumulation in GLE and GLE-free conditions. The scale bar shows 200 μm. Each experiment was repeated independently at least thrice. Significant differences were analyzed using the *t*-test method; [****] $p < 0.0001$ as compared with control.

2.5. Phytochemical Characterization in GLE

LC-MS was conducted for profiling the phytochemicals in GLE, and its results are presented in Figure 6. The chromatographic peaks were identified by comparing the MS data with databases based on the search of m/z values of molecular ion peaks in the positive mode $[M + H]^+$. Consequently, myricetin, coumestrin, chlorogenic acid, and hesperidin were detected as the major compounds (Table 4).

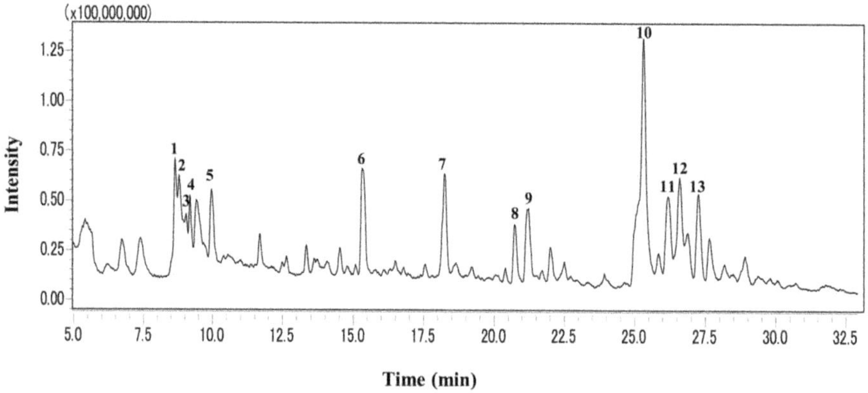

Figure 6. LC-MS profile of GLE. The total ion chromatogram was obtained by a triple quadrupole mass spectrometer operated in the positive electrospray ionization mode.

Table 4. Compounds identified from the chromatogram of GLE.

Peak	Retention Time (min)	$[M + H]^+$ (*m/z*)	Identified Compounds	Theoretical Mass	Mass Error (ppm)
1	8.7	431.0973	Coumestrin	430.0900	6
2	8.8	299.2005	All-trans-3,4-didehydro-retinoic acid	298.1933	1
3	9.0	248.2009	Lycopodine	247.1936	3
4	9.2	289.0707	2-Hydroxynaringenin	288.0634	10
5	10.0	166.1226	Hordenine	165.1154	1
6	15.3	355.1024	Chlorogenic acid	354.0952	7
7	18.3	611.1970	Hesperidin	610.1898	5
8	20.7	449.1078	Quercitrin	448.1006	17
9	21.2	459.0922	Epigallocatechin gallate	458.0849	3
10	25.3	319.0448	Myricetin	318.2370	5
11	26.2	465.1028	Isoquercitrin	464.0955	3
12	26.6	465.3575	Unknown	-	-
13	27.3	567.3038	Unknown	-	-

3. Discussion

Plant extracts are a rich source of natural bioactive compounds. Many studies have evaluated plant extracts used in Southeast Asian countries, including Thailand, where these extracts are components of folk medicine [13,14]. In this study, the extracts of 20 edible plant leaves from Thailand were screened, and *G. littorale* was selected for further investigation because it showed high phenol content, flavonoid content, and radical-scavenging activity. Several studies have investigated various species of the genus *Glochidion* [15–19]; however, there are few studies concerning the functional properties and constituents of *G. littorale*. Our data showed that GLE protected *C. elegans* against H_2O_2-induced oxidative stress by reducing intracellular ROS accumulation. This might have been due to the high content of phenolic compounds such as flavonoids, which are known to possess strong antioxidant activity [20]. These findings are similar to those obtained by Duangjan et al. (2019), who showed that *G. zeylanicum* leaf extracts can protect *C. elegans* against oxidative stress [21]. The insulin/insulin-like signaling (IIS) pathway regulates growth, stress responsiveness, and longevity in *C. elegans* [22,23]. We found that *daf-16* null mutant *C. elegans* treated with GLE were susceptible to oxidative stress. This result suggests that the antioxidant effect of GLE in reducing oxidative stress in nematodes is possibly involved in not only radical-scavenging activity but also the regulation of the DAF-16 transcription factor.

The protective effects of GLE against MPP$^+$-induced toxicity in *C. elegans* were examined. DA neurons in nematodes take up MPP$^+$ mainly via high-affinity DA transporters, which is similarly observed in mammals. The accumulation of MPP$^+$ inside the neurons inactivates the mitochondrial complex I of the respiratory chain and induces cell death [24–27]. GLE treatment was found to significantly reduce the lethality associated with MPP$^+$ treatment in wild-type worms. The IIS pathway is modulated by insulin-like peptides through the DAF-2 receptor in *C. elegans* [28]. Under normal conditions, the IIS pathway inhibits the phosphorylation of DAF-16 and prevents its nuclear translocation. In *daf-2* null mutants, the GLE-treated group survived longer than the control group. In contrast, no difference in survival was observed between the control group and the GLE-treated group containing *daf-16* null mutant worms. It is known that downregulated DAF-2 signaling facilitates the entry of DAF-16 into the nucleus, where it can upregulate the expression of target genes and control stress resistance and longevity [29]. This may explain why *daf-2* mutant worms treated with GLE showed a relatively higher survival. Furthermore, an increased nuclear accumulation of DAF-16 in worms treated with GLE was observed using transgenic TJ356 DAF-16::GFPC. *elegans*. Cumulatively, these results indicated that GLE might have exhibited its neuroprotective effects via activation of DAF-16.

LC-MS profiling led to the identification of 11 phytochemical compounds in GLE. Myricetin identified in the main peak of GLE is a flavonoid widely found in many plants and is well known to exhibit protective effects against oxidative stress. A previous study has demonstrated that myricetin extended the lifespan of *C. elegans* by diminishing stress-induced ROS accumulation and the pro-longevity effects of myricetin were dependent on DAF-16 [30]. Chlorogenic acid has also been reported to exhibit pro-longevity effects via the attenuation of oxidative stress in *C. elegans* [31]. Considering these findings, the neuroprotective effects of GLE were mainly induced by flavonoids such as myricetin, and GLE might be a suitable candidate for the management of neurodegenerative diseases.

In conclusion, our study demonstrated that GLE possessed strong antioxidant activity, which reduced oxidative stress in *C. elegans*. The extract also showed neuroprotective activity against MPP$^+$-induced neurotoxicity in *C. elegans*. Various experiments performed using different transgenic worms suggested the possible involvement of the DAF-16 transcription factor in the observed neuroprotection. The high content of phenolic compounds, including flavonoids present in GLE, may be responsible for the observed stress resistance and neuroprotective properties. Further studies should identify the target genes involved in the neuroprotection mechanism.

4. Materials and Methods

4.1. Materials

The leaves of 20 different plants (Table 1) were obtained from a local market in Chiang Rai, Thailand. All reagents were of analytical grade. DPPH and H$_2$DCF-DA were obtained from Sigma-Aldrich (St. Louis, MO, USA). All other chemicals were obtained from Wako Pure Chemical Industries (Osaka, Japan).

4.2. Preparation of Leaf Extracts

Leaf samples were frozen in liquid nitrogen, and then powdered samples (5 g) were mixed in 100 mL of distilled water at 45 °C for 30 min, following sonication using a Branson SLPe Sonifier (Branson, North Billerica, MA, USA) at 35 kHz. The extract was filtered and freeze-dried to obtain a powdered sample.

4.3. Total Phenolic Contents

The Folin-Ciocalteu method was used to determine the total phenolic content. Briefly, 11.4 μL of the extract (1 mg/mL) was mixed with 227.3 μL of 2% (*w/v*) Na$_2$CO$_3$ solution, and then the mixture was allowed to stand at room temperature for 2 min. After addition of 11.4 μL of 10% (*v/v*) Folin-Ciocalteu reagent. The incubation in the dark was conducted for 30 min. Subsequently, the absorbance was measured at 750 nm using a microplate reader (Nivo 3F Multimode Plate Reader, PerkinElmer, Waltham, MA, USA). Gallic acid was used as a standard for the calibration curve. The total phenolic content was expressed as gallic acid equivalents (mg gallic acid equivalent/g of plant extract).

4.4. Total Flavonoid Contents

The aluminum chloride colorimetric method was used to measure the total flavonoid content. Briefly, 25 µL of the extract (2 mg/mL) was mixed with 7.5 µL of 5% (*w/v*) $NaNO_2$ solution and 152.5 µL of distilled water. After 6 min, 15 µL of 10% (*w/v*) $AlCl_3$ solution was added and allowed to stand for 5 min. Then, 50 µL of 1 M NaOH solution was added to the mixture. Subsequently, the mixture was incubated in the dark for 15 min, and the absorbance was measured at 510 nm using a microplate reader. The total flavonoid content was calculated by generating a calibration curve using quercetin as a standard, and the results were expressed as quercetin equivalent (mg quercetin equivalent/g of plant extract).

4.5. Free Radical-Scavenging Activity

The capacity to scavenge free radicals was assessed using DPPH assays [32]. Briefly, 100 µL of the extract (1 mg/mL) were mixed with 100 µL of DPPH solution. After 30 min, the absorbance was measured at 517 nm using a microplate reader. The results were expressed as a percentage of inhibition of the DPPH radicals.

4.6. C. Elegans Maintenance

Wild-type N2, CF1038 (*daf-16(mu86) I*), CB1370 (*daf-2(e1370) III*), and TJ356 (*zIs356 [daf-16p::daf-16a/b::GFP + rol-6(su1006)]*) strains and their diet, *Escherichia coli* OP50, were obtained from the Caenorhabditis Genetics Center (Minneapolis, MN, USA). According to the standard protocols, N2, CF1038, and TJ356 strains and CF 1370 strain were maintained at 20 and 15 °C, respectively, on nematode growth medium (NGM) agar plates containing heat inactivated *E. coli* OP50 [33]. S-complete solution was prepared according to previously described literature [34].

4.7. Oxidative Stress Assays

Oxidative stress was induced by treating wild-type (N2) and *daf-16* mutant (CF1038) worms with H_2O_2. L1 larvae were added to 96-well plates at an average of 15 nematodes per well in a 40 µL solution containing *E. coli* OP50. Five mM of H_2O_2 solution and the tested GLE dissolved in S-complete solution were added to achieve a final volume of 50 µL per well. L1 larvae were incubated for 48 h with H_2O_2 alone or in the presence of various concentrations of GLE, and worm viability was visually inspected under a stereomicroscope. The results from the H_2O_2-treated groups were normalized and expressed as a percentage of normal controls. The results were obtained from three independent experiments (100–160 worms/treatment in each experiment).

4.8. Intracellular ROS Levels

Intracellular ROS levels were determined using the H_2DCF-DA probe. L1 larvae of wild-type N2 worms were treated with 5 mM H_2O_2 and GLE at different concentrations in S-solution for 48 h in black 96-well plates; each well comprised a minimum of 100 worms. Worms were subsequently incubated with 25 µM H_2DCF-DA in the dark at 20 °C for 1 h. After incubation, the fluorescence intensity was measured at wavelengths of 485/530 nm using a Powerscan HT microplate reader (DS Pharma Biomedical, Osaka, Japan).

4.9. Neurotoxicity Assay

Neurotoxicity was induced by treating wild-type (N2) and transgenic (CF1038 and CB1370) worms with MPP^+. L1 larvae were added to 96-well plates (15 worms/well) in a 40 µL solution containing *E. coli* OP50. Worms were then incubated with 50 µL of 0.75 mM MPP^+ alone or in the presence of different concentrations of the tested sample for 48 h. After incubation, worm viability was visually inspected under a stereomicroscope.

Live worms were counted every 12 h post treatment until no live worms remained. The results of the MPP$^+$-treated groups were normalized and expressed as a percentage of the normal controls. The results were obtained from three independent experiments (80–130 worms/treatment in each experiment).

4.10. Nuclear Localization of DAF-16

Transgenic *C. elegans* TJ356, which expresses a DAF-16-GFP fusion protein, was used to examine the intracellular distribution of DAF-16. L1 stage nematodes were treated with GLE for 48 h at 20 °C. The worms were then transferred to a 2% agarose pad on a glass slide and anesthetized by adding one drop (approximately 20 µL) of 25 µM sodium azide to the agarose pad. The expression of GFP was examined via fluorescence microscopy (EVOS fl; Advanced Microscopy Group, Bothell, WA, USA). The mean fluorescence intensity of DAF-16 in the nuclei was analyzed using Image J software (National Institutes of Health, Bethesda, MD, USA).

4.11. Phytochemical Profiling Using LC-MS

The leaf extract was analyzed using the LCMS-8040 (Shimadzu). Mass spectra were acquired over a range of m/z 50–1000 using the Q3 scan mode. The solution was injected onto an Inertsil ODS-3 (250 × 2.1 mm, 5 µm, GL Sciences, Tokyo Japan) at a column temperature at 40 °C using a gradient of (A) 0.1% formic acid and (B) acetonitrile/water (80/20) containing 0.1% formic acid. The following gradient with a flow rate of 0.2 mL/min was used: 0–100% B (0–45 min), 100% B (45–50 min), and 0% B (50–60 min). Compounds were putatively identified by matching the experimental m/z values to the library of theoretical calculated m/z values in databases, including the Human Metabolome Database and the METLIN database.

4.12. Statistical Analysis

Data were expressed as the mean ± standard deviation for each group. The significant difference between the two groups was assessed using the *t*-test, whereas the difference between three and more groups was assessed using one-way ANOVA, followed by Tukey's post-hoc comparison test. Statistical significance was set at $p < 0.001$ and $p < 0.0001$. For lifespan assays, C. elegans survival was plotted using Kaplan–Meier survival curves and analyzed via log-rank tests using GraphPad Prism software (version 9.01; GraphPad Software, San Diego, CA, USA).

Author Contributions: Conceptualization, S.R., D.A.V., and S.K.; investigation, A.F.B., Y.Z., M.N., S.R., H.-Y.P., D.A.V., and K.S.; writing—original draft preparation, A.F.B.; writing—review and editing, S.K.; supervision, S.N. All authors have read and agreed to the published version of the manuscript.

Funding: This research received no external funding.

Institutional Review Board Statement: Not applicable.

Informed Consent Statement: Not applicable.

Data Availability Statement: The data are available by the corresponding author upon reasonable request.

Conflicts of Interest: The authors declare no conflict of interest.

Sample Availability: Samples of the compounds are not available from the authors.

References

1. Pohl, F.; Lin, P.K.T. The Potential Use of Plant Natural Products and Plant Extracts with Antioxidant Properties for the Prevention/Treatment of Neurodegenerative Diseases: In Vitro, In Vivo and Clinical Trials. *Molecules* **2018**, *23*, 3283. [CrossRef] [PubMed]
2. Kim, G.H.; Kim, J.E.; Rhie, S.J.; Yoon, S. The Role of Oxidative Stress in Neurodegenerative Diseases. *Exp. Neurobiol.* **2015**, *24*, 325–340. [CrossRef] [PubMed]
3. Lu, X.-l.; Yao, X.-l.; Liu, Z.; Zhang, H.; Li, W.; Li, Z.; Wang, G.-L.; Pang, J.; Lin, Y.; Xu, Z. Protective effects of xyloketal B against MPP+-induced neurotoxicity in *Caenorhabditis elegans* and PC12 cells. *Brain Res.* **2010**, *1332*, 110–119. [CrossRef]
4. Trimmer, P.A.; Bennett, J.P., Jr. The cybrid model of sporadic Parkinson's disease. *Exp. Neurol.* **2009**, *218*, 320–325. [CrossRef]
5. Cheon, S.-M.; Jang, I.; Lee, M.H.; Kim, D.K.; Jeon, H.; Cha, D.S. Sorbus alnifolia protects dopaminergic neurodegeneration in *Caenorhabditis elegans*. *Pharm. Biol.* **2016**, *55*, 481–486. [CrossRef]
6. Schmidt, N.; Ferger, B. Neurochemical findings in the MPTP model of Parkinson's disease. *J. Neural Transm.* **2001**, *108*, 1263–1282. [CrossRef] [PubMed]
7. Harrington, A.; Yacoubian, T.A.; Slone, S.R.; Caldwell, K.; Caldwell, G. Functional Analysis of VPS41-Mediated Neuroprotection in *Caenorhabditis elegans* and Mammalian Models of Parkinson's Disease. *J. Neurosci.* **2012**, *32*, 2142–2153. [CrossRef] [PubMed]
8. Fu, R.-H.; Harn, H.-J.; Liu, S.-P.; Chen, C.-S.; Chang, W.-L.; Chen, Y.-M.; Huang, J.-E.; Li, R.-J.; Tsai, S.-Y.; Hung, H.-S.; et al. n-Butylidenephthalide Protects against Dopaminergic Neuron Degeneration and α-Synuclein Accumulation in *Caenorhabditis elegans* Models of Parkinson's Disease. *PLOS ONE* **2014**, *9*, e85305. [CrossRef]
9. Jadiya, P.; Khan, A.; Sammi, S.R.; Kaur, S.; Mir, S.S.; Nazir, A. Anti-Parkinsonian effects of Bacopa monnieri: Insights from transgenic and pharmacological *Caenorhabditis elegans* models of Parkinson's disease. *Biochem. Biophys. Res. Commun.* **2011**, *413*, 605–610. [CrossRef]
10. Andrade, J.M.D.M.; Fasolo, D. Polyphenol Antioxidants from Natural Sources and Contribution to Health Promotion. In *Polyphenols in Human Health and Disease*; Elsevier BV: Amsterdam, The Netherlands, 2014; pp. 253–265.
11. Hsu, A.-L.; Murphy, C.T.; Kenyon, C. Regulation of Aging and Age-Related Disease by DAF-16 and Heat-Shock Factor. *Science* **2003**, *300*, 1142–1145. [CrossRef]
12. Henderson, S.T.; Johnson, T.E. daf-16 integrates developmental and environmental inputs to mediate aging in the nematode *Caenorhabditis elegans*. *Curr. Biol.* **2001**, *11*, 1975–1980. [CrossRef]
13. Hutadilok-Towatana, N.; Chaiyamutti, P.; Panthong, K.; Mahabusarakam, W.; Rukachaisirikul, V. Antioxidative and Free Radical Scavenging Activities of Some Plants Used in Thai Folk Medicine. *Pharm. Biol.* **2006**, *44*, 221–228. [CrossRef]
14. Stewart, P.; Boonsiri, P.; Puthong, S.; Rojpibulstit, P. Antioxidant activity and ultrastructural changes in gastric cancer cell lines induced by Northeastern Thai edible folk plant extracts. *BMC Complement. Altern. Med.* **2013**, *13*, 60. [CrossRef]
15. Hui, W.; Lee, W.; Ng, K.; Chan, C. The occurrence of triterpenoids and steroids in three *Glochidion* species. *Phytochemistry* **1970**, *9*, 1099–1102. [CrossRef]
16. Takeda, Y.; Mima, C.; Masuda, T.; Hirata, E.; Takushi, A.; Otsuka, H. Glochidioboside, a glucoside of (7S,8R)-dihydrodehydrodiconiferyl alcohol from leaves of glochidion obovatum. *Phytochemistry* **1998**, *49*, 2137–2139. [CrossRef]
17. Zhang, X.; Chen, J.; Gao, K. Chemical constituents from *Glochidion* wrightii Benth. *Biochem. Syst. Ecol.* **2012**, *45*, 7–11. [CrossRef]
18. Tian, J.-M.; Fu, X.-Y.; Zhang, Q.; He, H.-P.; Gao, J.-M.; Hao, X.-J. Chemical constituents from *Glochidion assamicum*. *Biochem. Syst. Ecol.* **2013**, *48*, 288–292. [CrossRef]
19. Kongkachuichai, R.; Charoensiri, R.; Yakoh, K.; Kringkasemsee, A.; Insung, P. Nutrients value and antioxidant content of indigenous vegetables from Southern Thailand. *Food Chem.* **2015**, *173*, 838–846. [CrossRef]
20. Pietta, P.-G. Flavonoids as Antioxidants. *J. Nat. Prod.* **2000**, *63*, 1035–1042. [CrossRef] [PubMed]
21. Duangjan, C.; Rangsinth, P.; Gu, X.; Zhang, S.; Wink, M.; Tencomnao, T. *Glochidion zeylanicum* leaf extracts exhibit lifespan extending and oxidative stress resistance properties in *Caenorhabditis elegans* via DAF-16/FoxO and SKN-1/Nrf-2 signaling pathways. *Phytomedicine* **2019**, *64*, 153061. [CrossRef]
22. Jensen, V.L.; Gallo, M.; Riddle, D.L. Targets of DAF-16 involved in *Caenorhabditis elegans* adult longevity and dauer formation. *Exp. Gerontol.* **2006**, *41*, 922–927. [CrossRef] [PubMed]
23. Daitoku, H.; Fukamizu, A. FOXO Transcription Factors in the Regulatory Networks of Longevity. *J. Biochem.* **2007**, *141*, 769–774. [CrossRef]
24. Lakso, M.; Vartiainen, S.; Moilanen, A.-M.; Sirviö, J.; Thomas, J.H.; Nass, R.; Blakely, R.D.; Wong, G. Dopaminergic neuronal loss and motor deficits in *Caenorhabditis elegans* overexpressing human α-synuclein. *J. Neurochem.* **2004**, *86*, 165–172. [CrossRef] [PubMed]
25. Nass, R.; Hall, D.H.; Miller, D.M.; Blakely, R.D. Neurotoxin-induced degeneration of dopamine neurons in *Caenorhabditis elegans*. *Proc. Natl. Acad. Sci. USA* **2002**, *99*, 3264–3269. [CrossRef]
26. Wang, Y.-M.; Pu, P.; Le, W.-D. ATP depletion is the major cause of MPP+ induced dopamine neuronal death and worm lethality in α-synuclein transgenic *C. elegans*. *Neurosci. Bull.* **2007**, *23*, 329–335. [CrossRef]
27. Braungart, E.; Gerlach, M.; Riederer, P.; Baumeister, R.; Hoener, M. *Caenorhabditis elegans* MPP+ Model of Parkinson's Disease for High-Throughput Drug Screenings. *Neurodegener. Dis.* **2004**, *1*, 175–183. [CrossRef]
28. Panowski, S.H.; Dillin, A. Signals of youth: Endocrine regulation of aging in *Caenorhabditis elegans*. *Trends Endocrinol. Metab.* **2009**, *20*, 259–264. [CrossRef]

29. Mukhopadhyay, A.; Tissenbaum, H.A. Reproduction and longevity: Secrets revealed by *C. elegans*. *Trends Cell Biol.* **2007**, *17*, 65–71. [CrossRef]
30. Büchter, C.; Ackermann, D.; Havermann, S.; Honnen, S.; Chovolou, Y.; Fritz, G.; Kampkötter, A.; Wätjen, W. Myricetin-Mediated Lifespan Extension in *Caenorhabditis elegans* Is Modulated by DAF. *Int. J. Mol. Sci.* **2013**, *14*, 11895–11914. [CrossRef]
31. Zheng, S.Q.; Huang, X.B.; Xing, T.K.; Ding, A.J.; Wu, G.S.; Luo, H.R. Chlorogenic Acid Extends the Lifespan of *Caenorhabditis elegans* via Insulin/IGF-1 Signaling Pathway. *J. Gerontol. A Biol. Sci. Med. Sci.* **2017**, *72*, 464–472. [CrossRef] [PubMed]
32. Blois, M.S. Antioxidant Determinations by the Use of a Stable Free Radical. *Nat. Cell Biol.* **1958**, *181*, 1199–1200. [CrossRef]
33. Stiernagle, T. Maintenance of *C. elegans*. *WormBook* **2006**, *2*, 1–11. [CrossRef] [PubMed]
34. Amrit, F.R.G.; Ratnappan, R.; Keith, S.A.; Ghazi, A. The *C. elegans* lifespan assay toolkit. *Methods* **2014**, *68*, 465–475. [CrossRef] [PubMed]

 *molecules*

Article

The Anti-Inflammatory Effect of a γ-Lactone Isolated from Ostrich Oil of *Struthio camelus* (Ratite) and Its Formulated Nano-Emulsion in Formalin-Induced Paw Edema

Salah E. M. Eltom [1,*], Ahmed A. H. Abdellatif [2,3], Hamzah Maswadeh [2], Mohsen S. Al-Omar [1], Atef A. Abdel-Hafez [1], Hamdoon A. Mohammed [1,4,*], Eiman ME. Agabein [5], Ibrahim Alqasoomi [2], Salem A. Alrashidi [1], Mohammed S. M. Sajid [6] and Mugahid A. Mobark [7,8]

Citation: Eltom, S.E.M.; Abdellatif, A.A.H.; Maswadeh, H.; Al-Omar, M.S.; Abdel-Hafez, A.A.; Mohammed, H.A.; Agabein, E.M.; Alqasoomi, I.; Alrashidi, S.A.; Sajid, M.S.M.; et al. The Anti-Inflammatory Effect of a γ-Lactone Isolated from Ostrich Oil of *Struthio camelus* (Ratite) and Its Formulated Nano-Emulsion in Formalin-Induced Paw Edema. *Molecules* **2021**, *26*, 3701. https://doi.org/10.3390/molecules26123701

Academic Editors: Raffaele Pezzani and Sara Vitalini

Received: 2 May 2021
Accepted: 5 June 2021
Published: 17 June 2021

Publisher's Note: MDPI stays neutral with regard to jurisdictional claims in published maps and institutional affiliations.

[1] Department of Medicinal Chemistry and Pharmacognosy, College of Pharmacy, Qassim University, Buraydah 51452, Saudi Arabia; m.omar@qu.edu.sa (M.S.A.-O.); a.abdelgalil@qu.edu.sa (A.A.A.-H.); salem11098@gmail.com (S.A.A.)

[2] Department of Pharmaceutics, College of Pharmacy, Qassim University, Buraydah 51452, Saudi Arabia; A.Abdellatif@qu.edu.sa (A.A.H.A.); msodh@qu.edu.sa (H.M.); imq3911@gmail.com (I.A.)

[3] Department of Pharmaceutics and Industrial Pharmacy, Faculty of Pharmacy, Al-Azhar University, Assiut 71524, Egypt

[4] Department of Pharmacognosy, Faculty of Pharmacy, Al-Azhar University, Cairo 11371, Egypt

[5] Department of Histology, College of Medicine, Qassim University, Buraydah 51452, Saudi Arabia; eimanagabein@qumed.edu.sa

[6] Department of Pharmacology and Toxicology, College of Pharmacy, Qassim University, Qassim 51452, Saudi Arabia; su.mohammed@qu.edu.sa

[7] Department of Pharmacy Practice, College of Pharmacy, Qassim University, Buraydah 51452, Saudi Arabia; mu.mohammed@qu.edu.sa

[8] Department of Pathology, Faculty of Medicine and Health Sciences, University of Kordofan, El-Obied 157, Sudan

* Correspondence: S.Eltom@qu.edu.sa (S.E.M.E.); ham.mohammed@qu.edu.sa (H.A.M.)

Abstract: The ostrich oil of *Struthio camelus* (Ratite) found uses in folk medicine as an anti-inflammatory in eczema and contact dermatitis. The anti-inflammatory effect of a γ-lactone (5-hexyl-3H-furan-2-one) isolated from ostrich oil and its formulated nano-emulsion in formalin-induced paw edema was investigated in this study. Ostrich oil was saponified using a standard procedure; the aqueous residue was fractionated, purified, and characterized as γ-lactone (5-hexyl-3H-furan-2-one) through the interpretation of IR, NMR, and MS analyses. The γ-lactone was formulated as nano-emulsion using methylcellulose (MC) for oral solubilized form. The γ-lactone methylcellulose nanoparticles (γ-lactone-MC-NPs) were characterized for their size, shape, and encapsulation efficiency with a uniform size of 300 nm and 59.9% drug content. The γ-lactone was applied topically, while the formulated nanoparticles (NPs) were administered orally to rats. A non-steroidal anti-inflammatory drug (diclofenac gel) was used as a reference drug for topical use and ibuprofen suspension for oral administration. Edema was measured using the plethysmograph method. Both γ-lactone and γ-lactone-MC-NPs showed reduction of formalin-induced paw edema in rats and proved to be better than the reference drugs; diclofenac gel and ibuprofen emulsion. Histological examination of the skin tissue revealed increased skin thickness with subepidermal edema and mixed inflammatory cellular infiltration, which were significantly reduced by the γ-lactone compared to the positive control (p-value = 0.00013). Diuretic and toxicity studies of oral γ-lactone-MC-NPs were performed. No diuretic activity was observed. However, lethargy, drowsiness, and refusal to feeding observed may limit its oral administration.

Keywords: Diclofenac; γ-lactone; nano-emulsion; methylcellulose; Ostrich oil; *Struthio camelus*

1. Introduction

Natural products play a prominent role in the treatment of diseases and are of major concern in drug discovery [1–3]. Plants are the main renewable source of natural products; however, some other renewable sources such as animals and microorganisms can be used for natural product production [4,5]. Ostrich, *Struthio camelus* is a large flightless bird which belongs to the Ratite family. This family includes three types of birds, ostrich, emu, and rhea, the ostrich lives in African countries, mainly in the areas of rich savanna [6]. Ostrich oil has been used in folk medicine as topical treatment of eczema, psoriasis, dry skin, contact dermatitis, burns, hair growth, dry hair, bedsores, and muscular pain [7]. Emu oil, having similar composition as ostrich oil, has anti-inflammatory and skin desensitizing properties [4]. Topical application of emu oil was shown to reduce inflammation associated with reduced levels of interleukin 1-alpha, tumor necrosis factor-alpha, and other proinflammatory cytokines in a croton-oil-induced inflammation mouse model [4,8,9]; however, there is limited scientific research data on these properties with ostrich oil. The fatty acids' profile showed that the highest ratios include oleic acid 31.04% followed by palmitic acid 19.26%, then arachidonic acid 15.92%, and erucic acid 6.75% [10].

Although ostrich oil has similarities in chemical composition but not identical to emu oil which showed anti-inflammatory properties [7], it is expected that ostrich oil will have anti-inflammatory activity as emu oil [4]. As ostrich oil is almost triglyceride lipids and is free of phospholipids, consequently it could penetrate human skin. The triglyceride exists in the saponifiable fraction is responsible for the anti-inflammatory activity [11], eventually our search for the part of the saponifiable faction of ostrich oil was focused on, which might be responsible for the anti-inflammatory effect. Inflammation is an immediate and severe response by living tissue to injuries. The primary indicators for inflammations are four; pain, redness, warmness and swelling which is followed by raised blood circulation towards the inflamed area (redness).

The poor solubility of γ-lactone restrained its applications. As a good protection and oral delivery system, an optimal nano-emulsion can be developed by using low-energy emulsification method. These can increase the systematic bioavailability and biological activity. Also, the solubilized γ-lactone in nano-emulsion can be used as an oral delivery system dramatically with improved stability and solubility [12]. Nano-emulsion has been identified as an excellent delivery system for drugs. Nano-emulsion is a heterogeneous system composed of one immiscible liquid dispersed as droplets within a dispersion medium. Nano-emulsions consist of fine oil-in-water dispersions, the size range of covering droplets is 100–600 nm. Nano-emulsions droplets, usually spherical in shape of dispersed particles used for biopharmaceutical assistance [13]. The nano-emulsion drug-delivery system has an adequate storage capacity and thermodynamic stability for drug delivery. The nano-emulsion can contain large quantities of hydrophobic dissolved with their mutual compatibility and ability to get the drugs protected from hydrolysis and enzymatic degradation, making them ideal for parenteral transport [12,14]. There are a few factors that must be considered during the preparation of nano-emulsion, including: the careful choice of surfactant to achieve an ultralow interfacial tension, which is a crucial requirement to produce nano-emulsion. Also, concentration of the surfactant must be high enough to stabilize the microdroplets to create nano-emulsion. The surfactant must be flexible or fluid sufficient to promote the formation of nano-emulsion. Several nano-emulsion evaluation parameters must be kept in mind, such as, one droplet size analysis measured by a diffusion method using a light scattering and particle size-analyzer counter [15]. This study evaluated the potential anti-inflammatory effect of the γ-lactone isolated from ostrich oil and its nano-emulsion against formalin induced edema using diclofenac gel for topical application and ibuprofen for oral administration as standard drugs. Spectrophotometric analysis for both γ-lactone and its nano-emulsion together with the toxicity and diuretic activity of the oral γ-lactone-MC-NPs were studied. Also, skin tissues from rats treated with topical γ-lactone were histologically examined.

2. Results and Discussion

2.1. Isolation and Characterization of γ-Lactone (5-hexyl-3H-furan-2one)

The purified lactone of ostrich oil was subjected to spectroscopic analysis (IR, NMR, and MS) and shown to be 5-hexyl-3H-furan-2-one (Figure 1). The lactone was isolated as a colorless oil with the molecular formula $C_{10}H_{16}O_2$ as determined by EI-Ms $[M]^+$: $m/z = 168$ (100%). The existence of a γ-lactone was suggested by IR carbonyl absorption at 1708 cm^{-1} (C=O) (Figure S3 in Supplementary File) and the IR spectrum of the sodium salt of the open form of the lactone (Figure S4 in Supplementary File) formed by the hydrolysis of the respective lactone, revealed absorption at 3426, 2922, 1563, 1448, and 868 cm^{-1}. From which, absorption at 3426 cm^{-1} was assigned to the hydroxyl group and at 1563 and 1447 cm^{-1} for the carboxylate anion. The sodium salt of the open lactone was reverted back to the lactone when treated with dilute hydrochloric acid. ^{1}H-NMR spectrum showed (CDCl$_3$, 400 MHz, (Figure S1 in Supplementary File)) characteristic proton signals at δ H 0.9 (t, 3H, CH$_3$, J = 7.0 Hz), 1.31–1.41 (m, 6H, 3 × CH$_3$), 1.61 (quintet, 2H, J = 7.5 Hz), 2.31 (m, 2H), 3.21 (m, 2H, lactone CH$_2$), 5.23 (m, 1H, alkene-H). While ^{13}C-NMR spectrum displayed (CDCl$_3$, 100 MHz, (Figure S2 in Supplementary File)) carbon signals at δ C 14.2 (CH$_3$), 22.7, 25.7, 27.9, 28.9, 31.5 34.1 (6 CH$_2$), 98.0 (alkene CH), 157.4 (alkene quaternary), 178.1 (C=O). Based on the interpretation of the above-mentioned data, it was proposed that the structure of the isolated compound was assigned to be the lactone (Figure 1). 5-hexyl-3H-furan-2-one has been isolated for the first time from ostrich oil as was previously synthesized as a building block in enantioselective Bronsted acid catalyzed *N*-acyliminium cyclization cascade reactions [16].

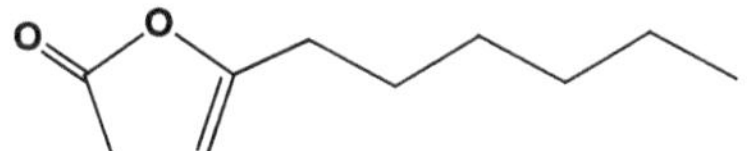

Figure 1. Chemical structure of 5-hexyl-3H-furan-2-one.

2.2. Nanoparticles Preparations

The size distribution for the prepared nanoparticles showed normal distribution with size diameter of 250 ± 13 nm (Figure 2A). γ-lactone-MC-NPs were formed in a stable nano-emulsion solution without any visible coalescence. Collectively, these results indicate a significant difference from γ-lactone itself which was completely immiscible in water. Nevertheless, the formulated γ-lactone-MC-NPs are completely dispersed in water (Figure 1). The entrapment of γ-lactone was confirmed by the change in color of MC to that of γ-lactone. Also, the size of γ-lactone-MC-NPs was increased to 250 ± 13 nm; this is attributed to the uniformity of γ-lactone-MC-NPs which showed only one peak without any aggregation. The particles were considered stable without aggregation after purification, and the formulated particles were also found to be relatively stable in size and did not form aggregates. The obtained data is considered as an intensity-weighted value and sensitive process for formulated smaller forms. This may be essential for aggregated or adulterated models [17–19].

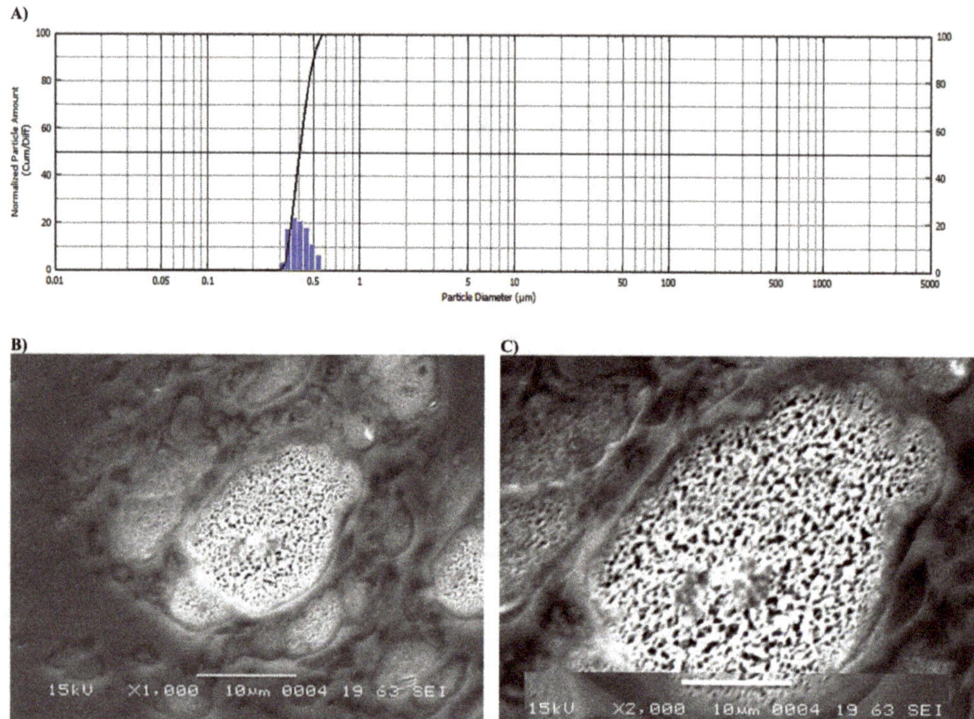

Figure 2. Characterization of the formulated γ-lactone-MC-NPs: (**A**) Particle size distribution, (**B**) Images taken with SEM at ×1000, and (**C**) Images taken with SEM at ×2000.

The γ-lactone-MC-NPs had a spherical shape and a particle size of 13–14 μm as shown by SEM. γ-lactone-MC-NPs showed highly spotted with more uniform and non-aggregated particles (Figure 2, image B). This result also confirmed the dynamic light scattering (DLS) study findings. Moreover, the γ-lactone-MC-NPs showed spotted dot inside the vesicle which are attributed to the γ-lactone itself (Figure 2; images C). The γ-lactone-MC-NPs showed difference in size as the particle diameters from DLS were lower than those obtained from SEM imaging and this was attributed to the presence of the layer coating of the γ-lactone-MC-NPs which further limits the total particle density [20]. The sizes of the γ-lactone-MC-NPs recorded by the DLS and SEM showed difference owing to different techniques. Our interpretation agrees with the size recorded with DLS and SEM. The data from DLS showed the average sizes of γ-lactone-MC-NPs varied significantly from that measured by the SEM technique. The DLS based average size estimations of the γ-lactone-MC-NPs were also varied with times and may not be reproducible wherein the nano-emulsion may aggregate and affect the average distribution of size [20,21].

2.3. Fourier-Transform Infrared Spectroscopy (FTIR)

The functional groups of γ-lactone, MC, and γ-lactone in a physical mixture with MC were checked by FTIR (Figure 3A) to examine the drug interaction between γ-lactone and MC. Absorption bands showed no prominent interaction between the two compounds in the physical mixture which had similar absorption bands as their raw powders. Moreover, the FTIR studies showed no interaction between the γ-lactone and MC, which is consistent with other results discussed previously that used FTIR to confirm the physical interaction of all components but found no interaction between γ-lactone and MC which means the γ-lactone can be released from the formulated γ-lactone-MC-NPs [22,23].

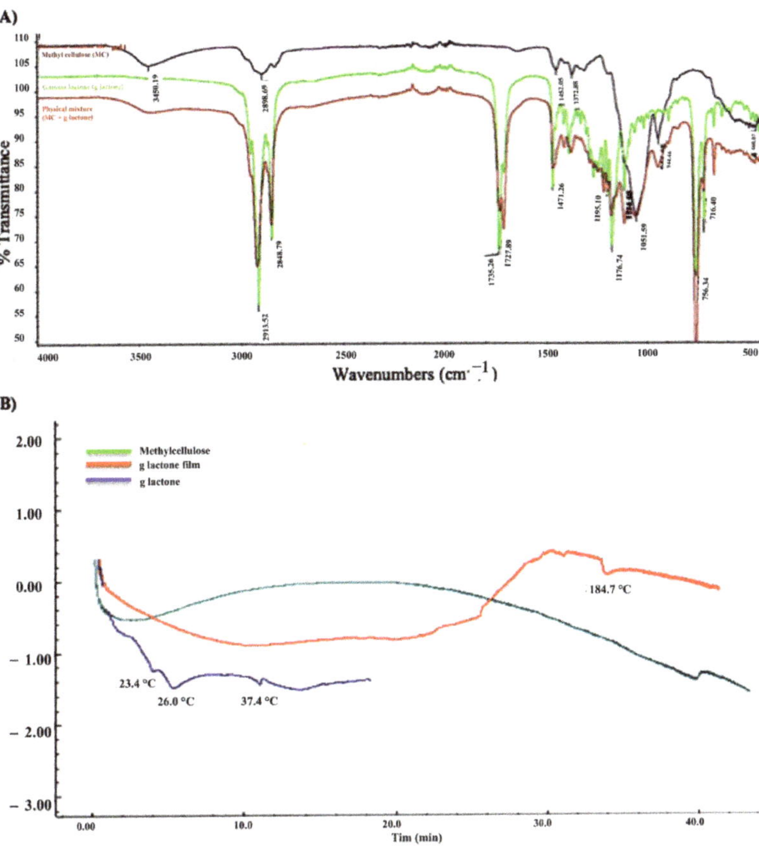

Figure 3. (**A**) FTIR spectra of γ-lactone (g lactone), methyl cellulose (MC), and γ-lactone in physical mixture with MC. The FTIR showed the typical peaks of γ-lactone without any foreign peaks due formulation. (**B**) Differential scanning calorimetry (DSC) for γ-lactone, MC, and γ-lactone in physical mixture with MC. Crosslinking of MC was detected forming plastic like film.

2.4. Differential Scanning Calorimetry (DSC)

As shown in the (Figure 3B) γ-lactone exhibits three endothermic peaks, at 23.4, 26, and 37.4 °C. The physical crosslinking of methylcellulose (MC) is a consequence of the formation of hydrophobic domains. Crosslinking kinetics of MC and its final stiffness depends on different parameters of the used product, such as the substitution degree or molecular weight, but it also might be adjusted using various concentrations and additives. Crosslinking of MC was observed after complete drying of the nanoparticle's formulation used in this study by forming plastic like film. The thermogram for the prepared nanoparticles showed a broad endothermic peak due to incorporation of γ-lactone into MC followed by an exothermic peak that indicates the formation of the crosslinking form of MC. A third endothermic peak at 184.7 °C was observed due to the melting of crosslinking MC [24].

2.5. Encapsulation Efficiency of γ-Lactone in Nanoparticles

The γ-lactone was incorporated in MC as γ-lactone-MC-NPs at 59.9%. This might be attributed to the low solubility of γ-lactone in water. Also, the presence of MC enhanced the entrapping of γ-lactone which was employed as an emulsion stabilizer in the distilled water (DW) emulsion.

2.6. Anti-Inflammatory Activity

In inflammations, cyclooxygenase (COX) is the key enzyme in the synthesis of autacoids [25]. Steroidal and non-steroidal anti-inflammatory drugs are currently the most widely used drugs in the treatment of acute inflammatory disorders, despite their renal and gastric negative secondary effects. As the result of long-term use of these drugs, the adverse effects become imminent such as gastric lesions and cardiovascular and renal failure [26]. Now, there is a need for new, safe, potent, and less toxic drugs. This stimulates our present study.

Previous studies on the 'oil' obtained from emu fat showed a very effective inhibition of chronic inflammation in rats when applied dermally (with a skin penetration enhancer) [27]. From our preliminary studies on saponification of ostrich oil, we noticed that the fatty acids obtained from ostrich oil showed similar composition to that of birds and rabbits and none of them showed anti-inflammatory property which is only rebutted for ostrich oil. The fatty acid profile showed: oleic acid (43.17%), palmitic acid (23.21%), and linoleic acid (16.88%), together with other fatty acids, in trace amounts: fatty acid, palmitoleic acid, linolenic acid, and lauric acid. Studies on the saponifiable fraction using chromatographic techniques furnished three compounds—A, B, and C. Compounds A and B from mass spectroscopy were shown to be high molecular weight hydrocarbons with simple functional groups, i.e., hydroxyl and ketone groups, respectively. Compound C showed to be the gamma lactone, which was focused on as the part which is responsible for the anti-inflammatory effect of ostrich oil. In light with similar results from the study on Emu oil, which showed the bulk of the anti-inflammatory activity, was present in the low triglyceride fraction [12]. This part of the fractionated oil was the concern of our present study.

The topical application of γ-lactone, 5-hexyl-3H-furan-2-one, isolated from ostrich oil showed reduction of formalin induced paw edema starting from the first hour and continued up to 24 h. At the first hour the reduction effect produced by γ-lactone was mild as the mean paw was 1.14 ± 0.09 cu.mm compared to 1.05 ± 0.06 cu.mm in diclofenac gel treated and 1.38 ± 0.14 cu.mm in positive control and the percent inhibition of edema was 56.8% while in diclofenac gel it was 67.8% (Table 1 and Figure 4).

Table 1. The anti-inflammatory activity of the γ-lactone-1 compared to that of the reference diclofenac gel expressed as percent inhibition of edema ($n = 30$) (Raw data available in the Tables S2–S4 in Supplementary File).

Test	Anti-Inflammatory Activity and Percentage Inhibition of Edema (Mean)				
	0 h	1 h	3 h	6 h	24 h
Control group	0	0	0	0	0
Diclofenac gel treated *	0%	67.8%	74.6%	75.2%	26.2%
Oral Ibuprofen suspension **	0%	81.8%	72.3%	71.98%	77.5%
γ-lactone treated *	0%	56.8%	79.3%	95.7%	75.6%
Oral γ-lactone-MC-NPs ***	0%	92.5%	88.3%	89.76%	77.9%

* Diclofenac gel and γ-lactone were administered at 300 mg diclofenac of 1% *w/w* and 30 mg γ-lactone. ** Ibuprofen suspension administered at 30 mg/kg; *** γ-lactone-MC-NPs was administrated at 10 mL/kg of γ-lactone (1 mL contains 223.5 µg).

After 3 h, γ-lactone induced significant reduction in paw edema resulting in mean paw volume of 0.95 ± 0.08 cu.mm compared to 1.03 ± 0.02 cu.mm in diclofenac gel and 1.64 ± 0.13 cu.mm in the positive control group with p-value < 0.05 indicating significant difference between groups. The percent inhibition of edema was also increased for both γ-lactone treated (79.3%) and diclofenac gel treated (74.6%). This study noticed that at three hours the edema reduction effect of γ-lactone started to increase and yield a greater effect than the reference diclofenac gel (Table 1 and Figure 4).

After 6 h, γ-lactone continued to show more anti-inflammatory activity than the reference diclofenac gel as the mean paw volume was 0.91 ± 0.06 cu.mm compared to 1.14 ± 0.05 cu.mm in diclofenac gel (p-value < 0.01) and the anti-inflammatory effect was clearer in the percent inhibition of edema: 95.7% for γ-lactone treated and 75.2% for diclofenac gel treated (Table 1 and Figure 4).

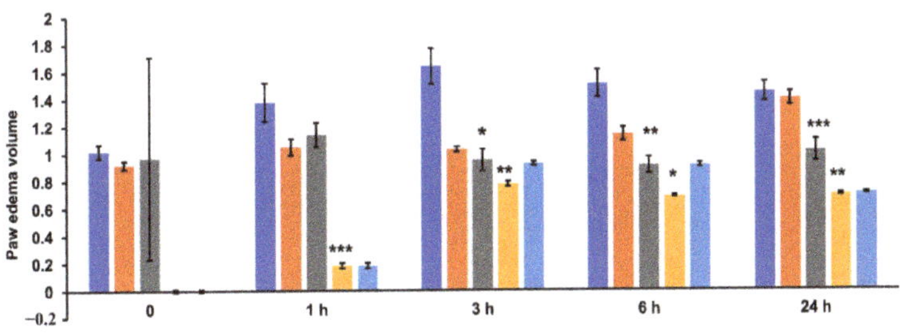

Figure 4. The percent of edema inhibition in cu.mm in the control group, topical diclofenac gel treated group, oral ibuprofen treated group, topical γ-lactone treated groups, and γ-lactone-MC-NPs treated group. Diclofenac gel and γ-lactone were administered at 300 mg diclofenac of 1% *w/w* and 30 mg γ-lactone. Ibuprofen suspension administered at 30 mg/kg, γ-lactone-MC-NPs was administrated at 10 mL/kg of γ-lactone (each 1 mL contains 223.5 µg). The results are expressed by Mean ± SEM ($n = 30$). One-way ANOVA followed by Tukey–Kramer multiple comparison test showed * $p < 0.05$ ** $p < 0.01$ and *** $p < 0.00$ as compared to the positive control group and the reference group.

After 24 h the diclofenac gel effect was nearly to be abolished as the mean paw volume was 1.40 ± 0.05 cu.mm approaching that of the positive control (1.45 ± 0.07 cu.mm), in contrast γ-lactone continued to show the anti-inflammatory effect even after 24 h as the mean paw volume was 1.02 ± 0.08 cu.mm (p-value < 0.001). These results confirmed that γ-lactone has prolonged anti-inflammatory effect that continued beyond 24 h with percent inhibition of edema volume of 75.6% (Table 1 and Figure 4).

The oral administration of γ-lactone-MC-NPs to the rats at the first hour showed potent paw edema inhibition of 92.5% compared to ibuprofen oral emulsion at 81.8%. Oral γ-lactone-MC-NPs continued to show effective reduction of paw edema at the third and sixth hour, while after 24 h the percent of edema inhibition was almost the same as that produced by ibuprofen (Table 1 and Figure 4).

These results showed a significant inhibition of edema in both topical γ-lactone, and oral γ-lactone-MC-NPs with a higher effect in the later (p-value < 0.001). The topical application of the γ-lactone and γ-lactone-MC-NPs were proved to be better than the reference drugs—diclofenac gel and ibuprofen emulsion. The anti-inflammatory activity lasts for up to 24 h as compared to the reference drugs. This result suggests a relatively potent and long-lasting effect for the topical γ-lactone and oral γ-lactone-MC-NPs over the reference drugs. The isolated γ-lactone; 5-hexyl-3H-furan-2-one is responsible for the anti-inflammatory activity of the ostrich oil.

2.7. The Histological Evaluation of Skin Tissues for Topical γ-Lactone

As shown in Table 2 and the raw data available in the Table S1 in Supplementary File, the histological assessment of skin thickness in a γ-lactone treated rats was compared to skin thickness in negative control, positive control and diclofenac treated groups. The results were expressed as mean ± SEM ($n = 24$). One-way ANOVA followed by the Tukey–Kramer multiple comparison test was performed and the comparison between groups was significant as the p-value is 0.00002 and the result was considered significant at $p < 0.05$. The comparison between the negative control and the positive control showed significant difference (p-value = 0.00003). The comparison between the negative control and the diclofenac treated rats was also significant (p-value = 0.023). While, the comparison between the negative control and the γ-lactone treated rats showed no difference as the p-value was insignificant (p-value = 0.899). The difference between the positive control and diclofenac treated and that between the positive control and γ-lactone treated were significant as the *p*-value was 0.0352 and 0.00013, respectively. However, there was insignificant difference between the diclofenac treated and γ-lactone treated rats, as the *p*-value was 0.096.

Table 2. The histological changes in skin-thickness measured in (mm) (n = 24). Diclofenac gel and γ-lactone were applied at 300 mg diclofenac of 1% *w/w* and 30 mg γ-lactone.

Number	Negative Control (Right Paw)	Positive Control (Left Paw)	Diclofenac Gel Treated	γ-Lactone Treated
1	1.0054	1.3361	0.9075	1.0451
2	0.9895	1.2779	1.1536	0.8281
3	0.8916	1.3758	1.2859	0.9789
4	0.7699	1.1906	1.0716	1.0477
5	0.9524	1.1986	1.1800	0.9842
6	0.8281	1.2912	0.9895	0.8043
Mean ± SEM	0.9062 ± 0.038	1.2784 ± 0.030	1.098 ± 0.056	0.9481 ± 0.036

The skin in positive control group showed sub epidermal edema with tortious dilated blood vessels, which lead to increase thickness of the skin. While in diclofenac treated and γ-lactone treated groups the skin showed significant reduction in thickness as compared to the positive control (Figure 5).

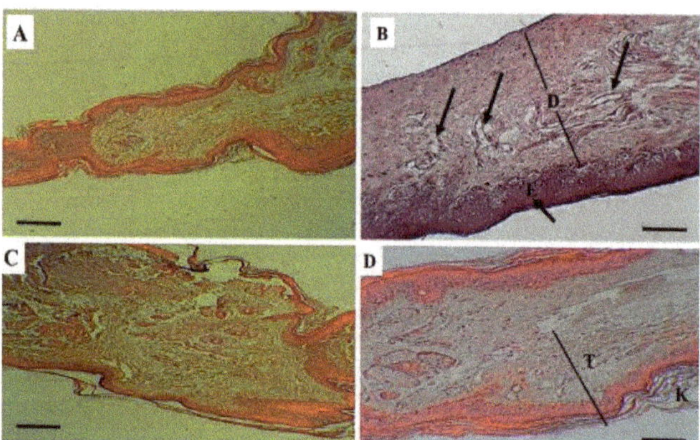

Figure 5. The histopathological assessment of skin thickness in rats: Photo (**A**): negative control, (**B**): Positive control, (**C**): Diclofenac treated, and (**D**): γ-lactone treated. E: epidermis (short arrow), D between two lines: Dermis, long arrow: loose tissue and dilated blood vessels, K: Keratinous layer, T: total thickness in lactone treated group. Scale bar: 100 μm. H&E satin.

The skin of positive control also showed mixed inflammatory cellular infiltration, which was more intense in the dermal layer with scattered inflammatory cells in the base of the epidermis. Figure 6.

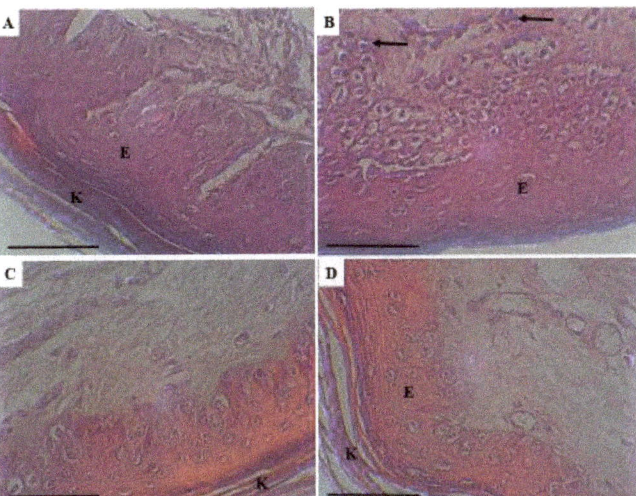

Figure 6. Histopathological assessment of inflammation in the skin of rats. Photo (**A**): negative control, (**B**): Positive control, (**C**): Diclofenac treated, and (**D**): γ-lactone treated. E: epidermis, arrows: inflammatory cells infiltration, K: Keratinous layer. Scale bar: 100 μm. H&E stain.

Histopathological evaluation showed a quantitative increase in the thickness of the skin in the positive control group as a result of edema and tortious dilatation of the blood vessels that developed after formalin injection. This increase in the thickness was statistically significant when compared to the negative control. The effect of formalin in tissue is well documented in previous studies [28,29].

This study also showed statistically significant anti-inflammatory effect of a γ-lactone compared to positive control and yielding a similar result to the diclofenac gel a recognized and widely used anti-inflammatory drug [30]. Moreover, the histopathological assessment revealed mixed infiltration of inflammatory cells, which was more prominent in the dermal layer and the basal part of the epidermis. The degree of inflammation was scored as intense inflammatory cellular infiltration (>40% of inflammatory cells in the sections) in the positive control group. Inversely, the degree of inflammation was markedly reduced and it was scored as slight inflammatory cellular infiltration (<20% of inflammatory cells in the sections) in both γ-lactone treated and diclofenac gel treated groups, which supports the hypothesis of anti-inflammatory effect of the γ-lactone against formalin-induced injury in rats.

2.8. Acute Toxicity and Weight Changes Study

From Table 3, the oral administration of γ-lactone-MC-NPs in a single dose of 5000 mg/kg at 20 mL/kg showed no toxic effect on the behavioral responses in the treated rats except for lethargy, drowsiness, and refusal of feeding observed in the first 4 h after dosing. There were no visual signs in skin, eyes, salivation, nor diarrhea in rats. After 24 h, no mortality occurred but significant weight loss (*p*-value 0.02 and 0.041) was observed. Apparently, the reason might be related to the refusal of feeding. After four days, 33.3% mortality in each group occurred. Although oral administration of the γ-lactone-MC-NPs showed potency in clearing paw edema in rats but lethargy, drowsiness, and refusal of feeding may limit its oral administration.

Table 3. Acute toxicity and weight changes study.

Groups	Weight of Rat in Grams in 0 h (Mean ± SD)	Weight of Rat in Grams in First 24 h (Mean ± SD)	*p*-Value	Initial Observation up to 4 h	Weight of Rat at Day 4
Control group ($n = 3$)	232.7 ± 13	232.7 ± 13	-	Normal	229.4 ± 11.2
Group 1 ($n = 3$)	168.7 ± 11.5	160.7 ± 11.4	0.02 *	Drowsiness, Sedation, Lethargic, and refusal of feeding	165.1 ± 12.6 One dead (33.3%)
Group 2 ($n = 3$)	261.0 ± 13.0	250 ± 13.0	0.041*	Drowsiness, Sedation, Lethargic, and refusal of feeding	258.9 ± 10.8 One dead (33.3%)

* *p*-value < 0.05 is considered significant. Oral γ-lactone-MC-NPs administered in a single dose of 5000 mg/kg at 20 mL/kg.

2.9. The Diuretic Effect of Oral γ-Lactone-MC-NPs

The parameters in Table 4 were calculated according to the formulae stated in the methods to compare the diuretic effects of oral γ-lactone-MC-NPs with a control and a standard diuretic. The diuretic action of oral γ-lactone-MC-NPs was found to be 1.01 which is similar to that of the control. Moreover, the diuretic activity of the oral γ-lactone-MC-NPs showed similar result to that of the control at 0.57 and 0.55, respectively. Diuretic activity in both the control and oral γ-lactone-MC-NPs apparently showed about 50% of the standard furosemide drug.

Table 4. The diuretic effect of oral γ-lactone-MC-NPs in rats.

Group	Volume of Urine (mL) Mean ± SD					Diuretic Action	Diuretic Activity
	1st Hour	2nd Hour	3rd Hour	4th Hour	5th Hour		
Control ($n = 3$)	0.61 ± 0.015	0.84 ± 0.025	0.93 ± 0.010	1.01 ± 0.015	1.05 ± 0.020	1	0.55
Furosemide ($n = 3$)	1.16 ± 0.03	1.44 ± 0.030	1.63 ± 0.030	1.77 ± 0.025	2.03 ± 0.015	1.81	1
γ-lactone-MC-NPs ($n = 3$)	0.64 ± 0.015	0.81 ± 0.006	0.95 ± 0.010	1.04 ± 0.006	1.03 ± 0.020	1.01	0.57

3. Materials and Methods

3.1. Extraction

The oil was obtained from traditional medicine shops in Sudan as a pale yellow oil that was saponified using a standard saponification procedure [31]. In which, 5 g of the oil was refluxed with 10% alcoholic KOH solution on water bath for 30 min. The alcohol was evaporated under vacuum, the residue was diluted with water, taken into a separatory funnel and extracted by dichloromethane to remove the higher hydrocarbons and then the aqueous layer was extracted with ethyl acetate. The ethyl acetate was evaporated under vacuum to yield semi-solid oil (3.2 g). The residue was fractionated through dry silica column using ethyl acetate as an eluent to give colorless semi-solid (2.9 g) as a major product. The residue was subjected to further separation as a result of the observation in a preliminary work that the anti-inflammatory activity was located in this extract.

3.2. Separation and Characterization

The crude ethyl acetate residue (500 mg) was purified using preparative thin layer chromatography using cyclohexane/ethyl acetate 60:40 as a developing solvent to yield a major compound (350 mg) as a low melting colorless solid. This was shown to be a lactone, from the ease of ring opening with 10% sodium hydroxide solution to give the sodium salt of hydroxy acid as a pale yellow solid, m.p. 245 °C. The lactone ring closed back on reaction with dilute hydrochloric acid, (λ max 1708 cm^{-1} for the lactone carbonyl stretch and at 3426 cm^{-1} for the hydroxyl group of the sodium salt). Further spectroscopic analyses showed this to be a γ-lactone, Figure 1. ^{1}H-NMR spectrum showed (CDCl$_3$, 400 MHz) characteristic proton signals at δ H 0.9 (t, 3H, CH$_3$, J = 7.0 Hz), 1.31–1.41 (m, 6H, 3 × CH$_3$), 1.61 (quintet, 2H, J = 7.5 Hz), 2.31 (m, 2H), 3.21 (m, 2H, lactone CH$_2$), 5.23 (m, 1H, alkene-H). While ^{13}C-NMR spectrum displayed (CDCl$_3$, 100 MHz) carbon signals at δ C 14.2 (CH$_3$), 22.7, 25.7, 27.9, 28.9, 31.5 34.1 (6 CH$_2$), 98.0 (alkene CH), 157.4 (alkene quaternary), 178.1 (C=O). Molecular formula C$_{10}$H$_{16}$O$_2$ [M]$^+$: m/z = 168 (100%). The γ-lactone was the subject of the anti-inflammatory activity in the present study.

3.3. Nanosuspension Preparation of γ-Lactone

γ-lactone was embedded in soluble nanoparticles (NPs) to increase the solubility of γ-lactone. These nanoparticles were made using MC polymer as the outer coating layer. γ-lactone-MC-NPs were prepared as the method stated previously with some modifications [32,33]. Briefly, 1 g of MC was dissolved in 100 mL distilled water and stirred continuously until a clear solution was obtained (Stock solution I). Further, 50 mg of γ-lactone were dissolved in 10 mL ethyl acetate (stock solution II). Moreover, dispersions containing 0.5 wt.% γ-lactone with respect to 1 g MC were prepared with 10 mL of ethyl acetate water and the dispersions were sonicated to obtain a uniform dispersion without agglomerates (Stock solution I and II). Both γ-lactone and MC mixture were mixed together and stirred continuously to let the ethyl acetate evaporate and a clear solution without bubbles was obtained and then left to stir for 12 h. The obtained nanoparticles were then characterized accordingly.

3.4. Size Measurements

Shimadzu particle size analyzer (SLAD-400, Shimadzu, Kyoto, Japan) was used to measure the size of γ-lactone-MC-NPs and the weight average of 10 of microparticles was announced. The samples were adjusted to 25 °C and exposed to a 633-nm laser beam at a 90° scattering angle. All samples have been conducted in an aqueous solution [34].

3.5. Determination of Surface Morphology

A volume of 20 µL of γ-lactone-MC-NPs solution were placed on the surface double-sided copper conductive tape and were let to dry. The γ-lactone-MC-NPs were then coated with a thin layer of platinum in a vacuum chamber (Zeny Vacuum pump, CA, USA) for 55 s at 25 mÅ, γ-lactone-MC-NPs solution were then sputter-coated (using Sputter coater, JOEL JFC-1300, Peabody, MA, USA) to create γ-lactone-MC-NPs electrically conductive before imaging in SEM. Both morphology and size of γ-lactone-MC-NPs were investigated using JEOL JSM-550 Scanning Electron Microscope (SEM) (Jeol, Akishima, Tokyo, Japan) [35].

3.6. Fourier Transform Infrared (FTIR) Spectroscopy

Fourier transform infrared (FTIR) γ-lactone, MC, and γ-lactone-MC physical mixture was performed using (Bruker, OPTIK GmbH, Type: Tensor 27, Ettlingen, Germany). The samples were scanned at the range of 400–4000 cm^{-1}. For purification, the γ-lactone-MC-NPs was centrifuged at 12500 rpm, for 20 min, and were washed with Millipore water three times [8].

3.7. Differential Scanning Calorimetry (DSC)

The thermal profiles of all materials and mixtures used in this study were measured by DSC-60 (Shimadzu, Kyoto, Japan) using 4–6 mg of sample in open aluminum pans, with an empty pan as a reference. The temperature increased with a heating rate of 20 °C/min for γ-lactone from 18 to 60 °C and from 18 to 220 °C for pure MC as well as for the final nanoparticle formulation prepared from the mixture of MC and γ-lactone (w/w) under a nitrogen gas flow.

3.8. Determination of Entrapment Efficiency %

The entrapment efficiency percent (EE%) of γ-lactone in γ-lactone-MC-NPs was determined spectrophotometrically. A volume of 10 mL of nanosuspension was mixed with 10 mL ethyl acetate, the mixture was stirred for 1 h until γ-lactone-MC-NPs dissolved and γ-lactone was completely released. The solution was measured for encapsulation efficiency of γ-lactone using a UV spectrophotometer (Varian Cary® 50 UV-Vis Spectrophotometer, Port Melbourne, Australia) at a wavelength of 388 nm. Drug content was computed using a calibration curve ($R^2 = 0.9998$) prepared from γ-lactone solutions with a concentration of 1–6 µg/mL. The EE% of γ-**lactone** was calculated according to the following equation:

$$EE\% = \frac{(Total\ amount\ of\ drug\ in\ particle)}{weight\ of\ particles\ taken} \times 100 \tag{1}$$

3.9. Animal Study

The study protocol was approved by the Ethics Committee for Animal Care and Use in the College of Pharmacy-Qassim University (Approval ID 2019-CP-9) [36].

3.9.1. Anti-Inflammatory Evaluation

Thirty male albino rats that were 12 to 16 weeks old, weighing 116–198 g, were obtained from the animal house, College of Pharmacy, Qassim University, Saudi Arabia and kept in standard conditions. The animals were divided randomly into five groups of six each.

3.9.2. Formalin-Induced Edema in the Left Paw of Rats

Edema was induced in the left hind paw of rats in all groups with sub planter injection of 0.1 mL of 5% formalin as described previously [36]. The right hind paw was maintained as a negative control. The anti-inflammatory activity was compared with control and treated groups using the plethysmograph method [36,37]. To all groups of rats, a mark at the tibia–tarsal junction was made with a permanent marker on both hind paws to ensure constant paw volume. In all groups, the left and right paw volume was measured at zero time (normal paw volume) and at 1, 3, 6, and 24 h after induction of inflammation.

The percent inhibition of edema was calculated as follow:

$$\%\ Edema\ inhibition = \frac{(VL - VR)control - (VL - VR)treated}{(VL - VR)control} \times 100 \tag{2}$$

where VL represents the mean of the left paw displacement volume and VR represents the mean of the right paw displacement volume.

3.9.3. Application of Diclofenac Gel and γ-Lactone

One group received 300 mg of 1% w/w diclofenac gel by applying it topically on the left hind paw 10 min before induction of inflammation [38]. Another group received the 30 mg of the neat γ-lactone isolated from the ethyl acetate extract of ostrich oil by applying it topically on the left hind paw 10 min before induction of inflammation. The last group was regarded as the positive and negative controls—R (Right paw)-negative control, L (Left paw) positive control.

3.9.4. Oral of Ibuprofen Suspension and γ-Lactone-MC-NPs

For the oral administration, ibuprofen was used as reference and medicated as γ-lactone-MC-NPs were orally administered at a volume of 5 mL/kg from one day at 0, 1, 3, 6, and 24 h. Ibuprofen suspension was administered at single daily doses of 30 mg/kg which was given orally 10 min before induction of the inflammation. The γ-lactone-MC-NPs was administrated at 10 mL/kg of γ-lactone (each 1 mL contains 223.5 µg) isolated from ostrich oil as a base nano-emulsion was given orally 10 min before induction of inflammation. γ-lactone and ibuprofen suspension, respectively, were administered 0.02 mL of 5% formalin by intra plantar injection in the left hind paw [39].

3.9.5. Acute Toxicity and Weight Changes Study

Acute oral toxicity was performed according to the Organization of Economic Cooperation and Development guidelines 420 for testing chemical compounds. Nine rats of both sexes aged 8 to 16 weeks-old were categorized in to groups 1, 2, and a control group. All rats were fasted for 18 h. γ-lactone-MC-NPs was administered as an oral gavage in a single dose of 5000 mg/kg at 20 mL/kg to rats in group 1 and 2 ($n = 6$) while deionized water was administered similarly to the control group ($n = 3$). All animals were allowed free-access to water before and after treatments but food was provided after 2 h of treatment. Observation like mortality, physical behavior, feeding, and other signs were carried out for four days.

3.9.6. Diuretic Activity of Oral γ-Lactone-MC-NPs

Diuretic activity was determined following the methods used by Lahlou et al. The Diuretic action and diuretic activity were calculated according to the Formulae (1) and (2), respectively.

$$1.\ \text{Diuretic action} = \frac{Urinary\ excretion\ of\ treatment\ group}{Urinary\ excretion\ of\ Control\ group}$$

$$2.\ \text{Diuretic activity} = \frac{Diuretic\ action\ of\ test\ drug}{Diuretic\ action\ of\ standard\ drug}$$

3.10. Histological Evaluation of Tissue

Three days after induction of inflammation, both right and left hind paws' slits were fixed in 10% formalin and left for 24 h for fixation. The formalin fixed tissue was further processed using automated tissue processor machine (Leica TP1020, Leica Microsystems GmbH, Wetzlar, Germany) and paraffin imbedded sections were prepared. Serial 3-μm sections were prepared using microtome (Leica RM2245, Leica Microsystems GmbH, Wetzlar, Germany) and stained by hematoxylin and eosin stain. All tissue sections were examined by light microscope (Olympus BX41, Olympus, Tokyo, Japan) using $4\times$, $10\times$, $20\times$, and $40\times$ magnifications and the image was taken via a digital image camera (5MP Binocular Microscope Electronic Eyepiece USB Video CMOS Camera for Image Capture, Zhejiang ULIRVISION Technology Co., Ltd, Zhejiang, China) and Top View image analyzer for more accurate measurement. The skin thickness quantification was measured in millimeters and the mean was derived from assessing eighteen sections form each group. Qualitative assessment of vascular tortuosity and dilatation together with edematous changes was performed. The inflammatory cellular infiltration was assessed in eighteen sections from each group and it was scored in to slight (presence of <20% of inflammatory cells in the section), moderate (presence of 20–40% of inflammatory cells in the section), and intense (presence of >40% of inflammatory cells in the section) as described in a previous study [40,41].

3.11. Statistical Analysis of Results

The volume measured by plethysmometer test and the thickness of the dermis were expressed as mean $\pm$ Standard Error of Mean. Paired t-test and One-Way ANOVA test followed by Tukey–Kramer multiple comparison tests were used to compare the mean between different groups and the result was considered significant at p-value < 0.05.

4. Conclusions

The present study showed that the γ-lactone, 5-hexyl-3H-furan-2-one, isolated from ostrich oil significantly lowers the paw edema in injured rats, which proved to be superior to the effect of diclofenac gel and ibuprofen suspension. The histological examination showed a significant reduction in skin thickness and inflammation. This study proved that the γ-lactone is solely responsible for the anti-inflammatory effect of ostrich oil, which agrees with a similar study on emu bird that showed this effect. These findings support its folkloric use.

Supplementary Materials: Figure S1: H-NMR spectrum of compound **1**; Figure S2: C-NMR spectrum of compound **1**; Figure S3: FT-IR spectrum of compound **1**; Figure S4: FT-IR spectra of the Na^+-salt of compound **1**; Table S1: The histological changes in skin—thickness measured in (mm).; Table S2: The anti-inflammatory activity result of control group in Plethysmometer test.; Table S3: The ant-inflammatory activity result of 30 mg/kg of oral ibuprofen suspension in Plethysmometer test.; Table S4: The anti-inflammatory activity result of 10 mL/kg of The γ-lactone nano emulsion base in Plethysmometer test.

Author Contributions: Conceptualization, S.E.M.E.; Formal analysis, S.E.M.E.; Investigation, H.M., E.M.A. and M.S.M.S.; Methodology, S.E.M.E., A.A.H.A., H.M., M.S.A.-O., A.A.A.-H., H.A.M., I.A., S.A.A. and M.A.M.; Software, M.S.A.-O. and E.M.A.; Writing—original draft, A.A.H.A., A.A.A.-H., H.A.M. and M.A.M.; Writing—review & editing, H.A.M. All authors have read and agreed to the published version of the manuscript.

Funding: This research received no external funding.

Institutional Review Board Statement: The study protocol was approved by the Ethics Committee for Animal Care and Use in the College of Pharmacy- Qassim University (Approval ID 2019-CP-9).

Informed Consent Statement: Not applicable.

Data Availability Statement: Data were provided in the text and supplementary file.

Acknowledgments: Researchers would like to thank the Deanship of Scientific Research, Qassim University for funding publication of this project.

Conflicts of Interest: The authors declare no conflict of interest.

Sample Availability: Sample of compound 1 is available from the first author.

References

1. Eldeeb, H.M.; Mohammed, H.A.; Sajid, M.S.M.; Eltom, S.E.M.; Al-Omar, M.S.; Mobark, M.A.; Ahmed, A.S. Effect of Roasted Date Palm Rich Oil Extracts in Liver Protection and Antioxidant Restoration in CCl_4-induced Hepato Toxicity in Rats. *Int. J. Pharmacol.* **2020**, *16*, 367–374. [CrossRef]

2. Mohammed, H.A.; Ba, L.A.; Burkholz, T.; Schumann, E.; Diesel, B.; Zapp, J.; Kiemer, A.K.; Ries, C.; Hartmann, R.W.; Hosny, M. Facile synthesis of chrysin-derivatives with promising activities as aromatase inhibitors. *Nat. Prod. Commun.* **2011**, *6*, 31–34. [CrossRef]

3. Mohammed, H.A.; Abdel-Aziz, M.M.; Hegazy, M.M. Anti-oral pathogens of tecoma stans (L.) and cassia javanica (l.) flower volatile oils in comparison with chlorhexidine in accordance with their folk medicinal uses. *Medicina* **2019**, *55*, 301. [CrossRef] [PubMed]

4. Yoganathan, S.; Nicolosi, R.; Wilson, T.; Handelman, G.; Scollin, P.; Tao, R.; Binford, P.; Orthoefer, F. Antagonism of croton oil inflammation by topical emu oil in CD-1 mice. *Lipids* **2003**, *38*, 603–607. [CrossRef] [PubMed]

5. Abdel-Aziz, M.M.; Al-Omar, M.S.; Mohammed, H.A.; Emam, T.M. In vitro and Ex vivo antibiofilm activity of a lipopeptide biosurfactant produced by the entomopathogenic Beauveria bassiana strain against microsporum canis. *Microorganisms* **2020**, *8*, 232. [CrossRef] [PubMed]

6. Horbańczuk, O.K.; Wierzbicka, A. Technological and nutritional properties of ostrich, emu, and rhea meat quality. *J. Vet. Res.* **2016**, *60*, 279–286. [CrossRef]

7. Basuny, A.; Arafat, S.; Soliman, H. Biological Evaluation of Ostrich Oil and Its Using for Production of Biscuit. *Egypt. J. Chem.* **2017**, *60*, 1091–1099.

8. Abdellatif, A.A.H.; Alawadh, S.H.; Bouazzaoui, A.; Alhowail, A.H.; Mohammed, H.A. Anthocyanins rich pomegranate cream as a topical formulation with anti-aging activity. *J. Dermatol. Treat.* **2020**, 1–8. [CrossRef]

9. Abdellatif, A.A.H.; Elgayed, S.H.; Afify, E.A.; Amin, H.A. Estrogenic Effect of Salvia officinalis Extract on Reproductive Function of Female Mice and Identification of Its Phenolic Content. *Comb. Chem. High Throughput Screen.* **2020**. [CrossRef] [PubMed]

10. Lopez, A.; Sims, D.E.; Ablett, R.F.; Skinner, R.E.; Léger, L.W.; Lariviere, C.M.; Jamieson, L.A.; Martínez-Burnes, J.; Zawadzka, G.G. Effect of emu oil on auricular inflammation induced with croton oil in mice. *Am. J. Vet. Res.* **1999**, *60*, 1558–1561. [PubMed]

11. Gunstone, F.D.; Russell, W.C. Animal fats. 3. The component acids of ostrich fat. *Biochem. J.* **1954**, *57*, 459. [CrossRef] [PubMed]

12. Zhang, Y.; Shang, Z.; Gao, C.; Du, M.; Xu, S.; Song, H.; Liu, T. Nanoemulsion for solubilization, stabilization, and in vitro release of pterostilbene for oral delivery. *AAPS PharmSciTech* **2014**, *15*, 1000–1008. [CrossRef]

13. Anton, N.; Benoit, J.-P.; Saulnier, P. Design and production of nanoparticles formulated from nano-emulsion templates—A review. *J. Control. Release* **2008**, *128*, 185–199. [CrossRef]

14. Lovelyn, C.; Attama, A.A. Current state of nanoemulsions in drug delivery. *J. Biomater. Nanobiotechnol.* **2011**, *2*, 626. [CrossRef]

15. Jaiswal, M.; Dudhe, R.; Sharma, P.K. Nanoemulsion: An advanced mode of drug delivery system. *3 Biotech* **2015**, *5*, 123–127. [CrossRef]
16. Muratore, M.E.; Holloway, C.A.; Pilling, A.W.; Storer, R.I.; Trevitt, G.; Dixon, D.J. Enantioselective Brønsted acid-catalyzed N-acyliminium cyclization cascades. *J. Am. Chem. Soc.* **2009**, *131*, 10796–10797. [CrossRef] [PubMed]
17. Abdellatif, A.A.H.; Abou-Taleb, H.A.; Abd El Ghany, A.A.; Lutz, I.; Bouazzaoui, A. Targeting of somatostatin receptors expressed in blood cells using quantum dots coated with vapreotide. *Saudi Pharm. J.* **2018**, *26*, 1162–1169. [CrossRef]
18. Abdellatif, A.A.H.; Tawfeek, H.M. Development and evaluation of fluorescent gold nanoparticles. *Drug Dev. Ind. Pharm.* **2018**, *44*, 1679–1684. [CrossRef] [PubMed]
19. Abdellatif, A.A.H.; Zayed, G.; El-Bakry, A.; Zaky, A.; Saleem, I.Y.; Tawfeek, H.M. Novel gold nanoparticles coated with somatostatin as a potential delivery system for targeting somatostatin receptors. *Drug Dev. Ind. Pharm.* **2016**, *42*, 1782–1791. [CrossRef]
20. Abdellatif, A.A.H.; Rasheed, Z.; Alhowail, A.H.; Alqasoumi, A.; Alsharidah, M.; Khan, R.A.; Aljohani, A.S.M.; Aldubayan, M.A.; Faisal, W. Silver citrate nanoparticles inhibit PMA-induced TNFα expression via deactivation of NF-κB activity in human cancer cell-lines, MCF-7. *Int. J. Nanomed.* **2020**, *15*, 8479. [CrossRef]
21. Zheng, T.; Bott, S.; Huo, Q. Techniques for accurate sizing of gold nanoparticles using dynamic light scattering with particular application to chemical and biological sensing based on aggregate formation. *ACS Appl. Mater. Interfaces* **2016**, *8*, 21585–21594. [CrossRef]
22. Zhang, H.-Y.; Xu, W.-Q.; Zheng, Y.; Omari-Siaw, E.; Zhu, Y.; Cao, X.; Tong, S.-S.; Yu, J.; Xu, X. Octreotide-periplocymarin conjugate prodrug for improving targetability and anti-tumor efficiency: Synthesis, in vitro and in vivo evaluation. *Oncotarget* **2016**, *7*, 86326. [CrossRef]
23. Nagpal, M.; Singh, S.K.; Mishra, D. Synthesis characterization and in vitro drug release from acrylamide and sodium alginate based superporous hydrogel devices. *Int. J. Pharm. Investig.* **2013**, *3*, 131.
24. Niemczyk-Soczynska, B.; Gradys, A.; Kolbuk, D.; Krzton-Maziopa, A.; Sajkiewicz, P. Crosslinking kinetics of methylcellulose aqueous solution and its potential as a scaffold for tissue engineering. *Polymers* **2019**, *11*, 1772. [CrossRef] [PubMed]
25. Tawfeek, H.M.; Abdellatif, A.A.H.; Dennison, T.J.; Mohammed, A.R.; Sadiq, Y.; Saleem, I.Y. Colonic delivery of indometacin loaded PGA-co-PDL microparticles coated with Eudragit L100-55 from fast disintegrating tablets. *Int. J. Pharm.* **2017**, *531*, 80–89. [CrossRef]
26. Pilotto, A.; Sancarlo, D.; Addante, F.; Scarcelli, C.; Franceschi, M. Non-steroidal anti-inflammatory drug use in the elderly. *Surg. Oncol.* **2010**, *19*, 167–172. [CrossRef] [PubMed]
27. Huerta, C.; Castellsague, J.; Varas-Lorenzo, C.; Rodríguez, L.A.G. Nonsteroidal anti-inflammatory drugs and risk of ARF in the general population. *Am. J. Kidney Dis.* **2005**, *45*, 531–539. [CrossRef]
28. Kim, H.-D.; Cho, K.-H.; Lee, B.-W.; Kwon, Y.-S.; Lee, H.-S.; Choi, S.-H.; Ku, S.-K. Effects of Magnetic Infrared Laser Irradiation on Formalin-Induced Chronic Paw Inflammation of Mice. *J. Phys. Ther. Sci.* **2010**, *22*, 395–404. [CrossRef]
29. Ahmad, N.S.; Waheed, A.; Farman, M.; Qayyum, A. Analgesic and anti-inflammatory effects of Pistacia integerrima extracts in mice. *J. Ethnopharmacol.* **2010**, *129*, 250–253. [CrossRef]
30. Chandra, S.; Chatterjee, P.; Dey, P.; Bhattacharya, S. Evaluation of anti-inflammatory effect of ashwagandha: A preliminary study in vitro. *Pharmacogn. J.* **2012**, *4*, 47–49. [CrossRef]
31. Al-Baidhani, A.M.; Al-Mossawi, A.H. The study of chemical content and physicochemical properties of ostrich (*Struthio camelus*) fat (local). In Proceedings of the IOP Conference Series: Earth and Environmental Science; IOP Publishing: Bristol, UK, 2019; 388, p. 12055.
32. Kuki, Á.; Nagy, L.; Zsuga, M.; Kéki, S. Fast identification of phthalic acid esters in poly (vinyl chloride) samples by direct analysis in real time (DART) tandem mass spectrometry. *Int. J. Mass Spectrom.* **2011**, *303*, 225–228. [CrossRef]
33. Orasugh, J.T.; Saha, N.R.; Sarkar, G.; Rana, D.; Mishra, R.; Mondal, D.; Ghosh, S.K.; Chattopadhyay, D. Synthesis of methylcellulose/cellulose nano-crystals nanocomposites: Material properties and study of sustained release of ketorolac tromethamine. *Carbohydr. Polym.* **2018**, *188*, 168–180. [CrossRef] [PubMed]
34. Mohammed, H.A.; Al-Omar, M.S.; El-Readi, M.Z.; Alhowail, A.H.; Aldubayan, M.A.; Abdellatif, A.A.H. Formulation of Ethyl Cellulose Microparticles Incorporated Pheophytin A Isolated from Suaeda vermiculata for Antioxidant and Cytotoxic Activities. *Molecules* **2019**, *24*, 1501. [CrossRef] [PubMed]
35. Tawfeek, H.M.; Abdellatif, A.A.H.; Abdel-Aleem, J.A.; Hassan, Y.A.; Fathalla, D. Transfersomal gel nanocarriers for enhancement the permeation of lornoxicam. *J. Drug Deliv. Sci. Technol.* **2020**, *56*, 101540. [CrossRef]
36. Soyocak, A.; Kurt, H.; Cosan, D.T.; Saydam, F.; Calis, I.U.; Kolac, U.K.; Koroglu, Z.O.; Degirmenci, I.; Mutlu, F.S.; Gunes, H.V. Tannic acid exhibits anti-inflammatory effects on formalin-induced paw edema model of inflammation in rats. *Hum. Exp. Toxicol.* **2019**, *38*, 1296–1301. [CrossRef]
37. Sepehri, Z.; Fereidoni, M.; Niazmand, S. Role of C-fibers during acute and chronic stress on formalin-induced paw edema in rats. *Indian J. Exp. Biol.* **2012**, *50*, 633–637.
38. Kaushik, D.; Kumar, A.; Kaushik, P.; Rana, A.C. Analgesic and Anti-Inflammatory Activity of Pinus roxburghii Sarg. *Adv. Pharmacol. Sci.* **2012**, *2012*, 245431. [PubMed]
39. Melarange, R.; Gentry, C.; O'Connell, C.; Blower, P.R. Anti-inflammatory efficacy and gastrointestinal irritancy: Comparative 1 month repeat oral dose studies in the rat with nabumetone, ibuprofen and diclofenac. *Agents Actions Suppl.* **1991**, *32*, 33–37.

40. Vilela, R.G.; Gjerde, K.; Frigo, L.; Junior, E.C.P.L.; Lopes-Martins, R.Á.B.; Kleine, B.M.; Prokopowitsch, I. Histomorphometric analysis of inflammatory response and necrosis in re-implanted central incisor of rats treated with low-level laser therapy. *Lasers Med. Sci.* **2012**, *27*, 551–557. [CrossRef] [PubMed]
41. Lahlou, S.; Tahraoui, A.; Israili, Z.; Lyoussi, B. Diuretic activity of the aqueous extracts of Carum carvi and Tanacetum vulgare in normal rats. *J. Ethnopharmacol.* **2007**, *110*, 458–463. [CrossRef]

molecules

Article

BDNF as a Putative Target for Standardized Extract of *Ginkgo biloba*-Induced Persistence of Object Recognition Memory

Beatriz G. Muratori [1], Cláudia R. Zamberlam [1], Thaís B. Mendes [2], Bruno H. N. Nozima [2], Janete M. Cerutti [2] and Suzete M. Cerutti [1,3,*]

[1] Cellular and Behavioral Pharmacology Laboratory, The Graduate Program in Biological Chemistry, Universidade Federal de São Paulo, São Paulo 09972-270, Brazil; muratoribeatriz@gmail.com (B.G.M.); crzamberlam@gmail.com (C.R.Z.)
[2] Genetic Bases of Thyroid Tumor Laboratory, Division of Genetics, Department of Morphology and Genetics, Federal University of Sao Paulo, São Paulo 04039-032, Brazil; thais_biude@hotmail.com (T.B.M.); bruno.nozima@gmail.com (B.H.N.N.); j.cerutti@unifesp.br (J.M.C.)
[3] Department of Biological Science, Universidade Federal de São Paulo, São Paulo 09972-270, Brazil
* Correspondence: smcerutti@unifesp.br; Tel.: +55-(11)-98229-3043

Abstract: Despite considerable progress on the study of the effect of standardized extract of *Gingko biloba* (EGb) on memory processes, our understanding of its role in the persistence of long-term memory (LTM) and the molecular mechanism underlying its effect, particularly episodic-like memory, is limited. We here investigated the effects of EGb on the long-term retention of recognition memory and its persistence and BDNF expression levels in the dorsal hippocampal formation (DHF). Adult male Wistar rats (n = 10/group) were handled for 10 min/5 day. On day 6, the animals were treated with vehicle or 0.4 mg/kg diazepam (control groups) or with EGb (250, 500 or 100 mg/kg) 30 min before the training session (TR1), in which the animals were exposed to two sample objects. On day 7, all rats underwent a second training session (TR2) as described in the TR1 but without drug treatment. Object recognition memory (ORM) was evaluated on day 8 (retention test, T1) and day 9 (persistence test, T2). At the end of T1or T2, animals were decapitated, and DHF samples were frozen at $-80\ ^{\circ}$C for analyses of the differential expression of BDNF by Western blotting. EGb-treated groups spent more time exploring the novel object in T2 and showed the highest recognition index (RI) values during the T1 and T2, which was associated with upregulation of BDNF expression in the DHF in a dose-and session-dependent manner. Our data reveal, for the first time, that EGb treatment before acquisition of ORM promotes persistence of LTM by BDNF differential expression.

Keywords: memory; object recognition; *Ginkgo biloba*; dorsal hippocampus formation; brain-derived neurotrophic factor

Citation: Muratori, B.G.; Zamberlam, C.R.; Mendes, T.B.; Nozima, B.H.N.; Cerutti, J.M.; Cerutti, S.M. BDNF as a Putative Target for Standardized Extract of *Ginkgo biloba*-Induced Persistence of Object Recognition Memory. *Molecules* **2021**, 26, 3326. https://doi.org/10.3390/molecules 26113326

Academic Editors: Raffaele Pezzani and Sara Vitalini

Received: 29 March 2021
Accepted: 10 May 2021
Published: 1 June 2021

Publisher's Note: MDPI stays neutral with regard to jurisdictional claims in published maps and institutional affiliations.

1. Introduction

Flavonoids (isoflavones, flavanones, anthocyanins, flavanols, flavones and flavonols) are bioactive molecules from plants or foods that, along with non-flavonoid compounds (phenolic acids and other phenols), are widely used to prevent or treat cognitive deficits. Extracts rich in flavonoids or flavonoid fractions have been associated with improvements in memory formation and stress-related memory processes [1–3]. A recent study showed that standardized extract of *Ginkgo biloba* (EGb), which is obtained from the dried leaves of the plant and is standardized to contain 24% to 27% ginkgo-flavonoglycosides and 6% ginkgo-terpenoid lactones [4], increases BDNF expression, which has a protective effect on neurons [5–7]. Standardized extract of *Ginkgo biloba* (EGb) has been used as a therapeutic alternative for the treatment of Alzheimer's disease and multi-infarct dementias and has significant positive effects on cognitive function in patients with Alzheimer's disease [8,9]. Similar results have been observed after Ginkgo terpene trilactone (GTTL) treatment. GT-TLs increase BDNF expression in the hippocampus, which protects hippocampal neurons against epilepsy and improves spatial memory, as evaluated by the Morris Water Maze (MWM) test [5,10].

Studies from our group have shown that acute or chronic treatment with EGb modulates fear memory formation, which is associated with increased levels of cAMP-dependent response element binding protein (CREB) in the hippocampal formation [11,12]. Conversely, our group reported an impairment in conditioned fear memory following diazepam treatment, which was associated with reduced CREB-1 expression (both mRNA and protein) in the dorsal hippocampus [11]. In addition, we showed that the mRNA and protein expression of the serotonin type-1A (5-HT$_{1A}$) receptor and the GluN2B and GluN2A subunits of the N-methyl-D-aspartate (NMDA)-type glutamate receptor in the dHF are modulated after EGb treatment and fear memory acquisition [13]. We also showed that flavones modulate fear memory formation and persistence through gamma-aminobutyric acid type A receptor (GABA$_A$R) modulation [14,15]. Furthermore, GABA$_A$R agonists impair the expression of object recognition memory (ORM), which seems to be correlated with BDNF expression and the persistence of LTM [16].

The ability of an animal to recognize a familiar object using the physical characteristics of the stimulus or the contextual information concerning the object (spatial and/or temporal order) involves discriminative memory formation. Recognition memory for an object or object-in-context in animals has been termed episodic-like memory and reflects an experience-dependent internal representation of what happened and where and when the event occurred. When the original memory is re-experienced, we recall details of the experience [17–20]. In this regard, recognition memory formation involves both familiarity and recollection of an episode, i.e., quantitative or qualitative information about a previously encountered item or event [21]. Both the familiarity and recollection components become less persistent over time [22,23].

The ORM task has been employed to investigate the effects of drug treatments on memory as well as the neuroanatomical, neurophysiological and neurochemical changes underlying spontaneous exploratory behavior in animals [24–26]. Numerous studies have explored the contribution of the perirhinal and parahippocampal cortices and the hippocampal formation (HF) to familiarity and recollection [18,27–31]. Recent advances have led to substantial improvements in our understanding of the role of the hippocampus in processing and integrating multisensory information and encoding long-term novel object recognition (NOR) memory, novel location memory (NLM) or temporal order memory (recency) [17,18,29,32–34], but the debate undoubtedly continues [35]. Moreover, the role of the hippocampus in the reconsolidation and persistence of ORM as well as molecular changes underlying these processes has been shown [16,36–38].

It was recently shown that reduced expression of brain-derived neurotrophic factor (BDNF) in the hippocampus is associated with impaired long-term ORM [39]. Activity-dependent secretion of BDNF has been associated with both hippocampus-dependent long-term memory formation and long-term biochemical and morphological dendritic changes [40,41] since it is crucial for long-term memory consolidation [42,43]. A growing body of evidence has shown that BDNF overexpression and activity modulate the differentiation, maturation, survival and maintenance of different populations of neurons and glial cells and has identified the main effects of BDNF on short- and long-term memory-related processes [44,45] and neurogenesis [46]. *De novo* synthesis of BDNF has been shown to be essential and sufficient for promoting late-phase long-term potentiation (L-LTP) [47] and improving long-term memory (LTM) [48]. Increased BDNF expression (mRNA and protein) in the hippocampus (1 or 12 h after training, respectively) is critical for LTM formation in a hippocampus-dependent acquisition task [24]. In addition, BDNF is a key molecule for the prevention of memory loss, the generation of persistent fear and long-lasting memory traces and the extended maintenance of LTM [22] and is also required for reconsolidation of NOR memory [16,49].

Together, these data show that BDNF/TrKB receptor-induced synaptic plasticity, survival and dendritic and axonal growth during hippocampal-dependent memory processes are mediated by distinct signalling pathways in both pre- and post-synaptic cells, which involve increases in extracellular signal-regulated protein kinase (ERK) expression and phosphorylation and activation of cyclic AMP response element-binding protein (CREB) [22,50,51].Despite the findings of these studies, little is known about novel therapeutic strategies for modulating the cellular and molecular processes that mediate the persistence of recognition memory. Flavonoids have long been targeted as prospective therapeutic agents for memory improvement [1].

Considering recent findings from our group describing the effects of EGb on the persistence of fear memory [52,53], we investigated the effects of the administration of EGb before the acquisition of object recognition memory (ORM), a non-aversive memory task, on the long-term retention of ORM and its persistence. Episodic-like memory loss is correlated with both early-stage Alzheimer's disease (AD) and normal aging. We wondered whether rats treated with EGb would have improved their original memory, i.e., does EGb affect memory consolidation and promote object recognition memory persistence? We analysed retrieval and within-session maintenance of ORM during the T1 phase, as well as the persistence of memory in the T2 phase. We also examine potential roles for diazepam, a $GABA_A R$ agonist, in modulating the acquisition of non-aversive memory and in mediating memory retention and persistence through alterations in BDNF expression. Last, given the pivotal role of BDNF in the neurochemical and morphological changes associated with LTM and the persistence of LTM, we analysed BDNF expression in the dorsal hippocampal formation of the control and EGb-treated groups after retention and persistence tests.

2. Results

2.1. Behavioral Analysis

To examine the cognition-enhancing effect of EGb on long-term memory for novel objects in rats, we employed two 10-min sessions (T1 and T2) of the NOR test, a non-aversive task, with an intertrial interval of 24 h. Memory formation and memory persistence were assessed by different analyses of total spontaneous exploration time for both objects in the T1 and T2 sessions (Figure 1).

Two-way RM ANOVA, indicated significant main effects of time exploring the familiar and novel objects ($F_{1.95} = 59.84$; $p < 0.0001$) and effects of group ($F_{4.95} = 2.967$; $p = 0.0234$) but no interaction between object and group ($F_{4.95} = 0.4771$; $p = 0.7524$). Comparative analyses of the mean total time spent (min) exploring each of the objects during the retention session (T1) between and within groups showed that the control groups (vehicle- and diazepam-treated group; n = 20 rats/group) and EGb-treated groups (250, 500 and 1000 mg/kg EGb-treated groups; n = 20 rats/group) spent more time exploring the novel object than the familiar object in the T1 session (Figure 1A). Based on our data, all rats recalled the familiar object and preferred to explore the novel object.

During T2 session when the animals were re-exposed to the objects, Two-way RM ANOVA indicated significant effects of time spent exploring the familiar object vs novel objects within groups ($F_{1.45} = 31.12$; $p < 0.0001$). Bonferroni's multiple comparisons test revealed the rats that had been treated with 1000 mg/kg EGb spent more time exploring the novel object than the familiar object compared with the vehicle- and diazepam-treated groups, as shown in Figure 1B. However, no interaction was observed between the factors group and object ($F_{1.45} = 1.385$; $p = 0.2549$) and no difference was observed between groups ($F_{4.45} = 1.144$; $p = 0.3496$).

Additionally, the time spent interacting with the sample object (A) and the novel object (C) were used to calculate the recognition index [RI = time spent exploring the novel object/total object exploration time (novel vs familiar)] [28,54,55]. RI represents a relative measure of discrimination between objects. The EI values in each session (T1 and T2) are shown in Figure 1C. The RI value in the first minute is shown for each session since the effects of novelty might decrease over time. The subsequent data *points* are the *mean RI value over a* three-minute period within a given session (three blocks of 3 min each as follows: trial 2–4 min; trial 5–7 min; and trial 8–10 min). Furthermore, within-session comparisons (min-by-min) at T1 using two-way RM ANOVA showed an interaction between groups and time ($F_{36, 855} = 1.756$; $p = 0.0043$, with a difference only in the first minute in relation to subsequent times (min 2–10) ($F_{7.9833, 758.4} = 1.914$; $p = 0.05$), as well as effects of the groups ($F_{4, 95} = 43.572$; $p = 0.0093$). Concerning the min-by-min analysis of the recognition index (RI) in T2, two-way RM ANOVA showed no interaction between groups and time ($F_{36, 405} = 1.411$; $p = 0.06$), no within-session difference (min-by-min), i.e., first minutes in relation to subsequent minutes (2–10) ($F_{7.706, 346.8} = 1.581$; $p = 0.1320$) and no effects of the groups ($F_{4, 45} = 0.5821$; $p = 0.6772$). Thus, this procedure evaluates the mean of a combination of 3 min to help reduce variance within sessions and has been systematically used in our lab since it allows us to evaluate retention of short- and long-term memory. An RI value of 1.0 indicates that the animal spent the majority of time exploring the novel object relative to the sample object, and a value of ≤ 0.5 indicates random exploration, i.e., no discrimination between objects or decreasing novelty. Two-way repeated-measures ANOVA followed by Tukey's multiple-comparison test revealed that all groups spent more time exploring the novel object than the previously explored (familiar) object (250 mg/kg EGb, $RI_1 = 0.66$; 500 mg/kg EGb, $RI_1 = 0.66$; 1000 mg/kg EGb, $RI_1 = 0.81$; vehicle, $RI_1 = 0.59$; 4 mg/kg diazepam, $RI_1 = 0.61$) There was an interaction between group and time ($F_{12.285} = 3.045$; $p = 0.0005$) and a main effect of time ($F_{3.285} = 11.76$; $p < 0.0001$) for the T1 session (Figure 1C, left panel). Moreover, the high-dose EGb-treated group spent more time exploring the novel object than the control groups (diazepam- and vehicle-treated groups). Furthermore, an intragroup comparison of the 1000 mg/kg EGb group showed that the RI in the first minute was significantly lower than that in any of the all three-minute blocks. Although our data showed difference between in first minute and last three minutes during T1, the RI was maintenance > 0.5, indicating that within-session the animals treated with 1000 mg/kg EGb spent more time exploring novel object. Because vehicle and EGb treatments were given only before TR1, our data suggest that 1000 mg/kg EGb improved the original memory, i.e., memory for early familiar object (object A) in relation to late object (object C). Our present data is consistent with previous results from our lab that show effects of EGb on persistence of fear memory [52,53].

Memory persistence was analyzed based on the data obtained in the T2 session (Figure 1C, right panel). Two-way RM ANOVA revealed no interaction between group or trial ($F_{12.135} = 1.49$; $p = 0.133$), but there was a main effect of group ($F_{4.45} = 2.932$; $p < 0.0308$). Rats treated with EGb (250 mg/kg, $RI_1 = 0.69$; 500 mg/kg, $RI_1 = 0.61$; 1000 mg/kg, $RI_1 = 0.71$) and vehicle ($RI_1 = 0.57$) spend more time exploring the novel object than the familiar object during the first minute. Conversely, rats treated with diazepam spend similar amounts of time exploring both objects ($RI_1 = 0.48$). Tukey's multiple comparison test revealed a significant decrease in the EI in the diazepam group compared to the 250 mg/kg EGb ($p < 0.05$) and 1000 mg/kg EGb ($p < 0.01$) groups. Analysis of the within-session data showed that the RI in the first minute was significantly lower than that in the second three-minute block for the 250 mg/kg EGb-treated group at ($p < 0.01$) and the third three-minute block for the 1000 mg/kg EGb group. The mean (and S.E.M) of the RI of the EGb-treated and control groups in both sessions are presented in the supplementary data (Table S1).

Moreover, to substantiate our results regarding the effects of EGb on the persistence of LTM, we also investigated recognition index (RI) between sessions by comparing the mean RIs from the first minute of the persistence test session with those from the last block of the three min of the retention test (Figure 1D). Two-way ANOVA revealed no interaction between group and session ($F_{4.45} = 2.095$; $p = 0.0969$), no effect of group ($F_{4.45} = 1.187$; $p = 0.3292$), and a main effect of time ($F_{1.45} = 6.249$; $p = 0.0161$).

Our present data corroborate previous findings and shows that rats treated with 1000 mg/kg EGb exhibit a higher RI during the first trial of the T2 session than during last three-minute block in the T1 session, indicating maintenance of the original memory, i.e., the information regarding the object appeared to be preserved (old vs. novel) 48 h after acquisition (T2).

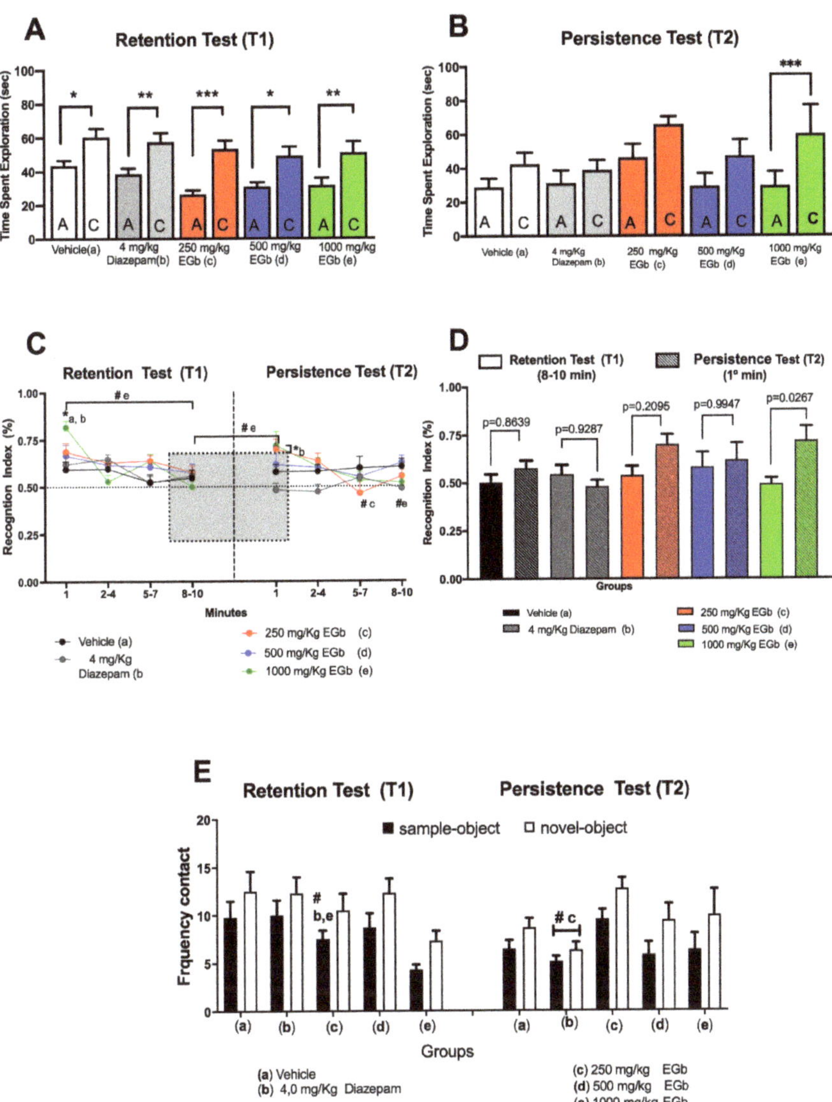

Figure 1. (**A**). Comparison of the mean of total time spent exploring/min the novel and sample objects in the retention test (day 8, n = 20/group), and the persistence test (**B**) (day 9, n = 10/group) between the control (diazepam- and vehicle-treated) and EGb-treated (250, 500 and 1000 mg/kg) groups. (**C**). The recognition index (RI) during the retention and the persistence sessions. The first points indicate the mean RI values during first min of the control (4 mg/kg diazepam- and vehicle-treated) and EGb-treated (250 mg/kg, 500 mg/kg and 1000 mg/kg) groups before the acquisition of ORM (n = 20/group) and the persistence test session (n = 10/group). The subsequent data points represent the mean DI values of nine min in blocks of 3 min (trials) each (2–4,5–7,8–10). EGb, diazepam and vehicle were administered orally 30 min prior to training session. No drugs were administered before retention test (T1) and persistence test (T2) sessions (**D**). Persistence of long-term memory (LTM) was inferred based on a comparison of the time spent spontaneously exploring sample object and that spent spontaneously exploring the novel object during the last three trial blocks at the end of the retention session and during the first trial of the initial persistence session. (**E**). Frequency of contact with the sample and novel objects during the retention test (n = 20/group) and persistence test (n = 10/group) for the groups treated with EGb (250 mg/kg, 500 mg/kg or 1000 mg/kg),

diazepam or vehicle (n = 10/group). Analyses of the sample vs novel object and group interaction, the time and group interaction (A, B, C, and D panels, respectively) and one random factor (rat) was performed using GraphPad Prism Software®. The values are presented as the means/min (± SEMs) (A, B, D, and E) and/or a block of three minutes (±SEMs) (C). * $p < 0.05$, ** $p < 0.01$ and *** $p < 0.001$; two-way repeated measures ANOVA with the *post hoc* Sidak multiple comparison test (A,B). Two-way repeated measures ANOVA was conducted for within- and between-group comparisons followed by Tukey's multiple comparisons test (C). Within-group comparison # $p < 0.05$; two-way ANOVA followed by Bonferroni's *post hoc* test (D,E). Comparisons between groups are presented * $p < 0.05$, ** $p < 0.01$ and *** $p < 0.001$.

Because a number of studies have shown that animals have preferences for any one object [56,57], we also examined the frequency of contact with each sample and novel object (Figure 1E). Two-way RM ANOVA revealed no interaction between the group and object in the T1 session ($F_{4.72} = 0.05093$; $p = 0.9950$) but showed main effects of frequency of contacting the object ($F_{1.18} = 9.996$; $p = 0.0055$) and treatment ($F_{3.237, 58.27} = 4.573$; $p = 0.0050$).

Our data showed that all groups contacted both objects with a similar frequency. Tukey's multiple comparisons test indicated a significant difference in the frequency of contact with the sample object between the 1000 mg/kg EGb group and the diazepam group. Comparative analyses of the frequency of contact with both the familiar and novel objects in the T2 session revealed no interaction between the group and object ($F_{4.36} = 1.082$; $p = 0.3800$) and no effect of treatment ($F_{4.36} = 2.454$; $p = 0.0633$), but a main effect of frequency of contact with both objects ($F_{1.9} = 36.46$; $p = 0.0002$). The EGb-treated groups exhibited a higher frequency of contact with the novel object than the familiar object. No differences were observed in the diazepam and vehicle groups.

Additionally, we investigated exploratory behaviors such as the amount of ambulation (lateral and central squares) and rearing and grooming of all animals during the retention test and persistence test (Figure 2). The data showed significantly increased ambulation in the lateral squares compared with the central squares in both the retention session ($F_{1.18} = 266$; $p < 0.0001$) and persistence session ($F_{1.18} = 6.663$; $p = 0.0191$). Two-way RM ANOVA revealed no interaction between group and area ($F_{4.72} = 0.4444$; $p = 0.7761$) and no effect of treatment ($F_{2963, 53.33} = 0.3472$; $p = 0.7875$) in the T1 session. Similarly, no interaction between group and area ($F_{4.72} = 0.6991$; $p = 0.5951$) and no effects of treatment ($F_{2.941, 5293} = 1.352$; $p = 0.2676$) were found in the T2 session (Figure 2).

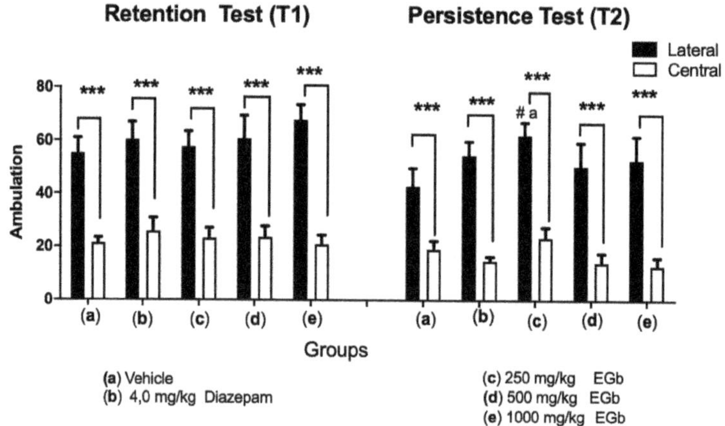

Figure 2. Number total of central and lateral square crossings (ambulation) in the control groups (4 mg/kg diazepam- and vehicle-treated group) and the groups treated with EGb (250 mg/kg; 500 mg/kg and 1000 mg/kg) during the retention test (n = 20/group) and persistence test (n = 10/group). The values are expressed as the mean values/ min (± SEMs). *** $p < 0.0001$; two-way ANOVA followed by post hoc Tukey's multiple comparison test. # $p < 0.05$ denotes a difference within groups. between sessions comparisons.

One-way RM ANOVA indicated that the total level of rearing behavior was similar for all groups in the T1 ($F_{4.45}$ = 2.424; p = 0.0619) and T2 session ($F_{4.45}$ = 0.8002; p = 0.5315). All data are shown as the means ± SEMs in Supplementary Table S2.

Finally, we examined the level of grooming, and the two-way RM ANOVA showed an effect of time ($F_{4.45}$ = 1647; p = 0.1790) in T1 and in T2 session ($F_{4.45}$ = 0.7617; p = 0.5557). All data are shown as the means ± SEMs in Supplementary Table S2.

2.2. Molecular Analysis

We also evaluated the levels of the BDNF protein in relation to tubulin in the hippocampal formation after the retention and persistence sessions. Compared with no treatment (naïve group), treatment with all doses of EGb before the acquisition of object recognition memory upregulated BDNF expression in animals subjected to ORM retrieval in the retention test. One-way ANOVA revealed that a main effect of the treatment ($F_{5.21}$ = 5.810; p = 0.0016) on performance in the retention test.

Furthermore, increased levels of BDNF were observed in the 250 and 1000 mg/kg EGb-treated groups compared with the naïve group in the T2 session ($F_{5, 23}$ = 4.005; p = 0.0093) (Figure 3). Our data in line with previous data about key role of BDNF on persistence of memory [18], since control and 500 mg/kg EGb groups didn't have differential expression of BDNF during retrieval of memory, which suggest that upregulation of BDNF during both sessions is required to persistence of memory.

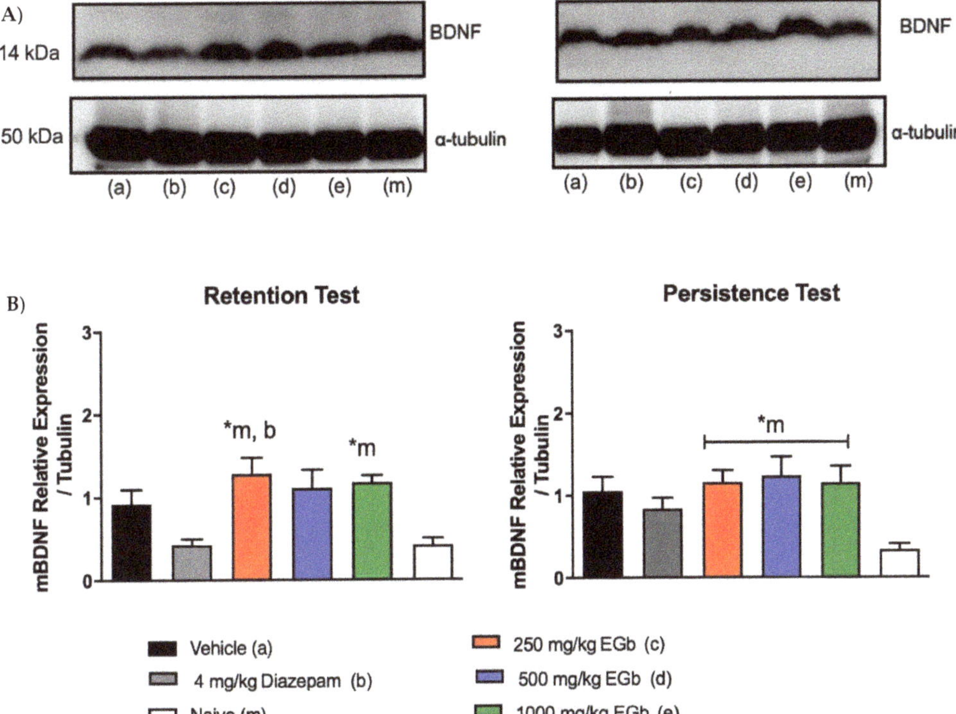

Figure 3. (**A**). Representative Western blot images. (**B**). Quantitative analysis of BDNF and α-tubulin expression in homogenates of the hippocampal formation obtained from the control (the 4 mg/kg diazepam (b), vehicle (a) and naïve (m) groups and the groups treated with EGb (250 mg/kg (c), 500 mg/kg (d) or 1000 mg/kg (e)), as determined by immunoblotting. The data were normalized to the level of α-tubulin in the retention and persistence test sessions (n = 5 rats/group). The values are presented as the mean values (± SEMs). * $p < 0.05$ one-way ANOVA followed by *post hoc* Tukey's multiple comparison test.

3. Discussion

Our present data show that all groups spent more time exploring the novel object than the familiar object in the retrieval (T1) session, but only the EGb-treated group at dose 1000 mg/kg spent more time exploring the novel object during the persistence session (T2). In addition, discrimination ability was evaluated within and between sessions by employing the recognition index (RI). Within-session analysis showed that RI was higher in the 1000 mg/kg EGb-treated group than the vehicle and diazepam groups during the first minute of the retrieval test. Similarly, the modulatory effect of EGb was observed during re-exposure of the animals to the same objects. Rats treated with 250 or 1000 mg/kg EGb had the highest RI values during the persistence session. The maintenance of memory for "novel" and familiar objects 96 h after acquisition of ORM was observed in the EGb-treated groups but not in the other groups.

Our data also revealed that rats treated with diazepam exhibit poor performance during the T2 session because the animals spent similar amounts of time exploring both the novel and familiar object, exhibiting an RI < 0.5. A reduced ability to recollect a previously experienced item suggests that diazepam prevented the persistence of ORM. To assess the effects of EGb on the persistence of ORM, we compared the RI data from the first min with those from the last three min of the retention test.

These data substantiate the findings regarding the total exploration time since the rats treated with the high dose of EGb exhibited an DI value similar to that observed during the first trial of the retention test. These effects suggest that EGb treatment led the animals to favor the innate novelty of the object, as reported in the literature [58]. Moreover, the number of line crosses in the central and lateral squares in the open field arena (Figure 2) and the levels of rearing and grooming (Supplementary Table S2) within sessions were evaluated to investigate whether or not the rats exhibited changes in spontaneous exploratory behavior or some aspects of emotionality including anxiety, which might lead to changes in object exploration and thus impair recognition memory. Based on the number of squares crossed, all groups of treated rats preferred to enter the lateral squares rather than the central squares, as the lateral squares elicited less anxiety than the central squares. Thus, EGb did not have an effect against anxiety. However, the size of open-field arena was chosen to reduce the levels of anxiety/aversion and to prevent potential interference with object exploration [59].

Because no difference in these behaviors was found between the treated groups and the control group (vehicle), it is unlikely that the findings observed following EGb or diazepam treatment were associated with motivation to explore or motor impairments. In addition, we evaluated the number of times (the frequency) the animals contacted each object since the number of contacts is a potential indicator of object recognition memory. During the T1 session, all groups had exhibited an increased frequency of contact with the novel object compared to the familiar object. However, during the T2 session, only the EGb-treated groups spent more time contacting the more recently encountered object than the familiar object.

Taken together, these results corroborate previous data from our group showing that that treatment with EGb improves long-term memory in a dose-dependent manner, as assessed by conditioned suppression of licking behavior as well as modulation of the spontaneous recovery of conditioned fear [11–13,52]. Here, we reveal a relationship between acute EGb treatment and the persistence of non-aversive memory for the first time. EGb-treated rats exhibit improvements in the recovery of memory for objects (what) and temporal order of the items (when). Moreover, we showed that maintenance of ORM caused by EGb treatment was associated with upregulation of BDNF in the dorsal hippocampal formation. However, no difference was detected in BDNF protein expression in the dHF between the vehicle, diazepam and naïve groups.

As mentioned in the introduction, novelty detection is associated with the hippocampus [60] and entorhinal cortex [61]. Here, we found that EGb treatment resulted in upregulation of BDNF expression in the dorsal hippocampal formation 72 and 96 hr after acquisition of ORM, which was associated with the persistence of ORM and was not seen in the control group [49,62,63].

Studies have shown that low levels of BDNF impair memory formation in animals subjected to an object recognition task. Reduced expression of BDNF in the hippocampus and parietal, entorhinal and frontal cortices results in noticeably reduced plasticity and cognitive impairment, which are correlated with progression of neurodegeneration and dementia [64]. In addition, it has been shown that EGb treatment can significantly enhance the levels of BDNF expression in the mouse hippocampus, leading to reversal of neural damage induced by acrylamide treatment [39]. The critical role of BDNF in the persistence of fear memory was previously described. Increased expression of BDNF may reverse memory deficits [22].

At the molecular level, BDNF is a modulator with a well-known effect on synaptic transmission and plasticity in different areas of the central nervous system [65], which contribute to the adaptative processes involved in consolidation of memory. CREB expression is modulated by multiple signalling pathways, including pathways associated with PKA-MAPKs, CaMK, and N-methyl-D-aspartate receptors (NMDA-Rs), which can converge to increase BDNF expression and modulate cytoskeleton protein synthesis and anti-apoptotic activity and are involved in molecular adaptation processes such as dendritic growth and ramification and the formation of stable long-term potentiated synapses necessary for consolidation processes.

Interactions of these different pathways, which converge on CREB and BDNF expression, are thought to be crucial for the effects of flavonoids on memory enhancement [1,66]. In previous studies from our group, we analysed the neural circuits and neurochemical changes following EGb treatment and fear memory formation [13,15,57,60] as well as the effects of crude extract of erythrina (CE), flavonoidic fraction or isolated compounds from CE on conditioned suppression [14,67]. EGb treatment prior to conditioning was subsequently evaluated after 72 or 96 h during the extinction retention test, and EGb was demonstrated to have effects on the spontaneous recovery of extinction. Furthermore, we showed that EGb treatment (acute or chronic) before conditioning increased the levels of CREB expression in the dorsal hippocampal formation. These data suggest that CREB expression might be correlated with the upregulation of BDNF observed in the present study. Although further studies are needed to better identify the mechanisms underlying object recognition memory, our data show for first time that BDNF in the dorsal hippocampal formation is an important target of EGb in promoting the persistence of object recognition memory.

4. Materials and Methods

4.1. Animals

A total of 120 experimentally naïve male Wistar rats (10 to 12 weeks old) were obtained from the Center for the Development of Experimental Medicine and Biology (Universidade Federal, São Paulo, Brazil). The animals were housed four per cage and provided food and water ad libitum during the acclimatization period (15 days) and experimental procedure. The animals were kept at a controlled temperature (21 °C $\pm$ 3) and relative humidity (55 $\pm$ 10) on a 12-h light/dark cycle (lights on at 6:00). The experiments were performed during the light cycle. All experimental procedures were approved by the local Committee Governing the Ethics on the use of Animal Experimentation of the Federal University of Sao Paulo (CEUA Unifesp 3447100417) and were conducted in accordance with the national animal care legislation and guidelines (Brazilian Law 11794/2008), as suggested by the APA Guidelines for Ethical Conduct in the Care. The animals were randomly assigned to control and EGb-treated groups. The animals in the control groups were distributed into the following three subgroups (n = 20/group): (I) the naïve group (no training/no treatment; protein expression control); (II) the vehicle group (0.9% saline; control) and, (III) diazepam-treated group (4 mg/kg, negative control). The animals in the EGb groups were further divided into subgroups according to the dose administered: (I) 250 mg/kg; (II) 500 mg/kg, or (III) 1000 mg/kg (n = 20/group).

4.2. Drug Administration

A previously described standardized *Ginkgo biloba* extract obtained from the green leaves of *Ginkgo biloba* (EGb) that contained 24% ginkgo-flavoglycosides and 6% ginkgo-terpenoid lactones [52] was used in all experiments. This extract has a formulation and composition identical to the products registered in Germany (Dr. Willmar Schwabe Pharmaceuticals, Karlsruhe, Germany). The dose range of EGb was chosen according to previous studies conducted in our laboratory [11,12,15,60]. According to previous data from our group, 4 mg/kg diazepam impairs the acquisition of fear memory in rats, and for this reason it was used as a negative control. EGb and diazepam were re-suspended in saline 0.9% (vehicle) and administered orally via an intragastric (IG) tube 30 min before the training session (TR1). No substances were administered in the other trial sessions.

4.3. Apparatus and Objects

The novel object recognition (NOR) task was conducted in a white square arena made of wood (40 × 40 × 40 cm), the floor of which was divided into different sections (central and lateral squares). A video camera fixed 100 cm above the apparatus was used to record all sessions for behavioral analysis. The objects used as the "sample objects" (old objects) for the training session and as the "novel object" (recent object) for the test session were chosen according to weight and size based on the literature [58]. The sample objects were always placed in the same corner (right or left) of the open field and positioned at the same place in the room to allow the use of distal cues. To minimize bias resulting from individual preference for the specific objects, we chose objects with a similar texture and colour pattern. In both the familiarization and test phases, rats were individually placed at the midpoint of the opposite wall of the box facing the wall to prevent coercing the rats to explore the objects. All experimental sessions were conducted using one rat from each group (control or treated). After each rat was tested, the apparatus and objects were carefully cleaned with 10% ethanol solution to remove odour cues [3,54,55].

4.4. Novel Object Recognition (NOR) Procedure

The testing protocol consisted of three phases (4 consecutive days): habituation, familiarization (two days) and the test phase (two days). All behavioral sessions were carried out between 8 a.m. and 1 p.m., including habituation of the animals to the testing room and to the researchers (the handling period).

4.4.1. Handling

Prior to beginning the experimental sessions, all animals were subjected to handling for a period of 20 min/day over 5 days (~2 min per rat each day) to avoid the effects of excessive handling after transport, as the novelty of handling has been shown to affect the acquisition of memory [58]. For the NOR test, the rats were individually transferred to and maintained in the adjacent room under light of a controlled intensity for 10 min before the sessions (Figure 1A).

4.4.2. Acquisition

On the 6th and 7th day all rats except those in the naïve group were allowed to freely explore the sample objects in order for 15 min to allow habituation to the objects (sample objects A and B) and the arena. This was considered the familiarization phase (F1 and F2) and is when the acquisition of object recognition memory occurs [68]. To prevent bias related to object exploration, the rats were placed against the centre of the opposite wall with their back to the objects. After the animals were placed in the open field arena, the experimenter left the room to avoid interfering with the animal's behavior (Figure 1B,C). Rationale for choosing of the familiarization procedure used in this study was based on previous data from the literature, such as the study by Shimoda and colleagues. [69], which showed no difference between the length of familiarization in the sample phase (5, 15 and 30 min) and subsequent novelty recognition performance at the test phase with a 24-h delay period. Complementarily, data from the study by Ozawa and colleagues suggest that longer exploration of the objects in the familiarization period may allow animals to make associations between each element of information, such as objects [70]. Moreover, another study revealed that a sample session of 10 min each for three days supports the encoding of strong object memory [60,71], and promotes persistence of object memory. Consistent with these data and previous data from our lab showing the effects of EGb as a cognitive enhancer [13,14,56,57] we hypothesized that associated effects, such as a longer sample phase and EGb treatment, would improve recognition memory formation and support the persistence of long-term recognition memory in rats in the object recognition test (ORT), since both the familiarity and recollection components become less persistent over time [22,23].

In addition, we counterbalanced the objects serving as A and B (e.g., a bottle served as object A and a statuette served as object B for half of the rats; the pattern was reversed for the remaining rats). Furthermore, the two objects had similar patterns (e.g., size and no odour cues). As defined by Ennaceur and Delacour [72], the exploration of an object was considered when the rat directed its nose at a distance ≥ 2 cm to the object and/or touched it with the nose, while turning around or sitting on the object was not considered an exploration.

4.4.3. Retention Test

All rats except those in the naïve group, were returned to the same chamber for analysis of long-term memory (LTM) for the objects (8th day) (n = 20/group) (Figure 1A–E). Each rat underwent a test trial that was identical to the second familiarization session except that one object was identical to the sample trial (object A or B) and one object was the novel object (object C). Twenty-four h after the end of the T1 session, half of the rats were euthanized (n = 10/group). The remaining half (n = 10/group) was subjected to the persistence test to evaluate the effects of EGb on the maintenance of memory.

4.4.4. Persistence Test

Previous data from the literature showed that long intervals of time lead to the deterioration of LTM for objects and that the hippocampus is required for object memory encoding after long intertrial intervals [73] and persistence of object recognition memory [74]. Thus, we hypothesized that treatment with a cognitive enhancer could result in the maintenance of LTM for objects over time. To evaluate the persistence of LTM (T2), an additional object recognition test session was performed 24 h following the T1 session. Similar to the T1 session, one sample object (object A) and one novel (object C) were placed in the opposite corners of the open field arena (right or left corner) in the T2 session. The session lasted 10 min (Figure 1A–E). Afterwards, the animal was removed from the test environment and returned to the vivarium (n = 10/group).

4.5. Molecular Analysis

4.5.1. Total Protein Extraction and Quantification by the Bradford Assay

Twenty-four hours after the T1 (day 8) or T2 (day 9) session ended, the animals (n = 5/group) were euthanized by decapitation. The dHF was rapidly removed, immediately frozen in liquid nitrogen and stored at $-80\,^{\circ}$C until extraction. Briefly, the samples were homogenized in 1 mL of extraction buffer containing 50 mM Tris HCl, 100 mM NaCl, 50 mM NaF, 1 mM NaVO$_4$, 0.5% NP-40, 10 µg/mL leupeptin, 10 µg/mL aprotinin and 0.5 mM PMSF and were incubated with the buffer for 20 min on ice. The sample lysates were centrifuged at 4 °C and 14,000 rpm for 15 min. The supernatants were collected, aliquoted and stored at $-80\,^{\circ}$C. Sodium orthovanadate (NaVO$_4$) was used as a phosphatase inhibitor, and leupeptin, aprotinin and PMSF were used as protease inhibitors. The protein concentrations of the supernatants were quantified with a NanoDrop 200c spectrophotometer (Thermo Scientific, Wilmington, DE, USA) using Bradford reagents

(Sigma-Aldrich, St. Louis, MO, USA) according to the manufacturer's recommendations and bovine serum albumin (BSA) as a standard.

4.5.2. Western Blotting

Homogenates (50 μg of protein) were subjected to SDS-PAGE on a 12% polyacrylamide gel (12% acrylamide, 0.1% *w/v* SDS, 0.1% (NH4) 2S2O8, Temed, 0.39 M Tris pH 8.8); 10% of the elution volume of each extract was mixed with 4× buffer (100 mM Tris-HCl pH 6.8, 10% *v/v* 2-mercaptoethanol, 4% *w/v* SDS, 0.02% *w/v* bromophenol blue, and 20% glycerol). Prior to being added to the gel, the samples were denatured by heating at 100 °C for 5 min. The potential difference applied was 200 V, and the current intensity was 150 mA. The running buffer contained 25 mM Tris-base, 250 mM glycine and 0.1% SDS. After electrophoresis, the proteins were transferred to a nitrocellulose membrane in transfer buffer (48 mM Tris, 39 mM glycine, 0.037% SDS and 20% methanol) and subjected to a potential difference of 60 V and a current intensity of 150 mA. The transfer was performed at 4 °C for 3 h. Nonspecific binding was then blocked by incubating the membrane with 5% skimmed milk powder in PBS-T (phosphate-buffered saline, 137 mM NaCl, 10 mM phosphate, 2.7 mM KCl, pH 7.4, and 0.1% Tween) for 1 h at room temperature under constant stirring. After blocking, the membrane was washed (3×) in PBS-T solution for 10 min and cut based on the size of the proteins to be evaluated. For BDNF detection, the membrane was incubated with a BDNF primary antibody (dilution, 1:700) (700 rabbit Abcam®, Inc., Cambridge, MA, USA) for 16 h at 4 °C under constant stirring. An anti-α-tubulin antibody) (dilution, 1: 10.000) (Sigma-Aldrich, St Louis, MO, USA) was used as the loading control. After incubation, the membrane was incubated with a conjugated anti-rabbit secondary antibody (Santa Cruz Biotechnology, Santa Cruz, CA, USA) diluted 1: 10.000 in 5% skimmed milk powder in PBS-T for 1 h at room temperature under constant agitation.

Finally, the membrane was washed (3×) times in PBS-T for 10 min. Immunoreactivity was evaluated with the Immobilon Western Chemiluminescent HRP Substrate kit (Millipore, Billerica, MA, USA). The bands were quantified using ImageQuant LAS 4000 (GE, Buckinghamshire, UK). The data were analysed using repeated-measures ANOVA followed by Dunnett's multiple comparisons.

5. Conclusions

Based on the findings from the present study, the persistence of memory is a process associated with mechanisms initiated during consolidation that persist over time. Furthermore, we reveal for the first time that EGb treatment before the acquisition of ORM increases BDNF expression in the hippocampus in a dose-dependent manner, which is associated with the persistence of object recognition memory. This finding was substantiated by the different levels of BDNF expression observed during the T1 and T2 sessions following treatment with 250 or 1000 mg/kg EGb. Conversely, rats treated with 500 mg/kg EGb showed upregulation during the T2 session, which was not sufficient to maintain the storage of recognition memory traces over time.

Supplementary Materials: The following are available online. Table S1: Recognition Index (RI) values in the retention and persistence tests. Table S2: The levels of rearing and grooming during analysis of long-term memory for objects sessions.

Author Contributions: Conceptualization, S.M.C. and J.M.C.; methodology, B.G.M., C.R.Z., T.B.M. and B.H.N.N.; formal analysis, B.G.M., C.R.Z., T.B.M. and B.H.N.N.; investigation, B.G.M., S.M.C. and J.M.C.; resources, B.G.M., C.R.Z., J.M.C. and S.M.C.; data curation, B.G.M., C.R.Z., T.B.M., B.H.N.N., J.M.C. and S.M.C.; writing—original draft preparation, B.G.M., J.M.C. and S.M.C.; writing—review and editing, B.G.M., J.M.C. and S.M.C.; supervision, S.M.C. and J.M.C.; project administration, S.M.C.; funding acquisition, S.M.C. All authors have read and agreed to the published version of the manuscript.

Funding: This study was supported by the São Paulo State Research Foundation (FAPESP) (grant **2016/18039-9** to SMC). BGM is a scholar from Coordenação de Aperfeiçoamento de Pessoal de Nível Superior-Brasil (CAPES)—Finance Code 001. CRZ is a Postdoctoral fellow from CNPq.

Institutional Review Board Statement: All procedures were approved by the Local Committee Governing the Ethics on the Use of Animal Experimentation of the Federal University of Sao Paulo (CEUA UNIFESP 2076240914) and were conducted in accordance with the Brazilian Law for the use of animals in scientific research (N°11.794) as suggested by the APA Guidelines for Ethical Conduct in the Care.

Informed Consent Statement: Not applicable for studies not involving humans.

Data Availability Statement: The data presented in this study are available on request from the corresponding author.

Conflicts of Interest: The authors declare no conflict of interest.

Sample Availability: Samples of the compounds are not available from the authors.

References

1. Spencer, J.P. Flavonoids: modulators of brain function? *Br. J. Nutr.* **2008**, *99*, ES60–ES77. [CrossRef] [PubMed]
2. Bakoyiannis, I.; Daskalopoulou, A.; Pergialiotis, V.; Perrea, D. Phytochemicals and cognitive health: Are flavonoids doing the trick? *Biomed. Pharmacother.* **2019**, *109*, 1488–1497. [CrossRef] [PubMed]
3. Walesiuk, A.; Trofimiuk, E.; Braszko, J.J. Ginkgo biloba extract diminishes stress-induced memory deficits in rats. *Pharmacol. Rep.* **2005**, *57*, 176–187. [PubMed]
4. Shen, Y.-F.; Chou, Y.-H.; Yang, Y.-L.; Lu, K.-T. The role of the dorsal hippocampus on the Ginkgo biloba facilitation effect of fear extinction as assessed with fear-potentiated startle. *Psychopharmacology* **2011**, *215*, 403–411. [CrossRef] [PubMed]
5. Sangiovanni, E.; Brivio, P.; Dell'Agli, M.; Calabrese, F. Botanicals as Modulators of Neuroplasticity: Focus on BDNF. *Neural Plast.* **2017**. [CrossRef]
6. Martin, Z.S.; Neugebauer, V.; Dineley, K.T.; Kayed, R.; Zhang, W.; Reese, L.C.; Taglialatela, G. α-Synuclein oligomers oppose long-term potentiation and impair memory through a calcineurin-dependent mechanism: Relevance to human synucleopathic diseases. *J. Neurochem.* **2012**. [CrossRef]
7. Zhao, L.; Wang, J.L.; Liu, R.; Li, X.X.; Li, J.F.; Zhang, L. Neuroprotective, anti-amyloidogenic and neurotrophic effects of apigenin in an Alzheimer's disease mouse model. *Molecules* **2013**, *18*, 9949–9965. [CrossRef]
8. Maurer, K.; Ihl, R.; Dierks, T.; Frölich, L. Clinical efficacy of Ginkgo biloba special extract EGb 761 in dementia of the Alzheimer type. *J. Psychiatr. Res.* **1997**, *31*, 645–655. [CrossRef]
9. Oken, B.S.; Storzbach, D.M.; Kaye, J.A. The efficacy of Ginkgo biloba on cognitive function in Alzheimer disease. *Arch. Neurol.* 1998. [CrossRef]
10. Chen, Y.; Feng, Z.; Shen, M.; Lin, W.; Wang, Y.; Wang, S.; Li, C.; Wang, S.; Chen, M.; Shan, W.; et al. Insight into Ginkgo biloba L. Extract on the Improved Spatial Learning and Memory by Chemogenomics Knowledgebase, Molecular Docking, Molecular Dynamics Simulation, and Bioassay Validations. *ACS Omega* **2020**. [CrossRef]
11. Oliveira, D.R.; Sanada, P.F.; Saragossa, F.A.C.; Innocenti, L.R.; Oler, G.; Cerutti, J.M.; Cerutti, S.M. Neuromodulatory property of standardized extract Ginkgo biloba L. (EGb 761) on memory: Behavioral and molecular evidence. *Brain Res.* **2009**, *1269*. [CrossRef]
12. Oliveira, D.R.; Sanada, P.F.; Filho, A.C.S.; Conceição, G.M.S.; Cerutti, J.M.; Cerutti, S.M. Long-term treatment with standardized extract of Ginkgo biloba L. enhances the conditioned suppression of licking in rats by the modulation of neuronal and glial cell function in the dorsal hippocampus and central amygdala. *Neuroscience* **2013**, *235*. [CrossRef] [PubMed]
13. Zamberlam, C.R.; Tilger, M.A.S.; Moraes, L.; Cerutti, J.M.; Cerutti, S.M. Ginkgo biloba treatments reverse the impairment of conditioned suppression acquisition induced by GluN2B-NMDA and 5-HT1A receptor blockade: Modulatory effects of the circuitry of the dorsal hippocampal formation. *Physiol. Behav.* **2019**. [CrossRef]
14. de Oliveira, D.R.; Zamberlam, C.R.; Rêgo, G.M.; Cavalheiro, A.; Cerutti, J.M.; Cerutti, S.M. Effects of a Flavonoid-Rich Fraction on the Acquisition and Extinction of Fear Memory: Pharmacological and Molecular Approaches. *Front. Behav. Neurosci.* **2016**, *9*. [CrossRef]
15. de Oliveira, D.R.; Todo, A.H.; Rêgo, G.M.; Cerutti, J.M.; Cavalheiro, A.J.; Rando, D.G.G.; Cerutti, S.M. Flavones-bound in benzodiazepine site on GABAAreceptor: Concomitant anxiolytic-like and cognitive-enhancing effects produced by Isovitexin and 6-C-glycoside-Diosmetin. *Eur. J. Pharmacol.* **2018**, *831*, 77–86. [CrossRef]
16. Radiske, A.; Rossato, J.I.; Gonzalez, M.C.; Köhler, C.A.; Bevilaqua, L.R.; Cammarota, M. BDNF controls object recognition memory reconsolidation. *Neurobiol. Learn. Mem.* **2017**. [CrossRef]
17. Barbosa, F.F.; de Oliveira Pontes, I.M.; Ribeiro, S.; Ribeiro, A.M.; Silva, R.H. Differential roles of the dorsal hippocampal regions in the acquisition of spatial and temporal aspects of episodic-like memory. *Behav. Brain Res.* **2012**. [CrossRef] [PubMed]

18. Eacott, M.J.; Gaffan, E.A. The roles of perirhinal cortex, postrhinal cortex, and the fornix in memory for objects, contexts, and events in the rat. *Q. J. Exp. Psychol. Sect. B* **2005**, *58*, 202–217. [CrossRef]

19. Dere, E.; Huston, J.P.; De Souza Silva, M.A. Episodic-like memory in mice: Simultaneous assessment of object, place and temporal order memory. *Brain Res. Protoc.* **2005**, *16*, 10–19. [CrossRef] [PubMed]

20. Crystal, J.D. Episodic-like memory in animals. *Behav. Brain Res.* **2010**, *215*, 235–243. [CrossRef] [PubMed]

21. Yonelinas, A.P.; Aly, M.; Wang, W.C.; Koen, J.D. Recollection and familiarity: Examining controversial assumptions and new directions. *Hippocampus* **2010**. [CrossRef]

22. Bekinschtein, P.; Cammarota, M.; Igaz, L.M.; Bevilaqua, L.R.M.; Izquierdo, I.; Medina, J.H. Persistence of Long-Term Memory Storage Requires a Late Protein Synthesis- and BDNF- Dependent Phase in the Hippocampus. *Neuron* **2007**, *53*, 261–277. [CrossRef]

23. Mandler, G. Familiarity Breeds Attempts: A Critical Review of Dual-Process Theories of Recognition. *Perspect. Psychol. Sci.* **2008**. [CrossRef] [PubMed]

24. Ennaceur, A.; Neave, N.; Aggleton, J.P. Spontaneous object recognition and object location memory in rats: The effects of lesions in the cingulate cortices, the medial prefrontal cortex, the cingulum bundle and the fornix. *Exp. Brain Res.* **1997**. [CrossRef] [PubMed]

25. Dere, E.; Huston, J.P.; De Souza Silva, M.A. Integrated memory for objects, places, and temporal order: Evidence for episodic-like memory in mice. *Neurobiol. Learn. Mem.* **2005**. [CrossRef] [PubMed]

26. Dere, E.; Kart-Teke, E.; Huston, J.P.; De Souza Silva, M.A. The case for episodic memory in animals. *Neurosci. Biobehav. Rev.* **2006**, *30*, 1206–1224. [CrossRef]

27. Langston, R.F.; Wood, E.R. Associative recognition and the hippocampus: Differential effects of hippocampal lesions on object-place, object-context and object-place-context memory. *Hippocampus* **2010**. [CrossRef]

28. Mumby, D.G.; Glenn, M.J.; Nesbitt, C.; Kyriazis, D.A. Dissociation in retrograde memory for object discriminations and object recognition in rats with perirhinal cortex damage. *Behav. Brain Res.* **2002**. [CrossRef]

29. Balderas, I.; Rodríguez-Ortiz, C.J.; Salgado-Tonda, P.; Chávez-Hurtado, J.; McGaugh, J.L.; Bermúdez-Rattoni, F. The consolidation of object and context recognition memory involve different regions of the temporal lobe. *Learn. Mem.* **2008**, *15*, 618–624. [CrossRef]

30. Miranda, M.; Morici, J.F.; Gallo, F.; Piromalli Girado, D.; Weisstaub, N.V.; Bekinschtein, P. Molecular mechanisms within the dentate gyrus and the perirhinal cortex interact during discrimination of similar nonspatial memories. *Hippocampus* **2021**. [CrossRef]

31. Callaghan, C.K.; Kelly, Á.M. Differential BDNF signaling in dentate gyrus and perirhinal cortex during consolidation of recognition memory in the rat. *Hippocampus* **2012**. [CrossRef] [PubMed]

32. Dees, R.L.; Kesner, R.P. The role of the dorsal dentate gyrus in object and object-context recognition. *Neurobiol. Learn. Mem.* **2013**. [CrossRef] [PubMed]

33. Squire, L.R.; Wixted, J.T.; Clark, R.E. Recognition memory and the medial temporal lobe: A new perspective. *Nat. Rev. Neurosci.* **2007**. [CrossRef] [PubMed]

34. Barker, G.R.I.; Warburton, E.C. When is the hippocampus involved in recognition memory? *J. Neurosci.* **2011**. [CrossRef]

35. Yonelinas, A.P.; Ranganath, C.; Ekstrom, A.D.; Wiltgen, B.J. A contextual binding theory of episodic memory: Systems consolidation reconsidered. *Nat. Rev. Neurosci.* **2019**. [CrossRef] [PubMed]

36. Rossato, J.I.; Bevilaqua, L.R.M.; Myskiw, J.C.; Medina, J.H.; Izquierdo, I.; Cammarota, M. On the role of hippocampal protein synthesis in the consolidation and reconsolidation of object recognition memory. *Learn. Mem.* **2007**. [CrossRef]

37. Kelly, A.; Laroche, S.; Davis, S. Activation of mitogen-activated protein kinase/extracellular signal-regulated kinase in hippocampal circuitry is required for consolidation and reconsolidation of recognition memory. *J. Neurosci.* **2003**. [CrossRef]

38. Martínez-Moreno, A.; Rodríguez-Durán, L.F.; Escobar, M.L. Late protein synthesis-dependent phases in CTA long-term memory: BDNF requirement. *Front. Behav. Neurosci.* **2011**. [CrossRef]

39. Wang, C.; Li, Z.; Han, H.; Luo, G.; Zhou, B.; Wang, S.; Wang, J. Impairment of object recognition memory by maternal bisphenol A exposure is associated with inhibition of Akt and ERK/CREB/BDNF pathway in the male offspring hippocampus. *Toxicology* **2016**. [CrossRef]

40. Alonso, M. ERK1/2 Activation Is Necessary for BDNF to Increase Dendritic Spine Density in Hippocampal CA1 Pyramidal Neurons. *Learn. Mem.* **2004**, *11*, 172–178. [CrossRef]

41. Leal, G.; Afonso, P.M.; Salazar, I.L.; Duarte, C.B. Regulation of hippocampal synaptic plasticity by BDNF. *Brain Res.* **2015**. [CrossRef] [PubMed]

42. Dudai, Y.; Eisenberg, M. Rites of passage of the engram: Reconsolidation and the lingering consolidation hypothesis. *Neuron* **2004**, *44*, 93–100. [CrossRef] [PubMed]

43. Tyler, W.J.; Alonso, M.; Bramham, C.R.; Pozzo-Miller, L.D. From acquisition to consolidation: On the role of brain-derived neurotrophic factor signaling in hippocampal-dependent learning. *Learn. Mem.* **2002**, *9*, 224–237. [CrossRef] [PubMed]

44. Alonso, M.; Vianna, M.R.M.; Depino, A.M.; Mello E Souza, T.; Pereira, P.; Szapiro, G.; Viola, H.; Pitossi, F.; Izquierdo, I.; Medina, J.H. BDNF-triggered events in the rat hippocampus are required for both short- and long-term memory formation. *Hippocampus* **2002**, *12*, 551–560. [CrossRef]

45. Kowiański, P.; Lietzau, G.; Czuba, E.; Waśkow, M.; Steliga, A.; Moryś, J. BDNF: A Key Factor with Multipotent Impact on Brain Signaling and Synaptic Plasticity. *Cell. Mol. Neurobiol.* **2018**, *38*, 579–593. [CrossRef] [PubMed]

46. Bhattarai, P.; Cosacak, M.I.; Mashkaryan, V.; Demir, S.; Popova, S.D.; Govindarajan, N.; Brandt, K.; Zhang, Y.; Chang, W.; Ampatzis, K.; et al. Neuron-glia interaction through Serotonin-BDNF-NGFR axis enables regenerative neurogenesis in Alzheimer's model of adult zebrafish brain. *PLoS Biol.* **2020**. [CrossRef] [PubMed]
47. Korte, M.; Kang, H.; Bonhoeffer, T.; Schuman, E. A role for BDNF in the late-phase of hippocampal long-term potentiation. *Neuropharmacology* **1998**, *37*, 553–559. [CrossRef]
48. Cunha, C.; Brambilla, R.; Thomas, K.L. A simple role for BDNF in learning and memory? *Front. Mol. Neurosci.* **2010**, *3*, 1. [CrossRef] [PubMed]
49. Mello-Carpes, P.B.; da Silva de Vargas, L.; Gayer, M.C.; Roehrs, R.; Izquierdo, I. Hippocampal noradrenergic activation is necessary for object recognition memory consolidation and can promote BDNF increase and memory persistence. *Neurobiol. Learn. Mem.* **2016**. [CrossRef] [PubMed]
50. Ying, S.W.; Futter, M.; Rosenblum, K.; Webber, M.J.; Hunt, S.P.; Bliss, T.V.P.; Bramham, C.R. Brain-derived neurotrophic factor induces long-term potentiation in intact adult hippocampus: Requirement for ERK activation coupled to CREB and upregulation of Arc synthesis. *J. Neurosci.* **2002**. [CrossRef]
51. Alonso, M.; Vianna, M.R.M.; Izquierdo, I.; Medina, J.H. Signaling mechanisms mediating BDNF modulation of memory formation in vivo in the hippocampus. *Cell. Mol. Neurobiol.* **2002**, *22*, 663–674. [CrossRef]
52. Zamberlam, C.R.; Vendrasco, N.C.; Oliveira, D.R.; Gaiardo, R.B.; Cerutti, S.M. Effects of standardized Ginkgo biloba extract on the acquisition, retrieval and extinction of conditioned suppression: Evidence that short-term memory and long-term memory are differentially modulated. *Physiol. Behav.* **2016**, *165*, 55–68. [CrossRef] [PubMed]
53. Gaiardo, R.B.; Abreu, T.F.; Tashima, A.K.; Telles, M.M.; Cerutti, S.M. Target proteins in the dorsal hippocampal formation sustain the memory-enhancing and neuroprotective effects of Ginkgo biloba. *Front. Pharmacol.* **2019**, *9*. [CrossRef]
54. Antunes, M.; Biala, G. The novel object recognition memory: Neurobiology, test procedure, and its modifications. *Cogn. Process.* **2012**. [CrossRef] [PubMed]
55. Broadbent, N.J.; Gaskin, S.; Squire, L.R.; Clark, R.E. Object recognition memory and the rodent hippocampus. *Learn. Mem.* **2010**, *17*, 5–11. [CrossRef]
56. Ennaceur, A. One-trial object recognition in rats and mice: Methodological and theoretical issues. *Behav. Brain Res.* **2010**, *215*, 244–254. [CrossRef] [PubMed]
57. Lueptow, L.M. Novel object recognition test for the investigation of learning and memory in mice. *J. Vis. Exp.* **2017**. [CrossRef] [PubMed]
58. Bevins, R.A.; Besheer, J. Object recognition in rats and mice: A one-trial non-matching-to-sample learning task to study "recognition memory". *Nat. Protoc.* **2006**. [CrossRef]
59. Steimer, T. Animal models of anxiety disorders in rats and mice: Some conceptual issues. *Dialogues Clin. Neurosci.* **2011**, *13*, 495. [PubMed]
60. Cohen, S.J.; Munchow, A.H.; Rios, L.M.; Zhang, G.; Ásgeirsdóttir, H.N.; Stackman, R.W. The rodent hippocampus is essential for nonspatial object memory. *Curr. Biol.* **2013**. [CrossRef]
61. VanElzakker, M.; Fevurly, R.D.; Breindel, T.; Spencer, R.L. Environmental novelty is associated with a selective increase in Fos expression in the output elements of the hippocampal formation and the perirhinal cortex. *Learn. Mem.* **2008**. [CrossRef] [PubMed]
62. Katche, C.; Cammarota, M.; Medina, J.H. Molecular signatures and mechanisms of long-lasting memory consolidation and storage. *Neurobiol. Learn. Mem.* **2013**, *106*, 40–47. [CrossRef]
63. Bekinschtein, P.; Cammarota, M.; Medina, J.H. BDNF and memory processing. *Neuropharmacology* **2014**, *76*, 677–683. [CrossRef]
64. Hock, C.; Heese, K.; Hulette, C.; Rosenberg, C.; Otten, U. Region-specific neurotrophin imbalances in Alzheimer disease: Decreased levels of brain-derived neurotrophic factor and increased levels of nerve growth factor in hippocampus and cortical areas. *Arch. Neurol.* **2000**. [CrossRef] [PubMed]
65. Bramham, C.R.; Messaoudi, E. BDNF function in adult synaptic plasticity: the synaptic consolidation hypothesis. *Prog. Neurobiol.* **2005**, *76*, 99–125. [CrossRef]
66. Vauzour, D.; Vafeiadou, K.; Rodriguez-Mateos, A.; Rendeiro, C.; Spencer, J.P.E. The neuroprotective potential of flavonoids: A multiplicity of effects. *Genes Nutr.* **2008**, *3*, 115–126. [CrossRef]
67. de Oliveira, D.R.; Zamberlam, C.R.; Gaiardo, R.B.; Rêgo, G.M.; Cerutti, J.M.; Cavalheiro, A.J.; Cerutti, S.M. Flavones from Erythrina falcata are modulators of fear memory. *BMC Complement. Altern. Med.* **2014**, *14*, 288. [CrossRef]
68. Rossato, J.I.; Radiske, A.; Kohler, C.A.; Gonzalez, C.; Bevilaqua, L.R.; Medina, J.H.; Cammarota, M. Consolidation of object recognition memory requires simultaneous activation of dopamine D1/D5 receptors in the amygdala and medial prefrontal cortex but not in the hippocampus. *Neurobiol. Learn. Mem.* **2013**. [CrossRef] [PubMed]
69. Shimoda, S.; Ozawa, T.; Ichitani, Y.; Yamada, K. Long-term associative memory in rats: effects of familiarization period in object-place-context recognition test. *bioRxiv* **2019**, 728295. [CrossRef]
70. Ozawa, T.; Yamada, K.; Ichitani, Y. Long-term object location memory in rats: Effects of sample phase and delay length in spontaneous place recognition test. *Neurosci. Lett.* **2011**. [CrossRef]
71. Cohen, S.J.; Stackman, R.W. Assessing rodent hippocampal involvement in the novel object recognition task. A review. *Behav. Brain Res.* **2015**. [CrossRef] [PubMed]

72. Ennaceur, A.; Delacour, J. A new one—Trial test for neurobiological studies of memory in rats. 1: Behavioral data. *Behav. Brain Res.* **1988**, *31*, 47–59. [CrossRef]
73. Hammond, R.S.; Tull, L.E.; Stackman, R.W. On the delay-dependent involvement of the hippocampus in object recognition memory. *Neurobiol. Learn. Mem.* **2004**. [CrossRef]
74. Vargas, L.S.; Ramires Lima, K.; Piaia Ramborger, B.; Roehrs, R.; Izquierdo, I.; Mello-Carpes, P.B. Catecholaminergic hippocampal activation is necessary for object recognition memory persistence induced by one-single physical exercise session. *Behav. Brain Res.* **2020**. [CrossRef] [PubMed]

molecules

Article

Astragaloside IV Suppresses Hepatic Proliferation in Regenerating Rat Liver after 70% Partial Hepatectomy via Down-Regulation of Cell Cycle Pathway and DNA Replication

Gyeong-Seok Lee [1], Hee-Yeon Jeong [1], Hyeon-Gung Yang [2], Young-Ran Seo [1], Eui-Gil Jung [3], Yong-Seok Lee [1], Kung-Woo Nam [1] and Wan-Jong Kim [1,*]

[1] Department of Life Science and Biotechnology, College of Natural Sciences, Soonchunhyang University, Asan 31538, Chungcheongnam-do, Korea; ssm4914@naver.com (G.-S.L.); youn6640@naver.com (H.-Y.J.); 94seoyoungran@gmail.com (Y.-R.S.); yslee@sch.ac.kr (Y.-S.L.); kwnam1@sch.ac.kr (K.-W.N.)

[2] Soonchunhyang Institute of Medi-bio Science (SIMS), Soonchunhyang University, Cheonan 31151, Chungcheongnam-do, Korea; yhg930205@naver.com

[3] Seoul Center, Korea Basic Science Institute, Seoul 02855, Korea; euigiljung@gmail.com

* Correspondence: wjkim56@sch.ac.kr; Tel.: +82-41-530-1251

Citation: Lee, G.-S.; Jeong, H.-Y.; Yang, H.-G.; Seo, Y.-R.; Jung, E.-G.; Lee, Y.-S.; Nam, K.-W.; Kim, W.-J. Astragaloside IV Suppresses Hepatic Proliferation in Regenerating Rat Liver after 70% Partial Hepatectomy via Down-Regulation of Cell Cycle Pathway and DNA Replication. *Molecules* **2021**, *26*, 2895. https://doi.org/10.3390/molecules26102895

Academic Editors: Raffaele Pezzani and Sara Vitalini

Received: 19 April 2021
Accepted: 11 May 2021
Published: 13 May 2021

Abstract: Astragaloside IV (AS-IV) is one of the major bio-active ingredients of huang qi which is the dried root of *Astragalus membranaceus* (a traditional Chinese medicinal plant). The pharmacological effects of AS-IV, including anti-oxidative, anti-cancer, and anti-diabetic effects have been actively studied, however, the effects of AS-IV on liver regeneration have not yet been fully described. Thus, the aim of this study was to explore the effects of AS-IV on regenerating liver after 70% partial hepatectomy (PHx) in rats. Differentially expressed mRNAs, proliferative marker and growth factors were analyzed. AS-IV (10 mg/kg) was administrated orally 2 h before surgery. We found 20 core genes showed effects of AS-IV, many of which were involved with functions related to DNA replication during cell division. AS-IV down-regulates MAPK signaling, PI3/Akt signaling, and cell cycle pathway. Hepatocyte growth factor (HGF) and cyclin D1 expression were also decreased by AS-IV administration. Transforming growth factor β1 (TGFβ1, growth regulation signal) was slightly increased. In short, AS-IV down-regulated proliferative signals and genes related to DNA replication. In conclusion, AS-IV showed anti-proliferative activity in regenerating liver tissue after 70% PHx.

Keywords: astragaloside IV; *Astragalus membranaceus*; huang qi; Astragali Radix; liver; liver regeneration; 70% partial hepatectomy; proliferation; rat

1. Introduction

For a long time, medicinal plants have played a key role in pharmacological research studies and drug development. *Astragalus membranaceus* is medicinal plant that has been used in traditional Chinese medicine throughout history; it is also called Astragali Radix. Huang qi (黃芪, in Chinese) is the name for the dried root of *Astragalus membranaceus* and it is mainly produced in China, Mongolia, and Korea [1]. The main components of huang qi are saponins, polysaccharides and flavonoids [2–4]. It has been used in traditional Chinese medicine for over 2000 years to treat various diseases including anemia, cardiovascular disorder, weakness, fatigue and fever [4,5]. In recent years, it has been reported that huang qi has anti-oxidative, anti-aging, anti-inflammatory, anti-diabetic, and anti-cancer properties [1–4].

Astragaloside IV (AS-IV) is a representative bio-active ingredient of huang qi [4,6]. AS-IV is classified as a tetra cyclic triterpenenoid, otherwise known as a protostane which means prototype of steroid (Figure 1) [7]. AS-IV has a similar carbon-tetra-cyclic structure to steroids which have a bio-activities as cholesterol. Recently, it was reported that long-term dietary cholesterol overload negatively affects liver regeneration after PHx. Cholesterol overload reduced hepatic DNA synthesis [8]. While, other steroid, estradiol accelerates liver regeneration [9]. Thus, it was expected that the AS-IV directly has an effects on liver regeneration after PHx. Many reports have shown that AS-IV has anti-oxidative properties and protective effects against hypoxic injury and ischemia-reperfusion injury [10–15]. The anti-cancer activities of AS-IV via inhibition of migration and invasion of cancer cells have also been demonstrated [16,17]. Moreover, AS-IV improves diabetic nephropathy, retinopathy, gastropathy, and wound healing [18–21] Hepato-protective effects against hepatic fibrosis, oxidative stress, and hepatitis B virus were also reported [10,22–25].

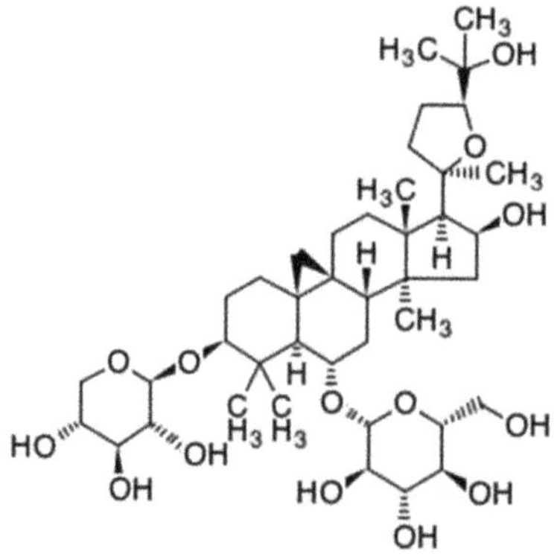

Figure 1. Chemical structure of astragaloside IV. ($C_{41}H_{68}O_{14}$, molecular weight: 794.97 g/mol).

The liver has ability to recover lost functional capacity after chemical or physical injury. After injury, the liver quickly regenerates to meet metabolic demand via the proliferation of hepatic cells. Liver regeneration involves a complex network of growth factors, signaling pathways, and transcriptional factors [26]. Uniquely, liver regeneration occurs without functional loss and through repopulation of mature cells rather than progenitor or stem cells [27].

The 70% partial hepatectomy (PHx) model is a commonly used model to investigate new aspects of liver regeneration [28]. It was first described by Higgins and Anderson in 1931. PHx is a surgical procedure to resect the median and left lateral lobes of the liver, which constitute about 70% of the liver mass [29–31]. Liver regeneration is especially rapid in small rodents; full-size restorations have been reported within 7 days in most rodents [27,32]. After PHx, remnant liver tissues undergo three phases. First is priming phase, which is characterized by the stimulation of hepatic mitogen. Hypoxic conditions after PHx and hemo-dynamic factors such as blood pressure are thought to be major stimulators of liver regeneration [33–36]. The liver is not alone in promoting its own regeneration, and cooperative signals for priming also come from the pancreas, spleen, duodenum, and adrenal glands [37,38]. Second is proliferating phase; DNA replication in hepatocytes started. The last step is growth termination. Transforming growth factor β1 (TGFβ1) is secreted by non-parenchymal cells including hepatic stellate cells, Kupffer cells, and platelets. TGFβ1 plays an important role in ending regeneration through suppression of hepatic proliferation [33,39,40].

The pharmacological effects of AS-IV have been actively studied in recent years and hepato-protective effects of AS-IV have been extensively reported. However, the effects of AS-IV on liver regeneration are not yet fully elucidated. In this study, the effects of AS-IV were investigated in a 70% PHx rat model through measurement of gene expression, the expression of hepatocyte growth factor (HGF, primary hepatic mitogen), and the expression of hepatic proliferation marker protein.

2. Results

2.1. mRNA Sequencing Analysis

To determine the effects of AS-IV on the gene expression of regenerating liver tissues after PHx, changes of gene expression were measured 12 h after PHx by mRNA sequencing analysis. The count of differentially expressed genes (DEGs) was 17,048; DEGs exhibiting changes more than 2-fold were considered significant. DEGs exhibiting changes more than 2-fold and 5-fold were 975 and 103 genes, respectively (Figure 2a,b). In the results of the two-fold DEGs, 370 DEGs were up-regulated and 605 DEGs were down-regulated by AS-IV (Figure 2c). Results from the 5-fold DEGs showed that 31 DEGs were up-regulated and 72 DEGs were down-regulated by AS-IV (Figure 2d).

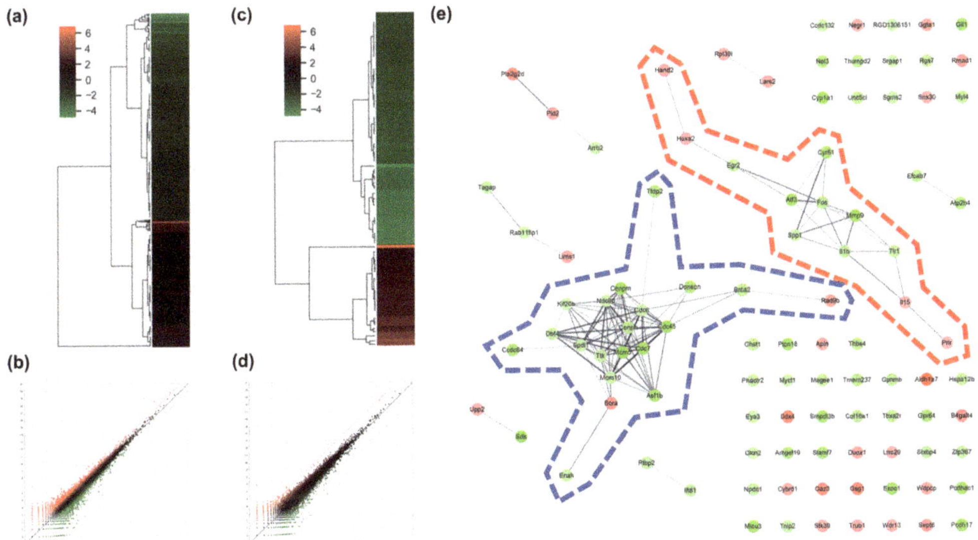

Figure 2. DEGs and gene network. (**a**) heatmap with hierarchical clustering of 2-fold changed DEGs, (**b**) expression pattern for 2-fold changed DEGs, (**c**) heatmap with hierarchical clustering of 5-fold changed DEGs, (**d**) expression pattern for 5-fold changed DEGs, (**e**) Gene network of 5-fold changed DEGs. 20 DEGs (in blue dotted line) and 12 DEGs (in red dotted line) were clustered. Markedly, DEGs within the blue dotted line showed strong relation. 14 other DEGs (not marked) were linked with only one or two genes. The green color in the figure indicates decreased expression; the red color in the figure indicates increased expression.

2.1.1. Key Gene Screening and Functional Annotation

To determine the pharmacological effects of AS-IV, we listed key genes from the result and investigated their function through corresponding protein of genes. We listed highly changed DEGs (more than 5-fold) and generated gene networks for the DEGs using a multiple protein search tool within the STRING database; two big cluster networks of DEGs were generated (Figure 2e). One network, which is marked by a blue dotted line in Figure 2e, consisted of 20 DEGs including *Mcm5, Cdc45, Cdc7, Cenpm, Asf1b, Donson, Tfdp2, Ccdc64, Cenpn, Brca2, Dbf4, Enahm, Mcm10, Ttk, Spdl1, Cdc6, Ndc80, Kif20b, Rad9b* and *Bora*. Another big network, marked by the red dotted line of Figure 2e, consisted of 12 genes including *Mmp9, Atf3, Cyr61, Fos, Il1b, Tlr1, Egr2, Spp1, Il15, Prlr, Hoxa2,* and *Hand2.* Functional annotation of DEGs in cluster network was conducted using the database for annotation, visualization and integrated discovery (DAVID) bioinformatic tool in three GO (gene ontology) categories including biological process, cellular component, molecular function (Table 1). The 20 DEGs in blue dotted line of Figure 2e were matched with

DNA replication initiation, cell division, double-strand break repair via break-induced replication, and DNA duplex unwinding, all of which fall within the biological process category. In the cellular component category, nucleoplasm was matched. In the molecular function category, DNA replication origin binding and chromatin binding were matched. Results from the functional annotation of the DEGs in this cluster (blue dotted line) showed that these genes were related to the DNA replication process of cell division. DEGs in the other clustered network within the red dotted line of Figure 2e were matched with positive regulation of angiogenesis, positive regulation of protein phosphorylation, and positive regulation of apoptotic process. They were matched only the biological process category (Table 1). From the changes in gene expression, strong interaction and function, we considered DEGs within the blue dotted line to be the key DEGs showing the effects of AS-IV.

Table 1. Functional annotation of clustered DEGs by DAVID in Figure 2e.

Gene network		
Clustered genes	*Mcm5, Cdc45, Cdc7, Cenpm, Asf1b, Donson, Tfdp2, Ccdc64, Cenpn, Brca2, Dbf4, Enahm, Mcm10, Ttk, Spdl1, Cdc6, Ndc80, Kif20b, Rad9b, Bora*	*Mmp9, Atf3, Cyr61, Fos, Il1b, Tlr1, Egr2, Spp1, Il15, Prlr, Hoxa2, Hand2*
GO category — Biological process	- DNA replication initiation - Cell division - Double-strand break repair via break-induced replication - DNA duplex unwinding	- Positive regulation of angiogenesis - Positive regulation of protein phosphorylation - Positive regulation of apoptotic process
GO category — Cellular component	- Nucleoplasm	(not matched)
GO category — Molecular function	- DNA replication origin binding - Chromatin binding	(not matched)

2.1.2. Pathway Mapping of DEGs

DEGs exhibiting 2-fold changes were further analyzed using the KEGG pathway database to evaluate the effect of AS-IV on signaling in regenerating liver tissues. Remarkably, DEGs were highly matched with three pathways including MAPK signaling, PI3K-Akt signaling, and cell cycle pathways (Table 2). The DEGs in these three pathways showed similar down-regulated patterns; overall, most matched DEGs were down-regulated (Figure 3).

Table 2. Results of pathway mapping for 2-fold changed DEGs by AS-IV.

KEGG Pathway	Count of Genes		
	Matched	Up-Regulated	Down-Regulated
MAPK signaling pathway	24	5	19
PI3K-Akt signaling pathway	22	5	17
Cell cycle pathway	22	1	21

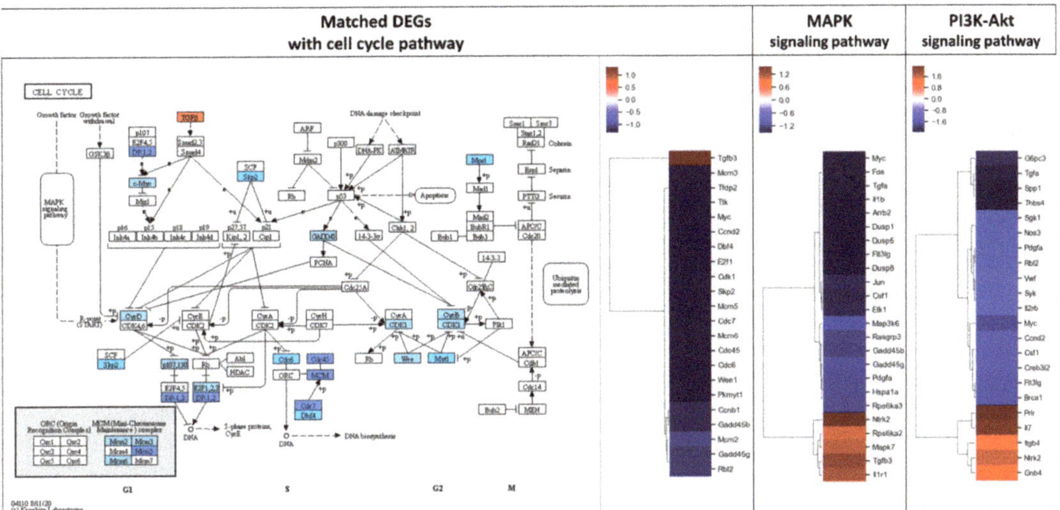

Figure 3. Results of KEGG pathway mapping. 2-fold changed DEGs were analyzed using the KEGG pathway database. 24, 22, and 22 DEGs were included in cell cycle, MAPK signaling, and PI3K-Akt signaling pathways, respectively. AS-IV showed down-regulatory effects on these pathways. Only a few DEGs were up-regulated in these three pathways. MAPK signaling and PI3K-Akt signaling pathways were upstream of the cell cycle pathway. The red and blue color in the figure indicate up-regulation and down-regulation of genes, respectively. Intensity of color is proportional to fold change.

2.2. Hepatic Proliferation of Regenerating Liver Tissue

Hepatic proliferation of regenerating liver tissue after AS-IV administration was evaluated via immunohistochemical staining and Western blot analysis for hepatocyte growth factor (HGF) (a mitogen for hepatocytes), cyclin D1 (a marker protein for proliferation), and transforming growth factor β1 (TGF β1) (a terminator for proliferation).

As HGF is a major stimulator of proliferation of hepatocyte during liver regeneration, the expression of HGF was evaluated to determine the hepato-proliferative signal after 70% PHx. Thus, the ratio of HGF positive cells increased immediately after 70% PHx and peaked 12 h after PHx. The ratio of immuno-positive cells in the AS-IV group showed a decrease when compared to controls (Figure 4). Remarkably, immediately after PHx, HGF expression decreased in half of the control group. Cyclin D1 was evaluated to determine the level of proliferation of hepatocyte after proliferative signal. It showed different expression. Cyclin D1 increased since 12 h after PHx with progression for cell cycle. In the AS-IV group, cyclin D1 expression was markedly decreased at 12 and 24 h after PHx (Figure 5). Results of Western blot analysis for cyclin D1 showed a decrease similar to that of the immunohistochemical results in Figure 5 (Figure 6). Cyclin D1 expression was dramatically decreased by AS-IV 24 h after PHx. TGF β1 was also determined by Western blot analysis to be a signal of growth regulation (Figure 6). The expression of TGFβ1 showed no difference 12 h after PHx, however, it was slightly increased with AS-IV treatment 24 h after PHx.

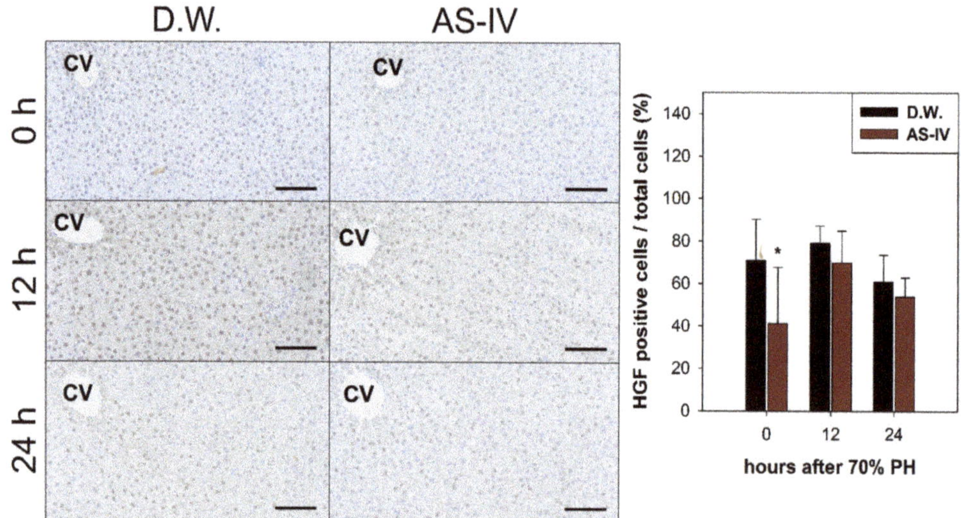

Figure 4. Immunostaining for HGF. HGF-immuno-positive reactions were a brown color after immunostaining. HGF peaked at 12 h after PHx. The AS-IV group showed a low expression of HGF. Hematoxylin was used as a counter stain. (Scale bar indicates 100 μm, CV: central vein, n = 3, mean ± standard deviation, *: $p < 0.05$).

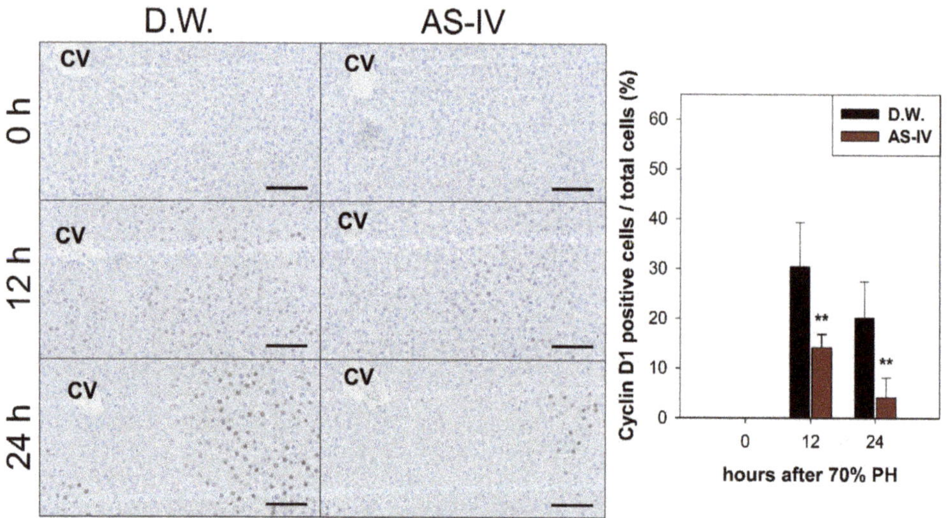

Figure 5. Immunostaining for cyclin D1. Cyclin D1 expression in regenerating liver tissues rapidly increased 12 h after PHx, peaking at 12 h after PHx then decreasing by 24 h. The AS-IV group showed a low expression of cyclin D1. Hematoxylin was used as a counter stain. (Scale bar indicates 100 μm, CV: central vein, n = 3, mean ± standard deviation, **: $p < 0.01$).

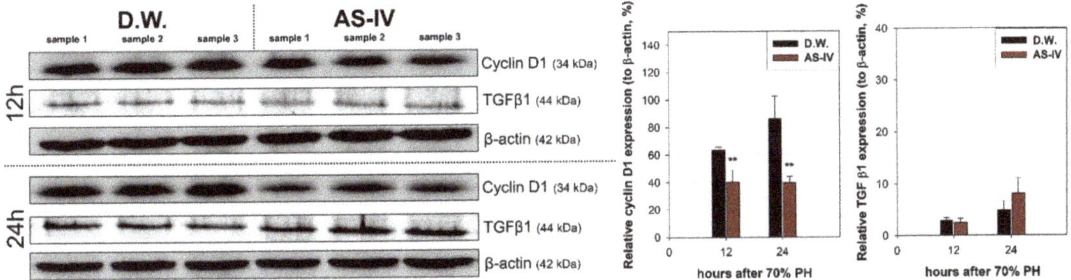

Figure 6. Relative expressions of cyclin D1 and TGF β1. Expression of cyclin D1 and TGFβ1 were analyzed by Western blot analysis. 12 h after PHx, cyclin D1 expression decreased in the experimental group and decreased further by 24 h while showing increased TGFβ1 expression. Relative protein expressions are presented as a ratio of the β-actin loading control. (mean ± standard deviation, *n* = 3, **: *p* < 0.01).

3. Discussion

Aastragalus membranaceus is one of the oldest known medicinal plants in traditional Chinese medicine. The dried root of this plant is called huang qi and an astragaloside IV (AS-IV) is the core bio-active ingredient of huang qi. Recently, it was reported that AS-IV showed anti-oxidative, anti-cancer, and hepato-protective activity in a variety of experiments [41–43]. Looking at recent research, it seems likely that AS-IV has many pharmacological potential. In 2019 Wei et al. reported that AS-IV could improve liver cirrhosis [44]. However, the effects of AS-IV on liver regeneration have not yet been elucidated. The 70% partial hepatectomy (PHx) model has been used in numerous studies of liver regeneration. In this study, we investigated the effects of AS-IV on liver regeneration in a 70% PHx model using mRNA sequencing, immunohistochemistry, and Western blot analysis. Our results suggest that AS-IV could suppress liver regeneration after 70% PHx. After oral administration of AS-IV, many genes changed their expression significantly (Figure 2). To determine the effects of AS-IV, we focused on the functions that are carried out by a group of differently expressed genes (DEGs) rather than the increases or decreases in expression of each individual gene. From the fold change, corresponding protein interactions, and function annotation of DEGs, we found 20 key DEGs including *Mcm5*, *Cdc45*, *Cdc7*, *Cenpm*, *Asf1b*, *Donson*, *Tfdp2*, *Ccdc64*, *Cenpn*, *Brca2*, *Dbf4*, *Enahm*, *Mcm10*, *Ttk*, *Spdl1*, *Cdc6*, *Ndc80*, *Kif20b*, *Rad9b*, and *Bora*, as shown in Figure 2e and Table 1. These genes showed strong functional relationship with cell division and our research suggests that these are the core genes responsible for the function of AS-IV in the regenerating liver. These results provide a potential mechanism for the therapeutic effects of AS-IV. Thus, as it relates to gene expression levels, by suppressing the genes required for molecular binding during DNA replication in the nucleus, AS-IV could inhibit DNA replication during cell division. These results suggest that AS-IV has a potent anti-proliferative effect in the regenerating liver.

DEGs were further analyzed via the KEGG pathway database to investigate the effects of AS-IV on proliferative signaling. Three signaling pathways (MAPK signaling pathway, PI3K-Akt signaling pathway, and cell cycle pathway) were down-regulated by AS-IV (Table 2, Figure 3). Other researchers have also reported similar down-regulation of MAPK signaling by AS-IV [45–47]. Decreased hepatocyte growth factor (HGF, hepato-proliferative signal molecule) and cyclin D1 (proliferation marker protein) expression also suggest the same anti-proliferative effects (Figures 4 and 5). Furthermore, AS-IV could affect the growth termination signal as well as could affect proliferative signal. Our extended study results show increased level of transforming growth factor β1 (TGF β1, growth termination signal) in the liver 24 h after PHx (Figure 6). In summary, AS-IV was revealed to have anti-proliferative effects in the regenerating liver via changes in gene expressions and protein expressions related to cell division and proliferative signals.

And So, AS-IV showed anti-proliferative effects in regenerating liver tissues. Thus, for the purpose of tissue regeneration through encouraging normal cell division, it is likely that AS-IV is unsuitable for liver regeneration. However, AS-IV could be applied for other purposes, such as for reducing oxidative injury after PHx surgery. After PHx, the liver is rapidly regenerated. In parallel, oxidative stress also rapidly increased. Some reports have shown that AS-IV has anti-oxidative and hepato-protective activities [48,49]. Additionally, our results could be used as basic study data for the anti-cancer activities of AS-IV with other studies. In recent, similar anti-cancer effects of AS-IV were reported including blocking of MAPK signal, decreased PCNA and Ki67 expression, and trigging G1 arrest in tumor cell [45,50].

4. Materials and Methods

4.1. Animals and Experimental Design

Male rats (SD strain, 8 weeks) were obtained from DBL Co., Ltd. (Eumseong, Korea). They were housed in an environmentally controlled room at 25 ± 1 °C, 12/12-light/dark cycle, and relative humidity $60 \pm 5\%$ with free access to standard food pellets and water (ad libitum). All animal experiments and procedures were performed in accordance with the guidelines for the care and use of laboratory animals of the national institutes of health (NIH) and after approval by the institutional animal care and use committee (IACUC) at the Sooonchunhyang University (permission No.: SCH20-0002).

Rats were randomly divided into control and experimental groups depending on treatment and sacrifice time. Both groups consisted of six rats. Three rats of each group were sacrificed 12 h after PHx and the other three rats were sacrificed 24 h after PHx. The rats in the experimental group received intragastric administration of astragaloside IV (AS-IV, 10 mg/kg) [4,13,19,22,51]. AS-IV was diluted in 1.5 mL of D.W. [52]. AS-IV (product No. #A3305, purity > 98.0%, HPLC) was purchased from the Tokyo Chemical Industry Co., Ltd. (Tokyo, Japan). The rats in the control group received the same volume of D.W. by intragastric administration. AS-IV and D.W. were administrated 2 h before surgery (Figure 7) [53–58].

To establish the liver regeneration model, a 70% partial hepatectomy (PHx) involving resection of the median and left lateral lobes was performed under anesthesia as previously described by Higgins and Anderson [29]. Animals were fasted for 12 h before surgery. Rats were sacrificed at 12 or 24 h after 70% PHx [39]. Regenerated remnant liver tissue was collected for analyses.

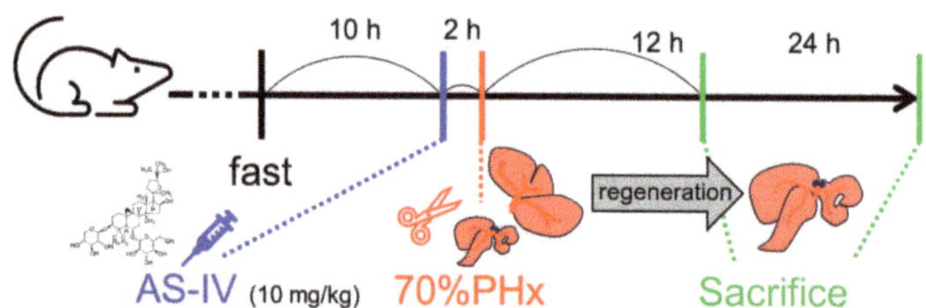

Figure 7. Administration of AS-IV and 70% PHx. Rats were received 10mg/kg of AS-IV diluted in D.W. (experimental group) or D.W. (control group) 2 h before 70% PHx and were sacrificed at 12 h or 24 h after PHx.

4.2. RNA Sequencing Analysis

TRIzol® reagent (Invitrogen, Carlsbad, CA, USA) was used for isolation of total RNA from liver tissue [59]. cDNA libraries were generated and purified using QuantSeq 3′ mRNA-Seq Library Prep Kit for Illumina (LEXOGEN, Vienna, Austria) according to manufacturer's instructions. High-throughput sequencing was performed as single-end 75 base pair sequencing using a NextSeq 500 (Illumina, Inc., San Diego, CA, USA). Raw reads were processed by BBDuk and aligned to the reference genome (rat, rn6, UCSC) using Bowtie2 [60]. The alignment file was assembled and estimated their abundances. Differentially expressed genes (DEGs) were determined based on counts from unique and multiple alignments with coverage in the Bedtools. The read count data were processed based on a quantile normalization method using EdgeR within R [61]. Genes were classified based on the database for annotation, visualization and integrated discovery (DAVID) and Medline databases (http://david.abcc.ncifcrf.gov/, accessed on 8 November 2020). DEGs exhibiting changes more than 2-fold were considered significant [59]. DEGs were also analyzed via the Kyoto Encyclopedia of Genes and Genomes (KEGG) mapper (address: https://www.genome.jp/kegg/tool/map_pathway2.html, accessed on 8 November 2020). DEGs were also analyzed based on protein-protein interactions via multiple protein searching tool within STRING database (ver. 11.0, address: http://string-db.org/db.org/, accessed on 8 November 2020).

4.3. Immunohistochemistry

Liver tissue was removed and immediately fixed in 10% neutral buffered formalin. It was then embedded in paraffin according to the routine process for light microscopy. The paraffin block was then cut into 4 μm thicknesses by rotary microtome (RM2235, Leica Biosystems, Germany). Antigen retrieval was conducted with sodium citrate buffer (10 mM, pH 6.0) at 95 °C. Heated sections were cooled for 30 min at room temperature. Sections were then incubated in 3% H_2O_2 (#1146, DUCSAN PURE CHEMICALS, Korea) and incubated in 5% bovine serum albumin (BSA, A7906, Merck KGaA, Darmstadt, Germany). Primary antibodies against Cyclin D1 (ab134175, Abcam plc., UK) and HGF (ab83760, Abcam plc., UK) were added. An HRP-conjugated secondary antibody (Thermo fisher scientific, Waltham, MA, USA) was then added. Immuno-detection was conducted with 3,3'-diaminobenzidine (SIGMAFAST™, Merck KGaA, Darmstadt, Germany). Counter staining was conducted using hematoxylin. All procedures were carried out in a humidified chamber to prevent the drying out of tissues. Tissues were observed using a microscope (CKX53, Olympus, Tokyo, Japan). Counts of positive reacted cells and total cells were measured by the color deconvolution tool within TMARKER (Ver. 2.146, open-source software) [62,63].

4.4. Western Blot Analysis

Total protein was extracted from liver tissues using protein extraction solution (PRO-PREP™, iNtRON BIOTECHNOLOGY, Seongnam, Korea) on ice and was determined by bicinchoninate (BCA) calorimetric assay kit (#23227, Thermo fisher scientific, USA) according to manufacturer's instruction. The protein was separated by 12% SDS-PAGE and transferred to PVDF (IPVH00010, Merck KGaA, Germany). The PVDF was then blocked with 5% skim milk (#232100, BD, Franklin Lakes, NJ, USA) and incubated with a primary antibody diluted in 0.5% skim milk overnight at 4 °C. The primary antibody against β-actin (A5316, Merck KGaA, Darmstadt, Germany) was diluted to 1:6000. Primary antibodies against cyclin D1 (ab134175) and TGF β1 (ab92486) were purchased from Abcam plc. (Cambridge, UK) and diluted to 1:3000. The membrane was further incubated with an HRP-conjugated secondary antibody (Thermo fisher scientific, USA) at room temperature for 1 h, followed by washing with PBS. Protein expression was detected by enhanced chemiluminescence (ECL) solution (K-12049-D50, Advansta Inc., San Jose, CA, USA) and chemiluminescence imaging system (GBox ichemi XL, Syngene, UK). β-actin was used as loading control. The relative expression level of target protein was expressed as ratio of β-actin [64].

4.5. Statistical Analysis

All quantitative data are presented as mean $\pm$ standard deviation (SD) from three independent experiments. Statistical analyses were performed using IBM SPSS statistics for windows (ver. 25, IBM, New York, NY, USA). Student's *t*-tests were used to analyze differences between control and the experimental group. Data with a *p*-value less than 0.05 were considered statistically significant. (*: *p*-value < 0.05, **: *p*-value < 0.01)

5. Conclusions

Astragaloside IV (AS-IV) is the major bio-active component of huang qi (the dried root of *Astragalus membranaceus*, a traditional Chinese medicinal plante). In present study, we demonstrated the pharmacological effects of AS-IV on regenerating rat liver tissue after 70% partial hepatectomy. AS-IV down-regulated proliferative signals, genes related to DNA replication, and cyclin D1 expression (Figure 8). In conclusion, AS-IV showed anti-proliferative activities in regenerating liver tissue.

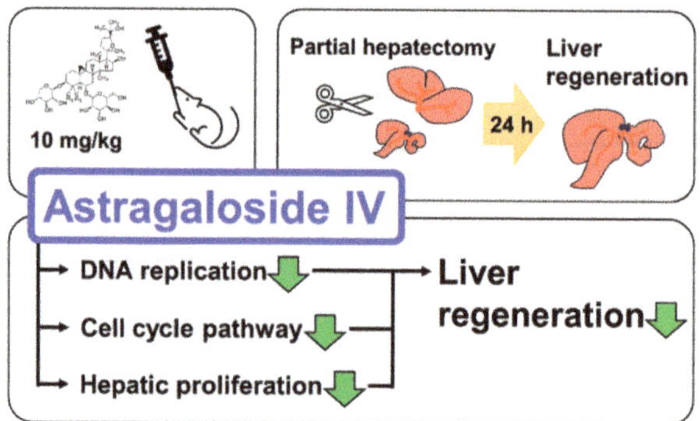

Figure 8. Graphical abstract of study. AS-IV suppressed hepatic proliferation of regenerating liver tissues after 70% PHx.

Author Contributions: Conceptualization: G.-S.L. and W.-J.K.; methodology: G.-S.L. and E.-G.J.; software: G.-S.L. and Y.-S.L.; validation: G.-S.L. and K.-W.N.; formal analysis: G.-S.L.; investigation: G.-S.L., H.-Y.J., H.-G.Y. and Y.-R.S.; resources: H.-Y.J., H.-G.Y. and Y.-R.S.; data curation: H.-Y.J. and H.-G.Y.; writing—original draft preparation: G.-S.L.; writing—review and editing: E.-G.J., Y.-S.L., K.-W.N. and W.-J.K.; visualization: G.-S.L., H.-Y.J. and H.-G.Y.; supervision: W.-J.K.; project administration: G.-S.L.; funding acquisition: E.-G.J., Y.-S.L., K.-W.N. and W.-J.K. All authors have read and agreed to the published version of the manuscript.

Funding: This work was supported by the Soonchunhyang University Research Fund.

Institutional Review Board Statement: The study was conducted according to the guidelines for the care and use of laboratory animal of the national institutes of health (NIH) and approved by the institutional animal care and use committee (IACUC) at the Soonchunhyang University (protocol No.: SCH20-0002, approved at February 2020).

Informed Consent Statement: Not applicable.

Data Availability Statement: We want to exclude this statement. Our study did not report public dataset.

Acknowledgments: This work was supported by the Soonchunhyang University Research Fund.

Conflicts of Interest: The authors declare no conflict of interest.

Sample Availability: Astragaloside IV is available from the authors or Tokyo Chemical Industry Co., Ltd. (Tokyo, Japan). Product No. #3305.

References

1. Fu, J.; Wang, Z.; Huang, L.; Zheng, S.; Wang, D.; Chen, S.; Zhang, H.; Yang, S. Review of the Botanical Characteristics, Phytochemistry, and Pharmacology of *Astragalus Membranaceus* (Huangqi). *Phytother. Res.* **2014**, *28*, 1275–1283. [CrossRef] [PubMed]
2. Liu, P.; Zhao, H.; Luo, Y. Anti-aging Implications of *Astragalus Membranaceus* (Huangqi): A Well-known Chinese Tonic. *Aging Dis.* **2017**, *8*, 868–886. [CrossRef] [PubMed]
3. Ma, X.Q.; Shi, Q.; Duan, J.A.; Dong, T.T.; Tsim, K.W. Chemical Analysis of *Radix Astragali* (Huangqi) in China: A Comparison with Its Adulterants and Seasonal Variations. *J. Agric. Food Chem.* **2002**, *50*, 4861–4866. [CrossRef] [PubMed]
4. Guo, Z.; Lou, Y.; Kong, M.; Luo, Q.; Liu, Z.; Wu, J. A Systematic Review of Phytochemistry, Pharmacology and Pharmacokinetics on *Astragali Radix*: Implications for *Astragali Radix* as a Personalized Medicine. *Int. J. Mol. Sci.* **2019**, *20*, 1463. [CrossRef]
5. Hao, X.; Wang, P.; Ren, Y.; Liu, G.; Zhang, J.; Leury, B.; Zhang, C. Effects of *Astragalus Membranaceus* Roots Supplementation on Growth Performance, Serum Antioxidant and Immune Response in Finishing Lambs. *Asian Australas J. Anim. Sci.* **2020**, *33*, 965–972. [CrossRef]
6. Ren, S.; Zhang, H.; Mu, Y.; Sun, M.; Liu, P. Pharmacological Effects of Astragaloside IV: A Literature Review. *J. Tradit. Chin. Med.* **2013**, *33*, 413–416. [CrossRef]
7. Zhao, M.; Gödecke, T.; Gunn, J.; Duan, J.A.; Che, C.T. Protostane and Fusidane Triterpenes: A Mini-Review. *Molecules* **2013**, *18*, 4054–4080. [CrossRef]
8. Živný, P.; Živná, H.; Palička, V.; Žaloudková, L.; Mocková, P.; Cermanová, J.; Mičuda, S. Modulation of Rat Liver Regeneration after Partial Hepatectomy by Dietary Cholesterol. *Acta Med.* **2018**, *61*, 22–28. [CrossRef]
9. Tsugawa, Y.; Natori, M.; Handa, H.; Imai, T. Estradiol Accelerates Liver Regeneration through Estrogen Receptor α. *Clin. Exp. Gastroenterol.* **2019**, *12*, 331–336. [CrossRef]
10. Li, X.; Wang, X.; Han, C.; Wang, X.; Xing, G.; Zhou, L.; Li, G.; Niu, Y. Astragaloside IV Suppresses Collagen Production of Activated Hepatic Stellate Cells via Oxidative Stress-mediated p38 MAPK Pathway. *Free Radic. Biol. Med.* **2013**, *60*, 168–176. [CrossRef]
11. Yu, W.; Lv, Z.; Zhang, L.; Gao, Z.; Chen, X.; Yang, X.; Zhong, M. Astragaloside IV Reduces the Hypoxia-Induced Injury in PC-12 Cells by Inhibiting Expression of miR-124. *Biomed. Pharmacother.* **2018**, *106*, 419–425. [CrossRef]
12. Gong, L.; Chang, H.; Zhang, J.; Guo, G.; Shi, J.; Xu, H. Astragaloside IV Protects Rat Cardiomyocytes from Hypoxia-Induced Injury by Down-Regulation of miR-23a and miR-92a. *Cell Physiol. Biochem.* **2018**, *49*, 2240–2253. [CrossRef]
13. Jiang, M.; Ni, J.; Cao, Y.; Xing, X.; Wu, Q.; Fan, G. Astragaloside IV Attenuates Myocardial Ischemia-Reperfusion Injury from Oxidative Stress by Regulating Succinate, Lysophospholipid Metabolism, and ROS Scavenging System. *Oxidative Med. Cell. Longev.* **2019**, *2019*, 9137654. [CrossRef]
14. Wei, D.; Xu, H.; Gai, X.; Jiang, Y. Astragaloside IV Alleviates Myocardial Ischemia-Reperfusion Injury in Rats through Regulating PI3K/AKT/GSK-3β Signaling Pathways. *Acta Cir. Bras.* **2019**, *34*, e201900708. [CrossRef]
15. He, Y.; Xi, J.; Zheng, H.; Zhang, Y.; Jin, Y.; Xu, Z. Astragaloside IV Inhibits Oxidative Stress-Induced Mitochondrial Permeability Transition Pore Opening by Inactivating GSK-3β via Nitric Oxide in H9c2 Cardiac Cells. *Oxidative Med. Cell Longev.* **2012**, *2012*, 935738. [CrossRef]
16. Jiang, K.; Lu, Q.; Li, Q.; Ji, Y.; Chen, W.; Xue, X. Astragaloside IV Inhibits Breast Cancer Cell Invasion by Suppressing Vav3 Mediated Rac1/MAPK Signaling. *Int. Immunopharmacol.* **2017**, *42*, 195–202. [CrossRef]
17. Cheng, X.; Gu, J.; Zhang, M.; Yuan, J.; Zhao, B.; Jiang, J.; Jia, X. Astragaloside IV Inhibits Migration and Invasion in Human Lung Cancer A549 Cells via Regulating PKC-α-ERK1/2-NF-κB Pathway. *Int. Immunopharmacol.* **2014**, *23*, 304–313. [CrossRef]
18. Wang, N.; Siu, F.; Zhang, Y. Effect of Astragaloside IV on Diabetic Gastric Mucosa in Vivo and in Vitro. *Am. J. Transl. Res.* **2017**, *9*, 4902–4913.
19. Fan, Y.; Fan, H.; Zhu, B.; Zhou, Y.; Liu, Q.; Li, P. Astragaloside IV Protects Against Diabetic Nephropathy via Activating eNOS in Streptozotocin Diabetes-Induced Rats. *BMC complementary Altern. Med.* **2019**, *19*, 355. [CrossRef]
20. Ding, Y.; Yuan, S.; Liu, X.; Mao, P.; Zhao, C.; Huang, Q.; Zhang, R.; Fang, Y.; Song, Q.; Yuan, D.; et al. Protective Effects of Astragaloside IV on db/db Mice with Diabetic Retinopathy. *PLoS ONE* **2014**, *9*, e112207. [CrossRef]
21. Luo, X.; Huang, P.; Yuan, B.; Liu, T.; Lan, F.; Lu, X.; Dai, L.; Liu, Y.; Yin, H. Astragaloside IV Enhances Diabetic Wound Healing Involving Upregulation of Alternatively Activated Macrophages. *Int. Immunopharmacol.* **2016**, *35*, 22–28. [CrossRef]
22. Zhao, X.M.; Zhang, J.; Liang, Y.N.; Niu, Y.C. Astragaloside IV Synergizes with Ferulic Acid to Alleviate Hepatic Fibrosis in Bile Duct-ligated Cirrhotic Rats. *Dig. Dis. Sci.* **2020**, *65*, 2925–2936. [CrossRef]
23. Liu, H.; Wei, W.; Sun, W.Y.; Li, X. Protective Effects of Astragaloside IV on Porcine-Serum-Induced Hepatic Fibrosis in Rats and in Vitro Effects on Hepatic Stellate Cells. *J. Ethnopharmacol.* **2009**, *122*, 502–508. [CrossRef]
24. Xie, D.; Zhou, P.; Liu, L.; Jiang, W.; Xie, H.; Zhang, L.; Xie, D. Protective Effect of Astragaloside IV on Hepatic Injury Induced by Iron Overload. *Biomed. Res. Int.* **2019**, *2019*, 3103946. [CrossRef]
25. Wang, S.; Li, J.; Huang, H.; Gao, W.; Zhuang, C.; Li, B.; Zhou, P.; Kong, D. Anti-hepatitis B Virus Activities of Astragaloside IV Isolated from *Radix Astragali*. *Biol. Pharm. Bull.* **2009**, *32*, 132–135. [CrossRef]
26. Yagi, S.; Hirata, M.; Miyachi, Y.; Uemoto, S. Liver Regeneration after Hepatectomy and Partial Liver Transplantation. *Int. J. Mol. Sci.* **2020**, *21*, 8414. [CrossRef]
27. Michalopoulos, G.K.; de Frances, M.C. Liver Regeneration. *Science* **1997**, *276*, 60–66. [CrossRef]

28. Andersen, K.J.; Knudsen, A.R.; Kannerup, A.S.; Sasanuma, H.; Nyengaard, J.R.; Hamilton-Dutoit, S.; Erlandsen, E.J.; Jørgensen, B.; Mortensen, F.V. The Natural History of Liver Regeneration in Rats: Description of an Animal Model for Liver Regeneration Studies. *Int. J. Surg.* **2013**, *11*, 903–908. [CrossRef]

29. Higgins, G.M.; Anderson, R.M. Experimental Pathology of Liver: Restoration of the Liver of the White Rat Following Partial Surgical Removal. *Arch. Pathol.* **1931**, *12*, 186–202.

30. Madrahimov, N.; Dirsch, O.; Broelsch, C.; Dahmen, U. Marginal Hepatectomy in the Rat: From Anatomy to Surgery. *Ann. Surg.* **2006**, *244*, 89–98. [CrossRef]

31. Huisman, F.; van Lienden, K.P.; Damude, S.; Hoekstra, L.T.; van Gulik, T.M. A Review of Animal Models for Portal Vein Embolization. *J. Surg. Res.* **2014**, *191*, 179–188. [CrossRef] [PubMed]

32. Taub, R. Liver Regeneration: From Myth to Mechanism. *Nat. Rev. Mol. Cell Biol.* **2004**, *5*, 836–847. [CrossRef] [PubMed]

33. Kwon, Y.J.; Lee, K.G.; Choi, D. Clinical Implications of Advances in Liver Regeneration. *Clin. Mol. Hepatol.* **2015**, *21*, 7–13. [CrossRef] [PubMed]

34. Cienfuegos, J.A.; Rotellar, F.; Baixauli, J.; Martínez-Regueira, F.; Pardo, F.; Hernández-Lizoáin, J.L. Liver Regeneration–The Best Kept Secret. A Model of Tissue Injury Response. *Rev. Esp. Enferm. Dig.* **2014**, *106*, 171–194.

35. Kang, L.I.; Mars, W.M.; Michalopoulos, G.K. Signals and Cells Involved in Regulating Liver Regeneration. *Cells* **2012**, *1*, 1261–1292. [CrossRef] [PubMed]

36. Hoffmann, K.; Nagel, A.J.; Tanabe, K.; Fuchs, J.; Dehlke, K.; Ghamarnejad, O.; Lemekhova, A.; Mehrabi, A. Markers of Liver Regeneration-the Role of Growth Factors and Cytokines: A Systematic Review. *BMC Surg.* **2020**, *20*, 31. [CrossRef] [PubMed]

37. Tarlá, M.R.; Ramalho, F.; Ramalho, L.N.; Silva Tde, C.; Brandão, D.F.; Ferreira, J.; Silva Ode, C.; Zucoloto, S. Cellular Aspects of Liver Regeneration. *Acta Cir. Bras.* **2006**, *21*, 63–66. [CrossRef]

38. Ozaki, M. Cellular and Molecular Mechanisms of Liver Regeneration: Proliferation, Growth, Death and Protection of Hepatocytes. *Semin. Cell Dev. Biol.* **2020**, *100*, 62–73. [CrossRef]

39. Rychtrmoc, D.; Libra, A.; Buncek, M.; Garnol, T.; Cervinková, Z. Studying Liver Regeneration by Means of Molecular Biology: How Far We are in Interpreting the Findings? *Acta Med.* **2009**, *52*, 91–99. [CrossRef]

40. Abu Rmilah, A.; Zhou, W.; Nelson, E.; Lin, L.; Amiot, B.; Nyberg, S.L. Understanding the Marvels Behind Liver Regeneration. *Wiley Interdiscip. Rev. Dev. Biol.* **2019**, *8*, e340. [CrossRef]

41. Yang, S.; Zhang, R.; Xing, B.; Zhou, L.; Zhang, P.; Song, L. Astragaloside IV Ameliorates Preeclampsia-Induced Oxidative Stress through the Nrf2/HO-1 Pathway in a Rat Model. *Am. J. Physiol. Endocrinol. Metab.* **2020**, *319*, E904–E911. [CrossRef]

42. Cui, X.; Jiang, X.; Wei, C.; Xing, Y.; Tong, G. Astragaloside IV Suppresses Development of Hepatocellular Carcinoma by Regulating miR-150-5p/β-catenin Axis. *Environ. Toxicol. Pharmacol.* **2020**, *78*, 103397. [CrossRef]

43. Han, L.; Li, J.; Lin, X.; Ma, Y.F.; Huang, Y.F. Protective Effect of Astragaloside IV on Oxidative Damages of Change Liver Cell Induced by Ethanol and H_2O_2. *Zhongguo Zhongyao Zazhi* **2014**, *39*, 4430–4435.

44. Wei, R.; Liu, H.; Chen, R.; Sheng, Y.; Liu, T. Astragaloside IV Combating Liver Cirrhosis through the PI3K/Akt/mTOR Signaling Pathway. *Exp. Ther. Med.* **2019**, *17*, 393–397. [CrossRef]

45. Li, B.; Wang, F.; Liu, N.; Shen, W.; Huang, T. Astragaloside IV Inhibits Progression of Glioma via Blocking MAPK/ERK Signaling Pathway. *Biochem. Biophys. Res. Commun.* **2017**, *491*, 98–103. [CrossRef]

46. Leng, B.; Li, C.; Sun, Y.; Zhao, K.; Zhang, L.; Lu, M.L.; Wang, H.X. Protective Effect of Astragaloside IV on High Glucose-Induced Endothelial Dysfunction via Inhibition of P2X7R Dependent P38 MAPK Signaling Pathway. *Oxidative Med. Cell Longev.* **2020**, *2020*, 5070415. [CrossRef]

47. Hsieh, H.L.; Liu, S.H.; Chen, Y.L.; Huang, C.Y.; Wu, S.J. Astragaloside IV Suppresses Inflammatory Response via Suppression of NF-κB, and MAPK Signalling in Human Bronchial Epithelial Cells. *Arch. Physiol. Biochem.* **2020**, *14*, 1–10. [CrossRef]

48. Zhu, Z.; Li, J.; Zhang, X. Astragaloside IV Protects Against Oxidized Low-Density Lipoprotein (ox-LDL)-Induced Endothelial Cell Injury by Reducing Oxidative Stress and Inflammation. *Med. Sci. Monit.* **2019**, *25*, 2132–2140. [CrossRef]

49. Nie, Q.; Zhu, L.; Zhang, L.; Leng, B.; Wang, H. Astragaloside IV Protects Against Hyperglycemia-Induced Vascular Endothelial Dysfunction by Inhibiting Oxidative Stress and Calpain-1 Activation. *Life Sci.* **2019**, *232*, 116662. [CrossRef]

50. Su, C.M.; Wang, H.C.; Hsu, F.T.; Lu, C.H.; Lai, C.K.; Chung, J.G.; Kuo, Y.C. Astragaloside IV Induces Apoptosis, G1-Phase Arrest and Inhibits Anti-apoptotic Signaling in Hepatocellular Carcinoma. *In Vivo* **2020**, *34*, 631–638. [CrossRef]

51. Lv, L.; Wu, S.Y.; Wang, G.F.; Zhang, J.J.; Pang, J.X.; Liu, Z.Q.; Xu, W.; Wu, S.G.; Rao, J.J. Effect of Astragaloside IV on Hepatic Glucose-Regulating Enzymes in Diabetic Mice Induced by a High-Fat Diet and Streptozotocin. *Phytother. Res.* **2010**, *24*, 219–224. [CrossRef]

52. Gad, S.C.; Cassidy, C.D.; Aubert, N.; Spainhour, B.; Robbe, H. Nonclinical Vehicle Use in Studies by Multiple Routes in Multiple Species. *Int. J. Toxicol.* **2006**, *25*, 499–521. [CrossRef]

53. Gu, Y.; Wang, G.; Fawcett, J.P. Determination of Astragaloside IV in Rat Plasma by Liquid Chromatography Electrospray Ionization Mass Spectrometry. *J. Chromatogr. B Analyt. Technol. Biomed. Life Sci.* **2004**, *801*, 285–288. [CrossRef]

54. Shi, X.; Tang, Y.; Zhu, H.; Li, W.; Li, Z.; Li, W.; Duan, J.A. Comparative Tissue Distribution Profiles of Five Major Bio-active Components in Normal and Blood Deficiency Rats after Oral Administration of Danggui Buxue Decoction by UPLC-TQ/MS. *J. Pharm. Biomed. Anal.* **2014**, *88*, 207–215. [CrossRef]

55. Liu, X.H.; Zhao, J.B.; Guo, L.; Yang, Y.L.; Hu, F.; Zhu, R.J.; Feng, S.L. Simultaneous Determination of Calycosin-7-O-β-D-Glucoside, Ononin, Calycosin, Formononetin, Astragaloside IV, and Astragaloside II in Rat Plasma after Oral Administration of Radix Astragali Extraction for Their Pharmacokinetic Studies by Ultra-pressure Liquid Chromatography with Tandem Mass Spectrometry. *Cell Biochem. Biophys.* **2014**, *70*, 677–686. [CrossRef]
56. Zhang, H.; Song, J.; Dai, H.; Liu, Y.; Wang, L. Effects of Puerarin on the Pharmacokinetics of Astragaloside IV in Rats and Its Potential Mechanism. *Pharm. Biol.* **2020**, *58*, 328–332. [CrossRef]
57. Chang, Y.X.; Sun, Y.G.; Li, J.; Zhang, Q.H.; Guo, X.R.; Zhang, B.L.; Jin, H.; Gao, X.M. The Experimental Study of *Astragalus Membranaceus* on Meridian Tropsim: The Distribution Study of Astragaloside IV in Rat Tissues. *J. Chromatogr. B Analyt. Technol. Biomed. Life Sci.* **2012**, *911*, 71–75. [CrossRef]
58. Zhang, W.D.; Zhang, C.; Liu, R.H.; Li, H.L.; Zhang, J.T.; Mao, C.; Moran, S.; Chen, C.L. Preclinical Pharmacokinetics and Tissue Distribution of a Natural Cardioprotective Agent Astragaloside IV in Rats and Dogs. *Life Sci.* **2006**, *79*, 808–815. [CrossRef]
59. Colak, D.; Al-Harazi, O.; Mustafa, O.M.; Meng, F.; Assiri, A.M.; Dhar, D.K.; Broering, D.C. RNA-Seq Transcriptome Profiling in Three Liver Regeneration Models in Rats: Comparative Analysis of Partial Hepatectomy, ALLPS, and PVL. *Sci. Rep.* **2020**, *10*, 5213. [CrossRef] [PubMed]
60. Langmead, B.; Salzberg, S.L. Fast Gapped-read Alignment with Bowtie 2. *Nat. Methods* **2012**, *9*, 357–359. [CrossRef]
61. Robinson, M.D.; McCarthy, D.J.; Smyth, G.K. EdgeR: A Bioconductor Package for Differential Expression Analysis of Digital Gene Expression Data. *Bioinformatics* **2010**, *26*, 139–140. [CrossRef] [PubMed]
62. Schüffler, P.J.; Fuchs, T.J.; Ong, C.S.; Wild, P.J.; Rupp, N.J.; Buhmann, J.M. TMARKER: A Free Software Toolkit for Histopathological Cell Counting and Staining Estimation. *J. Pathol. Inform.* **2013**, *4*, S2. [CrossRef] [PubMed]
63. Zhong, Q.; Rüschoff, J.H.; Guo, T.; Gabrani, M.; Schüffler, P.J.; Rechsteiner, M.; Liu, Y.; Fuchs, T.J.; Rupp, N.J.; Fankhauser, C.; et al. Image-based Computational Quantification and Visualization of Genetic Alterations and Tumour Heterogeneity. *Sci. Rep.* **2016**, *6*, 24146. [CrossRef] [PubMed]
64. Lee, G.S.; Yang, H.G.; Kim, J.H.; Ahn, Y.M.; Han, M.D.; Kim, W.J. Pine (*Pinus Densiflora*) Needle Extract Could Promote the Expression of PCNA and Ki-67 after Partial Hepatectomy in Rat. *Acta Cir. Bras.* **2019**, *34*, e201900606. [CrossRef]

Article

Protective Effects of Swertiamarin against Methylglyoxal-Induced Epithelial-Mesenchymal Transition by Improving Oxidative Stress in Rat Kidney Epithelial (NRK-52E) Cells

Kirti Parwani [†], **Farhin Patel** [†], **Dhara Patel** and **Palash Mandal** *

Department of Biological Sciences, P. D. Patel Institute of Applied Sciences,
Charotar University of Science and Technology, Changa 388421, India; kirtiparwani@gmail.com (K.P.);
farhinpatel1993@gmail.com (F.P.); dharapatel.phd@gmail.com (D.P.)
* Correspondence: palashmandal.bio@charusat.ac.in; Tel.: +91-9666164654
† These authors contributed equally to this work.

Abstract: Increased blood glucose in diabetic individuals results in the formation of advanced glycation end products (AGEs), causing various adverse effects on kidney cells, thereby leading to diabetic nephropathy (DN). In this study, the antiglycative potential of Swertiamarin (SM) isolated from the methanolic extract of *E. littorale* was explored. The effect of SM on protein glycation was studied by incubating bovine serum albumin with fructose at 60 °C in the presence and absence of different concentrations of swertiamarin for 24 h. For comparative analysis, metformin was also used at similar concentrations as SM. Further, to understand the role of SM in preventing DN, in vitro studies using NRK-52E cells were done by treating cells with methylglyoxal (MG) in the presence and absence of SM. SM showed better antiglycative potential as compared to metformin. In addition, SM could prevent the MG mediated pathogenesis in DN by reducing levels of argpyrimidine, oxidative stress and epithelial mesenchymal transition in kidney cells. SM also downregulated the expression of interleukin-6, tumor necrosis factor-α and interleukin-1β. This study, for the first time, reports the antiglycative potential of SM and also provides novel insights into the molecular mechanisms by which SM prevents toxicity of MG on rat kidney cells.

Keywords: advanced glycation end product (AGE); oxidative stress; epithelial to mesenchymal transition; AGE-inhibitor; swertiamarin; diabetic nephropathy

Citation: Parwani, K.; Patel, F.; Patel, D.; Mandal, P. Protective Effects of Swertiamarin against Methylglyoxal-Induced Epithelial-Mesenchymal Transition by Improving Oxidative Stress in Rat Kidney Epithelial (NRK-52E) Cells. *Molecules* **2021**, *26*, 2748. https://doi.org/10.3390/molecules26092748

Academic Editor: Raffaele Pezzani

Received: 11 March 2021
Accepted: 20 April 2021
Published: 7 May 2021

Publisher's Note: MDPI stays neutral with regard to jurisdictional claims in published maps and institutional affiliations.

1. Introduction

Type 2 Diabetes (T2D) is a metabolic syndrome, which results due to peripheral insulin resistance, affecting both metabolism and the disposal of glucose. The occurrence of diabetes is accelerating worldwide, with a consequent increase in diabetes associated complications like diabetic nephropathy (DN), neuropathy, retinopathy, and cardiovascular complications. Hyperglycaemia is one of the main reasons for causing DN, and various clinical trials have established that the progression to DN can be slowed and also reversed by good and strict glycemic controls [1–3]. Elevated blood glucose, which induces reactive oxygen species (ROS) formation, is understood to be one of the main reasons for the formation of advanced glycation end products (AGEs) in the intracellular and extracellular environment [4]. AGEs are the products of nonenzymatic glycation between free amino acids and reducing sugar via the Maillard reaction resulting into yellowish-brown fluorescent and insoluble adducts.

The formation of AGEs is divided into two stages. In the earlier stage, Amadori product, a stable compound is formed from an unstable Schiff base [5]. Amadori products can either form reactive dicarbonyls like glyoxal and methylglyoxal (MG) or undergo various chemical reactions like condensation, dehydration and oxidation to form AGEs.

The mechanism and the different stages involved in the formation of AGEs are explained in Supplementary Materials Figure S1. These adducts can alter the normal physiological functions of a protein upon glycation [6]. This can alter the half-life of the proteins and affect their physiological clearance. Patients with persistent T2D are established to have vastly higher levels of AGEs [7]. The role of AGEs in DN has been documented by reports suggesting the negative correlation between the accumulation of AGEs in plasma and renal function [8]. AGEs bind to their receptor called receptor for advanced glycation end products (RAGE) and primarily lead to the generation of ROS and oxidative stress which further leads to the damage of the renal tubular cells and mesangial cells leading to DN. Further, AGEs lead to an imbalance between synthesis and degradation of extracellular matrix components like collagen and cause their accumulation in the mesangium, tubular interstitial cells and glomerulus basement membrane leading to their hardening, which is a hallmark feature of DN [9].

Reduced antioxidant defenses have also been reported in diabetic patients thereby providing acceleration to the development of chronic complications [10]. It is reported that AGEs could alter the physiological functions of an antioxidant enzyme like superoxide dismutase (SOD) and could inactivate it [11]. Glycation has been found to induce aggregation and structural modifications in catalase by targeting lysine residues [12]. This reduced antioxidant machinery along with increased ROS due to hyperglycaemia can be detrimental for a cell. Many synthetic inhibitors that can restrain the formation of AGEs have been identified and studied but they have been withdrawn back due to their lower potency and side effects. Therefore, there arises a rationale to identify some inhibitor, which along with being potent and effective, should have low toxicity.

Traditional and herbal therapies are gaining importance slowly over conventional medicines due to their advantage of having lower or no side-effects. The molecules present in the plants are shown to be hypoglycaemic, hypolipidemic, and antioxidant in nature. Compounds in the plants like phenolics [13], polysaccharides [14] and carotenoids [15], etc. are shown to possess antiglycative properties. In addition, the regular consumption of edible products rich in antioxidants and polyphenolic compounds could be beneficial in the prevention of diabetes and its related complications [16]. *Enicostemma littorale* commonly called Mamejava has been shown to possess a hypoglycaemic effect in T2D rats [17]. *E. littorale* contains important phytoconstituents like gentianine, enicoflavine, gentiocrucine, swertiamarin (SM), etc., to name a few. SM has been shown to possess anticholinergic [18], antihyperlipidaemic [19], hypoglycaemic [20] and antioxidant effects [21]. SM in combination with quercetin, ameliorated T2D and oxidative stress in streptozotocin-treated Wistar rats [22].

Since glycation is a process associated with increased free radical formation, compounds with antioxidant activities could also act as an inhibitor for the formation of AGEs [23]. The antiglycation potential of SM has not been explored yet, although there exists a report where *Swertia Chirayata* extract, which contains various active molecules along with SM, showed inhibition of fructose mediated glycation. Given the important role of SM in diabetes mellitus and its antioxidant potential, the present study was executed to understand and assess its antiglycation potential. To our knowledge, this is the first report that shows inhibition of fructose mediated glycation by SM and also unravels the molecular mechanism involved in its protective role in MG-induced damage on kidney cells, thereby preventing DN.

2. Results

2.1. Isolation of SM from E. littorale and Characterization Using High-Performance Liquid Chromatography (HPLC), Fourier-Transform Infrared Spectroscopy (FTIR) and Liquid Chromatography-Mass Spectra(LC-MS)

Using the solvent chromatography on silica, SM was successfully isolated from *E. littorale* as confirmed by TLC. The presence of SM was confirmed with reference to the standard SM using the chloroform: methanol (8:2 *v/v*) as a mobile phase with R_f of 0.58. The visualization was done at 254 nm as shown in Figure 1a. By our isolation method, we report 7% yield of SM from the methanolic extract of *E. littorale*, which to our knowledge is the highest yield obtained until now. To analyze the purity of the isolated sample, the HPLC of the isolated compound was done and was compared with the standard SM. The HPLC profiles of the isolated SM (a) and the standard (b) are shown in Figure 1b. Both the standard and the lab isolated SM gave a characteristic peak at 238 nm, which is the absorbance maxima of SM as reported earlier. The retention time of the lab isolated SM was found to be 3.801 which fairly matched to 3.805 of that of the standard with the acetonitrile: water (10:90 *v/v*) as the mobile phase.

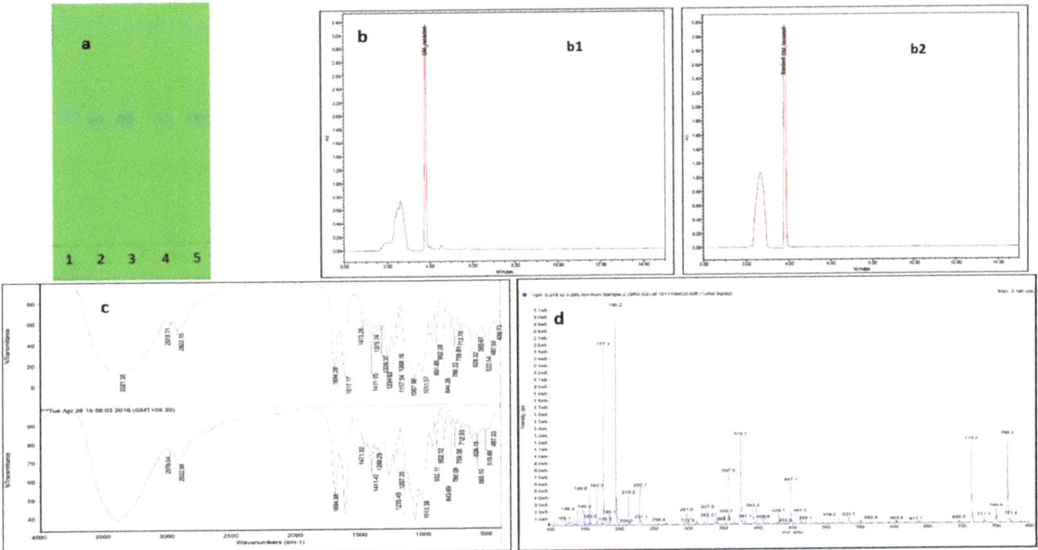

Figure 1. (**a**) TLC profile: Lane 1, 2: Fractions containing SM along with impurities. Lane 3: Standard Lane 4: Methanolic extract of *E. littorale* Lane 5: Fraction containing SM in the TLC mobile phase (Chloroform: methanol in 8:2 *v/v* proportion) (**b**) The HPLC chromatogram of SM isolated in the lab (**b1**) and of the standard SM (**b2**) at 1 mg/mL using acetonitrile: water (10:90) as the mobile phase. (**c**) An overlay of FT-IR spectrum of standard SM (red) and SM isolated in the lab (blue). (**d**) The mass spectrum (LC-MS) of isolated swertiamarin showing characteristic *m/z* of 375.1.

The analysis of the different functional groups present in the isolated sample was done using FTIR and was also compared with the standard SM. The lab isolated SM fairly matched the FTIR spectrum of the standard SM as shown in Figure 1c. FTIR spectra of both, isolated and standard swertiamarin showed definite peaks between wavenumbers 4000–500 cm^{-1}. The characteristic O-H stretch peak was found at 3381 cm^{-1}, C-H stretch at 2292 cm^{-1}, C=O stretch at 1694 cm^{-1}, C=C stretch at 1617 cm^{-1}, C-O-C at 1411 cm^{-1}, and C=CH$_2$ stretch at 844 cm^{-1}. This validates the purity of the isolated compound as the wavenumbers for both compounds (lab isolated SM and standard) match with each other.

The validation of the molecular weight of the isolated SM was done using LC-MS. The LC-MS spectrum of SM reported earlier showed a molecular ion peak at *m/z* of 374, and LC-MS spectrum for SM isolated in our lab as shown in Figure 1d, showed *m/z* of 375.1 in positive ion mode, indicating the addition of H$^+$ to the compound, thereby increasing molecular ion peak by 1, as shown in Merck Index. Together the characterization data using different methods confirmed the isolated compound to be SM and by our isolation method, we report 7% yield of SM from the methanolic extract of *E. littorale*, which to our knowledge is the highest yield obtained until now (Figure 1).

2.2. SM Inhibits the Formation of AGEs

The nonenzymatic glycation of the proteins leads to the formation of AGEs. AGEs possess characteristic property of fluorescence after glycation. The formation of AGEs was analyzed by checking the fluorescence at excitation and emission wavelengths of 370 nm and 440 nm, respectively. The fluorescence intensity was used as a measure of glycation, i.e., higher the intensity of fluorescence, the higher is the glycation of proteins. The intensity of fluorescence was significantly decreased ($p \leq 0.0001$) in SM treated samples as compared to untreated samples that contained neither SM nor metformin. In addition, SM could inhibit the formation of AGEs better than metformin ($p < 0.0001$) at similar doses (Figure 2). This indicates that SM could be a better glycation inhibitor and therefore can prevent the formation of endogenous AGEs due to persistent hyperglycaemia in diabetic subjects.

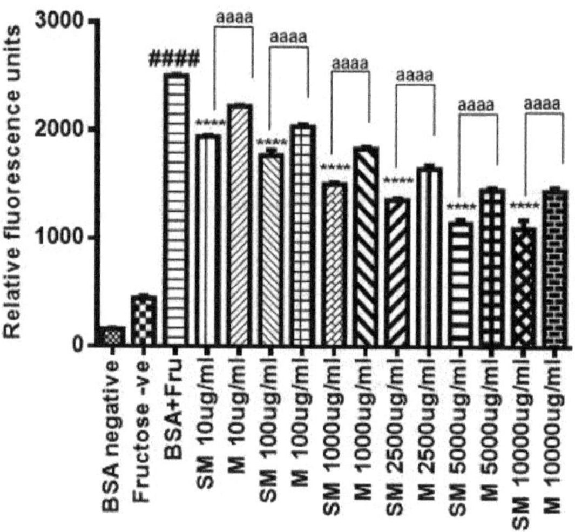

Figure 2. The intensity of fluorescence of AGEs in the presence and absence of swertiamarin (SM) and its comparison with metformin (M) at different concentrations measured using an excitation wavelength of 370 nm and an emission wavelength of 440 nm. The results are Mean ± SD of 3 individual experiments. #### $p < 0.0001$ represents comparison between bovine serum albumin (BSA) negative and BSA + Fructose (Fru) group. **** $p < 0.0001$ represents the statistical significance w.r.t the BSA + Fru group. aaaa $p < 0.0001$ represents the comparison between SM and M.

2.3. SM Reduces Fructose Mediated Hyperchromicity

Glycation of the proteins results in the structural changes of the proteins, which modifies their characteristic UV absorption spectrum. The changes in the structure of BSA and its absorbance were additionally analyzed by spectrophotometer (Figure 3a). As shown in Figure 3b, compared to normal BSA (Black line), glycated BSA (Blue line) shown increased absorbance and hyperchromicity (89%) due to fructose mediated changes in the structure of bovine serum albumin (BSA). SM could prevent the fructose mediated changes in the BSA (Green line) as depicted by reduced hyperchromicity (57.9%) indicating the inhibition of glycation ($p < 0.0054$) (Figure 3c).

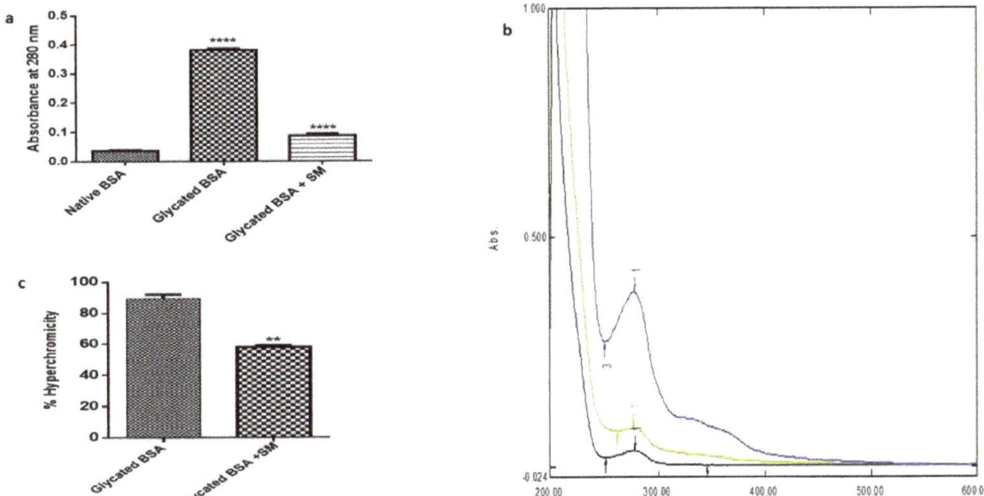

Figure 3. Antiglycative effect of swertiamarin at 100 µg/mL (SM 100) as shown by absorbance at 280 nm using UV-spectrophotometer (**a**), the structural changes as seen by the shift in the peaks due to glycation (**b**), suggesting hyperchromicity (**c**). The results are Mean ± SD of 3 individual experiments. **** $p < 0.0001$, ** $p < 0.054241$.

2.4. SM Reduces Carbonyl Content in the Glycated BSA

The carbonyl content was analysed as a marker for oxidation of protein during glycation. As shown in Figure 4, the glycated BSA showed remarkably higher carbonyl content upon the reaction with fructose, indicating the formation of radicals and dicarbonyls via Maillard reaction (*** $p < 0.001$). The treatment with SM could significantly limit the formation of carbonyls, which is due to its ability to scavenge free radicals as an antioxidant (** $p < 0.01$).

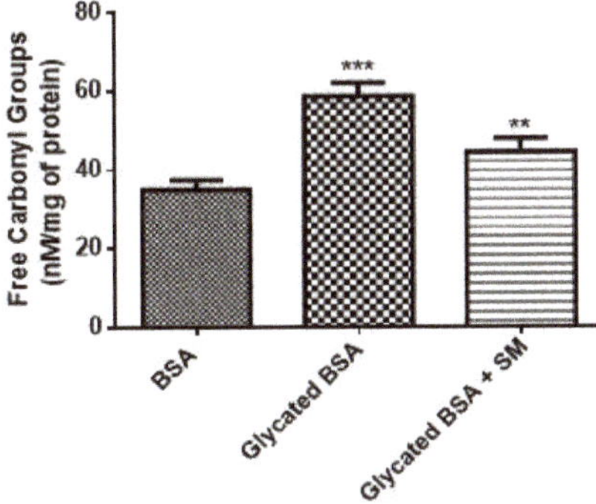

Figure 4. Estimation of carbonyl content. As compared to the native BSA, glycated BSA showed remarkably higher level of carbonyl content. The treatment with SM showed reduction in the carbonyl content, as compared to the glycated BSA. (*** $p < 0.001$ and ** $p < 0.01$). Results are mean ± SD of 3 individual experiments.

2.5. SM Prevents Fructose Mediated Side-Chain Modifications in BSA

The modifications in the functional group in the side chains of the BSA was analysed using FTIR. As shown in Figure 5, FTIR spectrum of the native BSA showed a peak at 1655 cm^{-1} corresponding to amide I, while red-shift was observed in case of glycated samples, where the amide I shifted to 1651 cm^{-1}. The treatment with SM showed a minor shift, indicating the relatively stable BSA molecule, as the amide I shift indicates the presence of α-helical structure in the protein. The amide III band shifted 9 cm^{-1}, from 1245 cm^{-1} to lower wavenumber 1236 cm^{-1} in the glycated BSA as compared to native BSA. However, the treatment of glycated BSA with SM did not show major shift in the wavenumber compared to native BSA with wavenumber 1242 cm^{-1}. Together these results indicate that glycated BSA exhibited various side-chain modifications, affecting the structure of BSA upon glycation. Table 1 shows the different bands and wavenumbers obtained in all the samples.

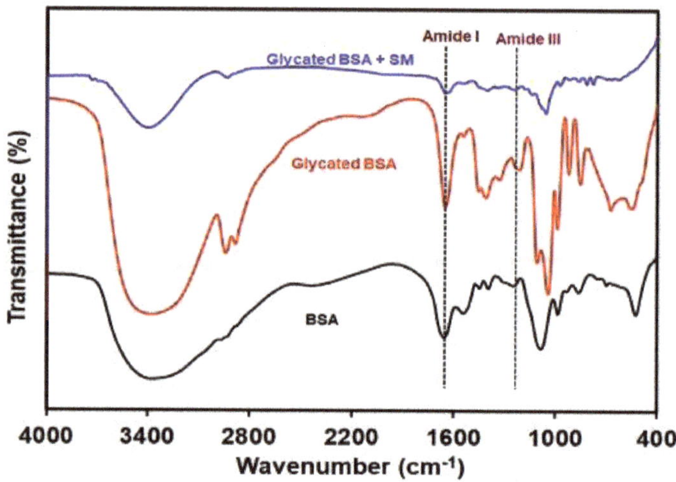

Figure 5. FTIR spectra of BSA (black line), glycated BSA (red line) and the glycated BSA treated with SM (blue line).

Table 1. Comparative analysis of bands and wavenumbers in all samples.

Samples	Bands and Wavenumber (cm^{-1})	
	Amide I	Amide III
Native BSA	1655	1245
Glycated BSA	1651	1236
Glycated BSA + SM	1654	1242

2.6. Effect of MG and SM on the Viability of NRK-52E Cells

The toxicity of MG on NRK-52E cells was analyzed for fixing the dose for further experiments on NRK-52E. The treatment with MG induced cell death of NRK-52E and the dose-dependent decrease in cell viability was observed with an increase in the concentration of MG ($p < 0.0001$) (Figure 6a). To determine the dose of MG to be used for further experiments with NRK-52E, we determined the IC50 value for MG, which was found to be 200 µM and therefore the cells were challenged with 200 µM MG for testing the protective effects of SM in presence of MG. However, SM did not show any toxicity on NRK-52E cells at all the different concentrations used for the experiment (Figure 6b).

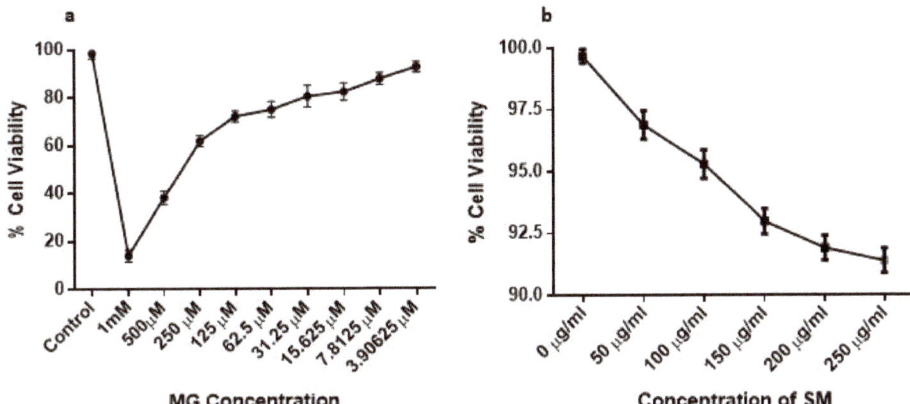

Figure 6. (a) The effect MG on the cell viability in NRK-52E cells. NRK-52E cells when treated with different concentrations of MG caused cell death in a dose-dependent manner as compared to control. (b) The cytotoxicity of SM was checked on NRK-52E cells. The treatment with SM did not show any toxicity on the NRK-52E cells at various concentrations. Results are Mean ± SD of 3 individual experiments.

2.7. SM Improves the Morphology of the NRK-52E Cells and Ameliorates against MG-Induced Damage

The effect of MG on the morphology of NRK-52E cells can be seen in Figure 7. MG treatment on the cells changed the cobblestone morphology of NRK-52E cells to a round and elongated fibroblast-like shape, indicating the cellular damage due to MG, which was prevented when the cells were treated with MG along with SM. The morphological changes are indicative of various inflammatory changes occurred due to the presence of MG, which was ameliorated by SM, as seen by the intact epithelial morphology of NRK-52E cells.

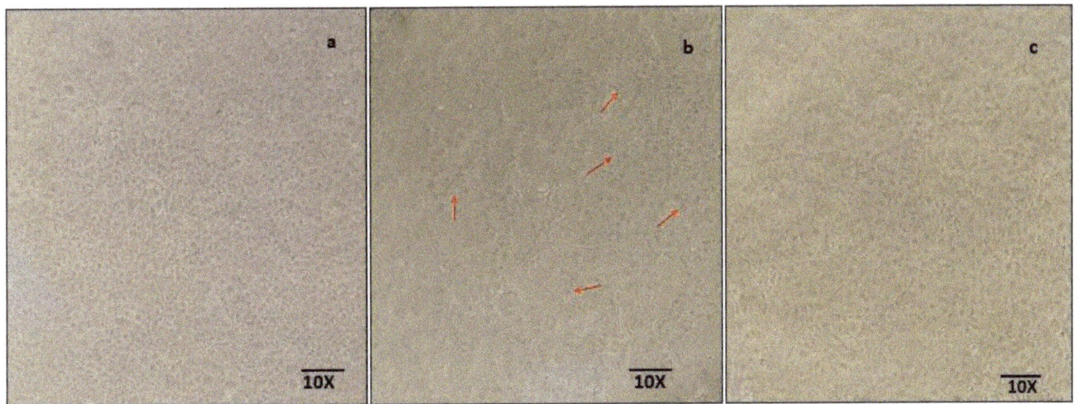

Figure 7. Morphology of NRK-52E cells in each treatment group at 24 h. The untreated NRK-52E cells showed its characteristic epithelial morphology (**a**). The treatment with 200 μM MG changed the epithelial morphology of NRK-52E cells to elongated fibroblast like morphology, indicated with arrows (**b**). Cotreatment with 100 μg/mL SM prevented MG—induced morphological changes in NRK-52E cells (**c**). Above images are representative microscopy images of each group under 10× objective.

2.8. SM Alleviates the Formation of Argpyrimidine Like AGEs in NRK-52E Cells

Dicarbonyls like MG react with arginine in the proteins to form AGEs. Argpyrimidine is a major modification of amino acid arginine by MG. As shown in Figure 8 NRK-52E cells when treated with MG showed higher levels of argpyrimidine as compared to the untreated cells. The treatment of SM along with MG alleviated the levels of argpyrimidine which could be due to the inhibition of modification of arginine by MG, thereby reducing the glycation induced damage to the cells.

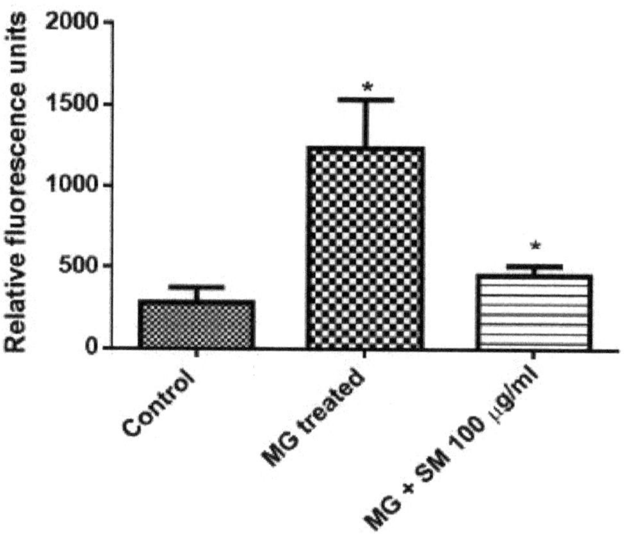

Figure 8. Detection of argpyrimidine levels. As compared to the control group, treatment with 200 µM MG in the NRK-52E cells after 24 h resulted in the formation of argpyrimidine by MG-induced modifications of arginine which could be prevented in MG and 100 µg/mL SM cotreatment group (* $p < 0.05$). Results are Mean ± SD of 3 individual experiments.

2.9. SM Alleviates the Formation of MG—Induced ROS and Mitigates the Oxidative Stress in NRK-52E Cells

The treatment with MG on NRK-52E cells resulted in the elevated oxidative stress, which could be due to the increased levels of AGEs formation in the presence of MG. Increased oxidative stress leads to the peroxidation of lipids resulting in the formation of malondialdehyde (MDA), a marker for oxidative stress in the cells. MG treatment in the NRK-52E cells exacerbated the levels of MDA as measured by HPLC (Figure 9a). This was ameliorated by the treatment of SM as seen by the reduced MDA levels. Oxidative stress further increased the formation of ROS in the cells, measured through fluorescence spectroscopy (Figure 9b). Treatment with SM in the presence of MG, prevented the formation of ROS, thereby unraveling the role of SM in preventing the MG—induced oxidative stress.

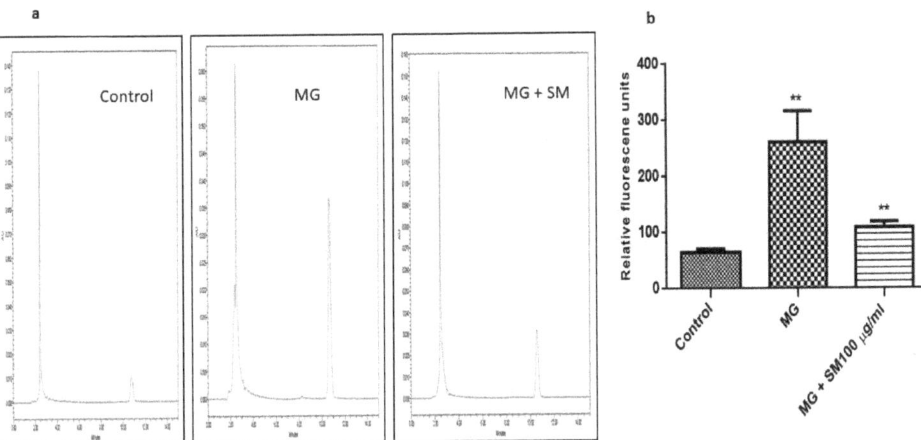

Figure 9. (**a**) HPLC chromatograms of MDA measured from NRK-52E cells after DNPH derivatization. The treatment with 200 µM MG in NRK-52E cells increased the levels of MDA measured using ODS2 reverse phase column in the presence of acetonitrile and milliQ water containing 0.2% acetic acid, with a ratio 38:62 respectively as the mobile phase. Treatment with 100 µg/mL SM in the presence of MG attenuated the production of MDA. (**b**) The levels of ROS were also measured using fluorescence spectroscopy with excitation and emission wavelengths 495 nm and 529 nm, respectively. Treatment with SM in the presence of MG could inhibit the elevation of ROS significantly (** $p < 0.01$) as shown in MG treated group (** $p < 0.01$), proving the antioxidative characteristic of SM. Results are Mean ± SD of 3 individual experiments.

2.10. SM Alleviates MG—Induced ER Stress in NRK-52E Cells

The treatment of the NRK-52E cells with MG leads to the production of AGEs which further results in the upregulation of ER stress by activating the unfolded protein response (UPR). The mRNA expression of ER stress markers like CCAAT-enhancer-binding protein homologous protein (CHOP) and glucose regulatory protein 78 (Grp78) were upregulated in the MG-treated NRK-52E cells. The treatment with SM along with MG could alleviate the upregulation of CHOP and Grp78, indicating the protection against ER stress (Figure 10).

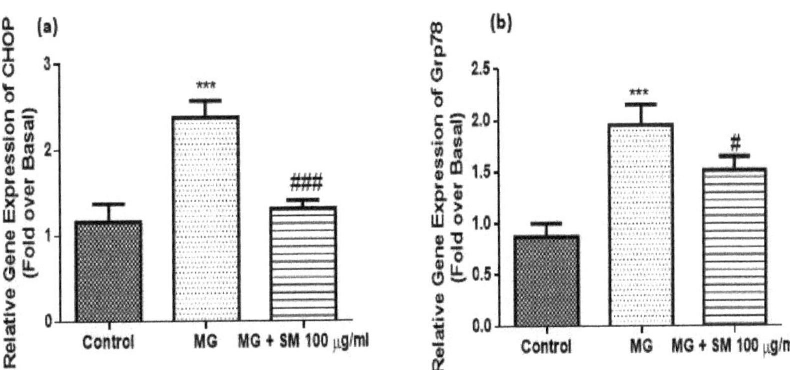

Figure 10. mRNA expression of ER stress genes (**a**) CHOP and (**b**) Grp-78 using qRT-PCR. The expression of CHOP and Grp78 was significantly upregulated in MG-exposed cells as compared to the untreated control NRK-52E cells after 24 h. The treatment with 100 µg/mL SM along with 200 µM MG could alleviate the upregulation of CHOP and Grp78, indicating the protection against ER stress. Results are Mean ± SD of 3 individual experiments analysed by ANOVA. The symbol *** $p < 0.001$ indicates the comparison of MG group w.r.t control and # $p < 0.05$, ### $p < 0.001$ indicates the comparison of SM group w.r.t. MG group.

2.11. Treatment with SM Improves the Inflammatory Response in MG Treated NRK-52E Cells

Cells when treated with 200 µM MG caused the upregulation of the mRNA levels of RAGE, that could be due to elevated levels of AGEs in MG-treated NRK-52E cells. The interaction of RAGE with AGEs and MG alone induced oxidative stress which resulted in the upregulation of ROS via increased NADPH oxidase levels in NRK-52E cells. The oxidative stress and the upregulated expression of RAGE induced the expression of various proinflammatory cytokines like interleukin-6 (IL-6), tumor necrosis factor-α (TNF-α), interleukin-1β (IL-1β) and inducible nitric oxide synthase (iNOS) in the cells treated with MG as compared to untreated control NRK-52E cells. This further exacerbated the expression of cell adhesion molecules like intracellular cell adhesion molecule-1 (ICAM-1) and monocyte chemoattractant protein-1 (MCP-1) in NRK-52E cells, leading to inflammation and cellular injury to kidney cells. Nonetheless, SM treatment at a concentration of 100 µg/mL to MG exposed cells (Figure 11) could prevent the upregulation of all the inflammatory cytokines under investigation (#### $p < 0.0001$; ### $p < 0.001$; ## $p < 0.01$), therefore explaining the preventive role of SM in inflammation of kidney cells.

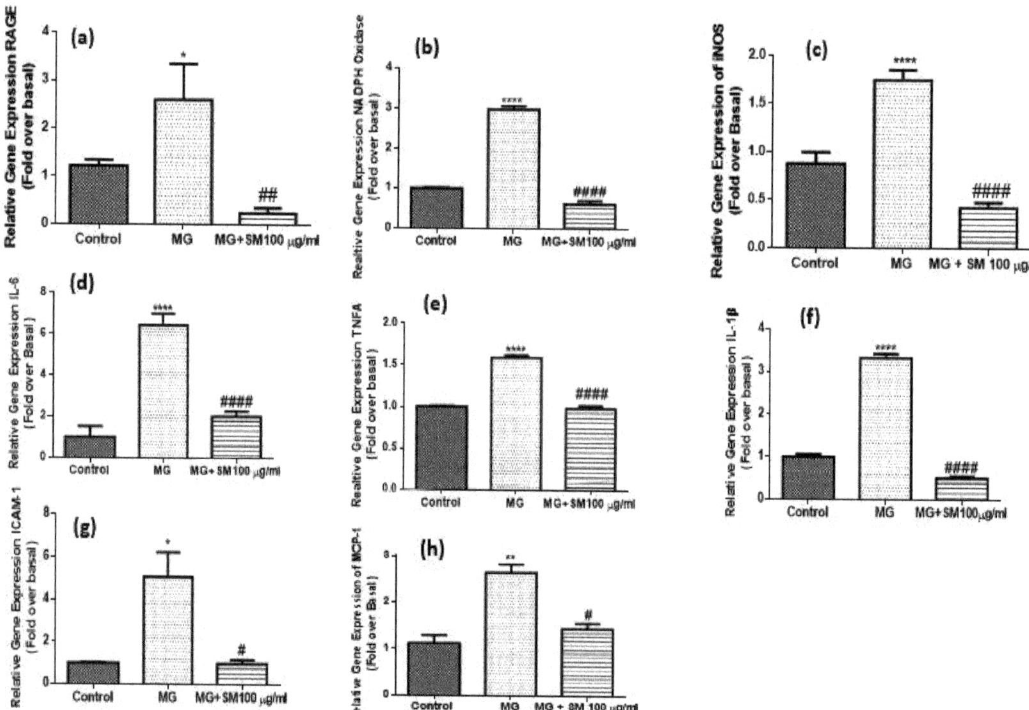

Figure 11. qRT-PCR analysis results of (**a**) RAGE (**b**) NADPH oxidase (**c**) iNOS (**d**) IL-6 (**e**) TNF-α (**f**) IL-1β (**g**) ICAM-1, and (**h**) MCP-1 in MG-treated NRK-52E cells alone and SM for 24 h. The gene expression levels were calculated after normalizing against housekeeping 18S rRNA and are presented as relative mRNA expression units. Values represent mean ± SD of 3 individual experiments. The symbol * $p < 0.05$, ** $p < 0.01$, **** $p < 0.0001$ indicate the comparison of MG group w.r.t control and # $p < 0.05$, ## $p < 0.01$, #### $p < 0.0001$ indicate the comparison between MG and SM group.

2.12. SM Improves MG-Induced Epithelial Mesenchymal Transition (EMT) in NRK-52E Cells

NRK-52E cells when treated with MG, showed morphological changes and the epithelial nature of cells changed to elongated, and fibroblast-like phenotype (Figure 10), possibly due to epithelial-mesenchymal transformation (EMT). SM treated group, did not show any change in the epithelial morphology of NRK-52E cells. MG induces the upregulation of p38 MAPK and also the downstream signaling by activating the transcription factors leading to various molecular changes in the kidney cells. It is previously reported that transforming growth factor- β (TGF-β) plays a very crucial role in the pathogenesis of DN. TGF-β induces the upregulation of fibroblast markers like alpha-smooth muscle actin (α-SMA), and fibronectin-1, along with downregulation of epithelial marker E-cadherin. Treatment with MG showed upregulated expression of these markers in NRK-52E cells which was ameliorated in the cells cotreated with MG and SM. These results indicate that MG induces the EMT changes in epithelial NRK-52E cells (Figure 12). As oxidative stress plays a very important role in the development of renal fibrosis via EMT, we also analyzed the levels of key intracellular antioxidant machinery like nuclear factor (erythroid-derived 2)-like 2 (Nrf-2) and heme oxygenase-1 (HO-1) expression levels. The treatment with MG in NRK-52E cells significantly attenuated the levels of Nrf-2 and HO-1, which were induced by the treatment with SM. Together these results indicate that SM ameliorated EMT in NRK-52E cells by inhibiting the expression of TGF-β via upregulation of HO-1 and Nrf-2 (Figure 12). As shown in Figure 13, MG treatment upregulated the protein levels of TGF-β and decreased the levels of HO-1, which was alleviated by SM treatment.

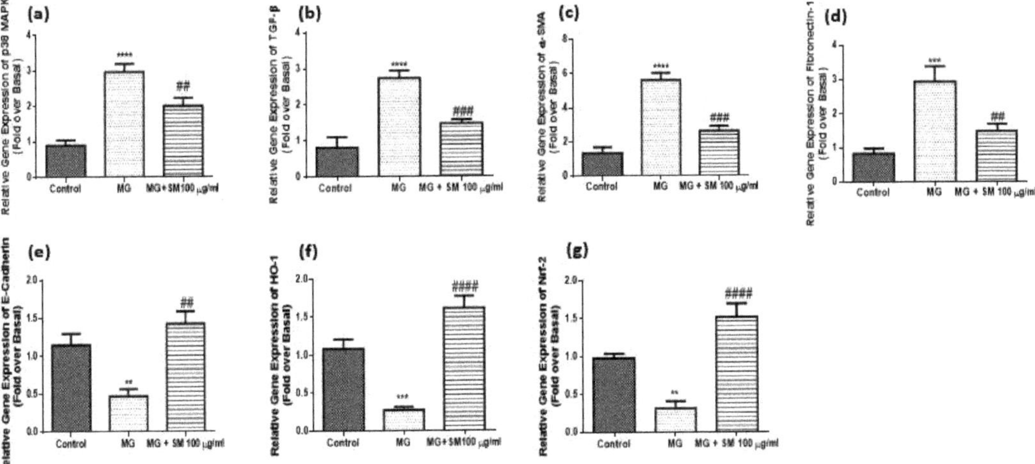

Figure 12. qRT-PCR analysis results of (**a**) p38 MAPK (**b**) TGF-β (**c**) α-SMA (**d**) Fibronectin-1 (**e**) E-cadherin (**f**) HO-1 and (**g**) Nrf-2 in MG-treated NRK-52E cells alone and SM for 24 h. The gene expression levels were calculated after normalizing against housekeeping 18S rRNA and are presented as relative mRNA expression units. Values represent mean ± SD of 3 individual experiments. The symbol ** $p < 0.01$, *** $p < 0.001$, **** $p < 0.0001$ indicate the comparison of MG group w.r.t control and ## $p < 0.01$, ### $p < 0.001$, #### $p < 0.0001$, indicate the comparison between MG and SM group.

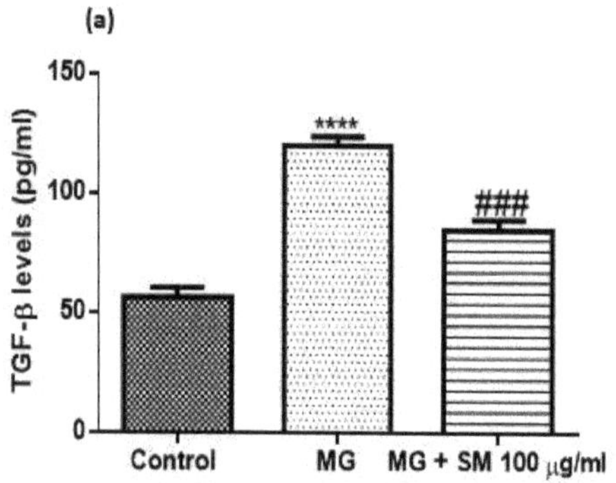
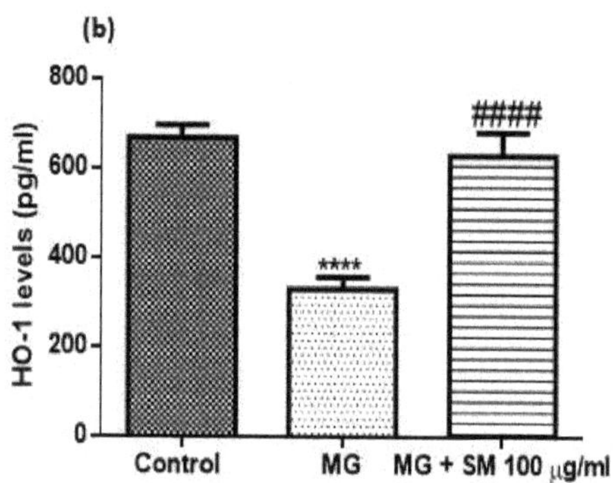

Figure 13. Protein levels of TGF-β and HO-1 in NRK-52E measured using ELISA. The treatment with 200 μM MG for 24 h, up-regulated the levels of TGF-β proteins which was prevented by the co-treatment with 100 μg/mL SM. The levels of HO-1 were declined in the MG treated group after 24 h which was alleviated by SM, proving the role of SM in inhibiting TGF-β expression by upregulating HO-1. Values represent mean ± SD of 3 individual experiments. **** $p < 0.0001$ indicate the comparison of MG group w.r.t control and ### $p < 0.001$, #### $p < 0.0001$, indicate the comparison between MG and SM group.

2.13. SM Shows Better Affinity and Binding with RAGE as Compared to Argpyrimidine

AGEs interact with their cellular receptor RAGE and initiate the inflammatory pathways by production of proinflammatory cytokines which further results in the oxidative stress. The selective inhibition of AGE-RAGE axis by preventing the binding of AGEs with RAGE can inhibit the deleterious effects observed due to their interaction. Therapeutic molecules that can bind with RAGE, thereby blocking the binding of AGEs with RAGE can be used as an approach to attenuate the progression of DN. In our study we have shown that SM not only reduces the expression of RAGE but also reduces the levels of AGEs like Argpyrimidine which is formed in the presence of MG. Therefore, to understand the possible role and mechanism of how SM prevents the MG-induced damage on NRK-52E cells, we adopted an in silico approach, where we checked if SM could interact with RAGE so as to prevent the binding of AGEs and therefore inhibit the activation of AGE-RAGE axis. Therefore, we docked both SM and Argpyrimidine with RAGE to check the affinity of both the ligands with RAGE. A total of 13 different molecules that encompassed different states of SM and Argpyrimidine were generated by LigPrep module. All the ligands and their various states were docked in the upstream C-Domain Type 1 using the Glide XP module. Further, the molecules were quantitatively evaluated based on their scores and qualitatively based on the type of interactions made with the residues of the amino acids. The chemical structures of the ligands are shown is Figure 14.

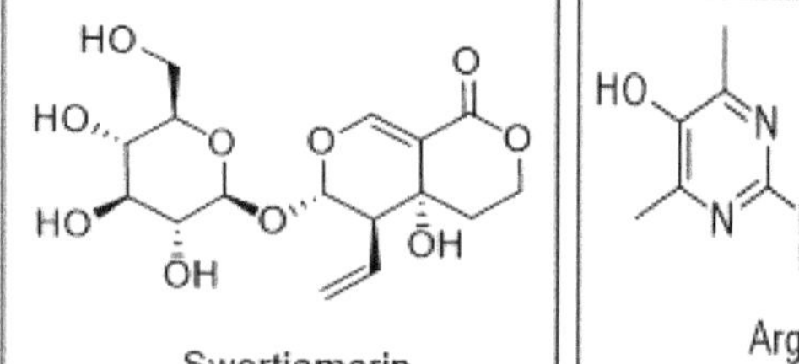

Figure 14. Chemical structure of swertiamarin and Argpyrimidine.

SM showed a G score of −4.7 kcal/mol while the docking score was −4.7 kcal/mol. The interactions formed by this ligand are as follows: OH acts as hydrogen bond acceptor for Leu 163 and Glu 174, and as a hydrogen bond donor to the amino acid Trp 156 (Figure 15a). Argpyrimidine on the other hand showed G score of −2.75 kcal/mol while the docking score is −2.73 kcal/mol. The interactions formed by this ligand are as follows: O⁻ acts as hydrogen bond acceptor for TRP 156, and hydrogen from amino group as a hydrogen bond donor to the amino acids Leu 154 and Glu 174. Further, the protonated nitrogen of argpyrimidine forms a salt bridge with GLU 174 (Figure 15b). Table 2 summarizes different interactions between SM-RAGE and Argpyrimidine-RAGE docking.

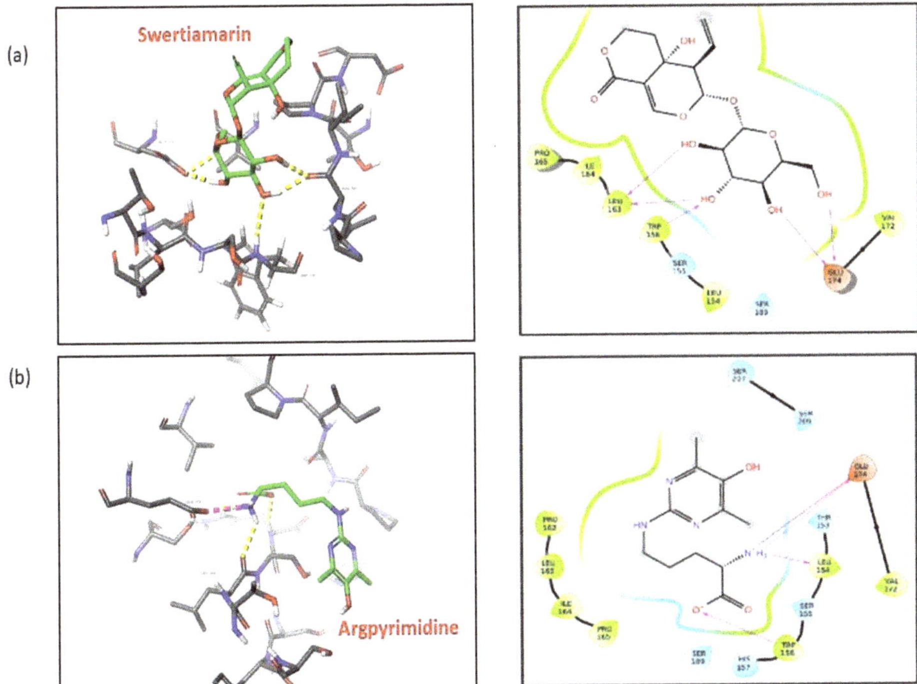

Figure 15. (**a**) Binding of SM with RAGE (left) and the amino acids involved in the interaction (right panel). (**b**) Binding of Argpyrimidine with RAGE (left) and the amino acid involved in the interaction (right panel).

Table 2. Interaction of ligands, type of interaction and total binding energies of RAGE with SM and Argpyrimidine.

Sr. No.	Interaction	Point Interaction	Donor Atom	Acceptor Atom	Type of Interaction	Bond Distance (Å)	Energy Binding (G Score) (kcal/mol)
1	RAGE: Swertiamarin	Trp 156–Swertiamarin	Trp 156: H	Swertiamarin: O	Hydrogen Bond	2.62	−4.7
		Leu 163–Swertiamarin	Swertiamarin: H	Leu 163: O	Hydrogen Bond	1.90	
		Leu 163–Swertiamarin	Swertiamarin: H	Leu 163: O	Hydrogen Bond	1.90	
		Glu 174–Swertiamarin	Swertiamarin: H	Glu 174: O	Hydrogen Bond	2.23	
		Glu 174–Swertiamarin	Swertiamarin: H	Glu 174: O	Hydrogen Bond	1.75	
2	RAGE: Argpyrimidine	Leu 154–Argpyrimidine	Argpyrimidine: H	Leu 154: O	Hydrogen Bond	2.21	−2.75
		Trp 156–Argpyrimidine	Trp 156: H	Argpyrimidine: O	Hydrogen Bond	2.38	
		Glu 174–Argpyrimidine	Argpyrimidine: H	Glu 174: O	Hydrogen Bond	1.82	
		Glu 174–Argpyrimidine	Argpyrimidine: N	Glu 174: O	Electrostatic Interaction	2.82	

3. Discussion

In the present study, we report isolation of SM and its role in inhibiting the formation of AGEs. The SM was isolated in the lab from *E. littorale* with better yield (7%) by column chromatography, which to our knowledge is the highest yield using silica of 60–120 mesh size as compared to the 5% yield obtained using silica based column chromatography [24], 0.6% using sephadex LH-20 column [25] and 2% using centrifugal partition chromatography [26]. The absorbance maxima of isolated SM matched with the absorbance maxima of the SM isolated by Rana et al. in the HPLC [26] and with the UV spectral analysis by Vishwakarma et al. to be at 238 nm [24,27]. The molecular ion peak of the isolated SM was 375.1 which matched with the results obtained in the literature along with their FT-IR fingerprint, proving the isolated compound to be SM [28]. Owing to its various properties like anticholinergic [18] antihyperlipidaemic [19] and given the important role of SM in diabetes mellitus [20] and its antioxidant potential [29], the present study was executed to understand and assess its antiglycation potential.

The persistent hyperglycaemia in the blood leads to the production of ROS which results in the oxidative stress. Oxidative stress results not only due to excessive ROS but may also result due to the defects or impairment in the antioxidant enzymes which otherwise can clear the ROS [30]. Increased blood glucose exacerbates the formation of ROS and is one of the prime reasons for AGEs formation, which further leads to various complications of T2D like DN [31–33]. Free radicals formed due to higher glucose levels can induce the formation of AGEs from Amadori products [34]. This is explained by the study which shows the increased levels of MG in the blood of diabetics as compared to nondiabetic individuals [35] and also supported by a study proving the accumulation of AGEs in most DN patients [36]. Under diabetic conditions, the excess of glucose leads to the activation of polyol pathway, resulting in the formation of fructose [37]. Fructose leads to the formation of AGEs and ROS by promoting the glycation of proteins and lipids, also aided by MG which is a dicarbonyl formed during glycolytic pathway for glucose metabolism [38]. Hence, preventing the glycation induced by fructose and MG could be one of the prime strategies to inhibit the formation of AGEs and prevent the diabetes related complications [39]. We, therefore assessed the potential of SM in inhibiting the fructose induced glycation. Fructose mediated glycation of proteins increases their fluorescence intensity at 440 nm which is due to formation of AGEs [40]. In the extant study, BSA glycated with fructose showed increased fluorescence intensity as compared to BSA treated with fructose in the presence of SM, which explains the role of SM in being able to prevent the formation of AGEs. To understand the potency of SM in being able to prevent the formation of AGEs, we compared it with metformin which is already known to prevent MG mediated glycation of albumin and also prevent renal tubular injury [41,42]. SM at similar dose to that of metformin could significantly inhibit the formation of AGEs much better as compared to metformin, which proves it to a better glycation inhibitor than metformin. The glycation of the BSA results in the increase in the carbonyl content, indicative of oxidative changes in the BSA [43]. In the present study, in accordance with the literature, we found increased carbonyl content in the glycated BSA. The treatment of SM in presence of glycated BSA, could reduce the carbonyl content, which possibly can be attributed to the antioxidative potential of SM.

The glycation induced by fructose leads to the modifications of various proteins, which renders the loss of function of proteins. This can be understood by hyperchromicity that can be attributed to the loss or fragmentation of the proteins, which causes the exposure of the aromatic amino acids responsible for its higher absorption at 280 nm [44]. It is of prime importance to understand the changes in the proteins caused due to their frutosylation, as the concentration of fructose increases up to as high as 5 mM in the kidneys and peripheral nerves due to polyol pathway [45] The structural changes as understood by hyperchromicity could be due to side chain modifications in the proteins due to glycation. Such modifications lead towards the structural as well as functional group modifications in the proteins, which can be analysed by FT-IR. We found the changes in the peaks corresponding to amide I from 1655 cm^{-1} of the native BSA to 1651 cm^{-1} in the glycated BSA. The treatment with SM showed a minor shift, indicating the relatively stable BSA molecule. The amide I band is a characteristic feature attributed to α-helix in the protein, and the decrease in the wavenumber indicates loss of α-helix in the glycated BSA, and therefore the structural change in the BSA due to glycation. Similarly, the amide III band of native BSA shifted from 1245 cm^{-1} to lower wavenumber 1236 cm^{-1} in the glycated BSA as compared to native BSA and the treatment of glycated BSA with SM did not show major shift in the wavenumber compared to native BSA with wavenumber 1242 cm^{-1}. The pronounced effect on shifting of amide III wavenumbers could be due to increase in the β-sheet and β-turns in the glycated BSA sample [46]. The treatment with SM however prevented the structural changes in the BSA as indicated by minor shift, suggesting the preventive role of SM in inhibiting the glycation induced structural changes in the protein. Our results show that the fructose mediated glycation was prevented when the glycation of BSA by fructose was carried out in presence of SM, as seen by reduced hyperchromicity which indicates the role of SM in preventing the fructose or AGEs mediated damage to kidney cells.

As discussed earlier, AGEs lead to the formation of ROS, which results in the oxidative stress. The formation of ROS and the repair of the damage caused due to ROS and oxidative stress is normally rescued in the cells by its antioxidant enzyme machinery comprising of antioxidant enzymes like catalase and SOD. However, glycation of enzymes like catalase and SOD lead to the loss of their activity, rendering an imbalance in homeostasis of oxidants and antioxidants in the cell [47]. ROS being toxic to kidney cells promotes cell death by inflammatory and fibrogenic reactions in the kidney cells [48]. Therefore, to understand the cellular effects of AGEs and how SM could prevent AGEs mediated progression of DN, we used an in vitro MG-induced model of DN. The concentration of MG to be used in the study was determined using MTT based cell viability assay. The dose-dependent decrease in the cell viability was observed, possibly because MG with increasing concentration may have induced apoptosis in the cells. Since nearly 50% cell death was observed at 199.5 μM concentration of MG, we chose 200 μM MG concentration for our further experiments, which was supported by the studies on rat mesangial cells, where 200 μM MG resulted in apoptosis of mesangial cells after 8 h of incubation with MG [49].

MG in particular can act as a source to the formation of a specific fluorescent AGE called Argpyrimidine, which is formed by a reaction of MG with the guanidine group of arginine [38]. Argpyrimidine was also found to be accumulated in the intima and media of small arteries of the kidneys of diabetic patients, which suggested its role in the progression of DN [50]. Approximately two to three-fold higher levels of argpyrimidine are found in diabetic patients as compared to nondiabetic individuals. The argpyrimidine levels also are found to positively correlate with glycosylated hemoglobin [51]. Higher levels of argpyrimidine were found to be present in the mesangial cells of rat kidney cultured in high glucose-containing medium [52]. Our study shows that SM can prevent the formation of AGEs and specifically could also inhibit the formation of MG mediated argpyrimidine formation in NRK-52 E cells. We also found increased levels of ROS in MG treated cells which as discussed earlier could be possibly due to the formation of AGEs [34]. However, the cells treated with SM along with MG showed reduced ROS, suggesting the antiglycative property of SM.

It is well established that oxidative stress and increased free radical can induce lipid peroxidation. This is because lipids are most vulnerable to the attack by ROS and reactive nitrogen species (RNS) [53]. This can generate increased levels of MDA, which is one of the major mutagenic and toxic products formed during the process of lipid peroxidation [54]. The present study shows that cotreatment of NRK-52E cells with MG and SM reduced the levels of ROS which led to reduced oxidative stress as evidenced by decreased production of MDA, which otherwise was higher due to lipid peroxidation in MG treated NRK-52E cells. Chronic hyperglycaemia, if unmanaged, not only culminates into various complications but also leads to pancreatic β-cell death. In a recent study, it was shown that MG leads to β-cell death by the formation of AGEs [55].

The MG mediated AGEs could readily interact with their membrane-bound receptor RAGE and result in the kidney dysfunction by inducing chronic inflammation. In a study on OVE26 mouse, it was shown that RAGE deletion could prevent renal function in diabetic mice, thereby explaining the crucial role played by RAGE in DN [56]. Therefore, elucidating the role of SM in MG-mediated RAGE expression became crucial, and we found that SM could decrease the mRNA expression of RAGE as compared to that of MG-treated cells. The activation of RAGE due to AGEs can lead to various inflammatory reactions as discussed earlier, by inducing the expressions of inflammatory markers like NADPH oxidase, TNF-α, IL-1β, IL-6, which gets accelerated due to the expression of cellular adhesion molecules like ICAM-1 [57]. In the current study, we found that treatment of MG not only induced the up regulated expression of proinflammatory cytokines like TNF-α, IL-1β, IL-6 but also of NADPH oxidase and ICAM-1, which was ameliorated in the cells treated with SM along with MG. The interaction of RAGE with AGEs can lead to the generation of ROS via NADPH oxidase which can promote the expression of TGF-β via NFκB, mitogen-activated protein kinase (MAPK), or PKC pathways in mesangial and renal tubulointerstitial cells in the kidneys [58]. We in the current study showed the MG mediated upregulation of p38 MAPK, which by further activation of inflammatory cascades led to the increase in the pro-inflammatory markers, causing the progression towards DN.

TGF-β being a pro-fibrotic cytokine plays a very crucial role in the renal fibrosis via EMT. EMT remains a very important phenomenon in the pathogenesis of DN, as it induces the expression of pro-fibrotic markers and mesenchymal markers like fibronectin 1 and α-SMA, inducing the tubulointerstitial fibrosis [59,60]. In addition, inhibition of particular stages involved in EMT, reduced the formation of fibrotic lesions in the kidney and therefore EMT is being considered to be significant in the development of DN [61–63]. We found that MG exposure changed the epithelial morphology of NRK-52E to more extended, fibroblast-like morphology, possibly due to higher mRNA expression of TGF-β as compared to lower expression of TGF-β in the cells co-treated with SM and MG. In the current study, MG induced the expression of fibronectin 1 and α-SMA along decreasing the epithelial marker e-cadherin, confirming the MG mediated EMT. These changes were prevented by SM. Our results are in agreement with a previous study, where TGF-β led to fibrosis of renal proximal tubules cells, leading to the death of the cells [64].

AGEs are known to induce pro-inflammatory signaling culminating in oxidative stress, further inducing apoptosis in the nephrons leading to renal fibrosis and therefore DN [65]. It is well established that ROS and oxidative stress plays a very important role in development of renal complications including EMT [66]. Therefore, the cellular antioxidants could be chosen as one of the prime targets for inhibiting EMT. The role of cellular antioxidant machinery like Nrf-2 and HO-1 has been demonstrated by investigations which showed that $Nrf2^{-/-}$ mice were highly sensitive to high glucose induced oxidative damage, and the HO-1 deficiency is associated with increased fibrosis and increased tubular expression of TGF-β [67,68]. In our study, we found that MG induced downregulation of HO-1 and Nrf-2 was upregulated in SM treated cells.

To understand the possible mechanism by which SM inhibited the expression of RAGE, we explored the in silico approach adopted by researchers [69] to analyze the role of SM in binding with RAGE, thereby blocking the binding of AGEs with RAGE. As MG leads to the production of Argpyrimidine, we docked both Argpyrimidine and SM with RAGE to check for their binding energies. Argpyrimidine is known to bind in the C-Domain type 1 present in the RAGE; therefore, we chose the region ranging from 123 to 219 amino acids for performing the docking [69]. SM showed better binding energy and docking score as compared to Argpyrimidine, which suggests that RAGE binds with more preference with SM than Argpyrimidine, thereby blocking the binding of Argpyrimidine with RAGE. The difference in binding energy is −1.95 kcal/mol. This docking result is complementary to the experimental results obtained in vivo, which also shows that binding of Swertiamarin results in reduction of the AGEs and the expression of its receptor RAGE.

In conclusion, we report that SM inhibits binding of AGEs with RAGE and hence prevents the activation of RAGE signaling and p38 MAPK, thereby inhibiting the downregulation of HO-1 and Nrf-2 to prevent the EMT in rat kidney cells.

4. Materials and Methods

Bovine serum albumin (BSA, fraction V) was acquired from Himedia (Mumbai, India). Potassium bromide (FTIR grade) was obtained from Sigma-Aldrich (St. Louis, MO, USA). Acetonitrile (HPLC grade), HPLC grade water for HPLC; hexane (AR grade), ethyl acetate (AR grade) and methanol (AR grade) were acquired from Merck (Darmstadt, Germany) for the isolation of SM using column chromatography. All the solvents were used in their pure form without any preprocessing.

4.1. Isolation of SM Using Silica Chromatography

The dry plant *E.littorale* was acquired from Saurashtra region, Gujarat, India. The plant was press dried and then powdered in a crusher. The powder was immersed overnight in double the volume of 70% methanol (SRL, Mumbai, India) and 30% water. The process was repeated three times for a single batch and then filtered. The filtrate was vacuum evaporated using a Heidolph rotary evaporator (Schwabach, Germany), to get a dry hydroalcoholic extract.

To obtain SM from the hydroalcoholic extract, silica based column chromatography was done. Furthermore, 1 g of hydroalcoholic extract was coated with 3 g silica. The extract was chromatographed using silica gel (60–120 mesh size, Merck, Germany) with hexane and then with hexane containing ethyl acetate (60–97%), followed by ethyl acetate and then with ethyl acetate mixed with methanol in the order of increasing polarity (0.5–1.5%). Fractions with different polarities were assessed for the presence of SM by thin-layer chromatography (TLC) using the solvent system of chloroform: methanol (8:2). The presence of SM was established by co-chromatography of the standard SM (TCI, Japan) along with different fractions. Fractions with SM were pooled and dried. They were further purified by precipitating the methanol dissolved residue with nonpolar solvents like diethyl ether (yield: 7%).

4.2. Characterization of Isolated SM for Its Purity

The characterization as well as the purity of the isolated compound was confirmed by various methods which included TLC, HPLC, LC-MS and FTIR.

4.2.1. Characterization Using High Performance Liquid Chromatography (HPLC)

For HPLC analysis, the solution of SM isolated in the lab was prepared to achieve a final concentration of 1 mg/mL in methanol. The samples and solvents to be used as mobile phases were filtered through 0.2 μm syringe filters (Axiva, India). HPLC was carried out using the previously described mobile phase for SM with slight modifications by using acetonitrile: water (10:90) as the mobile phase [70]. The flow rate was 1 mL/min and the column temperature was maintained at 25 °C. The detection wavelength and mode for SM was 238 nm and photodiode detector (PDA).

4.2.2. Characterization Using Fourier-Transform Infrared Spectroscopy (FTIR) and Mass Spectrometer (MS)

The isolated compound was analyzed by FTIR for the identification of functional groups present in the compound. The standard was also used as a reference for the isolated compound. A homogenous solution of both standard and the compound was individually prepared in potassium bromide (KBr) and subjected to FTIR spectrophotometer (Thermofisher Scientific, Waltham, MA, USA). The compound isolated was also subjected to positive-ion Electrospray Ionization, i.e., ESI using Perkin-Elmer Applied Biosystem Sciex API 2000 (Waltham, MA, USA) for identification of its characteristic molecular ion peak.

4.3. Analysis of Antiglycative Potential of SM Using In Vitro Glycation Assay

Antiglycative studies were done as suggested by McPherson et al. with slight modifications [71]. Bovine Serum Albumin (BSA, Himedia, Mumbai, India) and Fructose (Merck, Darmstadt, Germany) were used to induce glycation. In brief, in a 1.5 mL eppendorf, 100 µL of 60 mg/mL BSA (20 mg/mL final concentration) was incubated with 100 µL of 1.5 M fructose (0.5 M final concentration) and 100 µL of potassium phosphate buffer (pH 7.4). Negative control was incorporated as well, which consisted of 100 µL of 60 mg/mL BSA (20mg/mL final concentration) and 200 µL of potassium phosphate buffer (pH 7.4). To understand the antiglycative potential of the compound, the same reaction was carried out in the presence of varying concentrations of SM (1 µg/mL to 10 µg/mL). For the comparative studies, metformin was used as a positive antiglycative control at the same concentrations as of SM. The reaction was incubated at 60 °C for 24 h and samples were analyzed for fluorescence due to AGEs on a Perkin Elmer LS-55 spectrofluorometer (Waltham, MA, USA) by using an excitation wavelength of 370 nm and an emission wavelength of 440 nm.

4.4. Analysis of Protein Modifications Due to Glycation by UV Absorbance Spectroscopy

The native, glycated BSA and samples in which BSA was glycated in presence of SM were analyzed for their absorption spectra on Shimadzu UV–1700 spectrophotometer (Kyoto, Japan) in 200 to 800 nm wavelength range using a quartz cuvette. The samples were analyzed for the hyperchromic shift due to glycation by fructose as suggested by Allarakha et al. [45].

4.5. Estimation of Protein Carbonyl Content

The analysis of the carbonyl content as a marker for oxidative damage to protein due to glycation was done according to the protocol suggested by Meeprom et al. [43]. From each group, 100 µL of the sample was taken and mixed with 400 µL of 10 mM 2,4-dinitrophenylhydrazine (DNPH) prepared in 2.5 M HCl solution, followed by the incubation for 1 h in the dark. After the incubation, the protein was precipitated on ice using 500 µL of 20% *w/v* tri-choloroacetic acid (TCA) solution and subjected to centrifugation at $10,000 \times g$ for 10 min at 4 °C. The protein in the pellet was washed thrice with 500 µL of 1:1 solution of ethanol and ethyl acetate followed by resuspension in 250 µL of 6 M guanidium hydrochloride solution. The absorbance was recorded at 370 nm and the carbonyl content was measured using the molar extinction coefficient of DNPH, i.e., $\varepsilon = 22,000 \text{ M}^{-1} \text{ cm}^{-1}$. The results were expressed as nM carbonyl content/mg of protein.

4.6. Fourier Transform Infrared Spectroscopy of the Protein Samples

To check and analyze the functional and conformational changes in the BSA due to fructose mediated glycation, FTIR spectroscopy was done as suggested by Liu et al. [46]. The samples were lyophilized and then mixed with FTIR grade KBr in 1: 100 ratio (sample: KBr) to make a homogenous mixture and pressed to prepare a transparent pellet. The pellet was subjected to analysis in the transmission mode in the range of 400–4000 cm^{-1} on FTIR spectrophotometer (Thermofisher Scientific, Waltham, MA, USA).

4.7. Culturing of NRK-52 E Cells

Normal Rat Kidney (NRK-52E) cell line was acquired from National Centre for Cell Sciences (NCCS), Pune, India. The cells were cultured in Dulbecco's Modified Eagle Medium Low Glucose (5.5 mM/L) medium with 10% fetal bovine serum, 1% L-glutamine, and 1% penicillin/streptomycin (Thermofisher Scientific, Waltham, MA, USA) at 37 °C in a 5% CO_2 incubator. The cells were grown till 80–90% confluency, after which they were used for experiments with MG in the presence and absence of SM.

Cell Viability and Dose Determination of MG and SM Using MTT Assay

The concentration of MG (Sigma-Aldrich, St. Louis, MO, USA) used during the experiments was determined by checking MG toxicity on NRK-52E cells using 3-(4,5-dimethylthiazol-2-yl)-2,5-diphenyltetrazolium bromide (MTT) assay described by Riss et al. [72]. In brief, 1×10^4 cells were seeded in a 96-well plate. After 24 h, cells were exposed to different concentrations MG (1 mM, 500 μM, 250 μM, 125 μM, 62.5 μM, 31.25 μM, 15.625 μM, 7.812 μM, and 3.906 μM), along with an untreated control well, for 24 h. Similarly, the toxicity of SM was checked on NRK-52E. NRK-52E cells were treated with different concentrations of SM (50 μg/mL, 100 μg/mL, 150 μg/mL, 200 μg/mL and 250 μg/mL) along with a group left untreated with SM for 24 h. After treating them for 24 h, 10 μL of MTT solution was added to each well at a final concentration of 0.5 μg/mL and the cells were incubated at 37 °C in a 5% CO_2 atmosphere for 3 h in the dark. Following the incubation period, formazan crystals formed in each well were solubilized using 100 μL dimethyl sulfoxide (DMSO) solution. The absorbance was then measured at 570 nm to determine the viability of cells.

4.8. MG Stimulation and Different Treatment Groups

To check the effect of MG on NRK-52E cells, 8×10^5 cells were seeded in a 60 mm culture plates followed by incubation for 24 h. After this, cells received different treatments based on the respective groups like control (only growth medium), MG (medium containing 200 μM MG) and MG + SM 100 μg/mL (medium containing 200 μM MG in the presence of 100 μg/mL SM). The effect of MG-induced stress on kidney cells was assayed by checking the ROS production, lipid peroxidation, argpyrimidine levels and transcript levels of various genes involved in the progression of DN by qRT PCR.

4.9. Estimation of Argpyrimidine

To estimate the levels of Argpyrimidine, NRK-52E cells (8×10^6 cells/well) were seeded onto 60mm plates, and the cells were incubated for 24 h. After 24 h, cells were treated with MG, in the presence and absence of SM for 24 h. After the incubation, the cells were lysed using the lysis buffer followed by centrifugation at 16,000 rpm for 10 min at 4 °C. The supernatant was then analyzed using a Hitachi F-7000 fluorescence spectrophotometer (Tokyo, Japan) with an excitation wavelength of 330 nm and an emission wavelength of 380 nm for the presence of Argpyrimidine [51].

4.10. Estimation of Reactive Oxygen Species (ROS)

In order to estimate oxidative stress caused by MG in the presence or absence of SM, the cells were treated with 200 μM MG alone or along with SM and incubated for 24 h. After treatment, cells were incubated with 5(6)-carboxy-20,70-dichlorofluorescein Diacetate (Carboxy-H2-DCFDA) (Sigma-Aldrich, St. Louis, MO, USA) in the dark with a final concentration of 30 μM at 37 °C for 1 h. The cells were then harvested, washed with PBS, and resuspended in PBS. ROS production in the cells was measured by measuring the fluorescence of the sample at an excitation wavelength—485 nm and an emission wavelength—530 nm, using a fluorescence spectrophotometer (Perkin Elmer LS-55, Waltham, MA, USA [73].

4.11. Estimation of MDA as a Measure of Lipid Peroxidation by HPLC

Total malondialdehyde (MDA) in the cultured cells was estimated using HPLC by following the method described by Tukozkan et al. with some modifications [74]. Summarily, the medium was removed from the plates and the cells were rinsed with PBS. Cells were homogenized in cold 1.15% KCl to make 10% homogenate. 500 μl of the homogenate was then mixed with 100 μl of 6 M NaOH, and the samples were incubated in a water bath at 60 °C for 45 min. The hydrolyzed sample was then acidified with 250 μL of 35% perchloric acid. The samples were subjected to centrifugation at $15,000 \times g$ for 10 min. After centrifugation, 250 μL of the supernatant was collected and mixed with 25 μL of DNPH solution, followed by 10 min incubation in the dark. The samples were then analyzed by

HPLC in an ODS2 reverse column using acetonitrile: water (38:62) containing 0.2% acetic acid as a mobile phase. Isocratic conditions were maintained during HPLC with a flow rate of 1 mL/min and the MDA was detected in the samples at 310 nm with the UV detector with a retention time of about 10 min. The concentration of MDA was detected in the sample by comparing it with the standard curve prepared using 1,1,3,3 tetraethoxypropane.

4.12. Isolation of RNA, cDNA Synthesis and Analysis of Gene Expressions of Various Markers by qRT-PCR

Total RNA was isolated from the control and treated cells using TRIzol reagent. The purity and concentration of RNA in the sample were measured at 260 nm and A260/A280 ratio of the sample was determined using NanoDrop 2000 (Thermofisher Scientific, Waltham, MA, USA). A total of 1 µg RNA was used to synthesize cDNA using first strand cDNA synthesis kit. For qPCR, maxima SYBR Green/ROX qPCR master mix was used to quantify the mRNA expression levels for all genes under investigation on the Agilent Strategene Mx3005P system. The qRT-PCR system comprised of 1 µL cDNA, 0.5 µL of 10 µM FP (forward primers), 0.5 µL of 10 µM RP (reverse primers), 10 µL SYBR Green Mastermix, and 8.0 µL milli-Q water in a 20 µL reaction system. qRT-PCR cycling conditions were set according to Kema, V.H. et al. 2017 [75]. The $2^{-\Delta\Delta Ct}$ method (fold change over basal) was applied to evaluate mRNA expression levels in both control and treated cells. 18S rRNA was presented as an internal reference gene control. The list of primer sequences sets used are tabulated in Table 3.

Table 3. The list of *Rattus norvegicus* primers used for quantification of mRNA using qRT-PCR.

Sr. No.	Name of the Gene	Sense Primer Sequence	Antisense Primer Sequence
1.	18s	5′ACGGAAGGGCACCACCAGGA 3′	5′CACCACCACCCACGGAATCG 3′
2.	RAGE	5′ GGTACTGGTTCTTGCTCT 3′	5′ATTCTAGCTTCTGGGTTG 3′
3.	TNF-α	5′ CAAGGAGGAGAAGTTCCCAA 3′	5′CTCTGCTTGGTGGTTTGCTA 3′
4.	ICAM-1	5′CCCCACCTACATACATTCCTAC 3′	5′ACATTTTCTCCCAGGCATTC 3′
5.	NADPH oxidase	5′ GGCATCCCTTTACTCTGACCT 3′	5′ TGCTGCTCGAATATGAATGG 3′
6.	IL-6	5′ GCCCTTCAGGAACAGCTATGA 3′	5′ TGTCAACAACATCAGTCCCAAGA 3′
7.	IL-1β	5′ CCCTGCAGCTGGAGAGTGTGG 3′	5′ TGTGCTCTGCTTGAGAGGTGCT 3′
8.	TGF-β	5′ TGCTTCAGCTCCACAGAGAA 3′	5′ TGTGTTGGTTGTAGAGGGCA 3′
9.	iNOS	5′ TCACTGGGACAGCACAGAAT 3′	5′ TGTGTCTGCAGATGTGCTGA 3′
10.	Fibronectin	5′ CATGGCTTTAGGCGAACCA 3′	5′ CATCTACATTCGGCAGGTATGG 3′
11.	α-SMA	5′ GACCCTGAAGTATCCGATAGAACA 3′	5′CACGCGAAGCTCGTTATAGAAG 3′
12.	E-cadherin	5′ TGATGATGCCCCCAACACTC 3′	5′ CCAAGCCCTTGGCTGTTTTC 3′
13.	p38 MAPK	5′ CGAAATGACCGGCTACGTGG 3′	5′ CACTTCATCGTAGGTCAGGC 3′
14.	Grp78	5′ GAAACTGCCGAGGCGTAT 3′	5′ ATGTTCTTCTCTCCCTCTCTCTTT 3′
15.	CHOP	5′ GAAAGCAGAAACCGGTCCAAT 3′	5′ GGATGAGATATAGGTGCCCCC 3′
16.	MCP-1	5′ CCTCCACCACTATGCAGGTCTC 3′	5′ GCACGTGGATGCTACAGGC 3′
17.	Nrf-2	5′ CAGAGTTTCTTCGCCAGAGG 3′	5′ TGAGTGTGAGGACCCATCG 3′
18.	HO-1	5′ CAAATCCCACCTTGAACACA 3′	5′ CGACTGACTAATGGCAGCAG 3′

4.13. Estimation of Protein Levels of TGF-β and HO-1 by ELISA

The protein levels of TGF-β and HO-1 were analysed from the cultured NRK-52E cells after the treatment with 200 µM MG in the absence and presence of 100 µg/mL SM. Briefly, the cultured cells from different groups were removed using cell scraper after washing thrice with 1X PBS. The cells were then centrifuged at 1500 rpm for 10 min under cooling conditions. The collected cell free supernatant was analyzed as per the manufacturer's protocol (Abcam, Cambridge, UK). The levels of TGF-β and HO-1 were measured in pg/mL.

4.14. Molecular Docking Studies

4.14.1. Homology Modelling

Receptor for Advanced Glycation End Products (RAGE) from *Rattus norvegicus* consists of 402 amino acids [76]. The FASTA sequence for building up the 3D model was obtained from UniProt database (UniProt ID: Q63495) [77]. The 3D model of RAGE was constructed using RAGE (*Mus musculus*) (PDB ID: 4IM8) as the template as it showed a sequence identity of 92.04% [76,78]. The 3D model was created using SWISS-MODEL server which uses homology modelling method, wherein it aligns the amino acid sequence of the desired protein with the protein that has closest identity to our desired protein and whose structure has been solved, thus developing a model for the unknown desired protein whose sequence is known. The generated model was further used for docking studies. The molecular docking was performed using Glide module (XP) of Maestro software (Schrodinger) version 12.7.156 (New York, NY, USA) [79].

4.14.2. Protein Preparation Protocol

The protein structure of Receptor for Advanced Glycation End Products (RAGE) from *Rattus norvegicus* was constructed using SWISS-MODEL server. The structure so generated needs to be prepared using protein preparation wizard, which helps in pre-processing, optimizing hydrogen bonds and minimizing the protein, from its raw state to a better state which can be useful for the further calculations. The protein was preprocessed by assigning bond order, adding the missing hydrogen atoms, and converting the selenomethionine to methionine. The missing side chains and loops were filled in using Prime module, which is the protein refinement application. The termini were capped with N-acyl and N-methyl groups and finally, the water molecules which were more than 5 Å away from the het groups were deleted. A pH of 7.0 ± 2.0 was set in Epik to account for the possible tautomeric state of amino acid residues. The hydrogen bonds were optimized with PROPKA at a pH of 7. Lastly, the RAGE protein model was minimized using optimized potentials for liquid simulations 4 (OPLS4) force field [80–82].

4.14.3. Receptor Grid Generation

Receptor grid generation forms the first step of molecular docking protocol. A volume of the active site is represented by the generated grid file. The ligands of interest are then docked in this grid. Initially, the minimized protein was selected to generate the grid. It has been reported in the literature that Argpyrimidine binds to the upstream C-Domain Type 1 in RAGE [69]. Hence, the amino acids from 123 to 219, which represent the upstream C-Domain Type 1, were selected for generating the grid for docking SM and Argpyrimidine. The centroid was generated around these amino acids, and further criteria were set to default, to generate the grid for docking [80–82].

4.14.4. Glide Molecular Docking and Analysis of Docking

Glide is the molecular docking module in Maestro. It has a widespread use from knowing the ideal interactions between a particular ligand and the protein to screening large libraries of molecules for high throughput screening. The molecular docking using Glide starts with procuring the grid file and the Ligprep output file. Further, XP mode was selected for better accuracy and further criteria were set as default. The docking module was run and results were obtained and analysed using XP visualizer in Schrodinger [80–82].

4.14.5. Statistical Analysis

All respective experiments were performed at least thrice. Data of the replicates were calculated as mean $\pm$ SD. GraphPad Prism 7 software (GraphPad Software Inc., California

Corporation, San Diego, CA, USA) was used to analyze the results. The differences between all groups were analyzed using one way analysis of variance (ANOVA). Variations between groups were considered to be significant at p values less than or equal to 0.05.

5. Conclusions

The present study depicted the protective effect of SM in inhibiting the formation of fructose induced AGEs and its role in ameliorating the progression of DN. SM could prevent the fructose mediated conformational changes in the BSA. The formation of AGEs like argpyrimidine from MG upon interaction with RAGE led to the increased production of ROS and therefore resulted in the generation of oxidative stress. Moreover, treatment with SM in the presence of MG-treated NRK-52E cells prevented the oxidative stress and inflammation, by upregulating the antioxidants level of Nrf-2 and HO-1, thereby reducing the activation of TGF-β and prevented against the MG-induced EMT changes in NRK-52E. We for the first time report here that the inhibition of the of RAGE/ MAPK/ TGF-β pathway could be a possible mechanism contributing towards protecting effects of SM in preventing the DN induced changes in NRK-52E cells in presence of MG. In addition, the current study for the first time demonstrates that the molecular interactions between RAGE and SM can inhibit the binding of Argpyrimidine and block the AGE-RAGE axis, which also is supported by the cellular studies. Additionally, this may provide strong evidence for an in vivo study as a future prospect to unravel the mechanism of SM in inhibiting AGEs induced DN.

Supplementary Materials: The following are available online, Figure S1: The stages involved in the formation of advanced glycation end products.

Author Contributions: K.P. and F.P. contributed equally to this manuscript. The design and conception of the study was done by K.P., F.P. and P.M., K.P. did the acquisition of the data. Analysis and/or interpretation of data was done by K.P., F.P and P.M. The drafting of the manuscript was done by K.P. and F.P. while revision of the manuscript critically for important intellectual content was done by K.P., F.P., D.P. and P.M. All authors have read and agreed to the published version of the manuscript.

Funding: This research received no external funding.

Institutional Review Board Statement: Not applicable.

Informed Consent Statement: Not applicable.

Data Availability Statement: All data generated or analyzed during the current study are included in the manuscript.

Acknowledgments: The author would like to acknowledge Charotar University of Science and Technology for providing the author with CHARUSAT Ph.D. Scholar Fellowship (CPSF). Authors acknowledge Vaibhav Patel, Assistant Professor at P. D. Patel Institute of Applied Sciences, CHARUSAT for his help in the HPLC analysis. Authors would like to thank Mitesh Patel, Assistant Professor at P. D. Patel Institute of Applied Sciences, CHARUSAT for his help in doing FTIR of BSA. Authors are grateful to Vijay Thiruvenkatam, Pranav Bhagwat and Haritha D. for helping in the molecular docking studies and providing the facility at Indian Institute of Technology, Gandhinagar for the in silico analysis.

Conflicts of Interest: The authors declare no conflict of interest.

References

1. Writing Team for the Diabetes Control and Complications Trial/Epidemiology of Diabetes Interventions and Complications Research Group Sustained Effect of Intensive Treatment of Type 1 Diabetes Mellitus on Development and Progression of Diabetic Nephropathy. *JAMA* **2003**, *290*, 2159–2167. [CrossRef]
2. Fioretto, P.; Steffes, M.W.; Sutherland, D.E.; Goetz, F.C.; Mauer, M. Reversal of Lesions of Diabetic Nephropathy after Pancreas Transplantation. *N. Engl. J. Med.* **1998**, *339*, 69–75. [CrossRef] [PubMed]
3. Barbosa, J.; Steffes, M.W.; E Sutherland, D.; E Connett, J.; Rao, K.V.; Mauer, S.M. Effect of glycemic control on early diabetic renal lesions. A 5-year randomized controlled clinical trial of insulin-dependent diabetic kidney transplant recipients. *JAMA* **1994**, *272*, 600–606. [CrossRef] [PubMed]

4. Yao, D.; Brownlee, M. Hyperglycemia-Induced Reactive Oxygen Species Increase Expression of the Receptor for Advanced Glycation End Products (RAGE) and RAGE Ligands. *Diabetes* **2009**, *59*, 249–255. [CrossRef] [PubMed]

5. Neglia, C.; Cohen, H.J.; Garber, A.R.; Ellis, P.D.; Thorpe, S.R.; Baynes, J.W. 13C NMR investigation of nonenzymatic glucosylation of protein. Model studies using RNase A. *J. Biol. Chem.* **1983**, *258*, 14279–14283. [CrossRef]

6. Lapolla, A.; Traldi, P.; Fedele, D. Importance of measuring products of non-enzymatic glycation of proteins. *Clin. Biochem.* **2005**, *38*, 103–115. [CrossRef]

7. Kilhovd, B.; Giardino, I.; Torjesen, P.; Birkeland, K.; Berg, T.; Thornalley, P.; Brownlee, M.; Hanssen, K. Increased serum levels of the specific AGE-compound methylglyoxal-derived hydroimidazolone in patients with type 2 diabetes. *Metabolism* **2003**, *52*, 163–167. [CrossRef]

8. Cooper, M.E. Interaction of metabolic and haemodynamic factors in mediating experimental diabetic nephropathy. *Diabetologia* **2001**, *44*, 1957–1972. [CrossRef]

9. Pasupulati, A.K.; Chitra, P.S.; Reddy, G.B. Advanced glycation end products mediated cellular and molecular events in the pathology of diabetic nephropathy. *Biomol. Concepts* **2016**, *7*, 293–309. [CrossRef]

10. Ceriello, A.; Bortolotti, N.; Falleti, E.; Taboga, C.; Tonutti, L.; Crescentini, A.; Motz, E.; Lizzio, S.; Russo, A.; Bartoli, E. Total Radical-Trapping Antioxidant Parameter in NIDDM Patients. *Diabetes Care* **1997**, *20*, 194–197. [CrossRef]

11. Kang, J.H. Modification and inactivation of human Cu,Zn-superoxide dismutase by methylglyoxal. *Mol. Cells* **2003**, *15*, 194–199.

12. Najjar, F.M.; Taghavi, F.; Ghadari, R.; Sheibani, N.; Moosavi-Movahedi, A.A. Destructive effect of non-enzymatic glycation on catalase and remediation via curcumin. *Arch. Biochem. Biophys.* **2017**, *630*, 81–90. [CrossRef]

13. Choudhary, M.I.; Maher, S.; Begum, A.; Abbaskhan, A.; Ali, S.; Khan, A.; Rehman, S.-U.; Rahman, A.-U. Characterization and Antiglycation Activity of Phenolic Constituents from Viscum album (European Mistletoe). *Chem. Pharm. Bull.* **2010**, *58*, 980–982. [CrossRef]

14. Meng, G.; Zhu, H.; Yang, S.; Wu, F.; Zheng, H.; Chen, E.; Xu, J. Attenuating effects of Ganoderma lucidum polysaccharides on myocardial collagen cross-linking relates to advanced glycation end product and antioxidant enzymes in high-fat-diet and streptozotocin-induced diabetic rats. *Carbohydr. Polym.* **2011**, *84*, 180–185. [CrossRef]

15. Sun, Z.; Peng, X.; Liu, J.; Fan, K.W.; Wang, M.; Chen, F. Inhibitory effects of microalgal extracts on the formation of advanced glycation endproducts (AGEs). *Food Chem.* **2010**, *120*, 261–267. [CrossRef]

16. Yazdanparast, R.; Ardestani, A.; Jamshidi, S. Experimental diabetes treated with Achillea santolina: Effect on pancreatic oxidative parameters. *J. Ethnopharmacol.* **2007**, *112*, 13–18. [CrossRef]

17. Murali, B.; Upadhyaya, U.; Goyal, R. Effect of chronic treatment with Enicostemma littorale in non-insulin-dependent diabetic (NIDDM) rats. *J. Ethnopharmacol.* **2002**, *81*, 199–204. [CrossRef]

18. Yamahara, J.; Kobayashi, M.; Matsuda, H.; Aoki, S. Anticholinergic action of Swertia japonica and an active constituent. *J. Ethnopharmacol.* **1991**, *33*, 31–35. [CrossRef]

19. Vaidya, H.; Rajani, M.; Sudarsanam, V.; Padh, H.; Goyal, R. Antihyperlipidaemic activity of swertiamarin, a secoiridoid glycoside in poloxamer-407-induced hyperlipidaemic rats. *J. Nat. Med.* **2009**, *63*, 437–442. [CrossRef]

20. Vaidya, H.; Prajapati, A.; Rajani, M.; Sudarsanam, V.; Padh, H.; Goyal, R.K. Beneficial Effects of Swertiamarin on Dyslipidaemia in Streptozotocin-induced Type 2 Diabetic Rats. *Phytotherapy Res.* **2012**, *26*, 1259–1261. [CrossRef] [PubMed]

21. Jaishree, V.; Badami, S. Antioxidant and hepatoprotective effect of swertiamarin from Enicostemma axillare against d-galactosamine induced acute liver damage in rats. *J. Ethnopharmacol.* **2010**, *130*, 103–106. [CrossRef] [PubMed]

22. Jaishree, V.; Narsimha, S. Swertiamarin and quercetin combination ameliorates hyperglycemia, hyperlipidemia and oxidative stress in streptozotocin-induced type 2 diabetes mellitus in wistar rats. *Biomed. Pharmacother.* **2020**, *130*, 110561. [CrossRef]

23. Nakagawa, T.; Yokozawa, T.; Terasawa, K.; Shu, S.; Juneja, L.R. Protective Activity of Green Tea against Free Radical- and Glucose-Mediated Protein Damage. *J. Agric. Food Chem.* **2002**, *50*, 2418–2422. [CrossRef]

24. Vishwakarma, S.; Rajani, M.; Bagul, M.; Goyal, R. A Rapid Method for the Isolation of Swertiamarin from Enicostemma littorale. *Pharm. Biol.* **2004**, *42*, 400–403. [CrossRef]

25. Magora, H.B.; Rahman, M.; Gray, A.I.; Cole, M.D. Swertiamarin from Enicostemma axillare subsp. axillare (Gentianaceae). *Biochem. Syst. Ecol.* **2003**, *31*, 553–555. [CrossRef]

26. Rana, V.S. Separation and Identification of Swertiamarin from Enicostema axillare Lam. Raynal by Centrifugal Partition Chromatography and Nuclear Magnetic Resonance-Mass Spectrometry. *J. Pharm Sci. Emerg. Drugs* **2014**, *1*, 2.

27. Rana, V.S.; Dhanani, T.; Kumar, S. Improved and Rapid HPLC-PDA Method for Identification and Quantification of Swertiamarin in the Aerial Parts of Enicostemma Axillare. *Malaysian J. Pharma Sci.* **2012**, *10*, 1–10.

28. Kumar, S.; Jairaj, V. An Effective Method for Isolation of Pure Swertiamarin from Enicostemma littorale Blume. *Indo Glob. J. Pharm. Sci.* **2018**, *8*, 01–08. [CrossRef]

29. Wu, T.; Li, J.; Li, Y.; Song, H. Antioxidant and Hepatoprotective Effect of Swertiamarin on Carbon Tetrachloride-Induced Hepatotoxicity via the Nrf2/HO-1 Pathway. *Cell. Physiol. Biochem.* **2017**, *41*, 2242–2254. [CrossRef]

30. Nita, M.; Grzybowski, A. The Role of the Reactive Oxygen Species and Oxidative Stress in the Pathomechanism of the Age-Related Ocular Diseases and Other Pathologies of the Anterior and Posterior Eye Segments in Adults. *Oxidative Med. Cell. Longev.* **2016**, *2016*, 1–23. [CrossRef]

31. Fukami, K.; Yamagishi, S.-I.; Ueda, S.; Okuda, S. Role of AGEs in Diabetic Nephropathy. *Curr. Pharm. Des.* **2008**, *14*, 946–952. [CrossRef] [PubMed]

32. Ahmed, N. Advanced glycation endproducts—role in pathology of diabetic complications. *Diabetes Res. Clin. Pract.* **2005**, *67*, 3–21. [CrossRef]

33. Singh, V.P.; Bali, A.; Singh, N.; Jaggi, A.S. Advanced Glycation End Products and Diabetic Complications. *Korean J. Physiol. Pharmacol.* **2014**, *18*, 1–14. [CrossRef] [PubMed]

34. Lunceford, N.; Gugliucci, A. Ilex paraguariensis extracts inhibit AGE formation more efficiently than green tea. *Fitoter.* **2005**, *76*, 419–427. [CrossRef]

35. Lapolla, A.; Flamini, R.; Vedova, A.D.; Senesi, A.; Reitano, R.; Fedele, D.; Basso, E.; Seraglia, R.; Traldi, P. Glyoxal and Methylglyoxal Levels in Diabetic Patients: Quantitative Determination by a New GC/MS Method. *Clin. Chem. Lab. Med.* **2003**, *41*, 1166–1173. [CrossRef]

36. Stinghen, A.E.; Massy, Z.A.; Vlassara, H.; Striker, G.E.; Boullier, A. Uremic Toxicity of Advanced Glycation End Products in CKD. *J. Am. Soc. Nephrol.* **2015**, *27*, 354–370. [CrossRef]

37. Henning, C.; Liehr, K.; Girndt, M.; Ulrich, C.; Glomb, M.A. Extending the Spectrum of α-Dicarbonyl Compounds In Vivo. *J. Biol. Chem.* **2014**, *289*, 28676–28688. [CrossRef]

38. Wang, W.; Yagiz, Y.; Buran, T.J.; Nunes, C.D.N.; Gu, L. Phytochemicals from berries and grapes inhibited the formation of advanced glycation end-products by scavenging reactive carbonyls. *Food Res. Int.* **2011**, *44*, 2666–2673. [CrossRef]

39. Justino, A.B.; Franco, R.R.; Silva, H.C.G.; Saraiva, A.L.; Sousa, R.M.F.; Espindola, F.S. B procyanidins of Annona crassiflora fruit peel inhibited glycation, lipid peroxidation and protein-bound carbonyls, with protective effects on glycated catalase. *Sci. Rep.* **2019**, *9*, 1–15. [CrossRef]

40. Ardestani, A.; Yazdanparast, R. Cyperus rotundus suppresses AGE formation and protein oxidation in a model of fructose-mediated protein glycoxidation. *Int. J. Biol. Macromol.* **2007**, *41*, 572–578. [CrossRef]

41. Ahmad, S.; Shahab, U.; Baig, M.H.; Khan, M.S.; Khan, M.S.; Srivastava, A.K.; Saeed, M. Moinuddin Inhibitory Effect of Metformin and Pyridoxamine in the Formation of Early, Intermediate and Advanced Glycation End-Products. *PLoS ONE* **2013**, *8*, e72128. [CrossRef]

42. Ishibashi, Y.; Matsui, T.; Takeuchi, M.; Yamagishi, S. Metformin Inhibits Advanced Glycation End Products (AGEs)-induced Renal Tubular Cell Injury by Suppressing Reactive Oxygen Species Generation via Reducing Receptor for AGEs (RAGE) Expression. *Horm. Metab. Res.* **2012**, *44*, 891–895. [CrossRef]

43. Meeprom, A.; Sompong, W.; Chan, C.B.; Adisakwattana, S. Isoferulic Acid, a New Anti-Glycation Agent, Inhibits Fructose- and Glucose-Mediated Protein Glycation In Vitro. *Molecules* **2013**, *18*, 6439–6454. [CrossRef]

44. Jairajpuri, D.S.; Fatima, S.; Saleemuddin, M. Immunoglobulin glycation with fructose: A comparative study. *Clin. Chim. Acta* **2007**, *378*, 86–92. [CrossRef]

45. Allarakha, S.; Ahmad, P.; Ishtikhar, M.; Zaheer, M.S.; Siddiqi, S.S.; Moinuddin; Ali, A. Fructosylation generates neo-epitopes on human serum albumin. *IUBMB Life* **2015**, *67*, 338–347. [CrossRef]

46. Liu, J.; Xing, X.; Jing, H. Differentiation of glycated residue numbers on heat-induced structural changes of bovine serum albumin. *J. Sci. Food Agric.* **2018**, *98*, 2168–2175. [CrossRef] [PubMed]

47. Yan, H.; Harding, J.J. Glycation-induced inactivation and loss of antigenicity of catalase and superoxide dismutase. *Biochem. J.* **1997**, *328*, 599–605. [CrossRef]

48. Ha, H.; Hwang, I.-A.; Park, J.H.; Lee, H.B. Role of reactive oxygen species in the pathogenesis of diabetic nephropathy. *Diabetes Res. Clin. Pract.* **2008**, *82*, S42–S45. [CrossRef] [PubMed]

49. Liu, B.-F.; Miyata, S.; Hirota, Y.; Higo, S.; Miyazaki, H.; Fukunaga, M.; Hamada, Y.; Ueyama, S.; Muramoto, O.; Uriuhara, A.; et al. Methylglyoxal induces apoptosis through activation of p38 mitogen-activated protein kinase in rat mesangial cells. *Kidney Int.* **2003**, *63*, 947–957. [CrossRef]

50. Oya, T.; Hattori, N.; Mizuno, Y.; Miyata, S.; Maeda, S.; Osawa, T.; Uchida, K. Methylglyoxal Modification of Protein. *J. Biol. Chem.* **1999**, *274*, 18492–18502. [CrossRef]

51. Wilker, S.C.; Chellan, P.; Arnold, B.M.; Nagaraj, R.H. Chromatographic Quantification of Argpyrimidine, a Methylglyoxal-Derived Product in Tissue Proteins: Comparison with Pentosidine. *Anal. Biochem.* **2001**, *290*, 353–358. [CrossRef]

52. Padival, A.K.; Crabb, J.W.; Nagaraj, R.H. Methylglyoxal modifies heat shock protein 27 in glomerular mesangial cells. *FEBS Lett.* **2003**, *551*, 113–118. [CrossRef]

53. Niki, E. Lipid peroxidation products as oxidative stress biomarkers. *BioFactors* **2008**, *34*, 171–180. [CrossRef]

54. Ayala, A.; Muñoz, M.F.; Argüelles, S. Lipid Peroxidation: Production, Metabolism, and Signaling Mechanisms of Malondialdehyde and 4-Hydroxy-2-Nonenal. *Oxid. Med. Cell. Longev.* **2014**, *2014*, 1–31. [CrossRef]

55. Sompong, W.; Cheng, H.; Adisakwattana, S. Ferulic acid prevents methylglyoxal-induced protein glycation, DNA damage, and apoptosis in pancreatic β-cells. *J. Physiol. Biochem.* **2016**, *73*, 121–131. [CrossRef]

56. Reiniger, N.; Lau, K.; McCalla, D.; Eby, B.; Cheng, B.; Lu, Y.; Qu, W.; Quadri, N.; Ananthakrishnan, R.; Furmansky, M.; et al. Deletion of the Receptor for Advanced Glycation End Products Reduces Glomerulosclerosis and Preserves Renal Function in the Diabetic OVE26 Mouse. *Diabetes* **2010**, *59*, 2043–2054. [CrossRef]

57. Ye, S.D.; Zheng, M.; Zhao, L.L.; Qian, Y.; Yao, X.M.; Ren, A.; Li, S.M.; Jing, C.Y. Intensive insulin therapy decreases urinary MCP-1 and ICAM-1 excretions in incipient diabetic nephropathy. *Eur. J. Clin. Investig.* **2009**, *39*, 980–985. [CrossRef] [PubMed]

58. Yamagishi, S.-I.; Matsui, T. Advanced Glycation end Products, Oxidative Stress and Diabetic Nephropathy. *Oxidative Med. Cell. Longev.* **2010**, *3*, 101–108. [CrossRef]

59. Menon, M.C.; Ross, M.J. Epithelial-to-mesenchymal transition of tubular epithelial cells in renal fibrosis: A new twist on an old tale. *Kidney Int.* **2016**, *89*, 263–266. [CrossRef] [PubMed]
60. Zhang, X.; Liang, D.; Guo, L.; Liang, W.; Jiang, Y.; Li, H.; Zhao, Y.; Lu, S.; Chi, Z.-H. Curcumin protects renal tubular epithelial cells from high glucose-induced epithelial-to-mesenchymal transition through Nrf2-mediated upregulation of heme oxygenase-1. *Mol. Med. Rep.* **2012**, *12*, 1347–1355. [CrossRef] [PubMed]
61. Burns, W.C.; Twigg, S.M.; Forbes, J.M.; Pete, J.; Tikellis, C.; Thallas-Bonke, V.; Thomas, M.C.; Cooper, M.E.; Kantharidis, P. Connective Tissue Growth Factor Plays an Important Role in Advanced Glycation End Product–Induced Tubular Epithelial-to-Mesenchymal Transition: Implications for Diabetic Renal Disease. *J. Am. Soc. Nephrol.* **2006**, *17*, 2484–2494. [CrossRef]
62. Lv, Z.-M.; Wang, Q.; Wan, Q.; Lin, J.-G.; Hu, M.-S.; Liu, Y.-X.; Wang, R. The Role of the p38 MAPK Signaling Pathway in High Glucose-Induced Epithelial-Mesenchymal Transition of Cultured Human Renal Tubular Epithelial Cells. *PLoS ONE* **2011**, *6*, e22806. [CrossRef]
63. Lee, Y.J.; Han, H.J. Troglitazone ameliorates high glucose-induced EMT and dysfunction of SGLTs through PI3K/Akt, GSK-3β, Snail1, and β-catenin in renal proximal tubule cells. *Am. J. Physiol. Physiol.* **2010**, *298*, F1263–F1275. [CrossRef]
64. Hung, T.-J.; Chen, W.-M.; Liu, S.-F.; Liao, T.-N.; Lee, T.-C.; Chuang, L.-Y.; Guh, J.-Y.; Hung, C.-Y.; Hung, Y.-J.; Chen, P.-Y.; et al. 20-Hydroxyecdysone attenuates TGF-β1-induced renal cellular fibrosis in proximal tubule cells. *J. Diabetes Complicat.* **2012**, *26*, 463–469. [CrossRef]
65. Parwani, K.; Mandal, P. Role of advanced glycation end products and insulin resistance in diabetic nephropathy. *Arch. Physiol. Biochem.* **2020**, 1–13. [CrossRef]
66. Lu, Q.; Wang, W.; Zhang, M.; Ma, Z.; Qiu, X.; Shen, M.; Yin, X. ROS induces epithelial-mesenchymal transition via the TGF-β1/PI3K/Akt/mTOR pathway in diabetic nephropathy. *Exp. Ther. Med.* **2018**, *17*, 835–846. [CrossRef]
67. Jiang, T.; Huang, Z.; Lin, Y.; Zhang, Z.; Fang, D.; Zhang, D.D. The Protective Role of Nrf2 in Streptozotocin-Induced Diabetic Nephropathy. *Diabetes* **2010**, *59*, 850–860. [CrossRef]
68. Kie, J.H.; Kapturczak, M.H.; Traylor, A.; Agarwal, A.; Hill-Kapturczak, N. Heme Oxygenase-1 Deficiency Promotes Epithelial-Mesenchymal Transition and Renal Fibrosis. *J. Am. Soc. Nephrol.* **2008**, *19*, 1681–1691. [CrossRef]
69. Fatchiyah, F.; Hardiyanti, F.; Widodo, N. Selective Inhibition on RAGE-binding AGEs Required by Bioactive Peptide Alpha-S2 Case in Protein from Goat Ethawah Breed Milk: Study of Biological Modeling. *Acta Inform. Medica* **2015**, *23*, 90–96. [CrossRef]
70. Kshirsagar, P.R.; Pai, S.R.; Nimbalkar, M.S.; Gaikwad, N.B. RP-HPLC analysis of seco-iridoid glycoside swertiamarin from differentSwertiaspecies. *Nat. Prod. Res.* **2016**, *30*, 865–868. [CrossRef] [PubMed]
71. McPherson, J.D.; Shilton, B.H.; Walton, D.J. Role of fructose in glycation and cross-linking of proteins. *Biochemistry* **1988**, *27*, 1901–1907. [CrossRef]
72. Riss, T.L.; Moravec, R.A.; Niles, A.L.; Duellman, S.; Benink, H.A.; Worzella, T.J.; Minor, L. *Assay Guidance Manual*; Eli Lilly & Company and the National Center for Advancing Translational Sciences: Bethesda, MD, USA, 2004; pp. 296–302.
73. Karbowski, M.; Kurono, C.; Wozniak, M.; Ostrowski, M.; Teranishi, M.; Nishizawa, Y.; Usukura, J.; Soji, T.; Wakabayashi, T. Free radical–induced megamitochondria formation and apoptosis. *Free. Radic. Biol. Med.* **1999**, *26*, 396–409. [CrossRef]
74. Tukozkan, N.; Erdamar, H.; Seven, I. Measurement of total malondialdehyde in plasma and tissues by high-performance liquid chromatography and thiobarbituric acid assay. *Firat Tip Dergisi.* **2006**, *11*, 88–92.
75. Kema, V.H.; Khan, I.; Jamal, R.; Vishwakarma, S.K.; Reddy, C.L.; Parwani, K.; Patel, F.; Patel, D.; Khan, A.A.; Mandal, P. Protective Effects of Diallyl Sulfide Against Ethanol-Induced Injury in Rat Adipose Tissue and Primary Human Adipocytes. *Alcohol. Clin. Exp. Res.* **2017**, *41*, 1078–1092. [CrossRef]
76. Tsoporis, J.; Izhar, S.; Leong-Poi, H.; Desjardins, J.-F.; Huttunen, H.; Parker, T. S100B Interaction with the Receptor for Advanced Glycation End Products (RAGE). *Circ. Res.* **2010**, *106*, 93–101. [CrossRef]
77. Apweiler, R. UniProt: The Universal Protein knowledgebase. *Nucleic Acids Res.* **2004**, *32*, 115–119. [CrossRef]
78. Xu, D.; Young, J.H.; Krahn, J.M.; Song, D.; Corbett, K.D.; Chazin, W.J.; Pedersen, L.C.; Esko, J.D. Stable RAGE-Heparan Sulfate Complexes Are Essential for Signal Transduction. *ACS Chem. Biol.* **2013**, *8*, 1611–1620. [CrossRef]
79. *Schrödinger Release, Version 12.7.156*; Maestro Schrödinger LLC: New York, NY, USA, 2021.
80. Sastry, G.M.; Adzhigirey, M.; Day, T.; Annabhimoju, R.; Sherman, W. Protein and ligand preparation: Parameters, protocols, and influence on virtual screening enrichments. *J. Comput. Mol. Des.* **2013**, *27*, 221–234. [CrossRef]
81. Greenwood, J.R.; Calkins, D.; Sullivan, A.P.; Shelley, J.C. Towards the comprehensive, rapid, and accurate prediction of the favorable tautomeric states of drug-like molecules in aqueous solution. *J. Comput. Mol. Des.* **2010**, *24*, 591–604. [CrossRef] [PubMed]
82. Jacobson, M.P.; Pincus, D.L.; Rapp, C.S.; Day, T.J.F.; Honig, B.; Shaw, D.E.; Friesner, R.A. A hierarchical approach to all-atom protein loop prediction. *Proteins: Struct. Funct. Bioinform.* **2004**, *55*, 351–367. [CrossRef] [PubMed]

Article

Constituents of *Aquilaria sinensis* Leaves Upregulate the Expression of Matrix Metalloproteases 2 and 9

Sui-Wen Hsiao [1,†], Yu-Chin Wu [2,†], Hui-Ching Mei [3], Yu-Hsin Chen [4], George Hsiao [5,*] and Ching-Kuo Lee [1,2,6,7,*]

[1] Ph.D. Program in Drug Discovery and Development Industry, College of Pharmacy, Taipei Medical University, 250 Wu Xin Street, Taipei 11031, Taiwan; suifeng0506@gmail.com

[2] Graduate Institute of Pharmacognosy, Taipei Medical University, 250 Wu Xin Street, Taipei 11031, Taiwan; tmc761038@tmu.edu.tw

[3] Department of Science Education, National Taipei University of Education, Taipei 10671, Taiwan; hcmei@tea.ntue.edu.tw

[4] Taichung District Agricultural Research and Extension Station, Council of Agriculture, Executive Yuan, Taichung 42081, Taiwan; ychen@tdais.gov.tw

[5] Department of Pharmacology, Taipei Medical University, 250 Wu Xin Street, Taipei 110, Taiwan

[6] School of Pharmacy, Taipei Medical University, 250 Wu Xin Street, Taipei 110, Taiwan

[7] Department of Chemistry, Chung Yuan Christian University, Zhongbei Road, Zhongli District, Taoyuan City 320314, Taiwan

* Correspondence: geohsiao@tmu.edu.tw (G.H.); cklee@tmu.edu.tw (C.-K.L.); Tel.: +886-2-27361661 (ext. 6150) (C.-K.L.)

† These authors contributed equally to this work.

Citation: Hsiao, S.-W.; Wu, Y.-C.; Mei, H.-C.; Chen, Y.-H.; Hsiao, G.; Lee, C.-K. Constituents of *Aquilaria sinensis* Leaves Upregulate the Expression of Matrix Metalloproteases 2 and 9. *Molecules* **2021**, *26*, 2537. https://doi.org/10.3390/molecules26092537

Academic Editor: Raffaele Pezzani

Received: 20 March 2021
Accepted: 24 April 2021
Published: 26 April 2021

Publisher's Note: MDPI stays neutral with regard to jurisdictional claims in published maps and institutional affiliations.

Abstract: In this novel study, we isolated 28 compounds from the leaves of *Aquilaria sinensis* (Lour.) Gilg based on a bioassay-guided procedure and also discovered the possible matrix metalloprotease 2 (MMP-2) and 9 (MMP-9) modulatory effect of pheophorbide A (PA). To evaluate the regulatory activity on MMP-2 and MMP-9, the HT-1080 human fibrosarcoma cells were treated with various concentrations of extracted materials and isolated compounds. PA was extracted by methanol from the leaves of *A. sinensis* and separated from the fraction of the partitioned ethyl acetate layer. PA is believed to be an active component for MMP expression since it exhibited significant stimulation on MMP-2 and proMMP-9 activity. When treating with 50 μM of PA, the expression of MMP-2 and MMP-9 were increased 1.9-fold and 2.3-fold, respectively. PA also exhibited no cytotoxicity against HT-1080 cells when the cell viability was monitored. Furthermore, no significant MMP activity was observed when five PA analogues were evaluated. This study is the first to demonstrate that C-17 of PA is the deciding factor in determining the bioactivity of the compound. The MMP-2 and proMMP-9 modulatory activity of PA indicate its potential applications for reducing scar formation and comparative medical purposes.

Keywords: *Aquilaria sinensis*; pheophorbide A; MMP-2; MMP-9; HT-1080

1. Introduction

Aquilaria sinensis (Lour.) Gilg is an evergreen woody plant with high economic value that is widely distributed in the tropical area of China. Possessing analgesic, sedative, and antiemetic effects, the resinous heartwood of *A. sinensis*, which is the major source of agarwood in China, has been used as traditional medicine for centuries [1]. Apart from the agarwood, the non-medical parts of leaves and flowers are consumed as a healthy herbal tea as well. The bioactivity of agarwood has been highly valued for years, while studies on the leaves have still been limited. The major metabolic components of *A. sinensis* are polysaccharides, amino acids, flavonoids and their glycosides, phenols, and xanthones [2–5]. Previous studies have reported anti-inflammatory and analgesic activity [6,7], laxative activity [8], an α-glucosidase inhibitory effect [2], a nitric oxide inhibitory effect [9], anticancer activity [10,11], and blood glucose regulatory effects [12] of the isolated compounds and extracts of *A. sinensis* leaves. Given the abundance and availability of leaves, there can potentially be a variety of pharmacological applications of the leaves from *A. sinensis*.

Matrix metalloproteinases (MMPs) are a family of enzymes important in the regulation of developmental and homeostatic remodeling of the extracellular matrix (ECM). Family members of MMPs can proteolytically degrade all components of the ECM and are subdivided into collagenases (MMP-1, -8, -13), gelatinases (MMP-2, -9), stromelysins (MMP-3, -10, -11), matrilysin (MMP-7), elastase (MMP-12), and MT-MMPs (MMP-14, -15, -16, -17) [13]. The MMP-2 and MMP-9, also known as gelatinase A and gelatinase B, respectively, share some of the same substrates, such as gelatins, collagens IV and V, aggrecan, elastin, and vitronectin [14]. MMP-2 can also degrade collagens I, VII, X, XI, β-amyloid protein precursor, and other substrates [14]. To investigate the effect of components isolated from *A. sinensis* on MMPs expression activity, the widely used HT-1080 cell model was used in this study. The HT1080 cell line is derived from a human fibrosarcoma to mimic stromal fibroblasts and is sustained to express MMP-2 and MMP-9 [15,16].

In the present study, the phytochemical properties of the components isolated from *A. sinensis* leaves and their regulatory effects on MMP-2 and MMP-9 expression were investigated.

2. Results

2.1. Bioassay Guided Compound Isolation from A. sinensis Leaves

In a preliminary biological evaluation, methanolic crude extracts of *A. sinensis* leaves demonstrated the ability to promote MMP-2 and MMP-9 activity at a concentration of 100 μg/mL. An investigation of the active principles of this plant was thus undertaken by using a bioassay guided method. After being partitioned into water (91.34 g), *n*-butanol (77.58 g), and ethyl acetate (EA) layers (225.27 g), the effects of these layers (100 μg/mL) on MMPs expression were evaluated by gelatin zymography. The EA layer was the most potent (Figure 1).

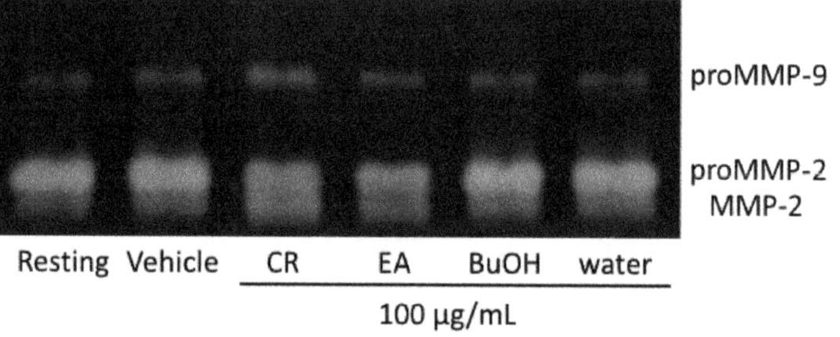

Figure 1. Effects of crude extracts and partitioned parts of *A. sinensis* leaves on MMP-2 and MMP-9 activity. Results of gelatin zymography. CR, crude extracts; EA, ethyl acetate layer; BuOH, butanol layer; water, water layer.

In order to find the constituents of *A. sinensis* leaves that regulate MMP-2 and MMP-9 activity, we carried out a biologically guided purification strategy. The EA layer in the methanol (MeOH) extract of *A. sinensis* leaves showed potential effects. The subsequent separation and identification of biologically active ingredients were thus focused on this layer. The most effective fractions were further fractionated and purified by repeated chromatography on silica gel columns and semi-preparative HPLC with refractive index detector to obtain twenty-eight compounds (Figure 2). The representative HPLC trace of compound AQ20 is shown in Figure 3.

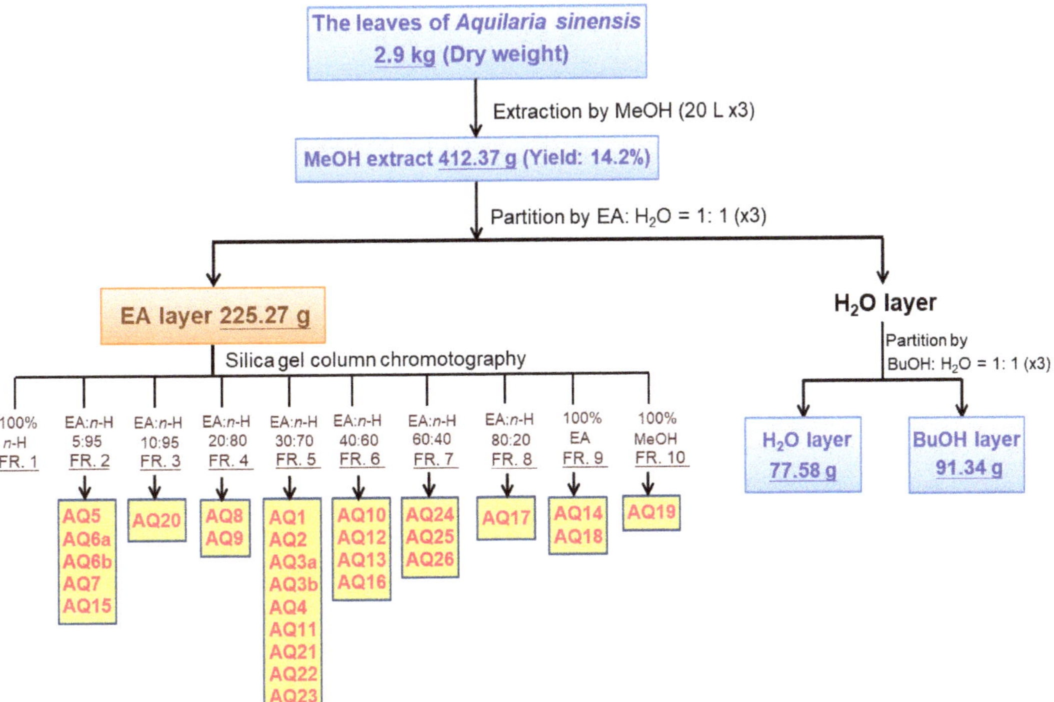

Figure 2. Flow diagram of bioassay-guided procedure. Bioassay-guided fractionation and isolation of the MeOH extract of the leaves of *A. sinensis* resulted in the isolation of 28 compounds.

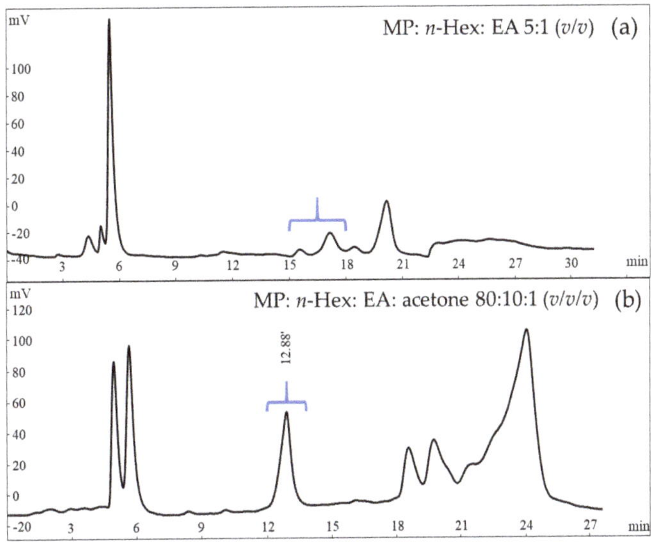

Figure 3. Representative HPLC trace of compound AQ20. (**a**) Eluent from RT 15-18 min was collected in one bottle and subjected to semi-preparative HPLC for further separation. (**b**) Peak at RT 12.9 min was collected and identified as AQ20. MP, mobile phase condition.

2.2. Compounds Isolated from the Ethyl Acetate Layer of of Aquilaria sinensis Leaves

To analyze the structures of the compounds separated from various fractions of the EA layer, the spectra from NMR and MS spectroscopic analysis were compared with the previously reported data. The classifications for identified isolated compounds were recognized as:

Terpenoids: loliolide (AQ4) [17], phytol (AQ7) [18], squalene (AQ15) [19]; Ligand: syringaresinol (AQ12) [20]; Phenolics: *p*-hydroxybenzoic acid (AQ8) [18], β-truxinic acid (AQ10) [21], *p*-hydroxyphenethyl *trans*-ferulate (AQ13) [22]; Lipids: methyl (6Z,9Z,12Z)-octadecatrienoate (AQ5) [23], palmitic acid and oleic acid (AQ6) [24,25], methyl-2-hydroxypentanedioate (AQ17) [26], glyceryl -9Z-octadecenate (AQ21) [27], glyceryl 1-monononadecylate (AQ22) [28], 1-[nonadeca-(9Z,12Z)-dienoyl]-sn-glycerol (AQ23) [29]; Chromones: 7-hydroxy-6-methoxy-2-[2-(4'-methoxyphenyl)ethyl] chromone (AQ11) [30], 6-hydroxy-7-methoxy-2-[2-(3'-hydroxy-4' -methoxyphenyl)ethyl] chromone (AQ16) [30]; Flavonoids: genkwanin (AQ1) [31], apigenin (AQ2) [32], velutin and pilloin (AQ3) [33,34];

Glycosides: 2-phenylethyl β-glucopyranoside (AQ14) [35], benzyl β-glucopyranoside (AQ18) [36], iriflophenone 2-*O*-α-rhamnoside (AQ19) [8], 1-*O*-caprylyl-2-*O*-linolenoyl-3-*O*-(β-D-glucopyranosyl)- *rac*-glycerol (AQ24) [37], 1-*O*-caprylyl-2-*O*-linoleoyl-3-*O*-(β-D-glucopyranosyl)- *rac*-glycerol (AQ25) [37], 1,2-*O*-dimyristoyl-3-*O*-(β-D-glucopyranosyl)-*rac*-glycerol (AQ26) [37]; Pheophorbides: methyl pheophorbide A (AQ9) [38], and pheophorbide A (PA) (AQ20) (Figure S1) [39].

2.3. Evaluation of Regulatory Effects on MMP-2 and MMP-9 Expression in HT-1080 Cells

The MMP-2 and MMP-9 modulating activity of the pure isolated compounds were measured at the concentrations of 50 μM in comparison with blank groups. Gelatin zymography was performed in triplicate, and the data were obtained from independent zymogram. As shown in Figure 4, the compound AQ20, which was identified as PA, distinctly increased the expression of MMP-2 and proMMP-9 by 1.9-fold and 2.3-fold elevation, respectively, with a 99.9% confidence level. While the protein levels of all other isolated compounds showed no significant difference between that of blank. To determine the cell toxicity of PA, HT-1080 cells were subjected to MTT cell viability assay in the presence of 50 μM PA. The cell viability of the HT-1080 cells was 94%, indicating that PA had no cell toxicity. With the exception of compound AQ3 where viability was 65%, all other compounds showed more than 75% cell viability.

Numerous concentrations were used to further investigate the dose–response effect of PA (Table 1 and Figure 5). The data on gelatin zymography were obtained from independent zymogram. For MMP-9, the expression levels were significantly elevated ($p < 0.001$) at concentrations of 25 and 50 μM. For MMP-2, the stimulating effect was observed at all of the tested concentrations. The *p*-value was lower than 0.05 at the concentration of 2.5 μM and lower than 0.001 at concentrations from 5 to 50 μM. Positive dose–response relationships between PA and the MMPs were clarified since higher expression levels were obtained along with increased concentrations of PA.

Along with the elevating effects of PA on MMP-2 and MMP-9 activity, the effects on MMP activity of five analogues of PA—bidenphytins A (20A), 13^2-hydroxypheophytin b (20B), (13^2S)-13^2-hydroxypheophytin a (20C), bidenphytins B (20D), and (13^2R)-13^2-hydroxypheophytin a (20E) (Figure 6) [40]—were also measured. The tested PA analogues were isolated in lab from leaves of *Biden pilosa* in previous study. However, the MMP-2 and MMP-9 expression levels showed no significant differences between each of the five PA analogues and blank, indicating that the five analogues had no MMP regulatory activity (Figure S2).

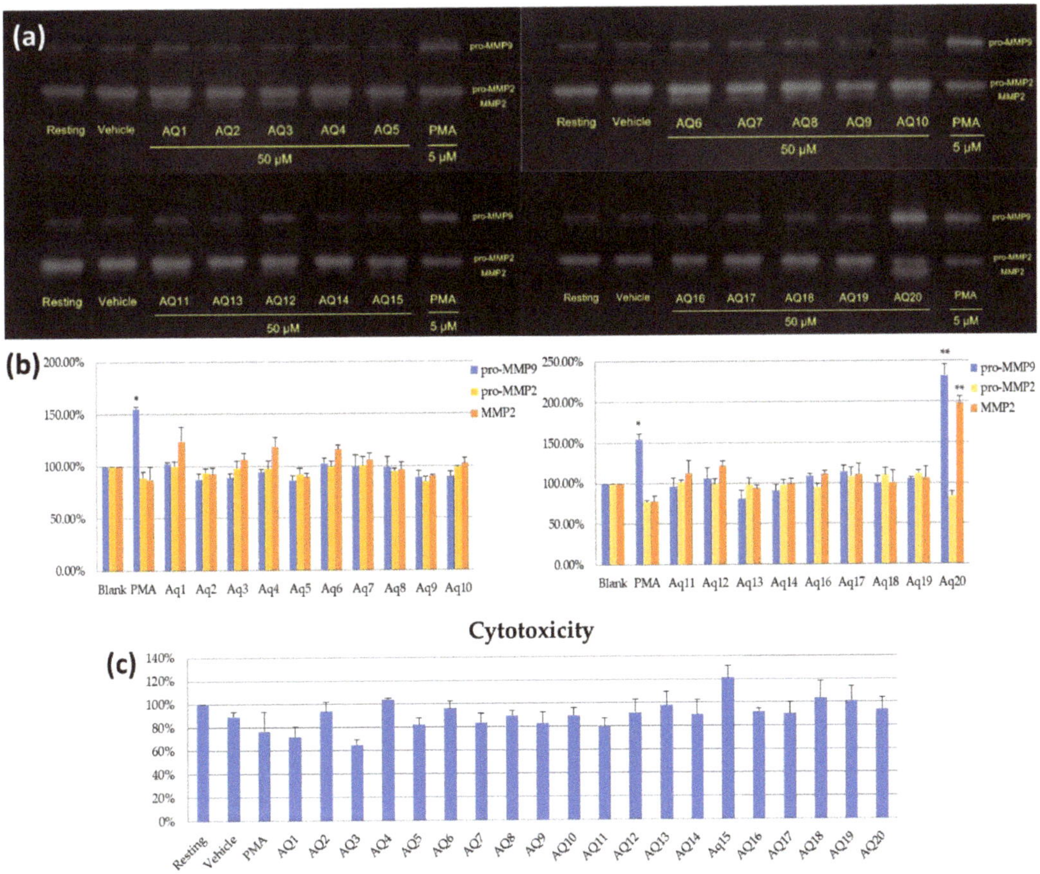

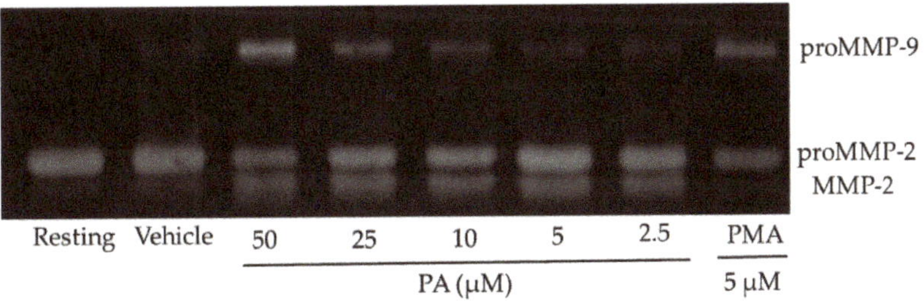

Figure 4. Effects of the isolated compounds AQ—AQ20 on MMP-2 and MMP-9 activity and cytotoxicity. (**a**) Results of gelatin zymography. (**b**) Activity of the compounds on MMPs expression. (**c**) Cell viability test. Data represent the mean ± SD ($n = 3$). PMA (5 µM); AQ1—AQ20 (50 µM); * $p < 0.05$, ** $p < 0.001$.

Figure 5. Effects of PA at various concentration.

Table 1. MMP-2 and MMP-9 activation by PA at various concentrations.

Concentration (μM)	Expression Level (%) [1]	
	pro-MMP9	MMP-2
50	230 ± 11.5 **	188 ± 6.5 **
25	146 ± 3 **	163 ± 2.5 **
10	118 ± 16	147 ± 7 **
5	104 ± 4	135 ± 2 **
2.5	104 ± 4	132 ± 11 *

[1] The expression level was compared with blank group (100%). * $p < 0.05$, ** $p < 0.001$.

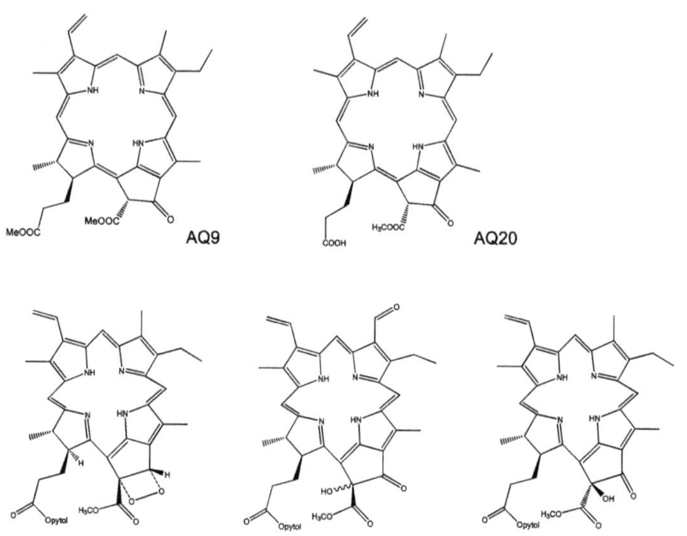

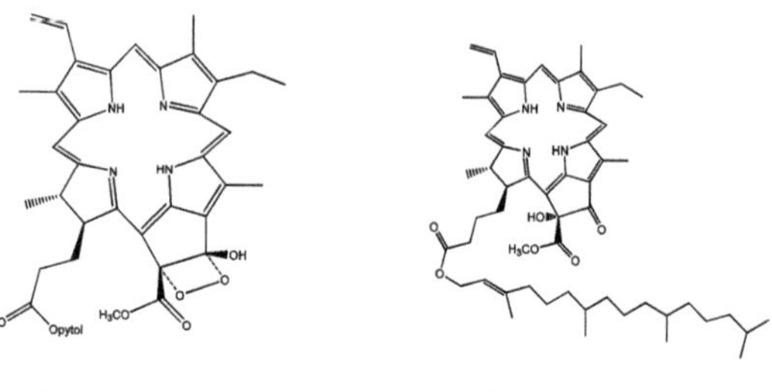

Figure 6. Structures of PA and its analogues. PA (AQ20), PA analogues (AQ9, 20A–E).

3. Discussion

In this study, 28 compounds were isolated from the leaves of *A. sinensis*, including three terpenoids, one ligand, three phenolics, seven lipids, two chromones, four flavonoids, six glycosides, and two pheophorbides. Among the 28 isolated pure compounds, PA presented effective promoting activity on MMP-2 and MMP-9.

Increased expression of MMP-2 and MMP-9 was reported to induce tumor angiogenesis, inflammation, and cancer metastasis [41]. Additionally, both MMP-2 and MMP-9 participate in the process of wound repair and tissue remodeling [42]. Prolonged and sustained MMP-2 expression during dermal wound repair suggested that MMP-2 may play a role in granulation and early remodeling phases of repair [43]. Upregulation of MMP-9 expression in the regeneration phase of wound healing can possibly prevent scar formation [44]. The MMP-2 and MMP-9 upregulatory activity revealed the potential scar reducing and comparative medical applications of PA.

PA is a metabolic breakdown product of chlorophyll in plants. Until now, various bioactivities of PA have been reported, including anticancer, antiviral, anti-inflammatory, antioxidant, immunostimulatory, and anti-parasite activity [45]. PA also serves as a good photosensitizer that absorbs energy from light to generate reactive oxygen species [46]. The regulatory effect of PA on MMP-2 and MMP-9 expression has also been reported. In ultraviolet (UV) B-exposed CCD-986sk human fibroblasts, PA and PA-derivatives including pyroPA and PA methyl ester reduced UVB-induced MMP-1 secretion and mRNA expression of MMP-1, -2, and -9 [47]. In the present study, expression of MMP-2 and MMP-9 were upregulated by PA in HT-1080 cells. The opposite result may due to the effect of UV irradiation on PA.

In contrast with PA, the five analogues showed no effect on MMP-2 and MMP-9 expression. In addition, the compound AQ9, identified as methyl PA, did not induce significant activity either. This can potentially be attributed to the phytol group of the analogues and the methyl group on C-17 of AQ9. Based on structure–activity relationships, the structure of the substituent on C-17 was suggested to be a crucial factor influencing the activity on MMP-2 and MMP-9. This result could potentially contribute to further investigations of the bioactivity of compounds similar in structure to PA.

4. Materials and Methods

4.1. General

Optical rotations were measured on a JASCO P-1020 digital polarimeter (Tokyo, Japan). ^{1}H- and ^{13}C-NMR were acquired on a Bruker DRX-500 SB spectrometer (Ettlingen, Germany). High-resolution and low-resolution mass spectra were obtained using an ABI Q-Star XLQ-TOF (Applied Biosystems, Foster, CA, USA) and ABI API 4000 Q-TRAP (Foster City, CA, USA), respectively. IR spectra were recorded on a JASCO FT/IR 4100 spectrometer (Tokyo, Japan). UV spectra were measured on a Thermo Helios μ spectrophotometer (Bellefonte, CA, USA). Silica gel (40–63 μm, Merck, Darmstadt, Germany) was used for gravity column chromatography. Pre-coated silica gel plates (Si 60 F$_{254}$, 0.2 mm, Merck, Darmstadt, Germany) were used for analytical TLC. Semi-preparative HPLC was performed on a Hitachi L-7110 HPLC with a refractive index detector (Thermo Separation Products, Sunnyvale, CA, USA). A Phenomenex® Luna silica column (5 μm, 10 × 250 mm, Phenomenex, CA, USA) and a VP 250/10 NUCLEODUR C18 HTec column (5 μm, 10 × 250 mm, MACHEREY-NAGEL, Düren, Germany) were used.

4.2. Plant Material

The leaves of *A. sinensis* were collected from Changhua County in June 2002 and were identified by Dr. S. Y. Chen at the Council of Agriculture, Executive Yuan. Voucher Specimens (No. CKL06012002) were deposited at the School of Pharmacy, Taipei Medical University, Taipei, Taiwan.

4.3. Extraction and Isolation

The dried leaves (2.9 kg) of *A. sinensis* were extracted three times with 20 L of MeOH, then partitioned between EA and water (1:1, *v/v*). Subsequently, the dried EA layer (190 g) was pre-coated using 250 g silica gel (70–230 mesh) applied onto a 2500 g silica gel open column (230–400 mesh) and eluted with mixtures of *n*-hexane (*n*-Hex), EA, and MeOH in a stepwise gradient mode. Ten conditions were used for elution and every 1000 mL of eluent was collected in one bottle. Eventually, 164 bottles were collected and combined into 10 fractions based on the elution condition. Each bottle of collected eluent was checked for its composition by TLC dipped in 10% H$_2$SO$_4$ in ethanol for the basis of further normal-phase or reversed-phase semi-preparative HPLC analysis. Purity of the compounds was checked by NMR and was higher than 80%. The structure of the compound was identified by NMR and MS.

The fraction 2 was eluted with *n*-Hex/EA (95:5, *v/v*) and 20 bottles were collected. The eluents were purified by HPLC on a semi-preparative normal-phase column at a flow rate of 3 mL/min and afforded **5** (59.0 mg, RT 22.2 min with *n*-Hex/EA (99:1, *v/v*)), **6a** + **6b** (323.4 mg, RT 18.6 min with *n*-Hex/EA (90:10, *v/v*)), **7** (57.2 mg, RT 27.9 min with *n*-Hex/EA (90:10, *v/v*)), and **15** (249.3 mg, RT 6.13 min with *n*-Hex/EA (99:1, *v/v*)).

The fraction 3 was eluted with *n*-Hex/EA (90:10, *v/v*) and 20 bottles were collected. The eluents were purified by HPLC on a semi-preparative normal-phase column at a flow rate of 3 mL/min and afforded **20** (19.3 mg, RT 12.9 min with *n*-Hex/EA/acetone (80:10:1, *v/v/v*)).

The fraction 4 was eluted with *n*-Hex/EA (80:20, *v/v*) and 20 bottles were collected. The eluents were purified by HPLC on a semi-preparative normal-phase column at a flow rate of 3 mL/min and afforded **8** (47.7 mg, RT 12.2 min with *n*-Hex/EA (70:30, *v/v*)) and **9** (13.5 mg, RT 21.7 min with *n*-Hex/EA (70:30, *v/v*)).

The fraction 5 was eluted with *n*-Hex/EA (70:30, *v/v*) and 19 bottles were collected. The eluents were purified by HPLC on a semi-preparative normal-phase column at a flow rate of 3 mL/min and afforded **1** (12.2 mg, RT 14.6 min with *n*-Hex/EA (60:40, *v/v*)), **2** (21.8 mg, RT 19.9 min with *n*-Hex/EA (60:40, *v/v*)), **3a** + **3b** (56.8 mg, RT 21.9 min with *n*-Hex/EA (60:40, *v/v*)), **4** (62.0 mg, RT 20.8 min with *n*-Hex/EA (50:50, *v/v*)), **11** (17.4 mg, RT 26.2 min with *n*-Hex/EA (30:70, *v/v*)), **21** (157.1 mg, RT 67.2 min with *n*-Hex/EA/acetone (71:28:1, *v/v/v*)), **22** (64.2 mg, RT 72.1 min with *n*-Hex/EA/acetone (71:28:1, *v/v/v*)), and **23** (206.0 mg, RT 77.7 min with *n*-Hex/EA/acetone (71:28:1, *v/v/v*)).

The fraction 6 was eluted with *n*-Hex/EA (60:40, *v/v*) and 17 bottles were collected. The eluents were purified by HPLC on a semi-preparative normal-phase column at a flow rate of 3 mL/min and afforded **10** (3.7 mg, RT 8.2 min with *n*-Hex/EA (30:70, *v/v*)), **12** (43.5 mg, RT 15.4 min with *n*-Hex/EA/acetone (64:32:4, *v/v/v*)), **13** (6.3 mg, RT 18.3 min with *n*-Hex/EA/acetone (64:32:4, *v/v/v*)), and **16** (20.2 mg, RT 24.3 min with *n*-Hex/EA/acetone (25:68:7, *v/v/v*)).

The fraction 7 was eluted with *n*-Hex/EA (40:60, *v/v*) and 20 bottles were collected. The eluents were purified by HPLC on a semi-preparative reversed-phase column at a flow rate of 2 mL/min and afforded **24** (42.7 mg, RT 8.2 min with MeOH/ACN (40:60, *v/v*)), **25** (39.1 mg, RT 37.8 min with MeOH/ACN (40:60, *v/v*)), and **26** (68.6 mg, RT 51.9 min with MeOH/ACN (40:60, *v/v*)).

The fraction 8 was eluted with *n*-Hex/EA (20:80, *v/v*) and 19 bottles were collected. The eluents were purified by HPLC on a semi-preparative reversed-phase column at a flow rate of 2 mL/min and afforded **17** (4.1 mg, RT 10.9 min with MeOH/H_2O (40:60, *v/v*)).

The fraction 9 was eluted with 100% EA and 11 bottles were collected. The eluents were purified by HPLC on a semi-preparative reversed-phase column at a flow rate of 2 mL/min and afforded **14** (5.7 mg, RT 33.1 min with MeOH/ACN/H_2O (12.5:12.6:75, *v/v/v*)) and **18** (11.8 mg, RT 24.3 min with MeOH/H_2O (1:3, *v/v*)).

The fraction 10 was eluted with 100% MeOH and 12 bottles were collected. The eluents were purified by HPLC on a semi-preparative reversed-phase column at a flow rate of 1.5 mL/min and afforded **19** (9.6 mg, RT 29.1 min with MeOH/ACN/H_2O (1:1:5, *v/v/v*)).

4.4. Cell Culture

HT1080 human fibrosarcoma cells were purchased from American Type Culture Collection (ATCC: CCL-121) and were maintained in a humidified incubator at 37 °C in 5% CO_2/95% air. The cells were cultured in RPMI-1640 medium (Gibco) supplemented with 10% heat-inactivated fetal bovine serum, 100 U/mL penicillin, and 100 mg/mL streptomycin.

4.5. Gelatin Zymography

Gelatin zymography was used to determine the expression of MMP-2 and MMP-9 [48]. HT1080 cell suspension (5×10^5 cells/mL) was seeded in 24-well cell culture plates using medium supplemented with 0.5% FBS for 24 h for cell adhesion and growth. Subsequently, the cells were treated with indicated concentrations of samples with vehicle (DMSO) as blank and phorbol 12-myristate 13-acetate (PMA, 5 μM) as positive control. After 24 h of incubation, conditioned medium was collected to analyze the activity of MMP-2 and MMP-9.

4.6. MTT Cell Viability Assay

HT1080 cell suspension (5×10^5 cells/mL) was seeded in 24-well cell culture plates for 24 h for cell adhesion and growth. After being treated with samples and incubated for 22 h, 5.5 mg/mL MTT were added to the cells and incubated for 2 h. DMSO was added for cell lysis and OD 550 nm was detected.

Supplementary Materials: The following are available online, Figure S1: NMR spectrum of AQ20; Figure S2. Effects of the five PA analogues on MMP-2 and MMP-9 expression.

Author Contributions: Conceptualization, C.-K.L. and Y.-H.C.; formal analysis, Y.-C.W. and G.H.; writing—original draft preparation, S.-W.H. and C.-K.L.; writing—review and editing, H.-C.M. All authors have read and agreed to the published version of the manuscript.

Funding: This research was funded by National Science Council, grant number NSC 98-2320-B-038-015-MY3.

Institutional Review Board Statement: Not applicable.

Informed Consent Statement: Not applicable.

Acknowledgments: We are grateful to Shwu-Huey Wang for the NMR data acquisition by the TMU Core Facility.

Conflicts of Interest: The authors declare no conflict of interest. The funders had no role in the design of the study; in the collection, analyses, or interpretation of data; in the writing of the manuscript; or in the decision to publish the results.

Sample Availability: Samples of the compounds are not available from the authors.

References

1. Yuan, H.-W.; Zhao, J.-P.; Liu, Y.-B.; Qiu, Y.-X.; Xie, Q.-L.; Li, M.-J.; Wang, W. Advance in studies on chemical constituents, pharmacology and quality control of Aquilaria sinensis. *Digit. Chin. Med.* **2018**, *1*, 316–330. [CrossRef]
2. Feng, J.; Yang, X.-W.; Wang, R.-F. Bio-assay guided isolation and identification of α-glucosidase inhibitors from the leaves of Aquilaria sinensis. *Phytochemistry* **2011**, *72*, 242–247. [CrossRef] [PubMed]
3. Lin, H.-Z.; Li, H.-N.; Mei, Q.-X. Study progress of Aquilaria sinensis leaves. *Pharm. Today* **2011**, *21*, 547–549.
4. Yang, M.-X.; Chen, H.-R. Antitumor constituents from the leaves of Aquilaria sinensis. In *Chinese Medicinal Chemistry Symposium*; Chinese Pharmaceutical Association: Guangzhou, Guangdong, 2011.
5. Yang, M.-X.; Liang, Y.-G.; Chen, H.-R. Chemical constituents from leaves of wild Aquilaria sinensis. *Zhong Cao Yao* **2014**, *45*, 1989–1992.
6. Zhou, M.-H.; Wang, H.-G.; Kou, J.-P.; Yu, B.-Y. Antinociceptive and anti-inflammatory activities of *Aquilaria sinensis (Lour.)* Gilg. Leaves extract. *J. Ethnopharmacol.* **2008**, *117*, 345–350. [CrossRef] [PubMed]
7. Qi, J.; Lu, J.-J.; Liu, J.-H.; Yu, B.-Y. Flavonoid and a rare benzophenone glycoside from the leaves of Aquilaria sinensis. *Chem. Pharm. Bull.* **2009**, *57*, 134–137. [CrossRef] [PubMed]
8. Hara, H.; Ise, Y.; Morimoto, N.; Shimazawa, M.; Ichihashi, K.; Ohyama, M.; Iinuma, M. Laxative effect of agarwood leaves and its mechanism. *Biosci. Biotechnol. Biochem.* **2008**, *72*, 335–345. [CrossRef]
9. Yang, X.-B.; Feng, J.; Yang, X.-W.; Zhao, B.; Liu, J.-X. Aquisiflavoside, a new nitric oxide production inhibitor from the leaves of Aquilaria sinensis. *J. Asian Nat. Prod. Res.* **2012**, *14*, 867–872. [CrossRef]
10. Cheng, J.-T.; Han, Y.-Q.; He, J.; De Wu, X.; Dong, L.-B.; Peng, L.-Y.; Li, Y.; Zhao, Q.-S. Two new tirucallane triterpenoids from the leaves of Aquilaria sinensis. *Arch. Pharm. Res.* **2013**, *36*, 1084–1089. [CrossRef]
11. Yang, M.-X.; Liang, Y.-G.; Chen, H.-R.; Huang, Y.-F.; Gong, H.-G.; Zhang, T.-Y.; Ito, Y. Isolation of flavonoids from wild Aquilaria sinensis leaves by an improved preparative high-speed counter-current chromatography apparatus. *J. Chromatogr. Sci.* **2018**, *56*, 18–24. [CrossRef] [PubMed]

12. Jiang, S.; Jiang, Y.; Guan, Y.-F.; Tu, P.-F.; Wang, K.-Y.; Chen, J.-M. Effects of 95% ethanol extract of Aquilaria sinensis leaves on hyperglycemia in diabetic db/db mice. *J. Chin. Pharm. Sci.* **2011**, *20*, 609. [CrossRef]

13. Chakrabarti, S.; Patel, K.D. Matrix metalloproteinase-2 (MMP-2) and MMP-9 in pulmonary pathology. *Exp. Lung Res.* **2005**, *31*, 599–621. [CrossRef] [PubMed]

14. Vincenti, M.P. The matrix metalloproteinase (MMP) and tissue inhibitor of metalloproteinase (TIMP) genes. In *Matrix Metalloproteinase Protocols*; Humana Press: Tortowa, NJ, USA, 2001; Volume 151, pp. 121–148.

15. Zhang, X.; Wang, X.; Zhong, W.; Ren, X.; Sha, X.; Fang, X. Matrix metalloproteinases-2/9-sensitive peptide-conjugated polymer micelles for site-specific release of drugs and enhancing tumor accumulation: Preparation and in vitro and in vivo evaluation. *Int. J. Nanomed.* **2016**, *11*, 1643.

16. Albright, C.F.; Graciani, N.; Han, W.; Yue, E.; Stein, R.; Lai, Z.; Diamond, M.; Dowling, R.; Grimminger, L.; Zhang, S.-Y. Matrix metalloproteinase–activated doxorubicin prodrugs inhibit HT1080 xenograft growth better than doxorubicin with less toxicity. *Mol. Cancer Ther.* **2005**, *4*, 751–760. [CrossRef]

17. Singh, A.; Chaudhuri, R.; Ghosal, S. Chemical constituents of Gentianaceae XX: Natural occurrence of (—)-loliolide in Canscora decussata. *J. Pharm. Sci.* **1976**, *65*, 1549–1551. [CrossRef]

18. Itoh, D.; Kawano, K.; Nabeta, K. Biosynthesis of chloroplastidic and extrachloroplastidic terpenoids in liverwort cultured cells: 13C serine as a probe of terpene biosynthesis via mevalonate and non-mevalonate pathways. *J. Nat. Prod.* **2003**, *66*, 332–336. [CrossRef] [PubMed]

19. Wu, T.S.; Chang, F.C.; Wu, P.L.; Kuoh, C.S.; Chen, I.S. Constituents of leaves of Tetradium glabrifolium. *J. Chin. Chem. Soc.* **1995**, *42*, 929–934. [CrossRef]

20. Ouyang, M.-A.; Wein, Y.-S.; Zhang, Z.-K.; Kuo, Y.-H. Inhibitory activity against tobacco mosaic virus (TMV) replication of pinoresinol and syringaresinol lignans and their glycosides from the root of Rhus javanica var. roxburghiana. *J. Agric. Food Chem.* **2007**, *55*, 6460–6465. [CrossRef] [PubMed]

21. Pattabiraman, M.; Kaanumalle, L.S.; Natarajan, A.; Ramamurthy, V. Regioselective photodimerization of cinnamic acids in water: Templation with cucurbiturils. *Langmuir* **2006**, *22*, 7605–7609. [CrossRef]

22. Darwish, F.M.; Reinecke, M.G. Ecdysteroids and other constituents from *Sida spinosa* L. *Phytochemistry* **2003**, *62*, 1179–1184. [CrossRef]

23. Durand, S.; Parrain, J.-L.; Santelli, M. A large scale and concise synthesis of γ-linolenic acid from 4-chlorobut-2-yn-1-ol. *Synthesis* **1998**, *1998*, 1015–1018. [CrossRef]

24. Barbosa, J.V.; Oliveira, F.; Moniz, J.; Magalhães, F.D.; Bastos, M.M. Synthesis and characterization of allyl fatty acid derivatives as reactive coalescing agents for latexes. *J. Am. Oil Chem. Soc.* **2012**, *89*, 2215–2226. [CrossRef]

25. Akita, C.; Kawaguchi, T.; Kaneko, F.; Yamamoto, H.; Suzuki, M. Solid-state13C NMR study on order disorder phase transition in oleic acid. *J. Phys. Chem. B* **2004**, *108*, 4862–4868. [CrossRef]

26. Utaka, M.; Nakatani, M.; Takeda, A. Dye-sensitized photooxygenation of 1, 2-cyclopentanediones in methanol. *J. Org. Chem.* **1986**, *51*, 1140–1141. [CrossRef]

27. Yu, C.-M.; Fathi-Afshar, Z.R.; Curtis, J.M.; Wright, J.L.; Ayer, S.W. An unusual fatty acid and its glyceride from the marine fungus Microsphaeropis olivacea. *Can. J. Chem.* **1996**, *74*, 730–735. [CrossRef]

28. Zhao, Q.; Mansoor, T.A.; Hong, J.; Lee, C.-O.; Im, K.S.; Lee, D.S.; Jung, J.H. New lysophosphatidylcholines and monoglycerides from the marine *sponge Stelletta* sp. *J. Nat. Prod.* **2003**, *66*, 725–728. [CrossRef]

29. Zeng, X.; Xiang, L.; Li, C.-Y.; Wang, Y.; Qiu, G.; Zhang, Z.-x.; He, X. Cytotoxic ceramides and glycerides from the roots of Livistona chinensis. *Fitoterapia* **2012**, *83*, 609–616. [CrossRef]

30. Yang, L.; Qiao, L.; Xie, D.; Yuan, Y.; Chen, N.; Dai, J.; Guo, S. 2-(2-Phenylethyl) chromones from Chinese eaglewood. *Phytochemistry* **2012**, *76*, 92–97. [CrossRef] [PubMed]

31. Isaev, I.; Agzamova, M.; Isaev, M. Genkwanin and iridoid glycosides from Leonurus turkestanicus. *Chem. Nat. Compd.* **2011**, *47*, 132–134. [CrossRef]

32. Bentamene, A.; Baz, M.; Boucheham, R.; Benayache, S.; Creche, J.; Benayache, F. Flavonoid aglycones from Centaurea sphaerocephala. *Chem. Nat. Compd.* **2008**, *44*, 234–235. [CrossRef]

33. Nunez-Alarcon, J. Pilloin, a new flavone from Ovidia pillo-pillo. *J. Org. Chem.* **1971**, *36*, 3829–3830. [CrossRef]

34. Herz, W.; Sosa, V.E. Sesquiterpene lactones and other constituents of Arnica acaulis. *Phytochemistry* **1988**, *27*, 155–159. [CrossRef]

35. Guo, Y.; Zhao, Y.; Zheng, C.; Meng, Y.; Yang, Y. Synthesis, biological activity of salidroside and its analogues. *Chem. Pharm. Bull.* **2010**, *58*, 1627–1629. [CrossRef]

36. Yoneda, Y.; Krainz, K.; Liebner, F.; Potthast, A.; Rosenau, T.; Karakawa, M.; Nakatsubo, F. "Furan endwise peeling" of celluloses: Mechanistic studies and application perspectives of a novel reaction. *Eur. J. Org. Chem.* **2008**, *2008*, 475–484. [CrossRef]

37. Cateni, F.; Bonivento, P.; Procida, G.; Zacchigna, M.; Favretto, L.G.; Scialino, G.; Banfi, E. Chemoenzymatic synthesis and antimicrobial activity evaluation of monoglucosyl diglycerides. *Bioorg. Med. Chem.* **2007**, *15*, 815–826. [CrossRef]

38. Hargus, J.A.; Fronczek, F.R.; Vicente, M.G.H.; Smith, K.M. Mono-(l)-aspartylchlorin-e6. *Photochem. Photobiol.* **2007**, *83*, 1006–1015. [CrossRef] [PubMed]

39. Ahn, M.-Y.; Kwon, S.-M.; Kim, Y.-C.; Ahn, S.-G.; Yoon, J.-H. Pheophorbide a-mediated photodynamic therapy induces apoptotic cell death in murine oral squamous cell carcinoma in vitro and in vivo. *Oncol. Rep.* **2012**, *27*, 1772–1778. [PubMed]

40. Lee, T.H.; Lu, C.K.; Kuo, Y.H.; Lo, J.M.; Lee, C.K. Unexpected novel pheophytin peroxides from the leaves of Biden pilosa. *Helv. Chim. Acta* **2008**, *91*, 79–84. [CrossRef]
41. Kessenbrock, K.; Plaks, V.; Werb, Z. Matrix metalloproteinases: Regulators of the tumor microenvironment. *Cell* **2010**, *141*, 52–67. [CrossRef]
42. Sawicki, G.; Marcoux, Y.; Sarkhosh, K.; Tredget, E.E.; Ghahary, A. Interaction of keratinocytes and fibroblasts modulates the expression of matrix metalloproteinases-2 and-9 and their inhibitors. *Mol. Cell Biochem.* **2005**, *269*, 209–216. [CrossRef]
43. Soo, C.; Shaw, W.W.; Zhang, X.; Longaker, M.T.; Howard, E.W.; Ting, K. Differential expression of matrix metalloproteinases and their tissue-derived inhibitors in cutaneous wound repair. *Plast. Reconstr. Surg.* **2000**, *105*, 638–647. [CrossRef] [PubMed]
44. Manuel, J.A.; Gawronska-Kozak, B. Matrix metalloproteinase 9 (MMP-9) is upregulated during scarless wound healing in athymic nude mice. *Matrix Biol.* **2006**, *25*, 505–514. [CrossRef]
45. Saide, A.; Lauritano, C.; Ianora, A. Pheophorbide a: State of the Art. *Mar. Drugs* **2020**, *18*, 257. [CrossRef] [PubMed]
46. Dolmans, D.E.; Fukumura, D.; Jain, R.K. Photodynamic therapy for cancer. *Nat. Rev. Cancer* **2003**, *3*, 380–387. [CrossRef] [PubMed]
47. Lee, H.; Park, H.-Y.; Jeong, T.-S. Pheophorbide a derivatives exert antiwrinkle effects on UVB-induced skin aging in human fibroblasts. *Life* **2021**, *11*, 147. [CrossRef]
48. Ra, H.-J.; Parks, W.C. Control of matrix metalloproteinase catalytic activity. *Matrix Biol.* **2007**, *26*, 587–596. [CrossRef]

 *molecules*

Article

Enhancement of β-Glucan Biological Activity Using a Modified Acid-Base Extraction Method from *Saccharomyces cerevisiae*

Enas Mahmoud Amer [1,†] , Saber H. Saber [2,†] , Ahmad Abo Markeb [3] , Amal A. Elkhawaga [4], Islam M. A. Mekhemer [3], Abdel-Naser A. Zohri [1] , Steve Harakeh [5,*] and Elham A. Abd-Allah [6]

[1] Botany and Microbiology Department, Faculty of Science, Assiut University, Assiut 71515, Egypt; anoss_86@yahoo.com (E.M.A.); zohriassiut@aun.edu.eg (A.-N.A.Z.)

[2] Laboratory of Molecular Cell Biology, Department of Zoology, Faculty of Science, Assiut University, Assiut 71515, Egypt; Saberhassan@aun.edu.eg

[3] Chemistry Department, Faculty of Science, Assiut University, Assiut 71515, Egypt; a_markeb@aun.edu.eg (A.A.M.); eslammekhemer236@gmail.com (I.M.A.M.)

[4] Medical Microbiology and Immunology Department, Faculty of Medicine, Assiut University, Assiut 71515, Egypt; amy.elkhawaga@aun.edu.eg

[5] Special Infectious Agents Unit, King Fahd Medical Research Center and Yousef Abdullatif Jameel Chair of Prophetic Medicine Application, Faculty of Medicine, King Abdulaziz University (KAU), Jeddah 21589, Saudi Arabia

[6] Zoology Department, Faculty of Science, New Valley University, El-Kharga 72511, Egypt; amira1422010@yahoo.com

* Correspondence: sharakeh@gmail.com or sharakeh@kau.edu.sa; Tel.: +96-6-55-939-2266

† These authors contributed equally to this work.

Citation: Mahmoud Amer, E.; Saber, S.H.; Abo Markeb, A.; Elkhawaga, A.A.; Mekhemer, I.M.A.; Zohri, A.-N.A.; Harakeh, S.; Abd-Allah, E.A. Enhancement of β-Glucan Biological Activity Using a Modified Acid-Base Extraction Method from *Saccharomyces cerevisiae*. *Molecules* **2021**, *26*, 2113. https://doi.org/10.3390/molecules26082113

Academic Editors: Raffaele Pezzani and Sara Vitalini

Received: 15 February 2021
Accepted: 19 March 2021
Published: 7 April 2021

Publisher's Note: MDPI stays neutral with regard to jurisdictional claims in published maps and institutional affiliations.

Abstract: Beta glucan (β-glucan) has promising bioactive properties. Consequently, the use of β-glucan as a food additive is favored with the dual-purpose potential of increasing the fiber content of food products and enhancing their health properties. Our aim was to evaluate the biological activity of β-glucan (antimicrobial, antitoxic, immunostimulatory, and anticancer) extracted from *Saccharomyces cerevisiae* using a modified acid-base extraction method. The results demonstrated that a modified acid-base extraction method gives a higher biological efficacy of β-glucan than in the water extraction method. Using 0.5 mg dry weight of acid-base extracted β-glucan (AB extracted) not only succeeded in removing 100% of aflatoxins, but also had a promising antimicrobial activity against multidrug-resistant bacteria, fungi, and yeast, with minimum inhibitory concentrations (MIC) of 0.39 and 0.19 mg/mL in the case of resistant *Staphylococcus aureus* (MRSA) and *Pseudomonas aeruginosa*, respectively. In addition, AB extract exhibited a positive immunomodulatory effect, mediated through the high induction of TNFα, IL-6, IFN-γ, and IL-2. Moreover, AB extract showed a greater anticancer effect against A549, MDA-MB-232, and HepG-2 cells compared to WI-38 cells, at high concentrations. By studying the cell death mechanism using flow-cytometry, AB extract was shown to induce apoptotic cell death at higher concentrations, as in the case of MDA-MB-231 and HePG-2 cells. In conclusion, the use of a modified AB for β-glucan from *Saccharomyces cerevisiae* exerted a promising antimicrobial, immunomodulatory efficacy, and anti-cancer potential. Future research should focus on evaluating β-glucan in various biological systems and elucidating the underlying mechanism of action.

Keywords: *Saccharomyces cerevisiae*; β-glucan; antimicrobial and anticancer activities; detoxification ability; immunomodulatory effect

1. Introduction

β-glucan (beta-glucans) is one of the most abundant forms of polysaccharide, with glucose polymers connected by a 1→3 linear β-glycosidic chain hub. The major branching group in β-glucan are 1→4 or 1→6 glycosidic chains, and its component has highly variable branches, while depending on the source, its physicochemical properties differ significantly [1]. β-Glucans occur in the cell walls of cereals and microorganisms (bacteria and fungi). Yeast cell wall β-glucan consists of 1→3 β-linked glucopyranosyl residues, with small number of 1→6 β-linked branches [2]. Several potential properties of β-glucans have been reported in the literature, such as being antioxidant, anti-inflammatory, anti-cholesterol, anti-ageing, a blood glucose regulator, and an antitumor agent [3,4]. Moreover, β-glucan can act as an immunomodulatory molecule, which acts via the modification of the host immune response by the activation of innate immune cells including macrophages, neutrophils, and granulocytes [5]. Based on both in vitro and in vivo studies, it has been reported that the innate immune cells, such as macrophages, neutrophils, monocytes, NK cells, and DCs, as well as T cells which function through Dectin-1, complement receptor 3 (CR3), and TLR-2/6 receptors can be activated by β-glucans [6–8]. More interestingly, the systemic injection of β-glucan contributed significantly to the regulation of tumor microenvironment (TME), resulting in the reduction of primary tumor growth and reduce metastasis, as shown in a preclinical mouse model [9]. Pelley et al. [10] reported that β-glucan molecules have an impressive characteristics when compared to other molecules, like being protein-, peptide-, virus-, and virus-like-particle-based immune regulators, in addition to their low cytotoxic effect. The in vivo tolerant dose was up to 10 mg/kg without any adverse side effects [11]. β-glucan has a specific immune-modulatory effect through a specific surface receptor, and it has an attractive chemical structure, with multiple aldehydes and hydroxyl groups, which highlight the opportunities for structural modifications and possible applications as a drug delivery molecule [12]. It was reported that β-glucans have a strong positive effect on the metabolic control of diabetes [13], wound healing induction [14], stress normalization, decrease of chronic fatigue syndrome [15], attenuating cholesterol levels [16], and anticancer activity [17]. In addition, β-glucan can be used as part of a vaccine for high-risk neuroblastoma [18]. Furthermore, many clinical studies have revealed the significant contribution of β-glucan to the treatment of chronic respiratory problems in children [19,20].

Most β-glucan supplements in the western world are obtained from baker's yeast. The European Food Safety Authority has approved *S. cerevisiae* as a new food ingredient [21]. A significant source of β-glucan is the *S. cerevisiae* cell wall, which contains about 55–65% β-glucan [22]. β-glucans have different functional characteristics that can be used in the food industry to make soups, sauces, drinks, and other food items, where they act as stabilizers, thickeners, and emulsifiers [23]. Natural sources of β-glucan are mushrooms including Shiitake mushroom (*Lentinus edodes*) in Japan and Lingzhi (*Ganoderma lucidum*) in China, which were used in traditional medicine as immune stimulators [24]. There are several techniques used to extract the strong β-glucan from microbial and plant sources, but the selection of a suitable extraction method plays a vital role in determining the structure, quality, and the functionality of the method [25].

Based on the above, the objectives of this work were to improve the biological activity of β-glucan by using a modified method for β-glucan extraction from *S. cerevisiae*, with good yield, and to determine its physical and chemical properties; detoxification; and its immune modulation abilities, as well as the anticancer, and antimicrobial activities of the extracted material.

2. Results

2.1. Extraction of β-Glucan

2.1.1. Acid-Base Extraction Method

In this method, the dry active yeast cells were propagated using yeast extract–glucose broth medium for 48 h. After collection of yeast cell pellets from the above step, they were mixed with five-fold 1 M NaOH, then heated in a water bath for 2 h at 80 °C, rewashed after centrifugation with distilled water, and lastly, recentrifuged to get the end yield, which was less than the starter by 40%. The lysed yeast cell pellets were collected and used for β-glucan extraction.

The extraction took place using five-fold acetic acid (CH_3COOH), to give a β-glucan yield at the end equal to 20% of the propagated yeast cells. Finally, the collected pellets of β-glucan were dried in an oven at 60 °C to remove any traces or organic-soluble materials, extra proteins, or other impurities that may exist, and then stored for further investigations. The pellets' weight after drying were 11.5% from the starter yeast cells.

2.1.2. Water Extraction Method

Lysis of yeast cells was done using sonication for 5 min in phosphate buffer at pH 8.0 (Bandelin, Sonopuls, Sicherungen, 2 X F2A, Berlin, Germany). The next steps after autolysis included shaking at 30 °C for 4 h, centrifugation at $5000 \times g$ for 15 min, washing twice with PBS, and then sodium dodecyl sulfate was added to separate any protein substances that may have been present. The resulting yeast cells weight was 80% of the starter yeast cells, which were treated by autoclaving at 121 °C for 4 h. The water-insoluble gradient was then centrifuged and air-dried to get 7% dry weight pellets from the starter yeast cells.

2.2. Characterization of β-Glucan
2.2.1. FTIR Spectroscopic Analysis

The recorded FTIR spectra of S1 (β-glucans sample extracted using the acid-base method) and S2 (β-glucans sample extracted using the water extraction method) samples is shown in Figure 1. The absorption bands at 3412 and 3445 cm^{-1} of samples S1 and S2, respectively, were attributed to the occurrence of a free hydroxyl group (O-H stretching). Whereas, the bands at 2921 and 2851 cm^{-1} were seen due to the appearance of C-H aliphatic in pyranoid rings (C-H and CH_2OH stretching bands). Moreover, the peaks of C-C and C-O stretching vibrations were confirmed by the existence of absorption in the region of 900–1200 cm^{-1}, which indicated the presence of polysaccharides as a major component. Furthermore, the strong absorption peak at 978.24 cm^{-1} indicated β-glycosidic bonds, i.e., (C1–H) deformation mode, and therefore indicated the presence of β-glucan. Interestingly, the FTIR bands at 1581 cm^{-1} of S1 and 1661 cm^{-1} of S2 were related to the presence of (C=O) amide band vibration. N-H stretching vibration was confirmed by the presence of the absorption band at 3661 cm^{-1}, which agrees with the results published by Hromádková [26].

2.2.2. HPLC Analysis

The HPLC technique was used to determine the quality and purity of the extracted β-glucan using the above two extraction methods. In addition, this was confirmed by injection of a β-glucan reference standard (S0). The HPLC chromatogram proved that the major peak of β-glucan appeared at 11 min, as shown in Figure 2. Furthermore, S1 and S2 showed the same retention time when compared with S0 as the major peak of β-glucan, which indicated the purity of the extracted β-glucan and the efficient method of the extraction.

2.2.3. Optical Properties

Figure 3 describes the UV-Vis spectrum of B-glucans; there was only a small hump at 390 nm for (S0), (S1), and (S2). This confirms the structure of extracts formed using the two different methods (i.e., acid-base and water methods) via the agreement of their absorption with the reference sample (S0), as show in Figure 3.

2.3. Detoxification Ability

In our study, serial amounts, 0.1, 0.2, 0.3, 0.4 and 0.5 mg dry weight, of the extracted β-glucan (acid-base extraction and water extraction) were tested for their ability to remove aflatoxins (AFs) (1000 ng/mL). As indicated in Figure 4, AF removal in case of using the S1 sample was more efficient than the S2 sample. In more details, in the case of S1, increasing the weight from 0.1 mg to 0.4 mg, the removal was up to 80%, while using 0.5 mg, the removal of the toxins was approximately 100%. However, while utilizing the S2 sample, the removal was about 40% of the toxins, using weights ranging from 0.1–0.3 mg. Moreover, 50% removal was achieved with 0.4 mg β-glucan dry weight, and finally around 80% of the toxins were removed at 0.5 mg dry weight of β-glucan. All results were compared with negative and positive controls.

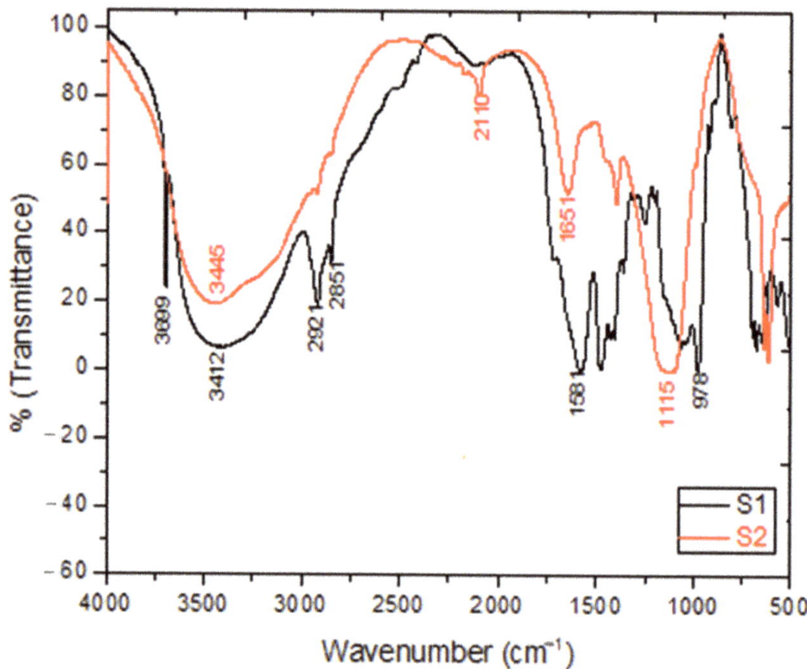

Figure 1. FTIR of β-glucan extracted from yeast cells by two methods: acid-base extraction method (S1), and water extraction method (S2).

2.4. Possible Mechanism for Interaction β-Glucan with AFs

Interaction of AFs with β-glucan was confirmed by FTIR, as illustrated in Figure 5. The absorption band of O-H stretching in S1 was shifted to 3473 cm^{-1}. Whereas the C-H and CH$_2$OH stretching bands at 2922 and 2852 cm^{-1} in pyranoid rings remained constant. Moreover, the FTIR bands at 1584 and 1653 cm^{-1} corresponding to the presence of (C=O) amide were slightly shifted, and the absence of 1744 cm^{-1} (ketone, C=O) in the AFs spectrum was attributed to the reduction of AFs via B-glucan. In addition, the absorption band of C1–H related to β-glycosidic bonds was decreased by the interaction with AFs. Therefore, the main interaction between AFs and β-glucan arose from the O-H and β-glycosidic bonds of β-glucan.

2.5. Antimicrobial Activity of β-Glucan

In the present study, the antimicrobial activity of β-glucan extracts S1 and S2 was screened by agar well diffusion assay using a concentration of 12.5 mg/mL. The results of the antimicrobial assay are presented in Table 1. It is clearly observed from the obtained data that the S1 sample showed a relatively high antimicrobial activity when compared to the S2 sample, particularly against multidrug resistant bacteria and fungi. The minimum inhibitory concentration (MIC) of the extracts S1 and S2 was determined. The MIC varied within the range of 12.5 mg/mL to 0.19 mg/mL. It was observed that S1 was potent against *S. aureus*, MRSA, and *P. aeruginosa*, with an MIC ranging from 0.39–0.19 mg/mL.

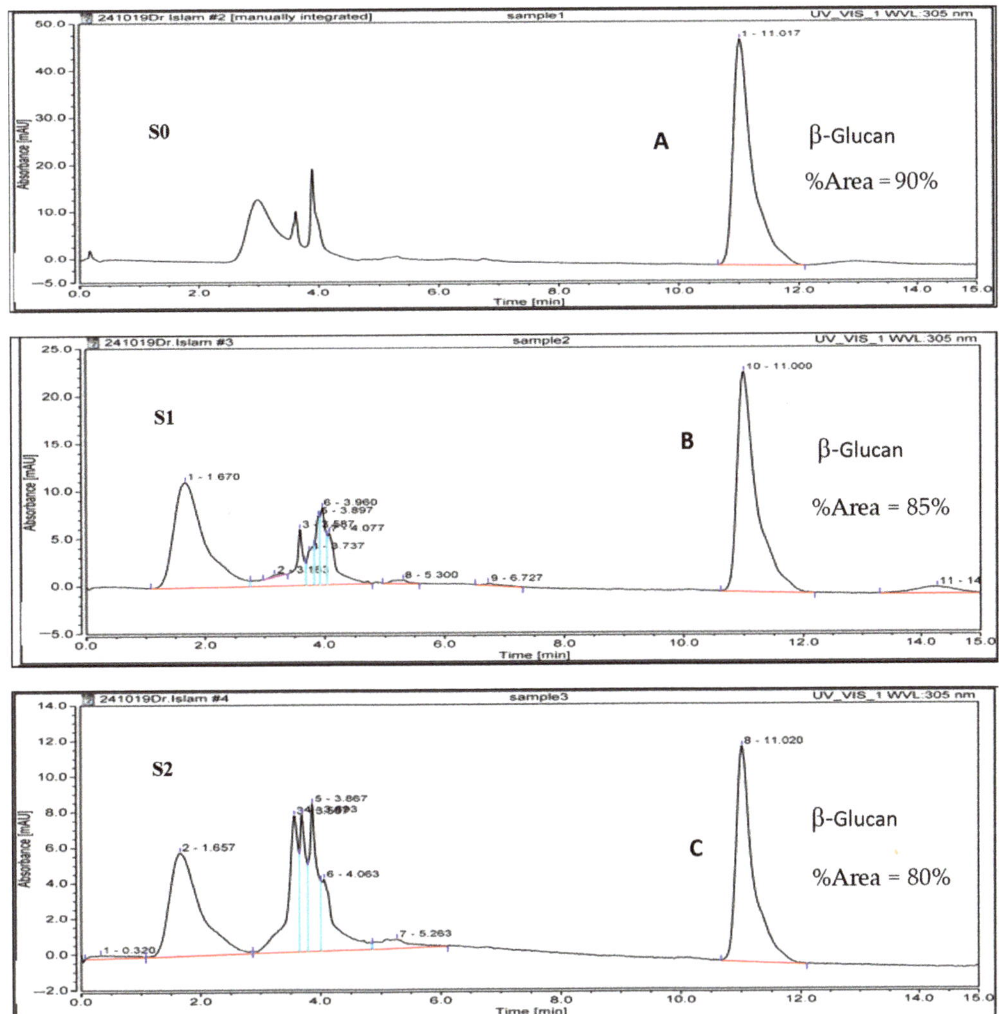

Figure 2. HPLC chromatogram of β-glucan standard (**A**), β-glucan extracted by acid-base extraction method (**B**), and β-glucan extracted by water extraction method (**C**).

2.6. β-Glucan as an Immune-Modulatory Molecule

The immune-modulatory effects of β-glucan were evaluated using phorbol-12-myriste-13-acetate (PMA) as a positive control [27] and non-treated peripheral blood mononuclear cells (PBMCs) media as a negative control. The levels of different cytokines, including IFN-γ, IL1, IL-2 and TNF-α, were measured. The data showed that the TNF-α level was significantly higher (about 2-fold) in β glucan treated PBMCs than the baseline level. Interestingly, the level of TNF-α was significantly higher in β glucan treated PBMCs than PMA treated PBMCs. Similarly, the levels of IL-6 and IFN-γ were significantly higher in β-glucan and PMA treated PBMCs than non-treated control cells. In the same way, β glucan induced more IL-2 secretion than PMA treated PBMCs, as shown in Figure 6.

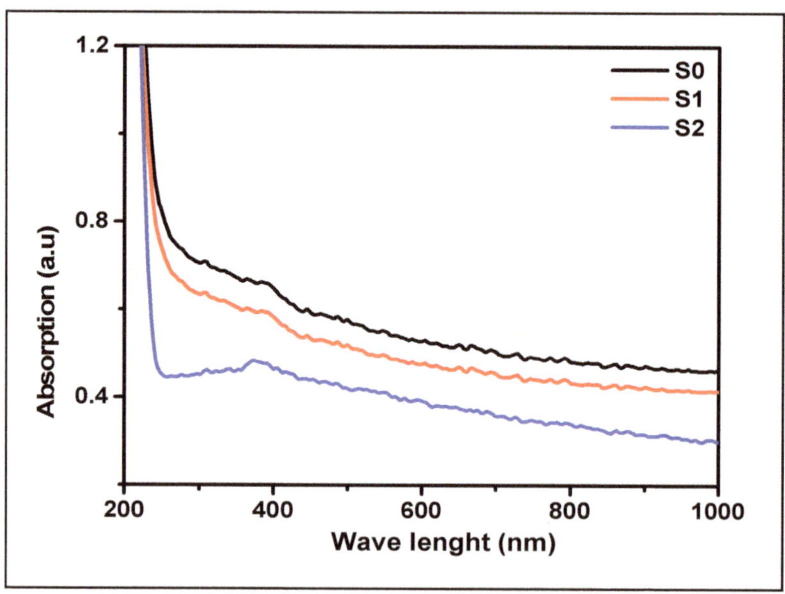

Figure 3. The optical properties of the two β-glucans extracted, S1 and S2, compared with the β-glucan standard, S0.

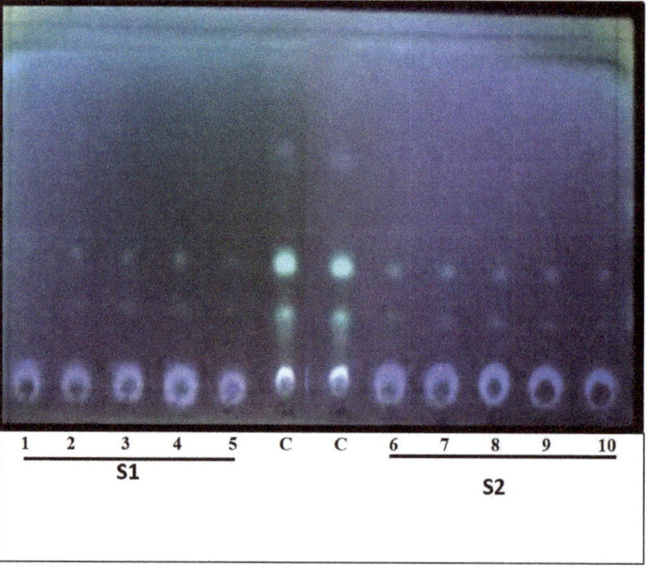

Figure 4. TLC plate showing decrease in Aflatoxin presence in accordance with β-glucan concentration increase, left points (1:5) are S1 samples, then two points of crude extract of aflatoxins (AFs), and finally the following five points (6:10) are samples of S2.

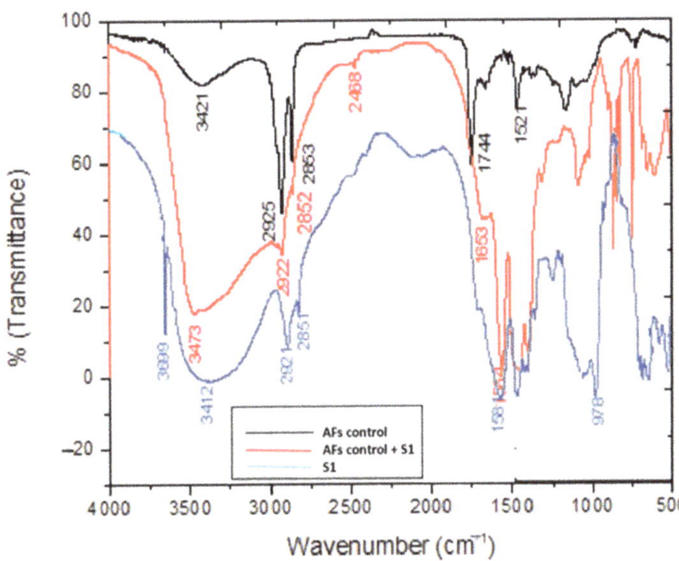

Figure 5. Possible mechanism of interaction between AFs and β-glucan extracted (S1).

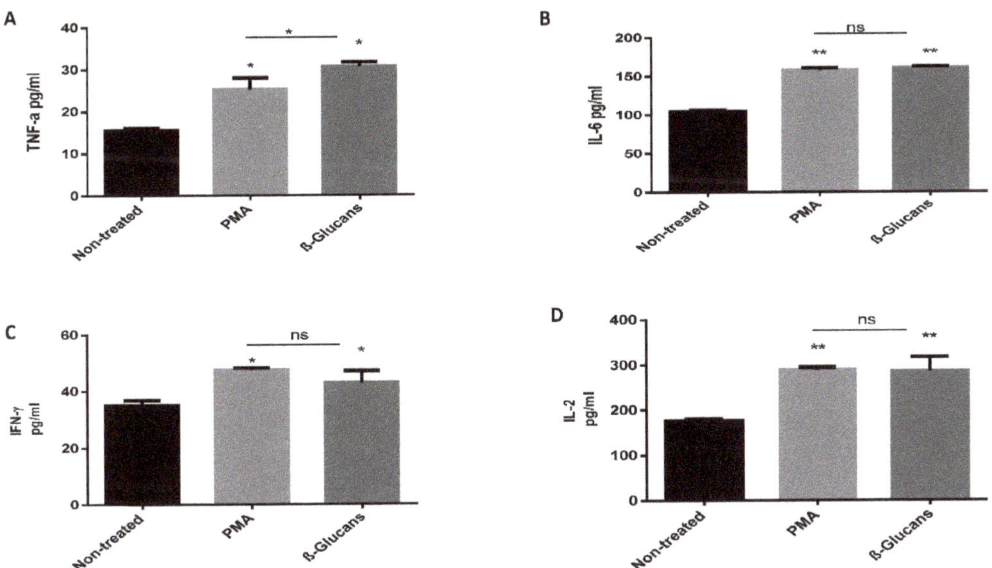

Figure 6. Assessment of the effect of β-glucan (S1) on the induction of cytokine release from peripheral blood mononuclear cells (PBMCs). Supernatants were collected from non-treated PBMCs, phorbol-12-myriste-13-acetate (PMA) treated PBMCs, and β-glucan treated PBMCs. The levels of TNF-α (**A**), IL-6 (**B**), IFN-γ (**C**), and IL-2 (**D**) were determined by ELISA. Data are represented as means ± SD of three separate experiments. * indicates $p \leq 0.05$, ** indicates $p \leq 0.01$ as assayed by two-tailed Student's *t*-test. ns = nonsignificant.

Table 1. Antimicrobial activity of β-glucans, S1 and S2, against the multidrug resistant tested microbial strains.

Tested Compounds	Test Strains/ZOI (mm) and MIC (mg/mL)																	
	S. aureus		MRSA		*S. pneumoniae*		*E. coli*		*K. pneumoniae*		*P. aeruginosa*		*A. flavus*		*A. niger*		*C. albicans*	
	ZOI	MIC	ZOI	MIC	ZOI	MIC	ZOI	MIC	ZOI	MIC	ZOI	MIC	ZOI	MIC	ZOI	MIC	ZOI	MIC
S1	11.9 ± 0.3	0.39	11 ± 0.3	0.39	5.1 ± 0.08	3.13	2.2 ± 0.1	6.25	10.2 ± 0.1	0.78	13.2 ± 0.1	0.19	2.6 ± 0.1	6.25	6.1 ± 0.07	3.12	7.7 ± 0.1	3.12
S2	9.2 ± 0.1	0.78	8.2 ± 0.2	1.56	1.6 ± 0.1	6.25	R	R	5.7 ± 0.2	3.125	11.1 ± 0.1	0.39	R	R	R	R	6.9 ± 0.4	3.12
PC	12 ± 0.3	0.78	11.3 ± 0.4	0.78	8.6 ± 0.1	0.31	16.9 ± 0.8	0.15	9.5 ± 0.3	0.31	17.2 ± 0.2	0.15	2.7 ± 0.1	0.78	5.1 ± 0.2	0.39	8.0 ± 0.3	0.39
DMSO	R		R		R		R	R	R	R	R	R	1.2 ± 0.05	R	4.2 ± 0.1	R	2.0 ± 0.04	R

ZOI = mean zone of inhibition in mm ± standard deviation (S.D.), R = (resistant, MIC = minimum inhibitory concentration, PC (μg/mL) = positive control (Vancomycin 50 μg/mL for Gram-positive bacteria, Gentamicin 10 μg/mL for Gram-negative bacteria), and fluconazole 25 μg/mL for fungi). R = resistant.

2.7. Anticancer Activity of β-Glucan

In the current study, we uncovered the antiproliferative effects of the two extracted β-glucans against normal and cancer cells. To be sure the effects of these samples reflected their potential anticancer properties, first, we screened different concentrations of water and methanol extract against WI-38 human normal lung fibroblast as control cells, A549 human lung cancer cells, MDA-MB-231 human breast cancer cells, and HePG-2 human liver cancer cells. Second, we investigated the mechanism of cell death using annexin (V) and propidium iodide (PI) and analyzed with FACS. The cytotoxic results using an MTT assay after 24 h showed that both the water and methanol extracts had a non-observed cytotoxic effect against normal and lung cancer cells with IC50 (1757, 1695, 1465, and 1494 µg/mL). In the case of the breast and liver cancer cells, both water and methanol extracts exerted a markedly cytotoxic effect at high concentration ($p < 0.05$) with IC50 769, 837, 723, and 703, respectively, as shown in Figure 7. The FACS results showed that using the IC_{50} of methanolic extract of β-glycan induced apoptotic cell death at a very low level (4.72%) of the total WI-38 normal cells. In the case of MDA-MB-231 breast cancer cells, methanol extract of β-glycan exerted a weak apoptotic cell death (11.9%), while inducing (23.1%) necrotic cell death of the total cell population. Finally, methanol extract induced a weak apoptotic cell death (12.3%) and weak necrotic cell death (8.28%) of total HePG-2 hepatic cells, as presented in Figure 7.

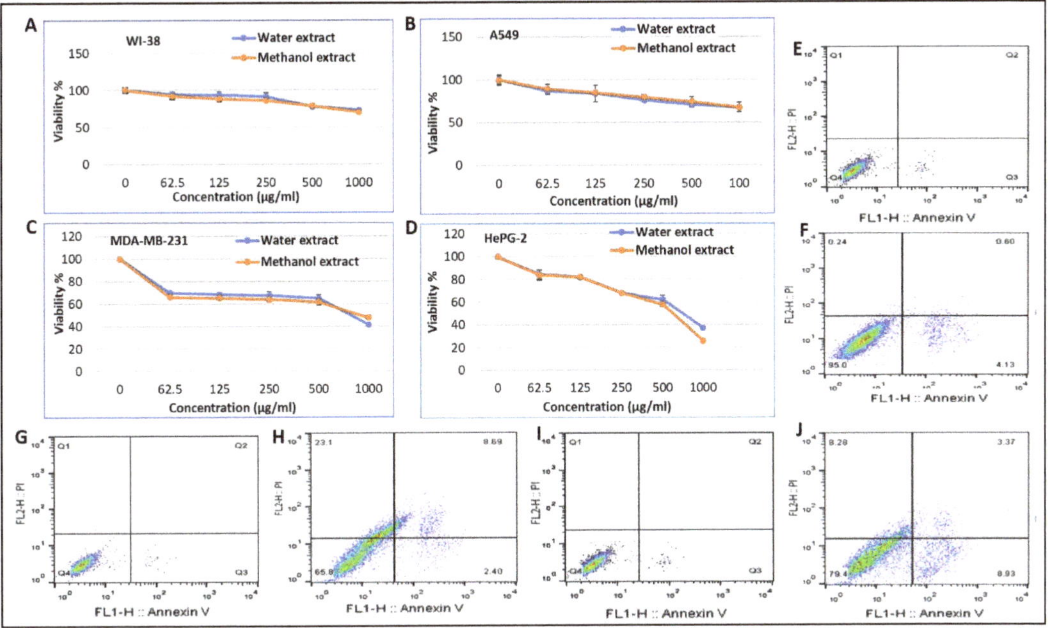

Figure 7. The cytotoxic effect of the indicated doses of water and methanol extracted β-glucan after 24 h treatment was assessed using MTT assay. (**A**) WI 38, (**B**) A549 cells, (**C**) MDA-MB-231 cells, and (**D**) HePG-2 cells treated with the indicated doses. Apoptotic and necrotic cell death were assessed using annexin (V) and propidium iodide (PI) staining and analyzed using a flow cytometer after 24 h treatment of methanol extracted β-glucan (**E**) WI-38 cells control, (**F**) WI-38 cell treated with 250 µg/mL of β-glucan, (**G**) MDA-MB-231 control treated cell, (**H**) MDA-MB-231 treated with 250 µg/mL of β-glucan, (**I**) HePG-2 control cells, (**J**) HePG-2 cells treated with 250 µg/mL of β-glucan.

3. Discussion

The first step in the extraction of β-glucan is (lysis) and according to the literature can be achieved by chemical (NaOH, HCl, acetic acid, citric acid), physical (sonication, high pressure) [28], and enzymatic (lytic enzymes) methods for yeast cells [29]. The second step can be done by using a number of extraction methods (acid-base, alkali-water, etc.).

For the acid-base extraction method, the dry active yeast cells were propagated; the propagation of the yeast bench player is a very important step to reach the stationary phase and for the autolysis to occur [30]. Autolysis of the yeast in the cells can be regarded as a lytic event. This is a process caused by intracellular yeast enzymes that is irreversible and associated with cell death [31]. It has been suggested that there are four yeast autolysis steps [32]. First, the structures of the cell endo degrade vascular proteases released into the cytoplasm. Second, cytoplasmic inhibitors originally inhibit the released proteases and then activate as a result of the degradation of these inhibitors, and third, intracellular polymer components are hydrolysed, with the hydrolysis products accumulating in the space restricted by the cell wall. Finally, the hydrolytic products are released when their molecular mass is low enough to cross pores in the cell wall.

Autolysis has been described as intracellular biopolymer hydrolysis under the influence of hydrolytic enzymes such as proteinases, ribonucleases, and glucanases [33]. During autolysis, the cell endo-structure degrades, releasing vascular proteases into the cytoplasm, then enzymes hydrolyse the intracellular polymer components, with the hydrolysis products accumulating in the space restricted by the cell wall, and finally the hydrolytic products are released when their molecular masses are low enough to cross pores in the cell wall [32]. The extraction was carried out using CH_3COOH. Acetic acid was preferred, due to its mild acidity, ease of handling, low toxicity, low cost, and availability. Generally, acids used for extraction should be mild enough to limit hydrolysis of the β-1, 3-linkages, and substantially only affect the β-1, 6-linked glucans, where the β-1, 6-glucan molecules interconnect the largest class of covalently linked cell wall proteins. The β-1, 3-linkages are resistant to hydrolysis by mild acids such as acetic acid [34].

The water extraction method using phosphate buffer sonication for cell disruption was used in this study instead of glass beads. Sonication gives efficient results for cell lysis in a short time, whereas glass beads need four hours to lyse cells, with less efficiency than sonication and with loss of cells during decantation of glass beads. According to Bzducha-Wróbel [35] after yeast cell autolysis combined with mechanical techniques, such as sonication, a higher amount of protein release was observed. The resulting yeast cells from the autolysis step were treated by autoclaving. According to Liu et al., (2013) the smallest effectiveness of nucleic acid release was defined by hot water extraction during autoclaving. At the same time, compared to cellular proteins, the hot water extraction method helps in the efficient solubilization of the genetic material [36].

The extracted β-glucan from two methods (acid-base and water methods) was characterized and confirmed using various spectroscopic and chromatographic techniques. First, UV-Vis spectrum was used to detect the absorptions in the chemical structure of β-glucans (S1 and S2) and these absorptions were compared with the standard sample to give the same pattern of spectral absorption as outlined in Figure 3. Furthermore, we confirmed all spectra (S1 and S2) via another comparison with the literature related spectrum [37]. The prepared formulation was validated as particulate β-glucan by the FTIR analysis, the structure of β-glucan was verified, together with the type of glycosidic bond through the presence of the characteristic functional groups by two extraction methods. In addition, the structure β-1,3-D-glucan was confirmed by comparison of our results with published data. To our delight, the FT-IR spectrums conformed to the same signal configuration characteristic for β-1,3-D-glucan extracted from yeast. Based on (HPLC) analysis, we proved that our modified acid-base method gave high quality and purity from β-glucan extracts (S1 and S2), as shown in Figure 2. All the above led to the success of the extraction and the perfect confirmation of β-glucan in this study.

Human beings are exposed to mycotoxins through various routes: directly through plant-based ingredients; through air (both indoors and outdoors); or indirectly through animal-based ingredients [38]. Aflatoxicosis is a disease that affects both humans and livestock that is induced by aflatoxins (AFs). These toxins belong to a group of compounds known as furocoumarins, which are generated by the filamentous fungi *Aspergillus flavus* and *Aspergillus parasitic* as secondary metabolites, and which are common in the environment [39,40]. These toxins affect various stages of the storage, sowing, and industrialization of agricultural and dairy products, while these metabolites can generate significant financial losses. A large amount of plants used for human and animal consumption may get contaminated [41].

In relation to its carcinogenic impact, particularly in the liver, it has a mutagenic impact in DNA adducts, DNA breaks, gene mutations, and the induction of DNA synthesis and inhibition of DNA repair [42]. Moreover, AFs have immunomodulatory impacts on the immune system, including macrophage alteration [43], resulting in compromising the ability of murine bone marrow cells to form colonies of myeloids, erythroids, and others [44]. In addition, AFs inhibit the development of nitric oxide (NO) in the murine bone marrow [45]. Several strategies have been attempted to minimize the financial and biological impacts resulting from AFs, B_1, contamination by implementing adsorbents, heat, irradiation, or chemical inactivation. The most promising techniques of decontamination include the use of microorganisms, in particular lactic acid bacteria. The latter have a very strong capacity to decrease the toxin content, which plays a major part in the adsorption cycle [46–51]. The yeasts, *S. cerevisiae*, are commonly used in baking, brewing, wine making, and distilling sectors in many biotechnological procedures. It has been reported that, the esterified form of β-glucan has a protective function toward individual or mixed aflatoxin B_1, ochratoxin A, and T-2 mycotoxicosis, which broiler chickens are subjected to [52]. There are many pathways for the binding of AFs with β-glucan, including ketone, hydroxyl, phenyl, and lactone groups, which are involved in the formation of both hydrogen bonds and van der Walls interactions [53–55]. Moreover, Eduardo et al. [56] explained the interaction between *AFB1 and* β-D-glucan as a supramolecular complex interaction. The β-glucan of yeast cell walls consists mainly of (1-3)-β-glucan and (1-6)-β-glucan; (1-3)-β-glucan is involved in both hydrogen and Van der Waals bonding between glucan and AFB_1; whereas (1-6)-β-glucan is involved in only Van der Waals bonding [54]. The (1-3)-β-glucan chains form triple helix three-dimensional structures, with spring like mechanical properties and are responsible for the strength of yeast cell walls [22] and their ability to bind toxins [53].

In this work, for the first time, the reduction of aflatoxin using β-glucan was confirmed by FT-IR analysis. Interestingly, the effect of the above interactions was confirmed through the absence of peaks, which are related to the ketone group at 1744 cm^{-1} in AFs. At this point, β-glucan was extracted from baker's yeast (Saccharomyces cerevisiae) to give the most effective binder (i.e., more exposure to B-glucan removes more AFs) [57].

Currently, the significant increase in microbial drug resistance represents a major problem worldwide [58]. The results of the antimicrobial assay in this work are presented in Table 1. Our results are consistent with Ginovyan and his group (2015) [59], as it is clearly observed from the obtained data that the S1 sample showed promising antimicrobial activity. Moreover, S1 gave the better results in the inhibition of different types of multidrug resistant bacteria and fungi than S2 [60,61]. This may be due to the variations in the extraction method and the solvents used in the extraction, as confirmed by Chamidah and Prihanto (2017) [1] through the determination of inhibition zones, which were found to be 11.9, 11, 10.2, 13.2, and 7.7 mm for *S. aureus*, MRSA, *K. pneumoniae*, *P. aeruginosa*, and *C. albicans*, respectively. The magnitude of this inhibition zone could be attributed to Gram-positive bacteria generally having a higher sensitivity to the antimicrobial compounds in comparison to Gram-negative bacteria. It was confirmed by Sudjaswadi (2006) that the effectiveness of a antimicrobial compound is influenced by the character of the wall or cell membrane of the bacteria [62]. The minimum inhibitory concentration (MIC) of extracts S1 and S2 was determined. Notably, the results agree with those of Chamida et al., (2017) [1] who reported an interesting antimicrobial activity for β-glucan against several bacterial and fungal strains. Finally, it is a clearly evident from our antimicrobial results that S1 exhibited dual activities as a promising antibacterial and antifungal agent.

To assess the immune-stimulatory effect of β-glucan on PBMCs, the levels of different cytokines after exposing these cells to β-glucan were measured compared to the control. It was reported that the immunostimulatory effect is common, and glucans in particular stimulate the release of different cytokines including IFN-γ, IL1, IL-2, and TNF-α [63]. The levels of these cytokines were comparable in both cells, suggesting that β glucan can stimulate the immune response in a comparable manner and magnitude to PMA. β-glucan stimulates IL-2 release from PBMCs. Based on these data, β-glucan has shown a potent immuno-stimulatory effect, which can induce the release of innate (such as TNFα, IL-6, and IFN-γ) and adaptive cytokines (IFN-γ and IL-2), which consequently modulate the immune response against pathogens. Our data are in agreement with that of Chan et al. (2009) [5], who reported that β-glucan stimulates diverse immune related receptors, in particularly Dectin-1 and CR3, and can trigger a wide spectrum of immune responses [64–69], not only the monocytes and macrophage, and β-glucan can also act on neutrophils, NK cells, and dendritic cells, and will polarize the T cells subset. All these data confirm the immune stimulation role of β-glucan. The presented results clearly show that the acid-base extraction method of β-glucan not only acts on cellular immunity, but on overall immunity.

Glucans have been used for decades in Japan to treat gastrointestinal tumors [70]. Much research has highlighted the anticancer activity of glucans [17,71,72]. It was reported that glucan administration can significantly reduce both colony formation and tumor size in lung and colon cancer models [73,74]. In the same context, the extracted β-glucan results support the previously published data, especially for breast and liver cancers [75]. It was reported that bacteria-extracted β-glucan induced apoptotic cell death in colon cancer, and upregulated apoptotic induced genes including Bax and Caspase-3 genes and down regulated Bcl-2 gene [76]. Consistently with the previous data, the new extracted fungal β-glucan induced apoptotic cell death in breast and liver cancer cells.

4. Materials and Methods

4.1. Materials

Packaged instant dry baker's yeast (*Saccharomyces cerevisiae*) were bought from local markets, transported to the laboratory on ice, and left in the refrigerator till used for the extraction of β-glucan.

4.2. β-Glucan Extraction Methods

Two methods of β-glucan extraction were employed, an acid-base extraction method with slight modification of the reported method 23, as described in SI.1.1, and a water extraction method as reported by Piotrowska and Masek [77] and other cited methods [78,79], with some modifications. Detailed information about the methodology is reported in the supplementary information (SI.1.2).

4.3. β-Glucan Characterization

The characteristics of the extracted β-glucan (S1 and S2) were confirmed with FTIR [80,81], HPLC [82,83], and optical analytical techniques [84], as described in SI.2.

4.4. Aflatoxins Removal

Aflatoxin (AFS) removal was carried out using the batch mode of experiment [85], as described in SI.3.

4.5. Antimicrobial Activity

Microorganisms were isolated using the reported method [86], and described in SI.4.1. The antimicrobial activities [59] and minimum inhibitory concentrations (MIC) [87] of the S1 and S2 β-glucan extracted were evaluated as cited in the SI.4.2 and SI.4.3 sections.

4.6. Immunomodulatory Effects

Cytokine assays of peripheral blood mononuclear cells (PBMCs) were cultures in RPMI-1640 complete growth media, followed by treatment with (50 µg/mL) β-glucan, and evaluated as described in SI.5.

4.7. Anticancer Properties

The efficiency of β-glucan against A549 human non-small lung cancer cells, MDA-MB-232 human breast cancer cells, and HepG-2 human hepatocarcinoma cells compared to WI-38 human lung fibroblast cancer cell lines was tested as reported in SI.6.

5. Conclusions

The combination of a strong base (NaOH) and weak acid (CH_3COOH) in the extraction of β-glucan from yeast cell wall could yield a high-quality and quantity β-glucan. On the other hand, although the water extraction method is easy, fast, and low cost, it is not efficient to obtain a pure substance. β-glucan extracted using the acid-base method showed a promising antimicrobial, antitoxic, immunostimulatory, and anticancer activity, and further studies are needed to better evaluate its efficacy in other biological systems.

Supplementary Materials: Supplementary data associated with this article can be found in the online version.

Author Contributions: E.M.A., S.H.S., A.A.E., A.-N.A.Z., S.H. and E.A.A.-A. designed and conducted the biological part. A.A.M., and I.M.A.M. carried out the chemistry part. All authors have read and agreed to the published version of the manuscript.

Funding: This work was financially supported by the academy of Scientific Research and Technology (ASRT), Egypt grant no 6371 under the project Science up, and ASRT is second affliation.

Institutional Review Board Statement: Not applicable.

Informed Consent Statement: Not applicable.

Data Availability Statement: The data presented in this study are available in supplementary material.

Conflicts of Interest: The authors declare no conflict of interest.

References

1. Volman, J.J.; Ramakers, J.D.; Plat, J. Dietary Modulation of Immune Function by β-Glucans. *Physiol. Behav.* **2008**, *94*, 276–284. [CrossRef]
2. Upadhyay, T.K.; Fatima, N.; Sharma, D.; Saravanakumar, V.; Sharma, R. Preparation and Characterization of Beta-Glucan Particles Containing a Payload of Nanoembedded Rifabutin for Enhanced Targeted Delivery to Macrophages. *Excli J.* **2017**, *16*, 210.
3. Chamidah, A.; Hardoko; Prihanto, A.A. Antibacterial Activities of β-Glucan (Laminaran) against Gram-Negative and Gram-Positive Bacteria. In *AIP Conference Proceedings*; AIP Publishing LLC, 2017; Volume 1844, p. 20011. Available online: https://aip.scitation.org/doi/abs/10.1063/1.4983422# (accessed on 23 March 2021).
4. Marasca, E.; Boulos, S.; Nyström, L. Bile Acid-Retention by Native and Modified Oat and Barley β-Glucan. *Carbohydr. Polym.* **2020**, *236*, 116034. [CrossRef] [PubMed]
5. Muta, T. Molecular Basis for Invertebrate Innate Immune Recognition of (1→3)-β-D-Glucan as a Pathogen-Associated Molecular Pattern. *Curr. Pharm. Des.* **2006**, *12*, 4155–4161. [CrossRef]
6. Gantner, B.N.; Simmons, R.M.; Canavera, S.J.; Akira, S.; Underhill, D.M. Collaborative Induction of Inflammatory Responses by Dectin-1 and Toll-like Receptor 2. *J. Exp. Med.* **2003**, *197*, 1107–1117. [CrossRef]
7. Chan, G.C.-F.; Chan, W.K.; Sze, D.M.-Y. The Effects of β-Glucan on Human Immune and Cancer Cells. *J. Hematol. Oncol.* **2009**, *2*, 1–11. [CrossRef] [PubMed]
8. Qi, C.; Cai, Y.; Gunn, L.; Ding, C.; Li, B.; Kloecker, G.; Qian, K.; Vasilakos, J.; Saijo, S.; Iwakura, Y. Differential Pathways Regulating Innate and Adaptive Antitumor Immune Responses by Particulate and Soluble Yeast-Derived β-Glucans. *Blood J. Am. Soc. Hematol.* **2011**, *117*, 6825–6836. [CrossRef] [PubMed]

9. Zhang, M.; Yan, L.; Kim, J.A. Modulating Mammary Tumor Growth, Metastasis and Immunosuppression by SiRNA-Induced MIF Reduction in Tumor Microenvironment. *Cancer Gene Ther.* **2015**, *22*, 463–474. [CrossRef]

10. Pelley, R.P.; Strickland, F.M. Plants, Polysaccharides, and the Treatment and Prevention of Neoplasia. *Crit. Rev. Oncog.* **2000**, *11*. [CrossRef]

11. Camilli, G.; Tabouret, G.; Quintin, J. The Complexity of Fungal β-Glucan in Health and Disease: Effects on the Mononuclear Phagocyte System. *Front. Immunol.* **2018**, *9*, 673. [CrossRef]

12. Zhang, M.; Kim, J.A.; Huang, A.Y.-C. Optimizing Tumor Microenvironment for Cancer Immunotherapy: β-Glucan-Based Nanoparticles. *Front. Immunol.* **2018**, *9*, 341. [CrossRef] [PubMed]

13. Würsch, P.; Pi-Sunyer, F.X. The Role of Viscous Soluble Fiber in the Metabolic Control of Diabetes: A Review with Special Emphasis on Cereals Rich in β-Glucan. *Diabetes Care* **1997**, *20*, 1774–1780. [CrossRef] [PubMed]

14. Browder, W.; Williams, D.; Lucore, P.; Pretus, H.; Jones, E.; McNamee, R. Effect of Enhanced Macrophage Function on Early Wound Healing. *Surgery* **1988**, *104*, 224–230. [PubMed]

15. Vetvicka, V.; Vetvickova, J. B-Glucan Attenuates Chronic Fatique Syndrome in Murine Model. *J. Nat. Sci.* **2015**, *1*, e112.

16. Braaten, J.T.; Wood, P.J.; Scott, F.W.; Wolynetz, M.S.; Lowe, M.K.; Bradley-White, P.; Collins, M.W. Oat Beta-Glucan Reduces Blood Cholesterol Concentration in Hypercholesterolemic Subjects. *Eur. J. Clin. Nutr.* **1994**, *48*, 465–474.

17. Sima, P.; Vannucci, L.; Vetvicka, V. Glucans and Cancer: Historical Perspective. *Cancer Transl. Med.* **2015**, *1*, 209.

18. Kushner, B.H.; Cheung, I.Y.; Modak, S.; Kramer, K.; Ragupathi, G.; Cheung, N.-K. V Phase I Trial of a Bivalent Gangliosides Vaccine in Combination with β-Glucan for High-Risk Neuroblastoma in Second or Later Remission. *Clin. Cancer Res.* **2014**, *20*, 1375–1382. [CrossRef] [PubMed]

19. Richter, J.; Svozil, V.; Král, V.; Dobiášová, L.R.; Stiborová, I.; Vetvicka, V. Clinical Trials of Yeast-Derived β-(1, 3) Glucan in Children: Effects on Innate Immunity. *Ann. Transl. Med.* **2014**, *2*, 15. [PubMed]

20. Vetvicka, V.; Richter, J.; Svozil, V.; Dobiášová, L.R.; Král, V. Placebo-Driven Clinical Trials of Yeast-Derived β-(1-3) Glucan in Children with Chronic Respiratory Problems. *Ann. Transl. Med.* **2013**, *1*, 26. [PubMed]

21. LEGRAS, J.; Merdinoglu, D.; CORNUET, J.; Karst, F. Bread, Beer and Wine: Saccharomyces Cerevisiae Diversity Reflects Human History. *Mol. Ecol.* **2007**, *16*, 2091–2102. [CrossRef] [PubMed]

22. Klis, F.M.; Mol, P.; Hellingwerf, K.; Brul, S. Dynamics of Cell Wall Structure in Saccharomyces Cerevisiae. *FEMS Microbiol. Rev.* **2002**, *26*, 239–256. [CrossRef]

23. Sánchez-Madrigal, M.Á.; Neder-Suárez, D.; Quintero-Ramos, A.; Ruiz-Gutiérrez, M.G.; Meléndez-Pizarro, C.O.; Piñón-Castillo, H.A.; Galicia-García, T.; Ramírez-Wong, B. Physicochemical Properties of Frozen Tortillas from Nixtamalized Maize Flours Enriched with β-Glucans. *Food Sci. Technol.* **2015**, *35*, 552–560. [CrossRef]

24. Stier, H.; Ebbeskotte, V.; Gruenwald, J. Immune-Modulatory Effects of Dietary Yeast Beta-1, 3/1, 6-D-Glucan. *Nutr. J.* **2014**, *13*, 38. [CrossRef] [PubMed]

25. Ahmad, A.; Anjum, F.M.; Zahoor, T.; Nawaz, H.; Dilshad, S.M.R. Beta Glucan: A Valuable Functional Ingredient in Foods. *Crit. Rev. Food Sci. Nutr.* **2012**, *52*, 201–212. [CrossRef]

26. Hromádková, Z.; Ebringerová, A.; Sasinková, V.; Šandula, J.; Hříbalová, V.; Omelková, J. Influence of the Drying Method on the Physical Properties and Immunomodulatory Activity of the Particulate (1→3)-β-D-Glucan from Saccharomyces Cerevisiae. *Carbohydr. Polym.* **2003**, *51*, 9–15. [CrossRef]

27. Allan, S.E.; Crome, S.Q.; Crellin, N.K.; Passerini, L.; Steiner, T.S.; Bacchetta, R.; Roncarolo, M.G.; Levings, M.K. Activation-Induced FOXP3 in Human T Effector Cells Does Not Suppress Proliferation or Cytokine Production. *Int. Immunol.* **2007**, *19*, 345–354. [CrossRef] [PubMed]

28. Wenger, M.D.; DePhillips, P.; Bracewell, D.G. A Microscale Yeast Cell Disruption Technique for Integrated Process Development Strategies. *Biotechnol. Prog.* **2008**, *24*, 606–614. [CrossRef] [PubMed]

29. Ferrer, P. Revisiting the Cellulosimicrobium Cellulans Yeast-Lytic β-1, 3-Glucanases Toolbox: A Review. *Microb. Cell Fact.* **2006**, *5*, 10. [CrossRef] [PubMed]

30. Pengkumsri, N.; Sivamaruthi, B.S.; Sirilun, S.; Peerajan, S.; Kesika, P.; Chaiyasut, K.; Chaiyasut, C. Extraction of β-Glucan from Saccharomyces Cerevisiae: Comparison of Different Extraction Methods and in Vivo Assessment of Immunomodulatory Effect in Mice. *Food Sci. Technol.* **2017**, *37*, 124–130. [CrossRef]

31. Babayan, T.L.; Bezrukov, M.G. Autolysis in Yeasts. *Acta Biotechnol.* **1985**, *5*, 129–136. [CrossRef]

32. Babayan, T.L.; Bezrukov, M.G.; Latov, V.K.; Belikov, V.M.; Belavtseva, E.M.; Titova, E.F. Induced Autolysis of Saccharomyces Cerevisiae: Morphological Effects, Rheological Effects, and Dynamics of Accumulation of Extracellular Hydrolysis Products. *Curr. Microbiol.* **1981**, *5*, 163–168. [CrossRef]

33. Charpentier, C.; Freyssinet, M. The Mechanism of Yeast Autolysis in Wine. *Yeast (Chichester)* **1989**, *5*, 181–186.

34. Zohri, A.N.A.; Moubasher, H.; Abdel-Hay, H.M.; Orban, M.A.I. Biotechnological β-Glucan Production from Returned Bakers Yeast and Yeast Remaining after Ethanol Fermentation. *Egypt. Sugar J.* **2019**, *4*, 5.

35. Bzducha-Wróbel, A.; Błażejak, S.; Kawarska, A.; Stasiak-Różańska, L.; Gientka, I.; Majewska, E. Evaluation of the Efficiency of Different Disruption Methods on Yeast Cell Wall Preparation for β-Glucan Isolation. *Molecules* **2014**, *19*, 20941–20961. [CrossRef] [PubMed]

36. Liu, D.; Zeng, X.-A.; Sun, D.-W.; Han, Z. Disruption and Protein Release by Ultrasonication of Yeast Cells. *Innov. Food Sci. Emerg. Technol.* **2013**, *18*, 132–137. [CrossRef]

37. Pattanayak, S.; Chakraborty, S.; Biswas, S.; Chattopadhyay, D.; Chakraborty, M. Degradation of Methyl Parathion, a Common Pesticide and Fluorescence Quenching of Rhodamine B, a Carcinogen Using β-d Glucan Stabilized Gold Nanoparticles. *J. Saudi Chem. Soc.* **2018**, *22*, 937–948. [CrossRef]
38. Reddy, K.R.N.; Salleh, B.; Saad, B.; Abbas, H.K.; Abel, C.A.; Shier, W.T. Erratum: An Overview of Mycotoxin Contamination in Foods and Its Implications for Human Health (Toxin Reviews (2010) 29 (326)). *Toxin Rev.* **2010**, *29*. [CrossRef]
39. Williams, J.H.; Phillips, T.D.; Jolly, P.E.; Stiles, J.K.; Jolly, C.M.; Aggarwal, D. Human Aflatoxicosis in Developing Countries: A Review of Toxicology, Exposure, Potential Health Consequences, and Interventions. *Am. J. Clin. Nutr.* **2004**, *80*, 1106–1122. [CrossRef] [PubMed]
40. Madrigal-Santillán, E.; Morales-González, J.A.; Vargas-Mendoza, N.; Reyes-Ramírez, P.; Cruz-Jaime, S.; Sumaya-Martínez, T.; Pérez-Pastén, R.; Madrigal-Bujaidar, E. Antigenotoxic Studies of Different Substances to Reduce the DNA Damage Induced by Aflatoxin B1 and Ochratoxin A. *Toxins* **2010**, *2*, 738–757. [CrossRef] [PubMed]
41. Cleveland, T.E.; Dowd, P.F.; Desjardins, A.E.; Bhatnagar, D.; Cotty, P.J. United States Department of Agriculture—Agricultural Research Service Research on Pre-harvest Prevention of Mycotoxins and Mycotoxigenic Fungi in US Crops. *Pest Manag. Sci. Former. Pestic. Sci.* **2003**, *59*, 629–642. [CrossRef] [PubMed]
42. Anttila, A.; Bhat, R.V.; Bond, J.A.; Borghoff, S.J.; Bosch, F.X.; Carlson, G.P.; Castegnaro, M.; Cruzan, G.; Gelderblom, W.C.A.; Hass, U. IARC Monographs on the Evaluation of Carcinogenic Risks to Humans: Some Traditional Herbal Medicines, Some Mycotoxins, Naphthalene and Styrene. *IARC Monogr. Eval. Carcinog. Risks Hum.* **2002**, *82*. Available online: https://ucdavis.pure.elsevier.com/en/publications/iarc-monographs-on-the-evaluation-of-carcinogenic-risks-to-humans (accessed on 1 April 2021).
43. Baptista, A.S.; Horii, J.; Calori-Domingues, M.A.; Da Glória, E.M.; Salgado, J.M.; Vizioli, M.R. The Capacity of Manno-Oligosaccharides, Thermolysed Yeast and Active Yeast to Attenuate Aflatoxicosis. *World J. Microbiol. Biotechnol.* **2004**, *20*, 475–481. [CrossRef]
44. Bianco, G.; Russo, R.; Marzocco, S.; Velotto, S.; Autore, G.; Severino, L. Modulation of Macrophage Activity by Aflatoxins B1 and B2 and Their Metabolites Aflatoxins M1 and M2. *Toxicon* **2012**, *59*, 644–650. [CrossRef]
45. Roda, E.; Coccini, T.; Acerbi, D.; Castoldi, A.F.; Manzo, L. Comparative in Vitro and Ex-Vivo Myelotoxicity of Aflatoxins B1 and M1 on Haematopoietic Progenitors (BFU-E, CFU-E, and CFU-GM): Species-Related Susceptibility. *Toxicol. In Vitro* **2010**, *24*, 217–223. [CrossRef]
46. Shetty, P.H.; Jespersen, L. Saccharomyces Cerevisiae and Lactic Acid Bacteria as Potential Mycotoxin Decontaminating Agents. *Trends Food Sci. Technol.* **2006**, *17*, 48–55. [CrossRef]
47. Dalié, D.K.D.; Deschamps, A.M.; Richard-Forget, F. Lactic Acid Bacteria—Potential for Control of Mould Growth and Mycotoxins: A Review. *Food Control* **2010**, *21*, 370–380. [CrossRef]
48. Piotrowska, M. Adsorption of Ochratoxin A by Saccharomyces Cerevisiae Living and Non-Living Cells. *Acta Aliment.* **2012**, *41*, 1–7. [CrossRef]
49. Piotrowska, M.; Nowak, A.; Czyzowska, A. Removal of Ochratoxin A by Wine Saccharomyces Cerevisiae Strains. *Eur. Food Res. Technol.* **2013**, *236*, 441–447. [CrossRef]
50. Petruzzi, L.; Bevilacqua, A.; Baiano, A.; Beneduce, L.; Corbo, M.R.; Sinigaglia, M. In Vitro Removal of Ochratoxin A by Two Strains of S Accharomyces Cerevisiae and Their Performances under Fermentative and Stressing Conditions. *J. Appl. Microbiol.* **2014**, *116*, 60–70. [CrossRef]
51. Zhao, L.; Jin, H.; Lan, J.; Zhang, R.; Ren, H.; Zhang, X.; Yu, G. Detoxification of Zearalenone by Three Strains of Lactobacillus Plantarum from Fermented Food in Vitro. *Food Control* **2015**, *54*, 158–164. [CrossRef]
52. Raju, M.; Devegowda, G. Influence of Esterified-Glucomannan on Performance and Organ Morphology, Serum Biochemistry and Haematology in Broilers Exposed to Individual and Combined Mycotoxicosis (Aflatoxin, Ochratoxin and T-2 Toxin). *Br. Poult. Sci.* **2000**, *41*, 640–650. [CrossRef]
53. Yiannikouris, A.; FRANCois, J.; Poughon, L.; Dussap, C.-G.; Bertin, G.; Jeminet, G.; Jouany, J.-P. Adsorption of Zearalenone by β-D-Glucans in the Saccharomyces Cerevisiae Cell Wall. *J. Food Prot.* **2004**, *67*, 1195–1200. [CrossRef]
54. Yiannikouris, A.; André, G.; Poughon, L.; François, J.; Dussap, C.-G.; Jeminet, G.; Bertin, G.; Jouany, J.-P. Chemical and Conformational Study of the Interactions Involved in Mycotoxin Complexation with β-D-Glucans. *Biomacromolecules* **2006**, *7*, 1147–1155. [CrossRef]
55. Pereyra, C.M.; Cavaglieri, L.R.; Chiacchiera, S.M.; Dalcero, A. The Corn Influence on the Adsorption Levels of Aflatoxin B 1 and Zearalenone by Yeast Cell Wall. *J. Appl. Microbiol.* **2013**, *114*, 655–662. [CrossRef] [PubMed]
56. Madrigal-Bujaidar, E.; Morales-González, J.A.; Sánchez-Gutiérrez, M.; Izquierdo-Vega, J.A.; Reyes-Arellano, A.; Álvarez-González, I.; Pérez-Pasten, R.; Madrigal-Santillán, E. Prevention of Aflatoxin B1-Induced Dna Breaks by β-D-Glucan. *Toxins* **2015**, *7*, 2145–2158. [CrossRef]
57. Aazami, M.H.; Nasri, M.H.F.; Mojtahedi, M.; Mohammadi, S.R. In Vitro Aflatoxin B1 Binding by the Cell Wall and (1→3)-β-d-Glucan of Baker's Yeast. *J. Food Prot.* **2018**, *81*, 670–676. [CrossRef]
58. Akhapkina, I.G.; Antropova, A.B.; Akhmatov, E.A.; Zheltikova, T.M. Effects of the Linear Fragments of Beta-(1→3)-Glucans on Cytokine Production in Vitro. *Biochemistry* **2018**, *83*, 1002–1006. [CrossRef] [PubMed]
59. Ginovyan, M.; Keryan, A.; Bazukyan, I.; Ghazaryan, P.; Trchounian, A. The Large Scale Antibacterial, Antifungal and Anti-Phage Efficiency of Petamcin-A: New Multicomponent Preparation for Skin Diseases Treatment. *Ann. Clin. Microbiol. Antimicrob.* **2015**, *14*, 1–7. [CrossRef] [PubMed]

60. Irshad, A.; Sarwar, N.; Sadia, H.; Malik, K.; Javed, I.; Irshad, A.; Afzal, M.; Abbas, M.; Rizvi, H. Comprehensive Facts on Dynamic Antimicrobial Properties of Polysaccharides and Biomolecules-Silver Nanoparticle Conjugate. *Int. J. Biol. Macromol.* **2020**, *145*, 189–196. [CrossRef] [PubMed]

61. Sun, X.; Gao, Y.; Ding, Z.; Zhao, Y.; Yang, Y.; Sun, Q.; Yang, X.; Ge, W.; Xu, X.; Cheng, R. Soluble Beta-Glucan Salecan Improves Vaginal Infection of Candida Albicans in Mice. *Int. J. Biol. Macromol.* **2020**, *148*, 1053–1060. [CrossRef]

62. Sudjaswadi, R. Increasing of the Bacteriostatic Effects of HCl Tetracycline–Polyethylene Glycol 6000–Tween 80 (PT). *Indones. J. Pharm.* **2006**, *17*, 98–103.

63. Javmen, A.; Nemeikaitė-Čėnienė, A.; Bratchikov, M.; Grigiškis, S.; Grigas, F.; Jonauskienė, I.; Zabulytė, D.; Mauricas, M. β-Glucan from Saccharomyces Cerevisiae Induces IFN-γ Production in Vivo in BALB/c Mice. *In Vivo* **2015**, *29*, 359–363. [PubMed]

64. Liang, J.; Melican, D.; Cafro, L.; Palace, G.; Fisette, L.; Armstrong, R.; Patchen, M.L. Enhanced Clearance of a Multiple Antibiotic Resistant Staphylococcus Aureus in Rats Treated with PGG-Glucan Is Associated with Increased Leukocyte Counts and Increased Neutrophil Oxidative Burst Activity. *Int. J. Immunopharmacol.* **1998**, *20*, 595–614. [CrossRef]

65. Soltys, J.; Quinn, M.T. Modulation of Endotoxin-and Enterotoxin-Induced Cytokine Release by in Vivo Treatment with β-(1, 6)-Branched β-(1, 3)-Glucan. *Infect. Immun.* **1999**, *67*, 244–252. [CrossRef]

66. Williams, D.L.; Browder, I.W.; Di Luzio, N.R. Immunotherapeutic Modification of Escherichia Coli-Induced Experimental Peritonitis and Bacteremia by Glucan. *Surgery* **1983**, *93*, 448–454.

67. Markova, N.; Kussovski, V.; Drandarska, I.; Nikolaeva, S.; Georgieva, N.; Radoucheva, T. Protective Activity of Lentinan in Experimental Tuberculosis. *Int. Immunopharmacol.* **2003**, *3*, 1557–1562. [CrossRef]

68. Korotchenko, E.; Schießl, V.; Scheiblhofer, S.; Schubert, M.; Dall, E.; Joubert, I.A.; Strandt, H.; Neuper, T.; Sarajlic, M.; Bauer, R. Laser-facilitated Epicutaneous Immunotherapy with Hypoallergenic Beta-glucan Neoglycoconjugates Suppresses Lung Inflammation and Avoids Local Side Effects in a Mouse Model of Allergic Asthma. *Allergy* **2021**, *76*, 210–222. [CrossRef] [PubMed]

69. Hansen, J.Ø.; Lagos, L.; Lei, P.; Reveco-Urzua, F.E.; Morales-Lange, B.; Hansen, L.D.; Schiavone, M.; Mydland, L.T.; Arntzen, M.Ø.; Mercado, L. Down-Stream Processing of Baker's Yeast (Saccharomyces Cerevisiae)—Effect on Nutrient Digestibility and Immune Response in Atlantic Salmon (Salmo Salar). *Aquaculture* **2021**, *530*, 735707. [CrossRef]

70. Ina, K.; Kataoka, T.; Ando, T. The Use of Lentinan for Treating Gastric Cancer. *Anti-Cancer Agents Med. Chem. (Formerly Curr. Med. Chem. Agents)* **2013**, *13*, 681–688. [CrossRef]

71. Demir, G.; Klein, H.O.; Mandel-Molinas, N.; Tuzuner, N. Beta Glucan Induces Proliferation and Activation of Monocytes in Peripheral Blood of Patients with Advanced Breast Cancer. *Int. Immunopharmacol.* **2007**, *7*, 113–116. [CrossRef]

72. Shomori, K.; Yamamoto, M.; Arifuku, I.; Teramachi, K.; Ito, H. Antitumor Effects of a Water-Soluble Extract from Maitake (Grifola Frondosa) on Human Gastric Cancer Cell Lines. *Oncol. Rep.* **2009**, *22*, 615–620. [CrossRef]

73. Vetvicka, V.; Vetvickova, J. Glucans and Cancer: Comparison of Commercially Available β-Glucans—Part IV. *Anticancer Res.* **2018**, *38*, 1327–1333.

74. Sambrani, R.; Abdolalizadeh, J.; Kohan, L.; Jafari, B. Recent Advances in the Application of Probiotic Yeasts, Particularly Saccharomyces, as an Adjuvant Therapy in the Management of Cancer with Focus on Colorectal Cancer. *Mol. Biol. Rep.* **2021**, *48*, 951–960.

75. Vetvicka, V.; Vetvickova, J. Anti-Infectious and Anti-Tumor Activities of β-Glucans. *Anticancer Res.* **2020**, *40*, 3139–3145. [CrossRef] [PubMed]

76. Kim, M.-J.; Hong, S.-Y.; Kim, S.-K.; Cheong, C.; Park, H.-J.; Chun, H.-K.; Jang, K.-H.; Yoon, B.-D.; Kim, C.-H.; Kang, S.A. β-Glucan Enhanced Apoptosis in Human Colon Cancer Cells SNU-C4. *Nutr. Res. Pract.* **2009**, *3*, 180–184. [CrossRef]

77. Piotrowska, M.; Masek, A. Saccharomyces Cerevisiae Cell Wall Components as Tools for Ochratoxin A Decontamination. *Toxins* **2015**, *7*, 1151–1162. [CrossRef]

78. Freimund, S.; Sauter, M.; Käppeli, O.; Dutler, H. A New Non-Degrading Isolation Process for 1, 3-β-D-Glucan of High Purity from Baker's Yeast Saccharomyces Cerevisiae. *Carbohydr. Polym.* **2003**, *54*, 159–171. [CrossRef]

79. Liu, X.-Y.; Wang, Q.; Cui, S.W.; Liu, H.-Z. A New Isolation Method of β-D-Glucans from Spent Yeast Saccharomyces Cerevisiae. *Food Hydrocoll.* **2008**, *22*, 239–247. [CrossRef]

80. Kesika, P.; Prasanth, M.I.; Balamurugan, K. Modulation of Caenorhabditis Elegans Immune Response and Modification of Shigella Endotoxin upon Interaction. *J. Basic Microbiol.* **2015**, *55*, 432–450. [CrossRef]

81. Sivamaruthi, B.S.; Prasanth, M.I.; Balamurugan, K. Alterations in Caenorhabditis Elegans and Cronobacter Sakazakii Lipopolysaccharide during Interaction. *Arch. Microbiol.* **2015**, *197*, 327–337. [CrossRef]

82. Bradford, M.M. A Rapid and Sensitive Method for the Quantitation of Microgram Quantities of Protein Utilizing the Principle of Protein-Dye Binding. *Anal. Biochem.* **1976**, *72*, 248–254. [CrossRef]

83. Dubois, M.; Gilles, K.A.; Hamilton, J.K.; Rebers, P.T.; Smith, F. Colorimetric Method for Determination of Sugars and Related Substances. *Anal. Chem.* **1956**, *28*, 350–356. [CrossRef]

84. Zaki, S.A.; Abd-Elrahman, M.I.; Abu-Sehly, A.A.; Shaalan, N.M.; Hafiz, M.M. Thermal Annealing of SnS Thin Film Induced Mixed Tin Sulfide Oxides-Sn2S3 for Gas Sensing: Optical and Electrical Properties. *Mater. Sci. Semicond. Process.* **2018**, *75*, 214–220. [CrossRef]

85. Zohri, A.N.; Aboul-Nasr, M.B.; Adam, M.; Mustafa, M.A.; Amer, E.M. Impact of Enzymes and Toxins Potentiality of Four Aspergillus Species to Cause Aspergillosis. *Biol. Med.* **2017**, *9*, 2.

86. Farhan, M.A.; Moharram, A.M.; Salah, T.; Shaaban, O.M. Types of Yeasts That Cause Vulvovaginal Candidiasis in Chronic Users of Corticosteroids. *Med. Mycol.* **2019**, *57*, 681–687. [CrossRef] [PubMed]
87. Wiegand, I.; Hilpert, K.; Hancock, R.E.W. Agar and Broth Dilution Methods to Determine the Minimal Inhibitory Concentration (MIC) of Antimicrobial Substances. *Nat. Protoc.* **2008**, *3*, 163–175. [CrossRef] [PubMed]

Article

Antioxidant and Polyphenol-Rich Ethanolic Extract of *Rubia tinctorum* L. Prevents Urolithiasis in an Ethylene Glycol Experimental Model in Rats

Fatima Zahra Marhoume [1,2], Rachida Aboufatima [3], Younes Zaid [4,5,6], Youness Limami [4,6], Raphaël E. Duval [7,*], Jawad Laadraoui [2], Anass Belbachir [8,9], Abderrahmane Chait [2] and Abdallah Bagri [1,*]

[1] Laboratory of Biochemistry and Neuroscience, Integrative and Computational Neuroscience Team, Faculty of Sciences and Technology, Hassan First University, Settat 26002, Morocco; marhoume.fatimazahra@gmail.com

[2] Laboratory of Neurobiology, Pharmacology and Behavior, Faculty of Sciences Semlalia, Cadi Ayad University, Marrakech 40000, Morocco; jawad.laadraoui@edu.uca.ac.ma (J.L.); chait@uca.ma (A.C.)

[3] Laboratory of Biological Engineering, Faculty of Sciences and Technology, Sultan Moulay Slimane University, Beni Mellal 23000, Morocco; raboufatima@gmail.com

[4] Research Center of Abulcasis University of Health Sciences, Rabat 10000, Morocco; younes_zaid@yahoo.ca (Y.Z.); youness.limami@gmail.com (Y.L.)

[5] Botany Laboratory, Biology Department, Faculty of Sciences, Mohammed V University in Rabat, Rabat 10000, Morocco

[6] Immunology and Biodiversity Laboratory, Department of Biology, Faculty of Sciences, Hassan II University, Casablanca 20000, Morocco

[7] Université de Lorraine, CNRS, L2CM, F-54000 Nancy, France

[8] Morpho-Science Research Laboratory, Faculty of Medicine and Pharmacy, Cadi Ayad University, Marrakech 40000, Morocco; belbachir@gmail.com

[9] Regenerative Medicine Center University Hospital Center of Mohammed VI Marrakech, Marrakech 40000, Morocco

* Correspondence: raphael.duval@univ-lorraine.fr (R.E.D.); abagri511@gmail.com (A.B.); Tel.: +33-3-7274-7218 (R.E.D.); +212-5-2340-0736 (A.B.)

Citation: Marhoume, F.Z.; Aboufatima, R.; Zaid, Y.; Limami, Y.; Duval, R.E.; Laadraoui, J.; Belbachir, A.; Chait, A.; Bagri, A. Antioxidant and Polyphenol-Rich Ethanolic Extract of *Rubia tinctorum* L. Prevents Urolithiasis in an Ethylene Glycol Experimental Model in Rats. *Molecules* **2021**, *26*, 1005. https://doi.org/10.3390/molecules26041005

Academic Editors: Raffaele Pezzani and Sara Vitalini

Received: 12 January 2021
Accepted: 11 February 2021
Published: 14 February 2021

Publisher's Note: MDPI stays neutral with regard to jurisdictional claims in published maps and institutional affiliations.

Abstract: Treatment of kidney stones is based on symptomatic medications which are associated with side effects such as gastrointestinal symptoms (e.g., nausea, vomiting) and hepatotoxicity. The search for effective plant extracts without the above side effects has demonstrated the involvement of antioxidants in the treatment of kidney stones. A local survey in Morocco has previously revealed the frequent use of *Rubia tinctorum* L. (RT) for the treatment of kidney stones. In this study, we first explored whether RT ethanolic (E-RT) and ethyl acetate (EA-RT) extracts of *Rubia tinctorum* L. could prevent the occurrence of urolithiasis in an experimental 0.75% ethylene glycol (EG) and 2% ammonium chloride (AC)-induced rat model. Secondly, we determined the potential antioxidant potency as well as the polyphenol composition of these extracts. An EG/AC regimen for 10 days induced the formation of bipyramid-shaped calcium oxalate crystals in the urine. Concomitantly, serum and urinary creatinine, urea, uric acid, phosphorus, calcium, sodium, potassium, and chloride were altered. The co-administration of both RT extracts prevented alterations in all these parameters. In the EG/AC-induced rat model, the antioxidants- and polyphenols-rich E-RT and EA-RT extracts significantly reduced the presence of calcium oxalate in the urine, and prevented serum and urinary biochemical alterations together with kidney tissue damage associated with urolithiasis. Moreover, we demonstrated that the beneficial preventive effects of E-RT co-administration were more pronounced than those obtained with EA-RT. The superiority of E-RT was associated with its more potent antioxidant effect, due to its high content in polyphenols.

Keywords: *Rubia tinctorum* L.; antioxidants; polyphenols; ethylene glycol; urolithiasis; histophatology

1. Introduction

Urolithiasis, or kidney stones, is a major health concern with increasing prevalence rates worldwide [1]. It results from free or attached mineral crystallizations in the renal calyces [2]. The crystalline and organic components are formed when the urine becomes supersaturated with minerals, then they grow, and aggregate before being retained [3]. About 80% of calculi are composed of calcium phosphate (CaP) mixed with calcium oxalate (CaOx) [4]. Multiple factors can cause urolithiasis, including a sedentary lifestyle, unhealthy diet, irregular food habits and obesity [5]. the treatment of kidney stones involves the administration of symptomatic drugs (diuretics, anti-inflammatory drugs), percutaneous, nephrolithotomy and lithotripsy [6]. However, these treatments are frequently associated with complications such as hemorrhage, hypertension, and tubular necrosis followed by subsequent fibrosis [7]. In addition, they are very expensive and to date, there is no promising drug for the treatment and prevention of recurrence [8].

Herbal remedies could be an alternative to anti-urolithiasic drugs due to their many active compounds that can act synergistically and have—most the time—minimal side effects [9]. It was suggested that plants with anti-urolithiasic properties induce their effect via antioxidant capacities that mitigate the toxicity caused by free radicals involved in the initiation and development of urolithiasis [9]. Many experimental studies support this hypothesis. Indeed, phenols and flavonoids have been shown to be effective in attenuating the process of calculus formation, both in animal models and in humans [10]. Polyphenols from grape seeds prevented renal papilla from calcium monohydrate oxalate, calculi formation and lesions induced by oxidant cytotoxic substances [11]. Furthermore, polyphenols-rich extract from *Quercus gilva* Blume showed an anti-urolithiasic effect associated with its antioxidant and anti-inflammatory properties [12].

Rubia tinctorum L. (madder root) (RT) is a plant belonging to the Rubiaceae family whose root is used as a folk medicine to cure various ailments, including kidney stones and bladder diseases in several countries in Asia, Russia and Europe [9,13]. The therapeutic properties of RT such as anti-inflammatory, antioxidant, hepatoprotective and antibacterial activities were confirmed in vivo and in vitro by experimental data [14]. A local survey in Morocco has previously revealed the frequent use of *Rubia tinctorum* L. (RT) for the treatment of kidney stones [15]. However, to our knowledge, no experimental study reported an anti-urolithiasic effect of RT. Therefore, the present study was conducted on an ethylene glycol (EG)/ammonium chloride (AC) experimental model of urolithiasis in rats. The aim of this study was to evaluate the potential protective effect of both ethanolic and ethyl acetate extracts of *Rubia tinctorum* L. (E-RT and EA-RT, respectively) in this model. Then, the objective was to assess the RT extract's antioxidant activity and to identify the polyphenols contained in E-RT and EA-RT extracts which could be linked to these effects.

2. Results

2.1. Pathophysiology of EG/AC-Induced Urolithiasis

The treatment of rats for 10 days with EG/AC resulted in urolithiasis formation and in alterations in urinary, serum and kidney tissue parameters. Analysis of the changes induced showed that our experimental design reproduced the pathophysiology of a form of urolithiasis in the rat. As shown in Table 1, the levels of serum urea, creatinine, uric acid, calcium and phosphorus of the lithiasic group (G2) were significantly increased (5.91 ± 0.17, 39.51 ± 4.86, 44.86 ± 3.00, 119 ± 1.59 and 107.16 ± 1.74, respectively) compared to the vehicle control group (G1) (0.26 ± 0.02, 5.49 ± 0.21, 10.71 ± 0.29, 103.77 ± 1.01 and 83.34 ± 0.33, respectively). The levels of urinary urea, creatinine, uric acid, calcium and phosphorus concentrations were also significantly increased in the urolithiasic group (317.90 ± 0.95, 116.28 ± 4.02, 230.5 ± 10.79, 217.08 ± 3.73 and 2169.41 ± 1.47, respectively) when compared to G1 (33.96 ± 0.95, 27.75 ± 1.66, 58.33 ± 3.48, 74.36 ± 1.36 and 12.83 ± 0.60, respectively) (Table 2). Similar changes were also observed for electrolyte concentrations in urine and serum (e.g., Na⁺, Cl⁻, K⁺) (Tables 1 and 2). Indeed, the urolithiasic group's urine contained significant higher concentrations of potassium, chloride and sodium (196.03 ± 0.81, 132.31 ± 0.75 and 18.33 ± 0.49, respectively) than G1 (6.13 ± 0.16, 36.45 ± 0.59 and 8.33 ± 0.21 respectively). Moreover, marked histological alterations (Table 3), including interstitial mononuclear cell infiltration and damage in glomeruli, were also observed in the urolithiasic group but not in the vehicle control group.

Table 1. Effects of ethanolic and ethyl acetate extracts of *Rubia tinctorum* L. (E-RT and EA-RT, respectively, 1 g/kg and 2 g/kg) on serum parameters on 10 days ethylene glycol and ammonium chloride (EG/AC)-induced urolithiasis rat model.

Parameters	G1 Vehicle Control	G2 Lithiasic Control	G3 E-RT 1 g/kg	G4 E-RT 2 g/kg	G5 EA-RT 1 g/kg	G6 EA-RT 2 g/kg
Urea (g/L)	0.26 ± 0.02	5.91 ± 0.17 [b]	0.40 ± 0.01 [a,c]	0.31 ± 0.02 [a]	0.45 ± 0.07 [a,b]	0.38 ± 0.04 [a,d]
Creatinine (mg/L)	5.49 ± 0.21	39.51 ± 4.86 [b]	5.65 ± 0.71 [a]	5.16 ± 0.79 [a]	7.08 ± 0.37 [a]	6.93 ± 0.83 [a]
Ca^{2+} (mg/L)	103.77 ± 1.01	119 ± 1.59 [b]	106.33 ± 0.98 [a]	104.50 ± 1.03 [a]	109.5 ± 2.09 [a]	107.7 ± 1.63 [a]
Uric acid (mg/L)	10.71 ± 0.29	44.86 ± 3.00 [b]	13.66 ± 0.95 [a]	11.00 ± 1.03 [a]	18.66 ± 1.28 [a,b]	14.66 ± 1.60 [a]
Phosphorus (mg/L)	83.34 ± 0.33	107.16 ± 1.74 [b]	85.44 ± 1.39 [a]	84.03 ± 3.40 [a]	92.76 ± 3.23 [a,d]	91.38 ± 1.26 [a]
Na^+ (mmol/L)	137.53 ± 0.29	182.55 ± 2.32 [b]	139.20 ± 0.54 [a]	137.63 ± 0.53 [a]	141.68 ± 0.67 [a,b]	139.97 ± 2.07 [a]
Cl^- (mmol/L)	93.68 ± 0.72	151.55 ± 1.30 [b]	94.49 ± 1.17 [a]	93.00 ± 0.81 [a]	96.48 ± 0.76 [a]	95.34 ± 0.96 [a]
K^+ (mmol/L)	8.24 ± 0.07	12.27 ± 0.34 [b]	8.55 ± 0.07 [a]	8.36 ± 0.09 [a]	9.11 ± 0.39 [a]	8.56 ± 0.25 [a]

Values are mean $\pm$ standard error (SEM), $n = 6$, [a] $p < 0.001$ significantly different compared to the lithiasic control, [b] $p < 0.001$, [c] $p < 0.01$, and [d] $p < 0.05$ significantly different compared to the vehicle control. E-RT: ethanolic extract of RT; EA-RT: ethyl acetate extract of RT.

Table 2. Effects of ethanolic and ethyl acetate extracts of *Rubia tinctorum* L. (E-RT and EA-RT, respectively), at different concentrations (i.e., 1 g/kg and 2 g/kg) on urine parameters on 10 days ethylene glycol and ammonium chloride (EG/AC)-induced urolithiasis rat model.

Parameters	G1 Vehicle Control	G2 Lithiasic Control	G3 E-RT 1 g/kg	G4 E-RT 2 g/kg	G5 EA-RT 1 g/kg	G6 EA-RT 2 g/kg
Urea (g/L)	33.96 ± 0.95	317.90 ± 0.95 [b]	27.15 ± 1.85 [a,c]	34.12 ± 1.27 [a]	28.53 ± 1.64 [a]	25.96 ± 0.26 [a,c]
Creatinine (mg/L)	27.75 ± 1.66	116.28 ± 4.02 [b]	35.98 ± 1.53 [a,c]	31.32 ± 1.07 [a]	38 ± 2.31 [a,b]	35.21 ± 1.66 [a,d]
Ca^{2+} (mg/L)	74.36 ± 1.36	217.08 ± 3.73 [b]	98.16 ± 2.65 [a,b]	73.57 ± 1.76 [a]	83.95 ± 3.79 [a]	77.66 ± 1.56 [a]
Uric acid (mg/L)	58.33 ± 3.48	230.5 ± 10.79 [b]	101.83 ± 0.70 [a,b]	70 ± 3.14 [a]	121.83 ± 0.94 [a,b]	149.83 ± 0.98 [a,b]
Phosphorus (mg/L)	12.83 ± 0.60	299.41 ± 1.47 [b]	34.98 ± 0.78 [a,b]	17.16 ± 0.47 [a,d]	234.25 ± 0.93 [a,b]	163.16 ± 0.94 [a,b]
Na^+ (mmol/L)	8.33 ± 0.21	18.33 ± 0.49 [b]	10 ± 0.57 [a]	9.83 ± 1.16 [a]	8.66 ± 0.49 [a]	12 ± 0.57 [a,c]
K^+ (mmol/L)	6.13 ± 0.16	196.03 ± 0.81 [b]	42.74 ± 0.73 [a,b]	12.67 ± 0 [a,b]	22.56 ± 0.29 [a,b]	17.38 ± 0.16 [a,b]
Cl^- (mmol/L)	36.45 ± 0.59	132.31 ± 0.75 [b]	99.42 ± 0.50 [a,b]	56.66 ± 0.34 [a,b]	107.55 ± 2.61 [a,b]	68.94 ± 0.98 [a,b]
Protein (mg/L)	0.28 ± 0.01	1.03 ± 0.04 [b]	0.36 ± 0.06 [a]	0.11 ± 0.01 [a,d]	0.78 ± 0.02 [a,b]	0.76 ± 0.01 [a,b]

Values are mean $\pm$ SEM, $n = 6$; [a] $p < 0.001$ significantly different compared to the lithiasic control. [b] $p < 0.001$, [c] $p < 0.01$, and [d] $p < 0.05$ significantly different compared to the vehicle control.

Table 3. Histopathological changes in the kidneys of urolithiasic rats treated or not treated with RT extracts.

	G1 Vehicle Control	G2 Lithiasic Control	G3 E-RT 1 g/kg	G4 E-RT 2 g/kg	G5 EA-RT 1 g/kg	G6 EA-RT 2 g/kg
Cloudy swelling	-	+++	-	-	++	+
Infiltration of mononuclear cells	-	+++	-	-	++	+
Kidney hemorrhage	-	+++	+	-	++	+
Morphological disorganization of tubules and glomeruli	-	+++	+	+	++	++

The average results of evaluation in each group ($n = 6$) were scored: -: no modifications (no cellular damage); +: slight alteration; ++: moderate alteration; +++: marked alteration.

2.2. Effect of RT Extracts on the Pathophysiology of EG/AC-Induced Urolithiasis

2.2.1. Body Weight

The mean body weights (MBW) of the six experimental groups (i.e., vehicle control (G1), lithiasic control (G2), E-RT 1 g/kg (G3), E-RT 2 g/kg (G4), EA-RT 1 g/kg (G5), EA-RT 2 g/kg (G6)) were similar at the beginning of the experiment (day 1). After 10 days of treatment, the MBW of five groups (G2, G3, G4, G5 and G6) were decreased when compared to (G1) (Figure 1). Figure 1 also shows that the group G2 (i.e., −41.08 g) had a greater decrease in MBW than the RT-treated groups (G3, G4, G5 and G6) (i.e., −13.52 g; −5.68 g; −28.53 g; −22.96 g, respectively) and that the MBW of the group treated with E-RT (1 and 2 g/kg) did not significantly differ from the untreated control G1.

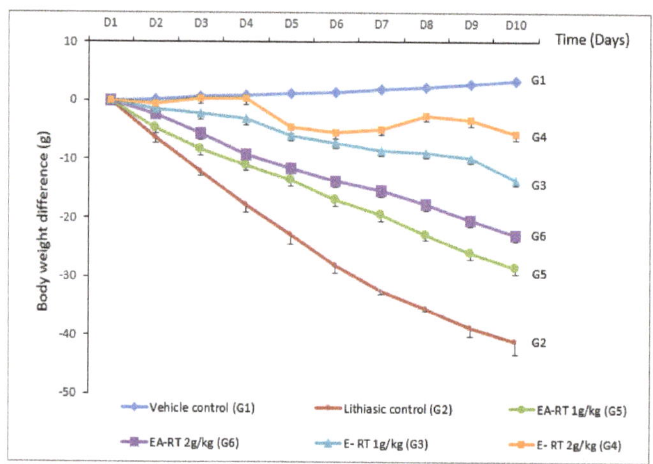

Figure 1. Cumulative body weight difference (mean ± SEM) in the 6 groups (6 rats/group) over the ten days of the treatment. Vehicle control (G1): received distilled water; Lithiasic control (G2): received 0.75% EG + 2% AC; E-RT 1 g/kg (G3): received concomitantly E-RT (1 g/kg) *plus* 0.75% EG + 2% AC; E-RT 2 g/kg (G4): received concomitantly E-RT (2 g/kg) *plus* 0.75% EG + 2% AC; EA-RT 1 g/kg (G5): received concomitantly EA-RT (1 g/kg) *plus* 0.75% EG + 2% AC; EA-RT 2 g/kg (G6): received concomitantly EA-RT (2 g/kg) *plus* 0.75% EG + 2% AC.

2.2.2. Urine Microscopic Analysis

Microscopic analysis of the urine showed that the vehicle control group did not contain CaOx crystals (Figure S1(G1)). However, EG/AC treatment resulted in bipyramidic CaOx crystal formation (Figure S1(G2)). Concomitant treatment with both RT extracts, regardless of dose, decreased the number and size of the crystals in a dose-dependent manner (Figure S1 panel G3–G6).

2.2.3. Serum and Urinary Analysis

As mentioned above, serum urea, creatinine, uric acid, calcium, phosphorus, chloride, sodium, and potassium levels were significantly higher in the lithiasic control group (G2) than in the vehicle control group (G1). These levels were significantly reduced by the preventive treatment of the two doses of E-RT: 1 g/kg and 2 g/kg and the higher dose of EA-RT (2 g/kg). The high dose of E-RT was found to be the most effective (Table 1).

In the same way, urinary concentrations of urea, creatinine, uric acid, calcium, phosphorus, chloride, sodium, potassium, and proteins were significantly higher in the lithiasic control group than in the vehicle control group. The treatment with E-RT and EA-RT significantly prevented an EG/AC-induced increase in the level of these concentrations. The high dose of E-RT (G4) was found to be the most effective fraction (Table 2).

2.2.4. Histopathological Analysis

Kidney sections from the vehicle control group (G1) showed normal cell structures in the glomeruli as well as in the renal tubules (T) and blood vessels (Figure 2(G1) and Table 3) while those obtained from the lithiasic control group (G2) showed severe and extensive tissue lesions. The alterations consisted of interstitial infiltration of the mononucleated cells, tubular dilatation, and lesions of the glomeruli (Figure 2(G2) and Table 3). The two doses (1 and 2 g/kg) of E-RT prevented these tissue alterations. Thus, only minor tissue damages in the blood vessels and in the tubular epithelial cells, but not in the glomeruli, were observed in the kidney sections from the E-RT groups (Figure 2 panel G3 and G4). EA-RT (1 and 2 g/kg) was also effective in preventing tissue damage that could be induced by the EG/AC treatment. Kidney histological sections from rats treated with EA-RT showed normal cell structures in the renal tubules, blood vessels, and glomeruli (Figure 2 panel G5 and G6) Furthermore, both RT extracts prevented the mononucleated cells' infiltration and hemorrhage but E-RT was more efficient than EA-RT (Table 3).

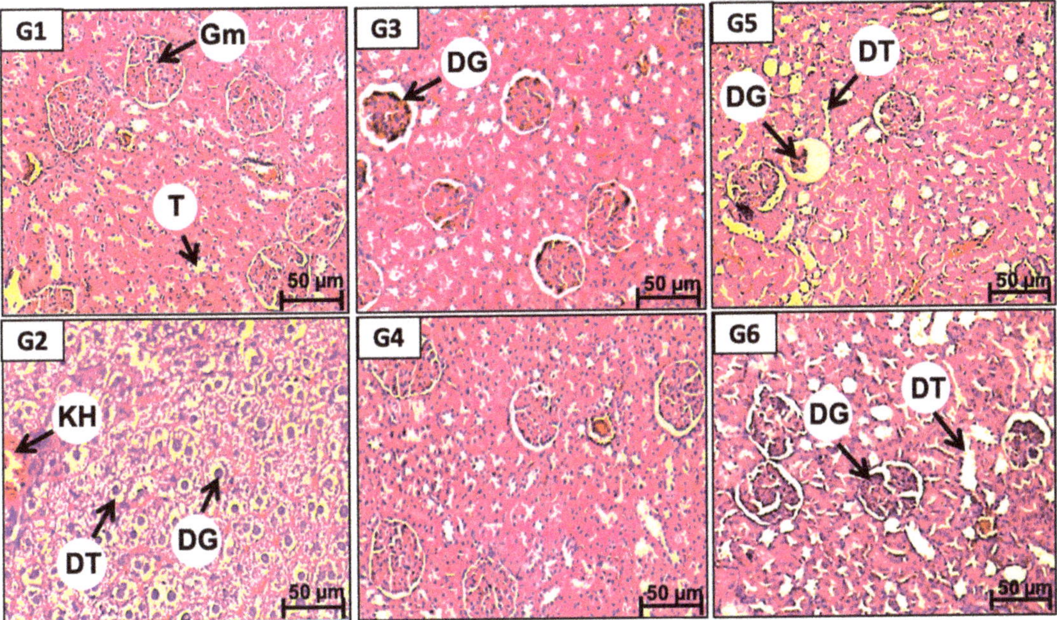

Figure 2. Representative microscopic images of hematoxylin-and-eosin-stained kidney sections observed under a light microscope from rats of: the vehicle control (**G1**), the lithiasic control (**G2**), treated with E-RT at 1 g/kg (**G3**) and 2 g/kg (**G4**), treated with EA-RT at 1 g/kg (**G5**) and 2 g/kg (**G6**). **Gm**: Glomeruli, **T**: Tubular, **DG**: Dilatation of glomeruli, **DT**: Dilatation of tubular, **KH**: kidney hemorrhage.

2.3. Antioxidant Properties of E-RT and EA-RT and Their Polyphenols Composition

2.3.1. Antioxidant Activity

Antioxidant activity of E-RT and EA-RT was assessed both in vitro (Tables 4 and 5) and in vivo (Figure 3).

Table 4. IC_{50} (µg/mL) values of E-RT and EA-RT extracts compared to gallic acid using the DPPH assay ($n = 3$).

	IC_{50} (µg/mL)
Gallic acid (GA)	64.50 ± 0.70
E-RT	156.44 ± 35.76 [a]
EA-RT	206.23 ± 90.68 [a]

Values are mean $\pm$ standard error (SEM), [a] $p < 0.001$ significantly different compared to gallic acid.

Table 5. IC$_{50}$ (µg/mL) values of E-RT and EA-RT extracts compared to reference antioxidants using the reducing power and ß-carotene assays (BHT and quercetin) (n = 3).

	IC$_{50}$ (µg/mL) Reducing Power	IC$_{50}$ (µg/mL) β-Carotene
Butylated hydroxytoluene (BHT)	0.12 ± 0.01	0.74 ± 0.02
Quercetin	0.07 ± 0.01	4.39 ± 0.05
E-RT	2.44 ± 0.02 [a,b]	75.61 ± 3.33 [a,b]
EA-RT	4.66 ± 0.04 [a,b]	101.64 ± 5.41 [a,b]

Values are expressed as mean ± standard error (SEM), [a] $p < 0.001$ significantly different compared to BHT, [b] $p < 0.001$, significantly different compared to quercetin.

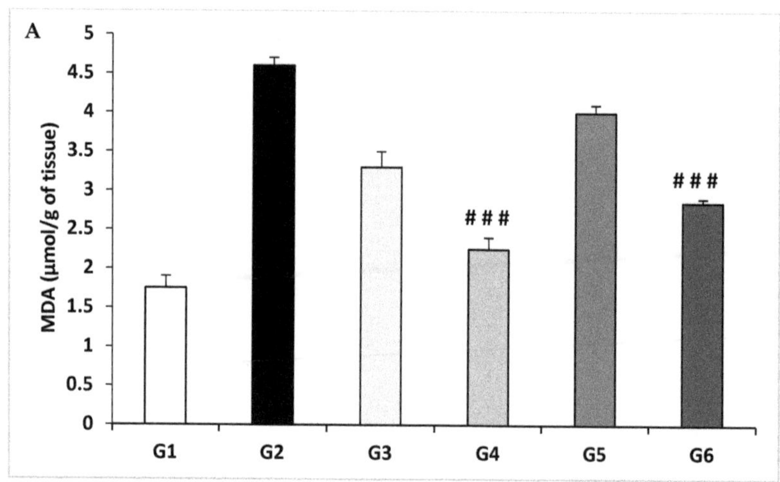

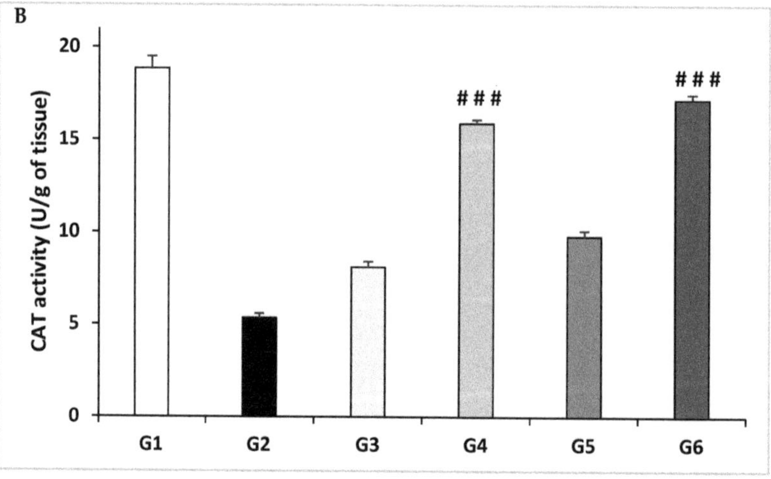

Figure 3. Evaluation of antioxidant activity in kidney homogenates using: (**A**) lipid peroxidation determination (MDA) and (**B**) catalase (CAT) enzyme activities. G1: Vehicle control; G2: Lithiasic control; G3: E-RT 1 g/kg; G4: E-RT 2 g/kg; G5: EA-RT 1 g/kg; G6: EA-RT 2 g/kg. The values shown are mean ± SEM of animals from each group (n = 6). ### $p \leq 0.001$ vs. lithiasic control.

In vitro, antioxidant activity was evaluated by DPPH (Table 4), the reducing power and β-carotene linoleic acid methods (Table 5). The antioxidant activity was determined by the value of the concentration (µg/mL) producing a 50% inhibition (IC$_{50}$) of free radicals. RT extracts showed higher IC$_{50}$ than gallic acid (64.5 ± 0.70 µg/mL) indicating that these extracts possess antioxidant activity that is lower than that of gallic acid (GA). IC_{50} of E-RT was 156.44 ± 35.76 µg/mL whereas IC_{50} of EA-RT was 206.23 ± 90.68 µg/mL, indicating that E-RT antioxidant potential was higher than that of EA-RT (Table 4).

As far as the reducing power and β-carotene linoleic acid methods are concerned, our data document that E-RT exhibited interesting IC_{50} antioxidant activity, respectively 2.44 ± 0.02 and 75.61 ± 3.33. IC_{50} of EA-RT was 4.66 ± 0.04 in reducing power assay, and 101.64 ± 5.41 for β-carotene assay. Overall, the antioxidant activity of the E-RT was higher than EA-RT. However, even if E-RT has an interesting antioxidant activity, it is less important than reference antioxidant agents such butylated hydroxytoluene (BHT) and quercetin.

In vivo, the antioxidant activity was evaluated in kidney homogenates using lipid peroxidation determination; malondialdehyde (MDA) concentration and catalase (CAT) activity (Figure 3).

EG/AC treatment increased significantly the level of MDA (4.60 ± 0.10 µmol/g of tissue) (Figure 3A) and decreased significantly the activity of catalase enzyme (5.35 ± 0.25 U/g of tissue) (Figure 3B) in kidneys of the untreated G2 compared to the vehicle control (G1, 1.75 ± 0.15 µmol/g and 18.85 ± 0.65 U/g, respectively). Treatments with *Rubia tinctorum* L. extracts (E-RT and EA-RT) reduced lipid peroxidation and protected against the oxidative stress induced by EG/AC treatment in G4 and G6 (2.25 ± 0.15 and 2.85 ± 0.05 µmol/g of tissue, respectively). CAT activity was significantly increased in the same groups (G4 and G6) (15.9 ± 0.20 and 17.2 ± 0.30 U/g of tissue, respectively).

2.3.2. Total Phenols Contents

The total phenol contents, determined in a sample volume of 20 µL by the Folin–Ciocalteu method, in the E-RT and EA-RT were 18.37 ± 0.58 and 0.61 ± 0.01 mg of GAE (gallic acid equivalent), respectively. The higher phenolic compound content found in E-RT is statistically different from that found in EA-RT.

2.3.3. Analysis of E-RT and EA-RT Polyphenols Contents

The polyphenol content of each extract, E-RT and EA-RT, respectively, were analyzed using HPLC (Figure S2). On the basis of standard compounds' retention times (RT, in min), the polyphenols identified in E-RT and EA-RT were syringic acid, rutin, ferulic acid, vanillin, rosmarinic acid, cinnamic acid, catechin, and quercetin. The respective amounts of each compound ranged from 12.14 to 26.74 mg/GAE/100 g DM (dry matter) (Table 6), according to the following order: vanillin > rosmarinic acid > quercetin. In EA-RT, quercetin was the most represented.

Table 6. Concentrations of the main phenolic compounds identified in E-RT and EA-RT expressed in mg GAE/100 g DM.

	Phenolic Compounds	Concentration (mg GAE/100 g DM)	Retention Time
E-RT	Syringic acid	13.79 ± 0.12	17.0
	Vanillin	26.74 ± 0.15	28.8
	Rosmarinic acid	15.61 ± 0.20	29.4
	Cinnamic acid	14.19 ± 0.10	41.8
	Catechin	13.78 ± 0.09	44.1
	Quercetin	14.90 ± 0.21	45.1
EA-RT	Rutin	12.32 ± 0.08	21.5
	Ferulic acid	12.14 ± 0.13	23.3
	Vanillin	12.32 ± 0.11	28.8
	Cinnamic acid	12.43 ± 0.09	41.8
	Quercetin	15.81 ± 0.12	45.1

Values are expressed as mean $\pm$ standard error (SEM) ($n = 3$).

3. Discussion

The results of the present study show that in response to 0.75% ethylene glycol (EG)/2% ammonium chloride (AC) oral administration over a 10-day period, young male rats developed kidney stones (or urolithiasis) mainly composed of calcium oxalate (CaOx). As shown in the Figure 1, treated rats (urolithiasis control group or G2) lost body weight because they drank less water and almost stopped eating. The EG/AC model used in the present this study is similar to the previously described experimental models of urolithiasis by Ravindra and colleagues [16]. The pathophysiological mechanisms responsible for the alterations elicited in this model could be related to an increase in the urinary oxalate (Ox) concentration. Indeed, EG is easily absorbed along the intestine and metabolized in the liver to Ox, leading to hyperoxaluria. Ox precipitates in the urine as CaOx because of its low solubility. High levels of Ox and CaOx crystals, particularly in the epithelial cells of the nephron, induce heterogeneous nucleation followed by crystal aggregation [17]. AC potentiates the action of EG and accelerates the phenomenon of urolithiasis [18].

Microscopic examination clearly showed that the treatment inducing urolithiasis led to the appearance of characteristic crystals of CaOx (i.e., with a bipyramid form) in the urine, while the urine of the untreated control group (G1) was free of these crystals. This result is similar to previously published data [16] and may be associated with a decreased urinary output, an elevated pH, hyperoxaluria, and hypercalciuria [19]. The biochemical analysis of the urine confirmed these results as lithiasic rats presented an increase in the excretion of phosphorus and calcium. A high concentration of phosphorus in the renal tubules could potentiate the Ox-induced lithiasis [20], whereas calcium would act as an important factor in the nucleation and precipitation of Ox in the form of CaOx [17] and in the resulting crystal growth [20]. An increase in urinary protein excretion has also been recorded indicating proximal tubular dysfunction [21]. Protein excretion could be related to severe lesions of the glomeruli and to tubular dilatation [22]. Another factor contributing to protein urea could be an interstitial inflammation attested by mononucleated cells' infiltration (Table 3).

Serum levels of urea, creatinine, and uric acid were significantly increased in the urolithiasic group (G2) compared to the untreated control group (G1), indicating renal damage (Table 1). These results are consistent with those of a previous study and indicate that the accumulation of nitrogenous substances in the serum may be a consequence of a decreased glomerular filtration rate (GFR) due to lithiasic obstruction [23]. Uric acid binds to CaOx and modulates its crystallization and solubility and also reduces the inhibitory activity of glycosaminoglycans [20]. Na^+, K^+, and Cl^- plasma concentrations were significantly increased in the lithiasic group (G2). Electrolyte imbalance disturbs the metabolism of the renal cells leading to the development of cell structure alterations [24].

The EG/AC treatment caused tubular dilation, glomeruli lesions, and mononucleocyte infiltration. These renal damages observed in the lithiasic group (G2) could be attributed to a peroxidative action on the renal epithelium resulting from the elevated rate of urinary Ox and its deposition in the tubules and glomeruli. Indeed, CaOx deposition was shown to induce oxidative stress, which could be responsible for papillary tissue lesions [2].

In the present study, we investigated the antilithiasic activities of ethanol and ethyl acetate extract of *Rubia tinctorum* L. (RT) on EG/AC-induced renal lithiasis in rats. RT is one of the several medicinal plants that are widely used in traditional medicine systems to cure various ailments. The plant has been extensively studied for its biological activities and therapeutic potentials such as anti-inflammatory, anti-aggregant, antioxidant and antibacterial properties [14,25,26]. In traditional medicine, RT dried roots are used for treatment of cardiovascular diseases including high blood pressure [15,27] liver pain, anemia and diarrhea [28,29]. Moreover, several studies have described the nephro-protective effects of plants [9]. However, to our knowledge, no previous studies have demonstrated an anti-urolithiasic activity of RT.

A previous toxicological experiment, conducted in our laboratory, demonstrated that up to 5 g/kg of the RT extracts, administered orally, triggered no major side effects [26]. Therefore, the selected doses of 1 g/kg and 2 g/kg were free of any toxicological effect.

Treatment with E-RT and EA-RT prevented the alterations induced by EG/AC to values close to the untreated control group (G1). This preventive effect concerned the formation of crystals in the urine and the biochemical parameters of the serum and urine. RT extracts also prevented the body weight loss induced by the lithiasic treatment in a dose-dependent manner (Figure 1). This finding is similar to that obtained with a standardized extract of fenugreek seed [30] that was linked to an improvement in diuresis which resulted in the dissolution of the formed calculus and an interruption in the process of aggregation and deposition of the additional crystals. E-RT and the higher dose of EA-RT prevented the increase in serum urea, creatinine, and uric acid levels, probably by preserving a normal GFR. Electrolyte (calcium, phosphorus, K^+, Na^+, and Cl^-) concentration enhancements were also inhibited by the RT extracts. Maintaining electrolytic balance may, therefore, result in the preservation of cell metabolism.

E-RT (2 g/kg) was the most effective in decreasing levels of urinary proteins and creatinine and restored EG/AC-induced low diuresis by improving the GFR. RT extracts also significantly reduced the levels of phosphorus, sodium, potassium, calcium, and uric acid. EA-RT (2 g/kg) had a lower beneficial effect and its action was limited to recovery of the GFR and the inhibition of stone formation. Of note, the lower dose (1 g/kg) of both EA-RT and E-RT was less effective than the higher dose 2 g/kg in inhibiting EG/AC-induced lithiasic effects as documented by biochemical changes in Tables 1 and 2.

Histopathological analyses showed concordant results with biochemical changes. RT extracts' dose consistently prevented the degenerative changes in kidney tissues that could be induced by EG/AC. Interestingly, RT extracts' preventive effects depended on the type of extract. E-RT (2 g/kg) was the most effective in protecting from EG/AC-induced disorganization in kidney architecture even though no differences were found between EA-RT and E-RT at 2 g/kg regarding their protective effect on calcium oxalate urolithiasis formation. Results obtained with RT extracts were similar to those previously obtained with *Acorus calamus* ethanolic extract [22], *Peucedanum grande* hydroalcoholic extract [31], and cystone [32]. Biochemical and histopathological improvements in lithiasic animals observed after treatment with several plant extracts were proposed to be directly related to their antioxidant capacity. Antioxidants could have an important action in preventing the formation of the intrapapillary calcifications induced by oxidative stress that lead to papillary calculi formation [11].

To verify the antioxidant preventing effect hypothesis, the antioxidant properties of RT extracts were evaluated both in vitro and in vivo.

The in vitro assays showed a promising antioxidant activity of both E-RT and EA-RT extracts. As documented in the DPPH scavenging, reducing power and ß-carotene assays, the E-RT extract has a better antioxidant activity compared to EA-RT extract. However, this effect is less important than reference antioxidant agents (Table 5).

In vivo, EG treatment increased the level of MDA and significantly decreased the activity of the catalase enzyme in kidneys compared to the untreated control group (G1) (Figure 3). Indeed, elevated free radical production, as observed by an increase in MDA level due to EG ingestion in the formation of nephrolithiasis, confirms that kidney tissue is under oxidative stress. This hypothesis is strengthened by the report that patients with kidney stones have less activity of antioxidant enzymes with increased lipid peroxidation [33].

Oxidative damages, as reflected by higher lipid peroxidation (MDA) and lower antioxidant enzymes activity such as catalase, deteriorate kidney structure and functions as observed in calculi-induced rats. Antioxidant and reactive oxygen scavengers have been shown to be effective for protecting the kidney in animals [33]. Here, treatment with E-RT and EA-RT showed an increase in catalase enzyme activity and a decrease in MDA levels in a dose-dependent manner in kidney homogenates.

Overall, our results present evidence that *Rubia tinctorum* L. extracts exhibited a marked protective effect against oxidative stress both in vitro and in vivo.

Rubia tinctorum L. extract contain large amounts of antioxidants which can play an important role in adsorbing and neutralizing free radicals, quenching oxygen, or decomposing peroxides. This antioxidant activity may be due, in a large part, to the specific polyphenolic composition identified by the HPLC analysis (Table 6). Indeed, a positive correlation exists between total phenolic contents and the antioxidant capacity [34]. This correlation is confirmed here, as qualitative and quantitative analysis of the RT extracts' composition showed that E-RT, which contained more various polyphenols at higher concentrations, has the most powerful antioxidant effect. The antioxidant potential of E-RT could be considered quite significant compared to the powerful phenol, gallic acid (this study) but it is important compared to extracts from other plants [35]. Analysis of the specific composition of each RT extract lead to the conclusion that E-RT antioxidant activity may be due to vanillin, rosmarinic acid, quercetin, catechin, syringic acid and cinnamic acid; while EA-RT antioxidant activity may be due to quercetin, cinnamic acid, vanillin and rutin. Several polyphenols with an antioxidant capacity were found to possess an antiurolithiasic preventing effect. As a matter of fact, quercetin antioxidant capacity was linked to a protective effect against oxidative stress associated with renal failure in the EG/AC model [29,36], to decreased oxidation of DNA bases [37] and to lead-induced DNA damage prevention and apoptosis [36]. There is also evidence that catechin has a preventive effect on renal calcium crystallization in vitro and in the EG model [38]. Vanillin and cinnamic acid may also contribute to an additional preventive effect as both compounds have antioxidant potential [39]. Therefore, it could be suggested that the superiority of E-RT over EA-RT related to their antioxidant action and prevention from urolithiasis is due to their richer polyphenolic constitution and their synergism.

These results are consistent with other studies that revealed that another species of the *Rubiaceae* family, *Rubia cordifolia* L., showed a great protective potential against different kidney and urinary disorders [21,40,41]. However, to our knowledge, for *Rubia tinctorum* L. this is the first time that a study reports such a preventive effective in urolithiasis.

4. Materials and Methods

4.1. Ethylene Glycol and Ammonium Chloride (EG/AC)-Induced Urolithiasis Model

4.1.1. Animals

Thirty-six male Sprague Dawley rats (5–6 months) weighing between 180 and 260 g were procured from the animal facility of the Faculty of Sciences Semlalia, Marrakech, Morocco. All animals were initially acclimated in their cages for 3 days before the experiments (Figure 4). Experiments were conducted in accordance with internationally accepted standard guidelines for the use of laboratory animals described in the Scientific Procedures on Living Animals ACT 1986 (European Council directive: 86/609/EEC) and approved by Semlalia Faculty of Sciences Ethic Committee (protocol code SEML/PR/2019-PR12). The rats had free access to drinking water and daily chow and were kept under a controlled 12 h light/dark cycle at 25 ± 2 °C. All the experiments were performed in the morning according to Zimmermann et al. [42].

4.1.2. Chemicals and Reagents

All chemicals used in this study were of analytical grade. Ethylene glycol (EG lot: BCBK 2604V), ethyl acetate (lot: SZBB184SV), gallic acid (lot: SLBM8746V), 1, 1-diphenyl-picrylhydrazyl (DPPH) (lot: 590790-379), and catechin (lot: WXBB6763V) were obtained from Sigma–Aldrich, Darmstadt, Germany. Ammonium chloride was purchased from Merck, Darmstadt, Germany (pro analysis, Lot: 9642642, ART. 1145). Ethanol (lot: 09L310512) and formaldehyde (lot: 11F200513) were procured from Prolabo VWR, Fontenay sous bois, France. PBS (Gibco, Ref 2285250) was obtained from Thermo Fischer Scientific, Waltham, MA USA 02451. Quercetin (≥95%) (Lot: SLBM7736V), rutin (≥95%) (Lot: VHS6475000V), ferulic acid (98%) (Lot: 1570363), rosmarinic acid (≥98%), cinnamic acid ($\geq$99%) (Lot: SZBF048AV) were purchased from Sigma–Aldrich, (Germany). Vanillin (99.8%) (Lot: 080417CE), catechin (lot: WXBB6763V) and Syringic acid ($\geq$95%) (Lot: 0478503CE) from Solvachim. Paraffin (lot: 1805052) was purchased from Leica biosystems, Nanterre, France and eosin (lot: 17060643) was purchased from Dako, Santa Clara, CA, USA. Kits used in this study for serum and urine dosage of urea (lot: 354868) (kinetic test with urease), creatinine (lot: 340775) (Jaffe reaction), uric acid (lot: 315894), phosphorus (lot: 325861), protein (lot: 307921) (Biuret Method), and Ca^{2+}, Na^+, K^+, and Cl^- reagent set (lot: 04522320) (indirect potentiometric) were procured from ROCHE diagnostics, Indianapolis, IN 46256, USA.

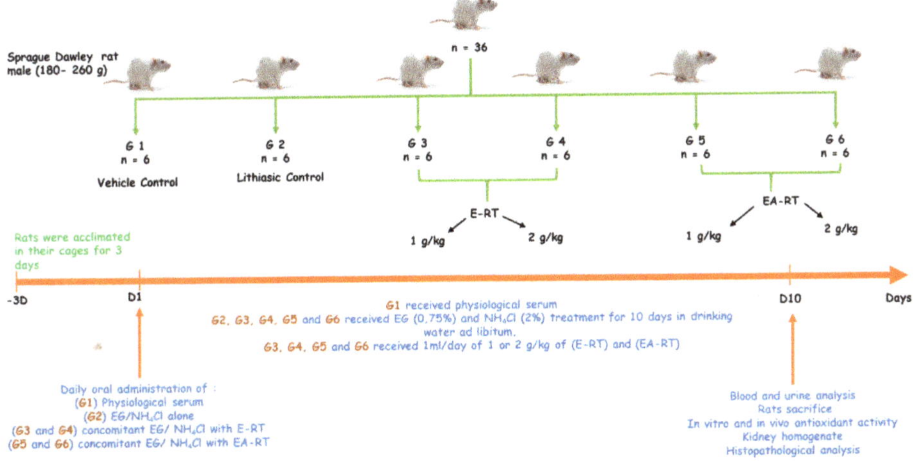

Figure 4. General experimental design diagram.

4.1.3. EG/AC Model Induction

The experimental urolithiasis rat model was induced according to the method described by Fan et al. [18]. Hyper-oxaluria and CaOx deposition in the kidney were induced by ethylene glycol (EG) being added to the drinking water to a final concentration of 0.75% for 10 days. To accelerate the process of lithiasis, 2% ammonium chloride (NH4CL or AC) was also added to the EG. Animals were weighed daily and divided into six groups of six each (Figure 4).

4.2. Preparation of RT Extracts and Assessment of Their Antilithiasic Effect

4.2.1. Collection of Plant Material

Rubia tinctorum L. (RT), locally named "el foua", was collected in June 2015 from the province of Azilal, Ait M'hamed village, geographic coordinates (31°51′00″ N, 6°30′48″ W), Morocco. The plant material was botanically classified and its correct botanical identification authenticated by Professor Ouhammou Ahmed Mohamed (Laboratory of Environment and Ecology (L 2 E, CNRST Associated Research Unit, URAC 32), Regional Herbarium MARK, Faculty of Sciences Semlalia, Cadi Ayad University, Marrakech, Morocco). A voucher specimen of the plant was deposited in the herbarium of the Semlalia Science Faculty, Cadi Ayad University (voucher number: 9825).

4.2.2. Preparation of the Extracts

The dried RT roots were coarsely powdered. Then, 303 g of powder were packed into a soxhlet column and extracted with 70% v/v ethanol in water at 75–79 °C for 15 h. The extract obtained was evaporated at 45 °C. The ethanol extract was successively separated by a series of increasing polar solvents (hexan, ethyle acetate, butanol and distilled water) according to the method previously published [26,43]. The fraction thus obtained was concentrated with a rotary evaporator to obtain the following proportions of yield: 12.75% of RT-ethanolic extract and 5.28% of RT-ethyl acetate extract.

4.2.3. Administration of the Extracts

A previous study did not show any toxicological effects of doses ranging between 0.5 to 5 g/kg; therefore, we selected the doses 1 and 2 g/kg [26].

After 3 days of acclimation, the 36 male Sprague Dawley rats were randomized in 6 groups of 6 individuals as follows (Figure 4):

- G1, the untreated control group (or vehicle control) received only distilled water during the ten days.
- G2, the lithiasic control group received distilled water supplemented with 0.75% ethylene glycol (EG) and 2% ammonium chloride (AC), during the ten days.
- G3 and G4 received distilled water supplemented with 0.75% ethylene glycol (EG) and 2% ammonium chloride (AC) and concomitantly, 1 mL/day of 1 or 2 g/kg of ethanolic extract of RT (E-RT), dissolved in distilled water, respectively, during the ten days.
- G5 and G6 received distilled water supplemented with 0.75% ethylene glycol (EG) and 2% ammonium chloride (AC) and concomitantly, 1 mL/day of 1 or 2 g/kg of ethyl acetate extract of RT (EA-RT), dissolved in distilled water, respectively, during the ten days.

4.3. Pathophysiological Parameters Evaluation

4.3.1. Urine Collection and Analysis

On the 9th day, rats from each group were individually placed in metabolic cages for 24 h for urine collection. The calcium oxalate crystals' shape and size were analyzed under light microscopy. After collection, urine samples were immediately analyzed for their volume, urea, creatinine, calcium, phosphorus, uric acid, Na, K, Cl, and total protein contents.

4.3.2. Serum Analysis

After the 10-day experimental period, the rats were anesthetized with an i.p. chloral hydrate (10 mg/kg) injection. A blood volume of 1–2 mL was collected from the jugular vein in two tubes without any additive. Serum was separated by centrifugation (350 G, for 10 min) and used for biochemical (creatinine, urea, phosphorus, uric acid, calcium, Na, K, and Cl) dosage. Substrates, minerals, and electrolyte concentrations were determined enzymatically by standard methods with a biochemical automat (Cobas C311 analyzer, Roche Diagnostics GmbH, D-68298, Mannheim, Germany).

4.3.3. Histopathological Analysis

After blood collection, the rats were sacrificed and both kidneys were carefully excised. Small slices of this freshly harvested tissue were fixed in a 10% formaldehyde solution buffered in 33 µM of sodium phosphate monobasic, dehydrated by serial ethanol solutions, diaphanized with ethanol-benzene, and enclosed in paraffin. Sections of 4 µm were sliced by a microtome (Leica, Microsystems Nussloch 6mbH, RM2125) before they were stained with haematoxylin and eosin and examined under a light microscope.

4.4. Antioxidant Activity

4.4.1. DPPH Assay

DPPH assay was realized according to the method previously described [44]. An aliquot of 1 mL of various sample concentrations of E-RT and EA-RT were added to a volume of 2 mL of DPPH. The reaction mixture was well stirred and incubated for 20 min at room temperature in the dark. The absorbance of the extracts was measured at 512 nm using ethanol and ethyl acetate as control. Gallic acid was used as a 50% positive control.

The percentage inhibition of the DPPH radical by the sample was calculated using the following equation:

$$I\% = [(A\ control - A\ sample)/A\ control] \times 100 \qquad (1)$$

"A control" is the absorbance of the control and "A sample" is the absorbance of the test sample. Extract concentration providing inhibition (IC_{50}) was extracted from the absorbance concentration graph by plotting the inhibition percentage against extract concentration. All tests were carried out in triplicate.

4.4.2. Lipid Peroxidation Determination

Tissue Samples Preparation

Kidneys tissues were homogenized with a polytron homogenizer in 1 mL of buffer containing PBS (Gibco pH = 7.4, Ref 2285250) and 0.04% Tween 80. Tissue samples were centrifuged at $10,000\times g$ for 10 min. After centrifugation, supernatants were kept with sera at $-20\ ^\circ$C until use.

Malondialdehyde Assay

The malondialdehyde (MDA) concentrations, as a product of lipid peroxidation, were evaluated by the thiobarbituric acid reactive substances method [45]. Briefly, the supernatant of the small intestine was mixed with 600 μL of 1% ortho-phosphoric acid and 200 μL of 0.6% thiobarbituric acid. They were then heated in a boiling water bath for 45 min. The completed reaction gave rise to a colorimetric product that will be extracted by the organic solvent n-butanol after centrifugation ($2000\times g$ during 15 min). The absorbance was read at 532 nm. MDA concentrations were expressed in terms of μmol of MDA per g of tissue (μmol/g of tissue) using a molar extinction coefficient of $1.56 \times 10^5\ M^{-1}\ cm^{-1}$ for MDA.

4.4.3. Catalase Activity Determination

The catalase activity (CAT) was initiated by adding 100 μL of each sample with 2.9 mL of the substrate hydrogen peroxide solution (H_2O_2, 0.2%) (Sigma–Aldrich, St. Louis, MO, USA) in a phosphate buffer (50 mM, pH 7.0) according to the method previously described [46]. Kinetics of degradation of H_2O_2 was followed for 3 min at 240 nm. The results were expressed in international unit per g of tissue (U/g of tissue).

4.4.4. Reducing Power Assay

Reducing power was assessed according to the method previously described [47]. 1 mL of various sample concentrations was added to 2.5 mL of phosphate buffer (200 mM, pH 6.6) and 2.5 mL of potassium ferricyanide (1%). The resulting solution was incubated for 20 min at 50 $^\circ$C, and after the incubation period 2.5 mL of 10% trichloroacetic acid (TCA) was mixed with the solution and then centrifuged at 200 G for 10 min. 2.5 mL of the upper layer solution was added to 2.5 mL of distilled water and 0.5 mL of 0.1% ferric chloride ($FeCl_3$). The measurement of coloration formed by the reduction of Fe^{3+} at 700 nm was used to determine the sample concentration providing 0.5 of absorbance (IC_{50}). BHT and quercetin were used as positive controls.

4.4.5. β-Carotene Linoleic Acid Method

This is one of the rapid methods to screen antioxidants, which is mainly based on the principle that linoleic acid, which is an unsaturated fatty acid, gets oxidized by "reactive oxygen species" (ROS) produced by oxygenated water [48]. The products formed will initiate the β-carotene oxidation, which will lead to discoloration. Antioxidants decrease the extent of discoloration, which was measured at 434 nm, and the activity was measured.

β-carotene (0.5 mg) in 1 mL of chloroform was added to 25 μL of linoleic acid and 200 mg of tween-80 emulsified mixture. Chloroform was evaporated at 40 $^\circ$C, 100 mL of distilled water saturated with oxygen was slowly added to the residue and the solution was vigorously agitated to form a stable emulsion. Four mL of this mixture was added into the test tubes containing 200 µL of the sample prepared in methanol. As soon as the emulsified solution was added to the tubes, zero-time absorbance was measured at 470 nm. The tubes were incubated for 2 h at 50 °C. Antioxidant activity is calculated as percentage of inhibition (I%) relative to the control using the following equation:

$$I\% = ((1 - (As - As120)/Ac - Ac120)) \tag{2}$$

where "As" was initial absorbance, "As120" was the absorbance of the sample at 120 min, "Ac" was initial absorbance of negative control and "Ac120" was the absorbance of the negative control at 120 min.

4.4.6. Determination of Total Phenolic Content

The total phenolic contents of the E-RT and EA-RT was determined using the Folin–Ciocalteu method with a slight modification [49]. Sample and standard readings were made using a spectrophotometer (VR-200 spectrophotometer) at 760 nm against the reagent blank. The test sample (20 µL) was mixed with 1.16 mL of distilled water and 100 µL of Folin–Ciocalteu's phenol reagent. After 8 min, 300 µL of saturated sodium carbonate solution (Na_2CO_3) 20% w/v in distilled water was added to the mixture. The reaction was kept in the dark for 40 min and after centrifugation, the absorbance of blue color from different samples was measured at 760 nm. The phenolic content was calculated as gallic acid equivalent (1–0.062 mg/mL, $Y = 0.981\,x + 0.003$, $R^2 = 0.9999$). All determinations were carried out in triplicate.

4.4.7. Determination of Phenolic Compounds

The EA-RT and E-RT were analyzed for their content of flavonoids and polyphenols by the HPLC method. Chromatography separations were performed on a reversed-phase (RP-18) column (Agilent Technologies 250 mm × 4.6 mm, 5.0 µm), protected by a pre-column (Agilent Technologies RP-18, 10 mm × 4.6 mm). Both columns were placed in a column oven set at 25 °C. The HPLC system consisted of a pump (SCL-10), an automatic injector (SIL-10AD) and a detector (SPD 10A UV-visible) set at a spectrum beginning at 200 nm and ending at 700 nm (Shimadzu, Japan). Data collection and analysis were performed using Shimadzu LC Solution chromatography data station software. Two solvents were used with a constant flow rate of 1 mL/min and injection volume of 10 µL. Solvent A consisted of 5% acetonitrile and 95% water, whereas solvent B consisted of a phosphate buffer dissolved in water (pH 2.6). HPLC analysis was performed with the standards, followed by the RT extracts, and then the samples' parameters were compared to the standards (syringic acid, vanillin, rosmarinic acid, catechin, cinnamic acid, quercetin, retinoic acid and ferulic acid).

4.5. Statistical Analysis

Statistical analyses were performed using the computer software Sigma Plot 12.5 for Windows. All data were expressed as mean ± standard error (SEM); p values less than 0.05, 0.01 and 0.001 were considered to be significant. Comparisons between different groups were performed using one-way analysis of variance (ANOVA). Significant differences between the control and experimental groups were assessed by Tukey's test.

5. Conclusions

In conclusion, the concomitant oral administration of RT extracts prevented the development of EG/AC-induced urolithiasis in the rat. E-RT was found to be more efficient than EA-RT. The superiority of the E-RT preventive effect over EA-RT was associated with its more powerful antioxidant effect, resulting from its specific and rich polyphenol composition. E-RT was constituted by vanillin, rosmarinic acid, quercetin, catechin, syringic acid and cinnamic acid, whereas that of EA-RT contained specifically quercetin, cinnamic acid, vanillin and rutin. Taken together, our results suggest that RT could be an important alternative to anti-urolithiasis drugs due to its efficiency and minor side effects.

Supplementary Materials: The following are available online, Figure S1: Typical calcium oxalate (CaOx) crystals, viewed under light microscope (50×), in 24h urine from rats of vehicle control (G1), lithiasic group (G2), groups treated with E-RT at 1 g/kg (G3) and 2 g/kg (G4) and groups treated with EA-RT at 1 g/kg (G5) and 2 g/kg (G6), Figure S2: HPLC chromatograms of E-RT (A) and EA-RT (B).

Author Contributions: A.B. (Abdallah Bagri), A.C., R.A., and F.Z.M.: the conception and design of study. F.Z.M. and J.L.: the acquisition of data and analysis and interpretation of the data. F.Z.M. and A.B. (Abdallah Bagri): drafting the manuscript. A.B. (Anass Belbachir), A.C., Y.Z., R.E.D. and Y.L.: critically revising the manuscript for important intellectual content. All authors have read and agreed to the published version of the manuscript.

Funding: This study was supported by the Faculty of Sciences and Technology, University Hassan First, Settat and the Faculty of Sciences, Semlalia, Cadi Ayad University Marrakech, Morocco.

Institutional Review Board Statement: Experiments were conducted in accordance with internationally accepted standard guidelines for the use of laboratory animals described in the Scientific Procedures on Living Animals ACT 1986 (European Council directive: 86/609/EEC) and approved by Semlalia Faculty of Sciences Ethic Committee (protocol code SEML/PR/2019-PR12).

Data Availability Statement: The data presented in this study are available upon reasonable request from the authors.

Acknowledgments: The authors are extremely grateful to Abderrazzak Regragui, an expert in animal laboratory handling, for his assistance in this study.

Conflicts of Interest: The authors declare no conflict of interest.

Sample Availability: Samples of the compounds are not available from the authors.

Abbreviations

AC	ammonium chloride
CaOx	calcium oxalate
Cl	chloride
DNA	deoxyribonucleic acid
DPPH	1, 1-diphenyl-2-picrylhydrazyl
EA-RT	ethyl acetate extract of *Rubia tinctorum* L.
EG	ethylene glycol
E-RT	ethanolic extract of *Rubia tinctorum* L.
GRF	glomerular filtration rate
HPLC	high performance liquid chromatography
IC_{50}	the half maximal inhibitory concentration
K	potassium
Na	sodium
Na_2CO_3	sodium carbonates solution
Ox	oxalate
RT	*Rubia tinctorum* L.

References

1. Romero, V.; Akpinar, H.; Assimos, D.G. Kidney stones: A global picture of prevalence, incidence, and associated risk factors. *Rev. Urol.* **2010**, *12*, e86–e96.
2. Khan, S.R.; Joshi, S.; Wang, W.; Peck, A.B. Regulation of macromolecular modulators of urinary stone formation by reactive oxygen species: Transcriptional study in an animal model of hyperoxaluria. *Am. J. Physiol. Renal Physiol.* **2014**, *306*, F1285–1295. [CrossRef]
3. Khan, S.R.; Pearle, M.S.; Robertson, W.G.; Gambaro, G.; Canales, B.K.; Doizi, S.; Traxer, O.; Tiselius, H.G. Kidney stones. *Nat. Rev. Dis. Primers* **2016**, *2*, 16008. [CrossRef]
4. Evan, A.P. Physiopathology and etiology of stone formation in the kidney and the urinary tract. *Pediatr. Nephrol.* **2010**, *25*, 831–841. [CrossRef]
5. Ahmed, A.; Wadud, A.; Jahan, N.; Bilal, A.; Hajera, S. Efficacy of *Adiantum capillus veneris Linn* in chemically induced urolithiasis in rats. *J. Ethnopharmacol.* **2013**, *146*, 411–416. [CrossRef]
6. Semins, M.J.; Matlaga, B.R. Medical evaluation and management of urolithiasis. *Ther. Adv. Urol.* **2010**, *2*, 3–9. [CrossRef]

7. Terlecki, R.P.; Triest, J.A. A contemporary evaluation of the auditory hazard of extracorporeal shock wave lithotripsy. *Urology* **2007**, *70*, 898–899. [CrossRef]
8. Butterweck, V.; Khan, S.R. Herbal medicines in the management of urolithiasis: Alternative or complementary? *Planta Med.* **2009**, *75*, 1095–1103. [CrossRef]
9. Agarwal, K.; Varma, R. Ethnobotanical study of antilithic plants of Bhopal district. *J. Ethnopharmacol.* **2015**, *174*, 17–24. [CrossRef]
10. Saha, S.; Shrivastav, P.S.; Verma, R.J. Antioxidative mechanism involved in the preventive efficacy of *Bergenia ciliata* rhizomes against experimental nephrolithiasis in rats. *Pharm. Biol.* **2014**, *52*, 712–722. [CrossRef]
11. Grases, F.; Prieto, R.M.; Fernandez-Cabot, R.A.; Costa-Bauza, A.; Tur, F.; Torres, J.J. Effects of polyphenols from grape seeds on renal lithiasis. *Oxid. Med. Cell Longev.* **2015**, *2015*, 813737. [CrossRef]
12. Youn, S.H.; Kwon, J.H.; Yin, J.; Tam, L.T.; Ahn, H.S.; Myung, S.C.; Lee, M.W. Anti-Inflammatory and Anti-Urolithiasis Effects of Polyphenolic Compounds from Quercus gilva Blume. *Molecules* **2017**, *22*, 1121. [CrossRef]
13. Shikov, A.N.; Narkevich, I.A.; Flisyuk, E.V.; Luzhanin, V.G.; Pozharitskaya, O.N. Medicinal plants from the 14th edition of the Russian Pharmacopoeia, recent updates. *J. Ethnopharmacol.* **2021**, *268*, 113685. [CrossRef]
14. Shilpa, P.N.; Venkatabalasubramanian, S.; Devaraj, S.N. Ameliorative effect of methanol extract of Rubia cordifolia in N-nitrosodiethylamine-induced hepatocellular carcinoma. *Pharm. Biol.* **2012**, *50*, 376–383. [CrossRef]
15. Bellakhdar, J.; Claisse, R.; Fleurentin, J.; Younos, C. Repertory of standard herbal drugs in the Moroccan pharmacopoea. *J. Ethnopharmacol.* **1991**, *35*, 123–143. [CrossRef]
16. Ravindra, D.H.; Sunil, S.J. Effect of hydro-alcoholic extract of Vernonia cinerea Less against ethylene glycol-induced urolithiasis in rats. *Indian J. Pharmacol.* **2016**, *48*, 434–440.
17. Scheid, C.R.; Cao, L.C.; Honeyman, T.; Jonassen, J.A. How elevated oxalate can promote kidney stone disease: Changes at the surface and in the cytosol of renal cells that promote crystal adherence and growth. *Front. Biosci.* **2004**, *9*, 797–808. [CrossRef]
18. Fan, J.; Glass, M.A.; Chandhoke, P.S. Impact of ammonium chloride administration on a rat ethylene glycol urolithiasis model. *Scanning Microsc.* **1999**, *13*, 299–306.
19. Leth, P.M.; Gregersen, M. Ethylene glycol poisoning. *Forensic Sci. Int.* **2005**, *155*, 179–184. [CrossRef]
20. Soundararajan, P.; Mahesh, R.; Ramesh, T.; Begum, V.H. Effect of *Aerva lanata* on calcium oxalate urolithiasis in rats. *Indian J. Exp. Biol.* **2006**, *44*, 981–986.
21. Divakar, K.; Pawar, A.T.; Chandrasekhar, S.B.; Dighe, S.B.; Divakar, G. Protective effect of the hydro-alcoholic extract of *Rubia cordifolia* roots against ethylene glycol induced urolithiasis in rats. *Food Chem. Toxicol.* **2010**, *48*, 1013–1018. [CrossRef]
22. Ghelani, H.; Chapala, M.; Jadav, P. Diuretic and antiurolithiatic activities of an ethanolic extract of *Acorus calamus* L. rhizome in experimental animal models. *J. Tradit. Complement. Med.* **2016**, *6*, 431–436. [CrossRef]
23. Pawar, A.T.; Vyawahare, N.S. Protective effect of ethyl acetate fraction of *Biophytum sensitivum* extract against sodium oxalate-induced urolithiasis in rats. *J. Tradit. Complement. Med.* **2017**, *7*, 476–486. [CrossRef]
24. Blachley, J.D.; Hill, J.B. Renal and electrolyte disturbances associated with cisplatin. *Ann. Intern. Med.* **1981**, *95*, 628–632. [CrossRef]
25. Kalyoncu, F.; Cetin, B.; Saglam, H. Antimicrobial activity of common madder (*Rubia tinctorum* L.). *Phytother. Res.* **2006**, *20*, 490–492. [CrossRef]
26. Marhoume, F.Z.; Laaradia, M.A.; Zaid, Y.; Laadraoui, J.; Oufquir, S.; Aboufatima, R.; Chait, A.; Bagri, A. Anti-aggregant effect of butanolic extract of *Rubia tinctorum* L on platelets in vitro and Ex Vivo. *J. Ethnopharmacol.* **2019**, *241*, 111971. [CrossRef]
27. Jouad, H.; Haloui, M.; Rhiouani, H.; El Hilaly, J.; Eddouks, M. Ethnobotanical survey of medicinal plants used for the treatment of diabetes, cardiac and renal diseases in the North centre region of Morocco (Fez-Boulemane). *J. Ethnopharmacol.* **2001**, *77*, 175–182. [CrossRef]
28. Eddouks, M.; Maghrani, M.; Lemhadri, A.; Ouahidi, M.L.; Jouad, H. Ethnopharmacological survey of medicinal plants used for the treatment of diabetes mellitus, hypertension and cardiac diseases in the south-east region of Morocco (Tafilalet). *J. Ethnopharmacol.* **2002**, *82*, 97–103. [CrossRef]
29. El Haouari, M.; Rosado, J.A. Medicinal Plants with Antiplatelet Activity. *Phytother. Res.* **2016**, *30*, 1059–1071. [CrossRef]
30. Kapase, C.U.; Bodhankar, S.L.; Mohan, V.; Thakurdesai, P.A. Therapeutic effects of standardized fenugreek seed extract on experimental urolithiasis in rats. *J. Appl. Pharm. Sci.* **2013**, *3*, 029–035. [CrossRef]
31. Kumar, B.N.; Wadud, A.; Jahan, N.; Sofi, G.; Bano, H.; Makbul, S.A.; Husain, S. Antilithiatic effect of *Peucedanum grande* C. B. Clarke in chemically induced urolithiasis in rats. *J. Ethnopharmacol.* **2016**, *194*, 1122–1129. [CrossRef]
32. Kasote, D.M.; Jagtap, S.D.; Thapa, D.; Khyade, M.S.; Russell, W.R. Herbal remedies for urinary stones used in India and China: A review. *J. Ethnopharmacol.* **2017**, *203*, 55–68. [CrossRef] [PubMed]
33. Selvam, R. Calcium oxalate stone disease: Role of lipid peroxidation and antioxidants. *Urol. Res.* **2002**, *30*, 35–47. [CrossRef] [PubMed]
34. Katalini, V.; Milpa, M.; Kuljaj, T.; Juki, M. Screening of 70 medicinal plant extracts for antioxidant capacity and total phenols. *Food Chem.* **2006**, *94*, 550–557. [CrossRef]
35. Naksuriya, O.; Okonogi, S. Comparison and combination effects on antioxidant power of curcumin with gallic acid, ascorbic acid, and xanthone. *Drug Discov. Ther.* **2015**, *9*, 136–141. [CrossRef]
36. Liu, C.M.; Sun, Y.Z.; Sun, J.M.; Ma, J.Q.; Cheng, C. Protective role of quercetin against lead-induced inflammatory response in rat kidney through the ROS-mediated MAPKs and NF-kappaB pathway. *Biochim. Biophys. Acta* **2012**, *1820*, 1693–1703. [CrossRef]

37. Calabrese, V.; Cornelius, C.; Dinkova-Kostova, A.T.; Calabrese, E.J.; Mattson, M.P. Cellular stress responses, the hormesis paradigm, and vitagenes: Novel targets for therapeutic intervention in neurodegenerative disorders. *Antioxid. Redox. Signal* **2010**, *13*, 1763–1811. [CrossRef]

38. Zhai, W.; Zheng, J.; Yao, X.; Peng, B.; Liu, M.; Huang, J.; Wang, G.; Xu, Y. Catechin prevents the calcium oxalate monohydrate induced renal calcium crystallization in NRK-52E cells and the ethylene glycol induced renal stone formation in rat. *BMC Complement. Altern. Med.* **2013**, *13*, 228. [CrossRef]

39. Stanely Mainzen Prince, P.; Rajakumar, S.; Dhanasekar, K. Protective effects of vanillic acid on electrocardiogram, lipid peroxidation, antioxidants, proinflammatory markers and histopathology in isoproterenol induced cardiotoxic rats. *Eur. J. Pharmacol.* **2011**, *668*, 233–240. [CrossRef]

40. Lodi, S.; Sharma, V.; Kansal, L. The protective effect of *Rubia cordifolia* against lead nitrate-induced immune response impairment and kidney oxidative damage. *Indian J. Pharmacol.* **2011**, *43*, 441–444. [CrossRef]

41. Son, J.K.; Jung, S.J.; Jung, J.H.; Fang, Z.; Lee, C.S.; Seo, C.S.; Moon, D.C.; Min, B.S.; Kim, M.R.; Woo, M.H. Anticancer constituents from the roots of *Rubia cordifolia* L. *Chem. Pharm. Bull.* **2008**, *56*, 213–216. [CrossRef]

42. Zimmermann, M. Ethical guidelines for investigations of experimental pain in conscious animals. *Pain* **1983**, *16*, 109–110. [CrossRef]

43. Elhabazi, K.; Aboufatima, R.; Benharref, A.; Zyad, A.; Chait, A.; Dalal, A. Study on the antinociceptive effects of *Thymus broussonetii Boiss* extracts in mice and rats. *J. Ethnopharmacol.* **2006**, *107*, 406–411. [CrossRef] [PubMed]

44. M'Sou, S.; Alifriqui, M.; Romane, A. Phytochemical study and biological effects of the essential oil of *Fraxinus dimorpha* Coss & Durieu (§). *Nat. Prod. Res.* **2017**, *31*, 2797–2800. [CrossRef]

45. Moreno, I.; Pichardo, S.; Jos, A.; Gomez-Amores, L.; Mate, A.; Vazquez, C.M.; Camean, A.M. Antioxidant enzyme activity and lipid peroxidation in liver and kidney of rats exposed to microcystin-LR administered intraperitoneally. *Toxicon* **2005**, *45*, 395–402. [CrossRef]

46. Hill, M.F.; Singal, P.K. Antioxidant and oxidative stress changes during heart failure subsequent to myocardial infarction in rats. *Am. J. Pathol.* **1996**, *148*, 291–300.

47. Bouimeja, B.; Yetongnon, K.H.; Touloun, O.; Berrougui, H.; Laaradia, M.A.; Ouanaimi, F.; Chait, A.; Boumezzough, A. Studies on antivenom activity of *Lactuca serriola* methanolic extract against *Buthus atlantis* scorpion venom by In Vivo methods. *S. Afr. J. Bot.* **2019**, *125*, 270–279. [CrossRef]

48. Miraliakbari, H.; Shahidi, F. Antioxidant activity of minor components of tree nut oils. *Food Chem.* **2008**, *111*, 421–427. [CrossRef]

49. Ainsworth, E.A.; Gillespie, K.M. Estimation of total phenolic content and other oxidation substrates in plant tissues using Folin-Ciocalteu reagent. *Nat. Protoc.* **2007**, *2*, 875–877. [CrossRef]

Article

Anti-Platelet Aggregation and Anti-Cyclooxygenase Activities for a Range of Coffee Extracts (*Coffea arabica*)

Nuntouchaporn Hutachok [1], Pongsak Angkasith [2], Chaiwat Chumpun [2], Suthat Fucharoen [3], Ian J. Mackie [4], John B. Porter [5] and Somdet Srichairatanakool [1,*]

[1]   Department of Biochemistry, Faculty of Medicine, Chiang Mai University, Chiang Mai 50200, Thailand; n.hutachok@hotmail.com
[2]   Royal Project Foundation, Chiang Mai 50200, Thailand; pongsak.a@cmu.ac.th (P.A.); ssomdet@hotmail.com (C.C.)
[3]   Thalassemia Research Center, Institute of Molecular Biosciences, Mahidol University Salaya Campus, Nakornpathom 71300, Thailand; suthat.fuc@mahidol.ac.th
[4]   Haemostasis Research Unit, Department of Haematology, University College London Medical School, London WC1E 6BT, UK; i.mackie@ucl.ac.uk
[5]   Red Cell Disorder Unit, Department of Haematology, University College London Medical School, London WC1E 6BT, UK; j.porter@ucl.ac.uk
*   Correspondence: somdet.s@cmu.ac.th; Tel.: +66-53935322; Fax: +66-53894031

Abstract: Coffee is rich in caffeine (CF), chlorogenic acid (CGA) and phenolics. Differing types of coffee beverages and brewing procedures may result in differences in total phenolic contents (TPC) and biological activities. Inflammation and increases of platelet activation and aggregation can lead to thrombosis. We focused on determining the chemical composition, antioxidant activity and inhibitory effects on agonist-induced platelet aggregation and cyclooxygenase (COX) of coffee beverages in relation to their preparation method. We prepared instant coffee and brewed coffee beverages using drip, espresso, and boiling techniques. Coffee extracts were assayed for their CF and CGA contents using HPLC, TPC using colorimetry, platelet aggregation with an aggregometer, and COX activity using ELISA. The findings have shown all coffee extracts, except the decaffeinated types, contained nearly equal amounts of CF, CGA, and TPC. Inhibitory effects of coffee extracts on platelet aggregation differed depending on the activation pathways induced by different agonists. All espresso, drip and boiled coffee extracts caused dose dependent inhibition of platelet aggregation induced by ADP, collagen, epinephrine, and arachidonic acid (ARA). The most marked inhibition was seen at low doses of collagen or ARA. Espresso and drip extracts inhibited collagen-induced platelet aggregation more than purified caffeine or CGA. Espresso, boiled and drip coffee extracts were also a more potent inhibitors of COX-1 and COX-2 than purified caffeine or CGA. We conclude that inhibition of platelet aggregation and COX-1 and COX-2 may contribute to anti-platelet and anti-inflammatory effects of espresso and drip coffee extracts.

Keywords: chlorogenic acid; coffee; cyclooxygenase; espresso; instant coffee; platelet aggregation

Citation: Hutachok, N.; Angkasith, P.; Chumpun, C.; Fucharoen, S.; Mackie, I.J.; Porter, J.B.; Srichairatanakool, S. Anti-Platelet Aggregation and Anti-Cyclooxygenase Activities for a Range of Coffee Extracts (*Coffea arabica*). *Molecules* **2021**, 26, 10. https://dx.doi.org/10.3390/molecules26010010

Received: 27 November 2020
Accepted: 17 December 2020
Published: 22 December 2020

1. Introduction

Arabica coffee (*Coffea arabica*) and Robusta coffee (*Coffea canephora*) are the most widely-consumed coffee beverages in the world. These coffee varieties differ in terms of cultivation area, their physical and chemical characteristics and the taste of the resulting coffee beverage [1]. Coffee beans contain lipids, carbohydrates, minerals, caffeine (CF) and phenolic acids such as chlorogenic acid (CGA) or 5-*O*-caffeoylquinic acid (CQA), caffeic acid (CA), ferulic acid (FA), quinic acid (QA), and feruloylquinic acid (FQA) (Figure 1), all of which contribute to their biological and pharmacological effects [2].

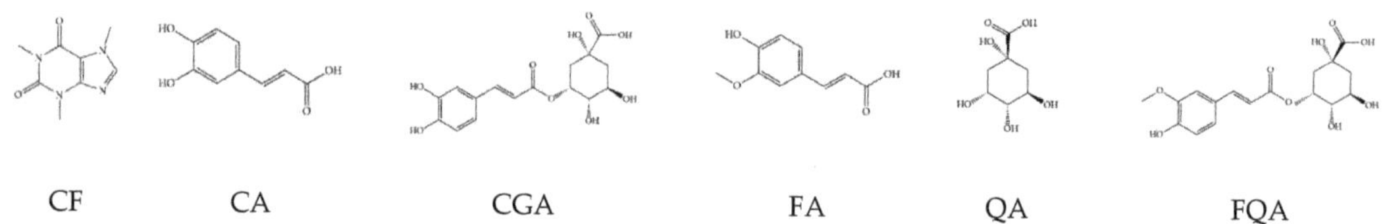

CF   CA   CGA   FA   QA   FQA

Figure 1. Chemical structures of caffeine (CF), caffeic acid (CA), chlorogenic acid (CGA), ferulic acid (FA), quinic acid (QA), and feruloylquinic acid (FQA) present in coffee [2].

Boiled coffee is the earliest and most commonly prepared form of coffee brew and is commonly referred to as Turkish coffee. Espresso coffee is a complex and much appreciated beverage among coffee consumers in Southern Europe, Central America and other areas. Indeed, differing varieties of coffee drink preparations will result in differences in aromatic compositions, bioactive compounds such as CF, CGA, CA, and other phenolic compounds, as well as other resulting pharmacological properties.

Importantly, coffee consumption provides hepato- and cardio-protective effects, displays antimicrobial activity and lowers the risk of type 2 diabetes mellitus [2]. In healthy people, moderate caffeine consumption (400 mg/day) neither induces cardiac arrhythmia nor increases serum cholesterol levels [3]. However, excessive coffee consumption is known to result in mineral deficiency, hypertension, hypercholesterolemia, insomnia, tremors, nausea, polyuria, diarrhea, and polyphagia [2–6]. Possibly, the positive effects of coffee are caused by certain compounds such as CF and CGA [3,7]. CGA (or caffeoylquinic acid, CQA) together with feruloylquinic (FQA) and dicaffeoylquinic (diCQA) acids are phenolic acids [7–9], and CF is recognized as a remarkable alkaloid [2,3]. These compounds are notably predominant in green coffee beans. Interestingly, the contents of these compounds in coffee are varied among tree species and can vary according to processing techniques [2,10–13], both of which can lead to differing degrees of antioxidant and radical scavenging activities. Nevertheless, degradation of CGA during the roasting process did not ultimately affect the antioxidant activity of the coffee beverage [14].

The hypercoagulable state is one of the most common hematological complications that can be caused by platelet activation, red cell disorders, the presence of reactive oxygen species (ROS) and/or decreased natural anticoagulants [15–17]. Regular blood transfusions and iron chelation therapy have become the main prophylactic methods employed in haemoglobinopathies (commonly in β-thalassaemia intermedia patients) to eliminate these multi-factors and prevent secondary complications [18,19]. In addition, platelet antagonists such as eptifibatide and prasugrel were found to reduce platelet activation and aggregation in patients with sickle cell diseases [20,21]. Notably, coffee is known to exhibit anti-thrombotic, nitric oxide (NO$^\bullet$) modulating and anti-platelet effects [22]. Accordingly, high and chronic consumption habits of coffee could alter the response of platelets to adenosine actions, decrease platelet aggregation and lower the risk of cardiovascular disease [23]. Moreover, cyclooxygenases (COXs), which are identified as constitutive COX-1 and inducible COX-2, can catalyze the synthesis of prostaglandins (PGs) and thromboanes (TXs) and are inhibited by acetylsalicylic acid (ASA). Specifically, COX-2 inhibitor drugs inhibit the synthesis of platelet-inhibitory prostacyclin (PGI$_2$); nonetheless, NS-398 was found to induce certain side effects such as increased vascular platelet adhesion, leading to thrombosis and hypercoagulation in patients with cardiovascular diseases [24]. Notably, they have been reported as being able to promote atherothrombosis by inhibiting the vascular formation of PGI$_2$ and increasing thrombotic-associated risks [25]. We have hypothesized that coffee preparations with different amounts of CGA and CF could help to prevent a hypercoagulable state, by inhibiting platelet aggregation and COXs and thereby, assist in the inhibition of platelet aggregation and COX activities to differing degrees. The aims of this study were to prepare different roasted Arabica coffee drinks using boiled (brewed coffee) and pressurized (espresso beverage) extraction methods and then evaluate their inhibitory effects on platelet aggregation and COXs activity.

2. Results

2.1. Chemical Compositions and Antioxidant Activity in Coffee Preparations

HPLC/DAD analysis revealed that authentic 125 µg/mL caffeine (CF), 125 µg/mL chlorogenic acid (CGA) and 1 mg/mL caffeic acid (CA) were eluted at the specific retention time (T_R) values of 11.72 min, 12.50 and 17.18 min, respectively (Figure 2). Consistently, free forms of CF and CGA at the corresponding time points were almost exclusively detected in extracts of instant coffee (both regular and decaffeinated formulas), as well as in extracts of espresso, drip and boiled coffees. However, CA was notably not detectable. In addition, the chromatographic profiles of CF and CGA were very similar among these coffee extracts with the exception of the decaffeinated coffee. Not surprisingly, nearly all of the CF was eliminated from the decaffeinated instant coffee. With regard to espresso, drip, boiled and instant (regular) coffee extracts, the peaks eluted between 2 and 10 min were observed to be evenly predominant. This outcome was possibly attributed to the presence of other phenolic/flavonoid compounds.

Results presented in Table 1 reveal differences of CGA, CF, TPC, and AA contents among five coffee extract samples, among which the CF and CGA contents are notable. Decaffeinated coffee is usually prepared by employing a treatment using an organic solvent or carbon dioxide to remove intact caffeine from regular coffee, reaching 1–2% of the original content of the coffee. Herein, instant decaffeinated coffee was found to contain higher CGA content but less CF content than instant regular coffee. Additionally, it displayed higher CGA, TPC, and antioxidant activity than other coffee extracts.

Table 1. Average values of chlorogenic acid, caffeine, total phenolic contents and antioxidant activity in coffee extracts.

Extract	CGA (µg/mg)	CF (µg/mg)	CA (µg/mg)	TPC (mg GAE/g)	Antioxidant Activity (mg TE/g)
Espresso Coffee	8.60	48.83	ND	113	125
Drip Coffee	8.81	45.73	ND	104	160
Boiled Coffee	9.45	49.58	ND	114	118
Instant (Regular) Coffee	5.15	39.81	ND	135	156
Instant (Decaffeinate) Coffee	10.61	0.53	ND	166	163

Abbreviations: CA = caffeic acid, CF = caffeine, CGA = chlorogenic acid, GAE = gallic acid equivalent, ND = not detectable, TE = Trolox equivalent, TPC = total phenolic content.

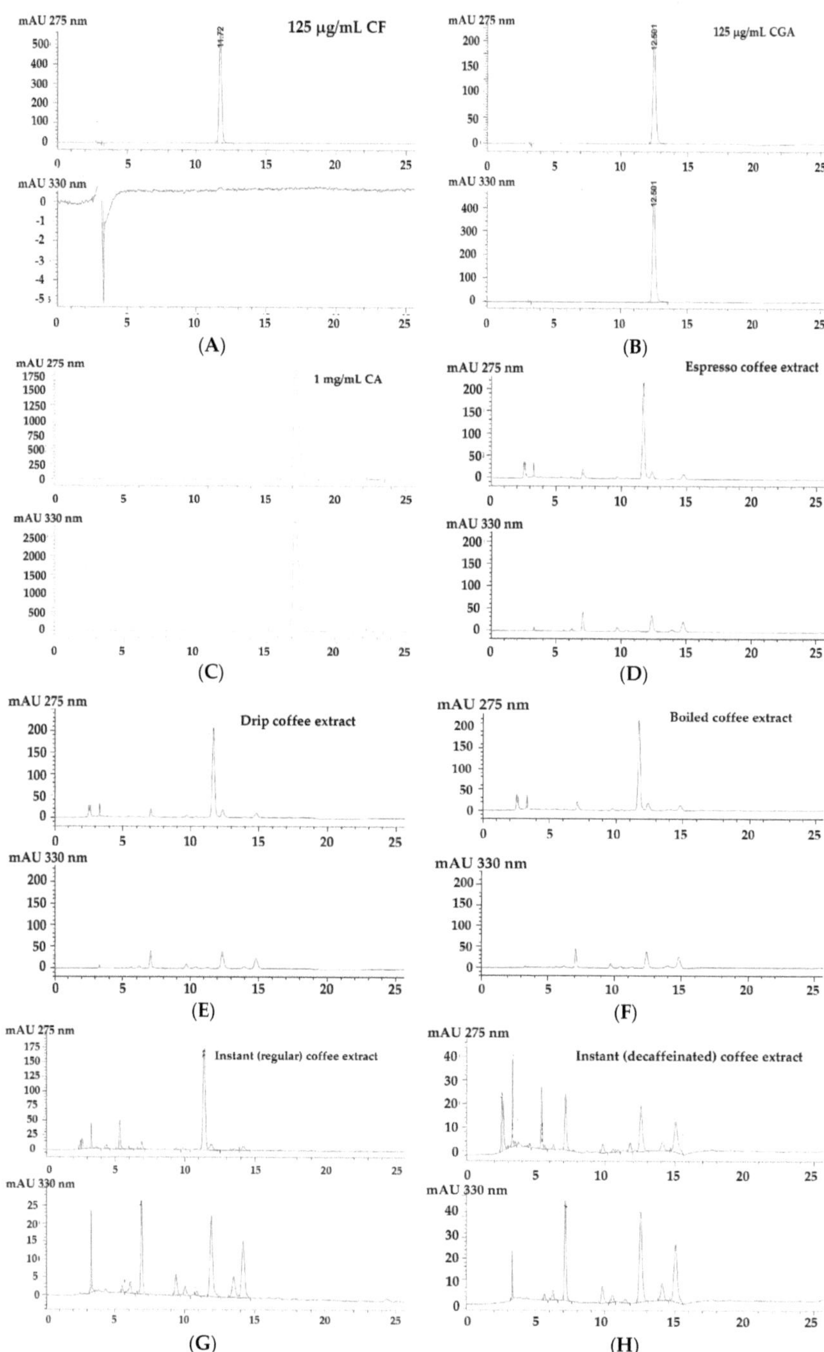

Figure 2. HPLC/DAD profiles of authentic standards including caffeic acid (CA), caffeine (CF), and chlorogenic acid (CGA); espresso, drip, boiled and instant coffee extracts (**A–H**).

2.2. Inhibitory Effect of Coffee Extracts on Platelet Aggregation

In this study, the platelets present in platelet rich plasma (PRP), were treated with different coffee extract preparations before adding different platelet agonists in vitro and monitoring platelet aggregation using an aggregometer instrument. All espresso, drip and boiled coffee extracts tended to cause dose dependent inhibition of platelet aggregation induced by ADP, collagen, epinephrine, and arachidonic acid (ARA), in which the espresso and drip coffee extract at the dose of 1 mg/mL showed significant inhibition of platelet aggregation induced by ADP (5 μM) and collagen (1 μg/mL). However, the inhibitory effects became weaker when the doses of the agonists were increased. The most marked inhibition was seen with low doses of collagen and with ARA. In addition, the inhibitory effects of the coffee extracts on the platelet aggregation to U46619 were weak (only occurring in 1/5 donors) and there was no inhibition of aggregation to TRAP (Figure 3). These results indicate that the inhibitory effects of coffee extracts on platelet aggregation differ depending on the particular pathways and mechanisms leading to platelet aggregation that are activated by different agonists.

As shown in Figure 4, CF (54.17 mg/mL) was found to be more effective than CGA (18.23 mg/mL) in inhibiting the aggregation of platelets induced by collagen. However, they showed less inhibitory effects on platelet aggregation than the coffee extracts, even though their concentrations were slightly higher than those of espresso (48.83 μg CF and 8.60 μg CGA equivalent/mL), boiled coffee (49.58 μg CF and 9.45 μg CGA equivalent/mL), drip coffee (45.73 μg CF and 8.81 μg CGA equivalent/mL), instant coffee (39.81 μg CF and 5.15 μg CGA equivalent/mL), and decaffeinated instant coffee (0.53 μg CF and 10.61 μg CGA equivalent/mL). There were no clear differences in the degree of inhibition between the coffee types at equal concentrations (1 mg/mL). Similarly, CF and CGA slightly inhibited the aggregation of platelets induced by collagen and showed much lower inhibitory effects than the coffee extracts. Therefore, the inhibitory effects of Arabica coffee and instant coffee on platelet aggregation were interesting, but will require further investigation.

In time course experiments, coffee extracts (1 mg/mL) were mixed with PRP in the cuvette and incubated for indicated periods of time. Collagen and ARA agonist were then added and platelet aggregation were immediately measured using an aggregometer instrument. Espresso coffee slowly inhibited platelet aggregation while boiled and drip coffee extracts caused immediate (within 1 min) inhibition of platelet aggregation to collagen (1 μg/mL). Likewise, the degree of inhibition was more potent at higher concentrations and over longer incubation periods. Consistently, the degree of inhibition fell in the following order: Boiled coffee ~ drip coffee > espresso coffee (Figure 5A–C). In contrast, when using ARA (1 mM) as an agonist, all of the coffee extracts completely inhibited platelet aggregation within 1–2 min regardless of concentration. However, the degree of inhibition fell in the order of boiled coffee ~ drip coffee > espresso coffee (Figure 5D–F).

Moreover, the inhibitory effects on the agonists-induced platelet aggregation seemed to be dependent upon the incubation time of the coffee extracts. Consequently, the inhibition potency was found to be in the order of drip coffee > boiled coffee > espresso coffee with regard to ARA stimulation, whereas it was not observed to be different with regard to collagen stimulation (Figure 6A–B).

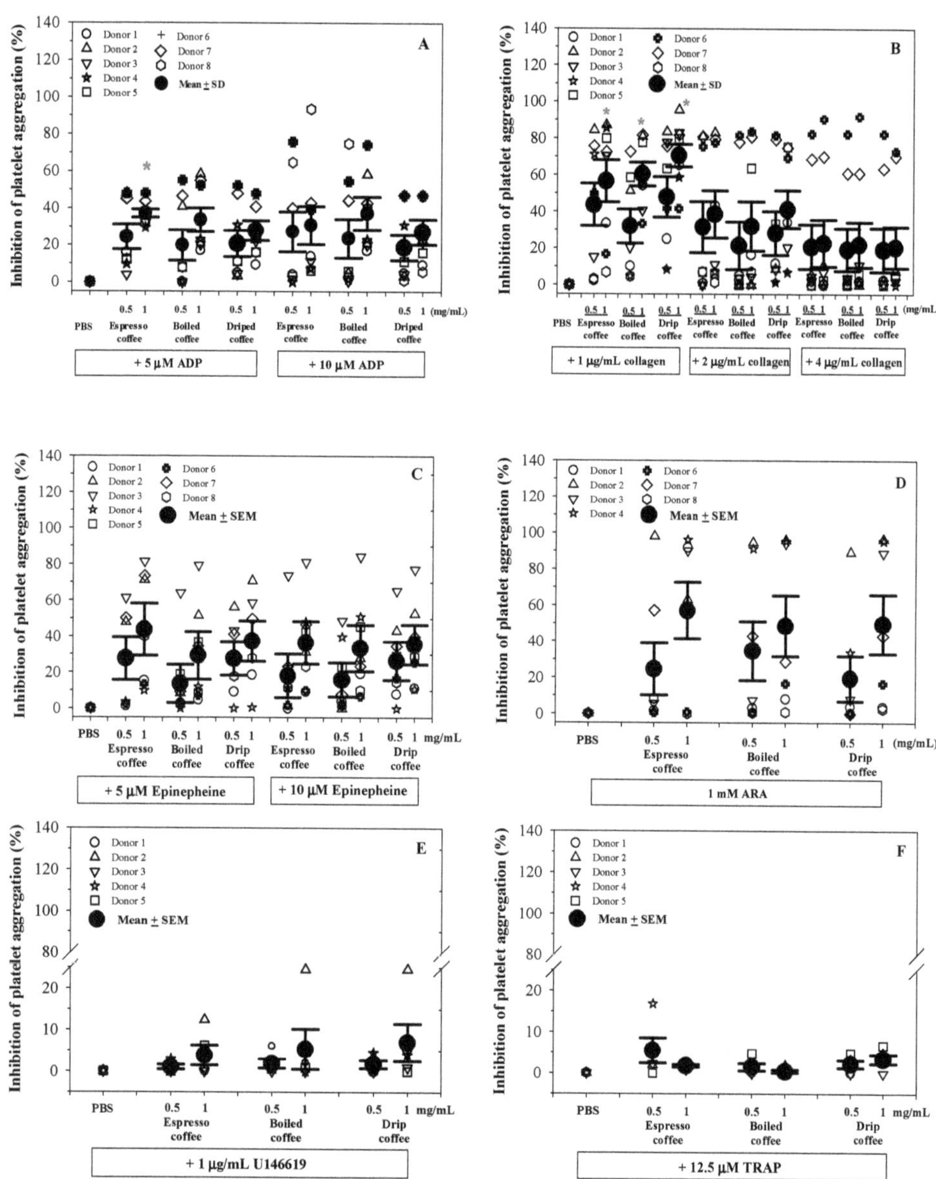

Figure 3. Inhibitory effects of phosphate-buffered saline (PBS) and coffee extracts (0.5 and 1 mg/mL each) on aggregation of platelets induced by adenosine diphosphate (ADP) (**A**), collagen (**B**), epinephrine (**C**), sodium arachidonate (ARA) (**D**), U46619 (**E**) and thrombin receptor agonist peptide (TRAP) (**F**). Data obtained from different blood samples are expressed as mean ± SEM values. * $p < 0.05$ when compared without the coffee extract treatments (PBS).

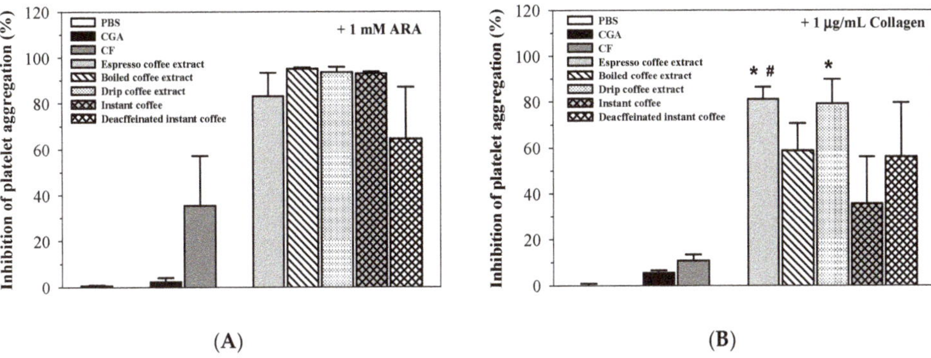

Figure 4. Inhibitory effects of phosphate-buffered saline (PBS), coffee extracts (0.5 and 1 mg/mL each), 18.23 mg/mL chlorogenic acid (CGA) and 54.17 mg/mL caffeine (CF) on aggregation of platelets induced by collagen (**B**) and sodium arachidonate (ARA) (**A**). Data obtained from five blood samples are expressed as mean + SEM values. * $p < 0.05$ when compared with the CGA; # $p < 0.05$ when compared to the CF.

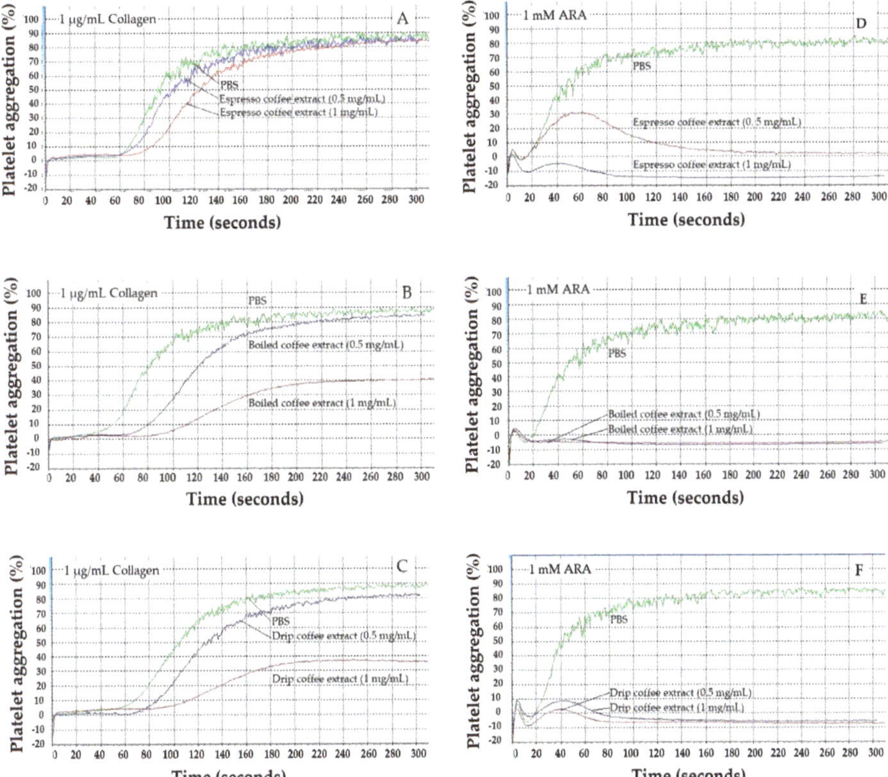

Figure 5. Representative aggregation traces of collagen (**A–C**) and sodium arachidonate (ARA)(**D–F**)-induced platelet aggregation by espresso, boiled, and drip coffee extracts (1 mg/mL each). Data were obtained from three platelet-rich plasma samples and are presented as mean ± SEM values.

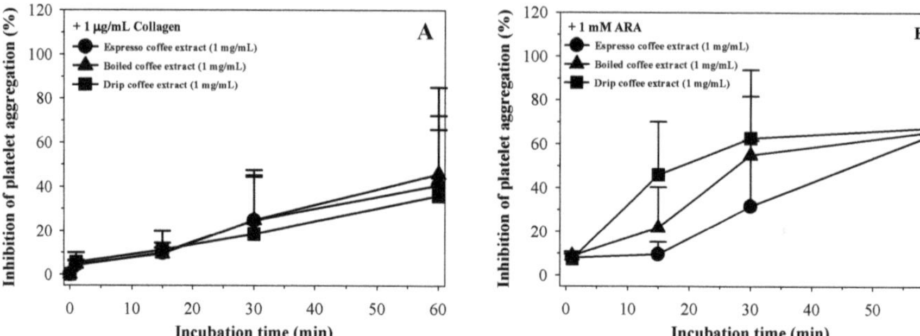

Figure 6. Time-course experiments of collagen (**A**) and sodium arachidonate (ARA) (**B**)-induced platelet aggregation by espresso, boiled and drip coffee extracts (1 mg/mL each). Data were obtained from three platelet-rich plasma samples and are presented as mean ± SEM values.

2.3. Effect of Coffee Extracts and Instant Coffee on Cyclooxygenase Activities

Treatments with espresso, boiled and drip coffee extracts (2 mg/mL equivalent to 97.7, 99.2, and 91.4 µg CF, 17.2, 18.9, and 17.6 µg CGA, respectively) and indomethacin (100 mg/mL) considerably inhibited COX-1 activity ($p < 0.05$), while CF (100 µg) and CGA (20 µg) treatments were found to be less active. Similar to COX-1 inhibition, the coffee extracts revealed significant inhibitory effects on COX-2 activity that were noticeably weaker than indomethacin (100 mg/mL) but stronger than CF and CGA (Figure 7). The inhibitory effects of the three coffee extracts compared with CF and CGA on COX-1 and COX-2 are broadly similar. The results suggest that with respect to inhibition of COX-1 and COX-2, that there are compounds present in coffee extracts, which are relatively lacking in CGA or CF. Although these differences appear marked for both COX-1 and COX-2 (Figure 7), further experiments would be required to establish statistical significance under a range of conditions.

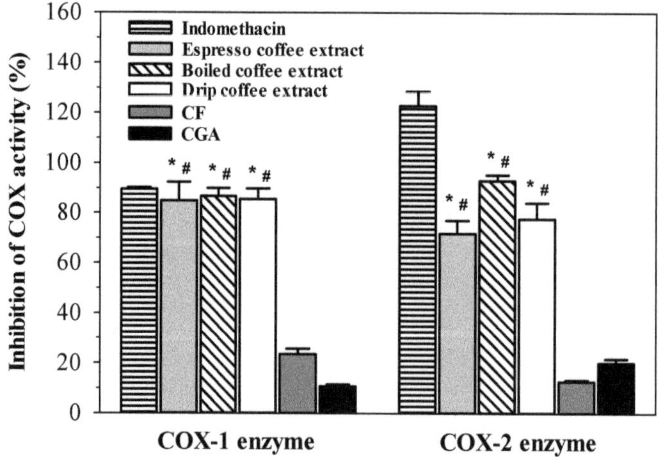

Figure 7. Inhibition of cyclooxygenase 1 (COX-1) and cyclooxygenase 2 (COX-2) activities by indomethacin (100 mg/mL), caffeine (CF, 100 µg/mL), chlorogenic acid (CGA, 20 µg/mL), espresso coffee, boiled coffee, and drip coffee extracts (2 mg/mL each). Data obtained from five separate duplicate measurements are expressed as mean ± SEM values. * $p < 0.05$ when compared with the CGA; # $p < 0.05$ when compared to the CF.

As is shown in Figure 8, the results demonstrate that espresso, boiled and drip coffee extracts exerted almost equal degrees of inhibition of COX-1 and COX-2 activities in a concentration-dependent manner, while CF and CGA were found to have exerted lesser degrees of inhibition. In the case of a constitutive COX-1, the degree of inhibition was minimal at a dose of 0.15 mg/mL and considerable at doses of 1–2 mg/mL. In the case of an inducible COX-2, the inhibitions were increasingly obvious as the doses were increased. Even with less CF and CGA equivalent concentrations, all the coffee extracts still displayed more potent degree of inhibition of COX-1 and COX-2 activities than the authentic standards.

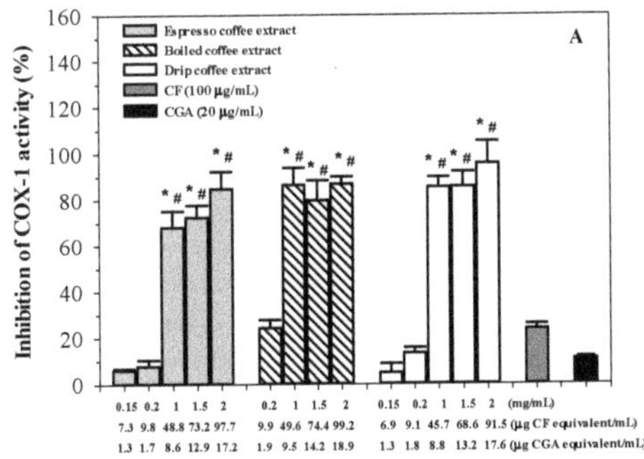
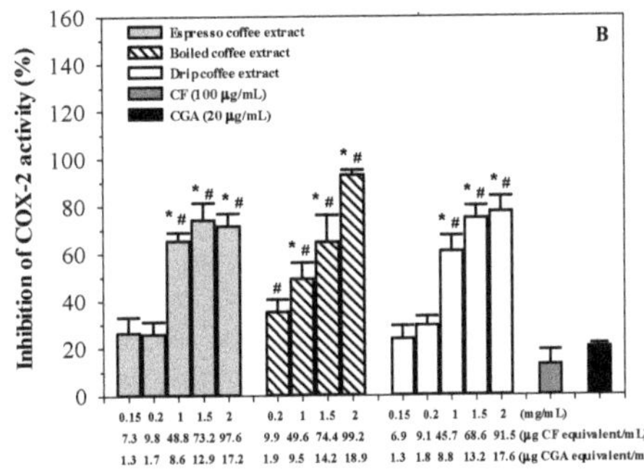

Figure 8. Inhibitory effects of espresso, boiled and drip coffee extracts (0.15–2 mg/mL each), caffeine 100 µg/mL) and chlorogenic acid (CGA, 20 µg/mL) on cyclooxygenase 1 (COX-1) (**A**) and cyclooxygenase 2 (COX-2) (**B**) activities. Data obtained from five separate duplicate measurements are expressed as mean ± SEM values. * $p < 0.05$ when compared with the CGA; # $p < 0.05$ when compared to the CF.

In this study, the coffee extracts are likely to contain other bioactive phenolic/flavonoid compounds (such as QA, FA, and FQA) besides CF and CGA that have synergistic anti-platelet aggregation and health benefits.

3. Discussion

Natural bioactive compounds in coffee beans help promote human health and wellness; however, the varying classes of phenolic compounds, sources of antioxidants, extraction methods, thermal processing and sample storage may influence the biological and pharmacological effects of coffee drinks. Generally, instant coffee is prepared by extracting the soluble and volatile contents of coffee beans or pulp with pressurized hot water (approximately 175 °C) and then through varying degree of concentration using either evaporation or spray drying methods of preparation. Instant coffee has been reported to possess anti-oxidative and anti-peroxidation properties; however, it may also act as a pro-oxidant at high concentrations, probably by converting an inactive ferric ion (Fe^{3+}) to a redox-active ferrous ion (Fe^{2+}) in a Fenton reaction [26]. Importantly, coffee contains significantly higher amounts of free phenolic compounds and displays greater antioxidant activity levels than cocoa and tea, of which their anti-oxidation yields are highly related to their TPC content; particularly with regard to their free phenolic compound content. This outcome is possibly due to the presence of the main content of CGA, particularly 5-caffeoylquinic acid (5-CQA) [27]. Natella et al. [28] reported on the presence of CGA, but not of CA, *p*-coumaric acid (CMA) and ferulic acid (FA) in American-style filtered coffee (60 g roasted coffee/L water) that had been prepared from a commercial automatic brewing machine. On the contrary, CA, CMA, and FA, but not CGA, were detectable in the brewed coffee after alkaline hydrolysis. In the present study, we found only the free form CGA, but not CA, with regard to the authentic standards in all of our coffee extracts. This outcome suggests that most or all of the CA in coffee would be esterified with QA and eventually form caffeoylquinic acid or CGA. In terms of roasting, 4-vinylcatechol of the CA moieties in green coffee usually oligomerizes to form bitter compounds [29].

The extraction phase of espresso coffee combines physical and chemical variables in a very short period of time. These variables directly affect its flavor, aroma, quality, and the resulting coffee beverage's bioactivities [30]. When compared with green coffee beans, roasted coffee beans are known to contain considerably lower contents of phytochemicals. This was confirmed by a 68% decrease of CGA. However, in this study, higher TPC and AA contents and unchanged CF contents were recorded [31]. Additionally, secondary metabolites, particularly CGAs, that are present in green beans were found to be degraded during the roasting process into other phenolic compounds. This is ultimately responsible for the bitterness of the coffee beverage. The present findings have revealed that all the coffee extracts exhibited differences in TPC and AA as a result of the differing coffee processes. Consistently, the TPC and AA levels detected in instant decaffeinated coffee and espresso coffee beverage were directly related to the CGA content, but were inversely related to CF content. Moreover, technological factors such as those associated with the decaffeination process, the brew volume and the method of preparation test, as well as the coffee species and the degree of roast, may all affect antioxidant activity and phenolic content in coffee drinks. Aguilera and colleagues have reported that heat/acidity-assisted extraction of coffee has produced higher concentrations of CGA, vanillic acid (VA), protocatechuic acid (PCA) and CMA than conventional methods [32]. In addition, abundant contents of CF (approximately 1200 ± 57 mg/kg spent coffee residual) and CGA (approximately 1700 ± 90 mg/kg spent coffee residual), some phenolic acids (e.g., CA, FA, PCA, VA, gallic acid, syringic acid and *trans*-cinnamic acid), and some flavonoids (e.g., rutin, cyanidin 3-glucoside and quercetin) were also present in the final coffee beverage [33]. The findings of Aguilera and colleagues have supported those of our study by confirming greater TPC and AA levels in the decaffeinated coffee than in the coffee brew extractions. With regard to free radical-scavenging activity and TPC, espresso (decaffeinated) coffee (30 mL cup) displayed significantly greater effects than espresso (regular), while a long-extraction espresso coffee beverage (70 mL cup) was found to result in greater radical-scavenging activity when compared to a short-extraction espresso coffee beverage (20 mL cup). Furthermore, Robusta brewed coffee revealed significantly greater radical-scavenging activity than Arabica brewed coffee [34]. In addition, at least 23 hydroxycinnamic derivatives including CGA, lactones and cinnamoyl-amino acid conjugates were detected in the coffee brews [34].

Thermal processing (such as steaming, boiling or frying) helps release bound phenolic compounds from the structural matrices and glycosides in coffee beans, but tends to reduce antioxidant activity. However, high-pressure processing tends to better retain the antioxidant properties of coffee drinks when compared to thermal treatments [35]. CGA, CA and FA are the three main anti-oxidative phenolic compounds present in both roasted coffee grounds and spent coffee grounds. Of these, CGA is the most abundant compound [36]. In the present study, CA was not detectable in all coffee extracts, while authentic CA was detected in the chromatogram. In modifications, feruloyl esterase obtained from microbial substances may be included in the preparation of coffee to release anti-oxidative phenolic acids such as CA, PCA, and FA from the coffee pulp and beans and would then contribute in the way of health benefits [37]. Moreover, water coffee extracts can be prepared using manual methods (such as boiling) and automatic coffee makers (such as percolation, espresso, and drip) depending upon the facility used and the customers' taste and degree of satisfaction. According to both the aroma and flavor, roasted coffee drinks are extremely popular when compared to both green coffee and instant coffee; nonetheless, roasting may influence the nutraceutical, biological, and pharmacological properties of the coffee beverage. Interestingly, medium roasted coffee displayed the highest TPC content, while light roast coffee exhibited the highest 2,2-diphenyl-1-picrylhydrazyl (DPPH) radical scavenging activity. Lastly, dark roasted coffee revealed both the lowest TPC content and the lowest degree of radical-scavenging activity [38]. We found that coffee extracts were capable of inhibiting platelet aggregation induced by ADP, epinephrine, collagen, and ARA, for which the most remarkable degree of inhibition took place with the ARA agonist, while inhibition was more marked at lower doses of collagen (which are thought to activate platelets mainly through the cyclo-oxygenase pathway). According to the results, high inter-individual variability, coupled with the fact that the sensitivity to inhibition is influenced by the strength of the aggregation response in the presence of PBS in each individual (i.e., if the response is at the lower end of normal, there is a more limited scope for inhibition than if the response was upper normal, since aggregation responses below zero cannot be recorded). Additionally, Disaggregation is thought to occur when the level of stimulus is below a certain threshold for full, irreversible aggregation, i.e., only limited amounts of TXA2 are generated and are compensated for by the inhibitory pathways within the platelet (e.g., cAMP generation). The concentration of TXA2 is also likely to be insufficient to provoke ADP release from the dense granules. Surprisingly, we found that inhibition was not influenced by CGA and CF when compared to the standard. Importantly, CF augmented the activity of NSAIDs such as aspirin, indomethacin and ketoprofen; nevertheless, it did not influence NSAIDs-induced inhibition of platelet thromboxanes (TXs) [39]. Likewise, there is increasing evidence that the consumption of phenolic-rich beverages has the potential to lower and prevent the risk of developing thrombosis. For example, catechol, hydroquinone and their derivatives, which are the antioxidants present in coffee and other plant products, were found to inhibit the production of prostaglandin E_3 (PGE$_3$) and TXB$_2$, as well as the aggregation of platelets induced by ARA but not by U46619 [40,41]. In addition, CGA has displayed inhibitory effects on collagen-induced platelet aggregation and TXA$_2$ production in a concentration dependent manner, along with increases in microsomal platelet cyclic adenosine-5′-monophosphate (cAMP) and cyclic guanosine-5′-monophosphate (cGMP) levels [42]. In contrast, CA was not detectable in all brewed coffee and instant coffee samples in this study, even though it had been previously identified as an intense bitter compound in strongly roasted espresso coffee [29]. With regard to CA composition, the brewed and instant coffee beverages displayed proper inhibitory effects on platelet aggregation and COXs activities. This was possibly due to the actions of other bioactive compounds. Interestingly, espresso coffee contains the highest amount of CGAs, mainly 5-CQA, (approximately 5 mM) and much lower amount of CA among all commonly consumed beverages [43]. With regard to its bioavailability, CGAs per se may directly interact with platelets in blood circulation or can be metabolized extensively by colonic microbiota into degradation products including CA, FA, hydrocaffeic acid, dihydroferulic acid and 3-(3′-hydroxyphenyl)propionic acid before being absorbed from the gut lumen [44]. Nevertheless, consumption of instant coffee (16 g/d, equivalent to 520 mg CF) for a 3-wk period did not change the plasma and urinary levels of TXB$_2$ and PGE$_2$ [45]. Furthermore, antioxidant activity of coffee products has been reported in the order of Robusta coffee, coffee cherry, Indian green coffee (72%) > CA (71%) > CGAs (70%) > dicaffeoylquinic acids (69%) > Nescafe Espresso (49%) [46].

Lastly, we have identified the mechanism of the coffee extracts on ARA-induced platelet aggregation by estimating the degree of COX activity through quantitation of prostaglandin. We found that the extracts were more potent than CGA or CF alone in inhibiting the aggregation by suppressing both COX-1 and COX-2 in the PGE$_2$ pathway; which the inhibition would possibly be interactions of CGA with CF. The two compounds with other phenolic compounds (such as quercetin), demonstrating the synergistic effects on inhibiting pro-inflammatory responses. Similarly, combination of quercetin and tea saponin extract exerted synergistic effect on inhibiting PGE$_2$ production through the COXs/PGE$_2$ pathway [47]. Naito and colleagues have previously demonstrated that hot water extracts of Blue Mountain, Yunnan and Kilimanjaro coffee beans containing CGA, CF, QA, and CA exhibited weak anti-platelet aggregation activity [48]. Moreover, CGA (5 mg/kg) and CF (0.5 g/kg) treatment nearly 60% decreased platelet aggregation in streptozotocin-induced diabetic rats [49,50]. Nonetheless, decaffeinated lyophilized coffee which contained higher phenolic compounds and CGA and integral coffee beverage did not change the hemostatic parameters in normal and high fat-fed rats [51]. Interestingly, an anti-oxidative diterpene "kahweol" found in unfiltered coffee has been shown to have a significant anti-inflammatory effect on the inhibition of inducible COX-2 and nitric oxide synthase (iNOS) activities and monocyte chemoattractant protein-1secretion [52–54]. Surprisingly, CGA has been reported to bind to the active site of the adenosine A_{2A} receptor and consequently attenuates the anti-platelet effect. Moreover, coffee extracts and authentic CGA, GA, and kaempferol consistently suppressed the expression of the COX-2 gene in macrophage (RAW264.7-Mϕ) cells [55] and in regular coffee beverages, while CF inhibited COX-2 gene expression in rats [56]. Furthermore, a recent study has reported that the anti-inflammatory effect on COX-2 gene suppression could be attributed to the extracts of the coffee leaves, which contain CGA, CF, mangiferin, rutin, and other bioactive molecules [57]. Synergistically inhibitory effect of agonists-induced platelet aggregation, which is related to antioxidant activity and reported in extracts of plants (such as *Salvia* species), is possibly caused by the interaction of persisting phenolic compounds [58–60]. Taken together, our findings suggest the potential inhibition of platelet aggregation and COXs activities in the following order: Drip coffee > espresso coffee > boiled coffee > instant coffee > CF, CGA, suggesting that the degree of inhibition may not be influenced by the anti-oxidative compounds and caffeine present in the coffee extracts.

4. Materials and Methods

4.1. Chemicals and Reagents

Accordingly, 2,2′-Azino-bis(3-ethylbenzothiazoline-6-sulfonic acid) diammonium salt (ABTS), Dulbecco's minimal essential medium (DMEM), RPMI1640 medium, phosphate buffered saline pH 7.0 (PBS), fetal bovine serum (FBS) and other materials for cell culturing were purchased from Gibco, Life Technologies (Eugene, OR. USA). Histopaque®-1077, dimethyl sulfoxide solution (DMSO), adenosine 5′-diphosphate (ADP) sodium salt, epinephrine, sodium arachidonate (AA), Thrombin receptor agonist peptide (TRAP), 9,11-dideoxy-11α,9α-epoxymethanoprostaglandin $F_{2\alpha}$ (U46619), gallic acid (GA), penicillin-streptomycin, 6-hydroxy-2,5,7,8-tetramethylchroman-2-carboxylic acid (Trolox), chlorogenic acid, and caffeine were purchased from Sigma-Aldrich Chemicals Company (Gillingham, UK). Acetonitrile, phosphoric acid, Folin-Ciocalteu reagent, potassium thiosulfate ($K_2S_5O_8$) and sodium carbonate (Na_2CO_3) were purchased from Merck KGaA (Darmstadt, Germany). Collagen suspension was obtained from Helena Biosciences (Gateshead, UK). Competitive enzyme linked immunosorbent assay (cELISA) kit for ovine/human cyclooxygenase (COX) (No.560131) was purchased from Cayman Chemicals Company (Ann Arbor, MI, USA). UltraPure water and other chemicals and solvents that were used in this study were of the highest analytical grade.

4.2. Coffee Extract Preparation

Green coffee beans (Coffea arabica) were kindly provided by the Royal Project Foundation, Chiang Mai, Thailand. Coffee beans were peeled, dried at room temperature until humidity reached <13% of their original weight, roasted in a coffee roaster at 200–250 °C for 15–20 min, ground using a milling machine, and kept in aluminum-foil bags at room temperature. Coffee extracts were prepared in coffee brewing devices using the boiled/brewed process and the pressurized espresso and drip methods according to the manufacturer's instructions and the description established by Caprioli et al. [61]. For the boiled and brewed method, ground roasted coffee (10 g) and boiled water (100 mL, 90 °C) were combined in a cylindrical vessel. The coffee was then left to brew for a few min. A circular filter was then tightly fitted to the cylinder and a fixed plunger was pushed down from the top through the liquid to force the grounds to the extract bottom containing most of the active substances. This process makes the coffee beverage stronger and leaves the spent coffee grounds in the vessel. Finally, the boiled coffee brew was then poured from the container, and this was what was used in our experiments. By using manual drip coffee equipment (Brand Delonghi, Model BCO421.S, Treviso, Italy), boiled water (100 mL, 90 °C) was poured into a container filled with ground coffee (10 g) with a perforated filter base. The hot water was then allowed to seep for approximately 5 min (15 bar pressure) through the filter. The resulting coffee then dripped into the container below. For the espresso method, a limited amount of hot water (90 °C) was pressurized at 12 bars using a portable espresso coffee machine (Brand MINIMEX, Model Miniespresso, Bangkok, Thailand). The hot water would then percolate through the ground roasted coffee beans (10 g) in a relatively short period of time to yield a small cup of concentrated foamy coffee extract. Coffee brew extracts were filtrated through filter paper (cellulose type, Whatman's No. 1, Merck KGaA, Darmstadt, Germany), centrifuged at 3000 rpm for 15 min, freeze-dried using a lyophilisation machine and kept at −20 °C until being analyzed and studied. In this experiment, instant (regular and decaffeinated) coffee (Tesco Gold, Tesco Stores Limited, United Kingdom) (10 g) was freshly prepared by reconstituting the coffee product with boiled water (100 mL, 90 °C) and then filtering the resulting beverage through Whatman's No. 1 filter paper.

4.3. Chemical Analysis of Coffee Preparations

CGA and CF were quantified using high-performance liquid chromatography/diode array detection (HPLC/DAD) [62]. The conditions included a column (C18-type, 4.6 mm × 250 mm, 5 μm particle size, Agilent Technologies, Santa Clara, California, United States), isocratic mobile-phase solvent containing 0.2% phosphoric acid and acetonitrile (90:10, v/v), a flow rate of 1.0 mL/min and a wavelength detection of 275 nm for CF and 330 nm for CGA. Data were recorded and integrated using Millennium 32 HPLC Software. CGA and CF were identified by comparison with the specific TR of the authentic standards and determined concentrations were established from the standard curves constructed from different concentrations.

Total phenolic compounds (TPC) of the coffee extracts were determined using the Folin–Ciocalteu method [63]. Briefly, the coffee extract (100 μL) was incubated with 10% Folin–Ciocalteu reagent (200 μL) at room temperature for 4 min and then incubated with 700 mM Na2CO3 (800 μL) for 30 min. The optical density (OD) was then measured at 765 nm against the reagent blank. GA (6.25–200 μg/mL) was used to generate a calibration curve that was used to calculate TPC and identify the gallic acid equivalent (GAE).

Antioxidant activity was determined using the ABTS radical cation (ABTS$^{\bullet}$+) decolorization method [64]. In the assay, ten microliters of absolute ethanol (blank) and coffee extracts (62.5–4000 μg/mL) or Trolox (6.25–800 μg/mL) were incubated with a freshly prepared solution consisting of 3.5 mM ABTS$^{\bullet}$+ and 1.22 mM K2SsO8 in the dark at room temperature for exactly 6 min. The results were photometrically measured at 764 nm. Results were expressed as the percentage of inhibition of ABTS$^{\bullet}$+ production and were reported in terms of mg trolox equivalent antioxidant capacity (TEAC)/g extracts.

4.4. Determination of Anti-Platelet Aggregation Activity

4.4.1. Isolation of Platelet-Rich Plasma

Blood collection was permitted by the director of Maharaj Nakorn Chiang Mai, Faculty of Medicine Chiang Mai University, Chiang Mai, Thailand (Reference Number 8393(8).9/436) and approved by the Research Ethics Committee for Human Study, Faculty of Medicine, Chiang Mai University, Chiang Mai, Thailand (Research ID: 7575/Study Code: BIO-2563-07575/Date of Approval: 28th September 2020). Informed consent was provided by healthy blood volunteers. Subjects were asked to avoid drinking caffeine-containing beverages for 24 h prior to blood collection. At 9.00 am, blood samples of non-fasting volunteers were collected directly from veins and deposited into tubes containing 0.106 M tri-sodium citrate. The tubes were mixed gently and centrifuged at 170 g at an ambient temperature (25 °C) for 10 min. Platelet-rich plasma (PRP) supernatant was then transferred to capped polypropylene tubes. Residual blood was centrifuged at 2000 g for 15 min and platelet-poor plasma (PPP) supernatant was separated using a plastic pipette then stored in a capped polypropylene tube for use as a PRP diluent and for setting the transmission blank on the aggregometer. PRP platelet counts were obtained using a KX-21 cell counter (Sysmex Corp, Kobe, Japan).

4.4.2. Platelet Aggregation Assay

Platelet aggregation was conducted using a light transmission aggregometer (AggRAM™, Helena Biosciences, Gateshead, UK) [65,66]. PRP (150–600×10^9 platelet/L) was stirred in an aggregometer cuvette at $37\,^\circ C$, and different coffee extracts (0.5 mg/mL and 1 mg/mL, 25 μL) were added. After incubation for 15 min, 25 μL of PBS (control) or agonists including: ADP (1–10 μM), epinephrine (1–10 μM), collagen (1–4 μg/mL), TRAP (12.5 μM), ARA (1 mM) and U46619 (1 μg/mL) were added to the mixture, and it was further incubated for 5 min. Aggregation was monitored as light transmission at 650 nm for 300 sec. The percentage of platelet aggregation inhibition was calculated using the following formula; $100 \times (OD_{PBS} - OD_{PRP})/OD_{PBS}$.

In a time-course study, PRP was incubated with different coffee extracts (1 mg/mL each, 25 μL) in an aggregometer cuvette at $37\,^\circ C$ for 15, 30, and 60 min. Finally, PBS or agonists including 1 μg/mL collagen and 1 mM ARA (25 μL) were added to the mixture and platelet aggregation was monitored as has been mentioned above. The percentage of platelet aggregation inhibition was calculated using the following formula: $100 \times (OD_{PBS} - OD_{PRP})/OD_{PBS}$.

4.5. Determination of Cyclooxygenase Activity

Biologically, COXs (or PGH synthase) that exhibit both COX and peroxidase (POD) activities catalyze the conversion of ARA to hydroperoxy endoperoxide (or PGG_2), which will subsequently be reduced by the POD component to PGH_2, a precursor of PGs, thromboxane and prostacyclins. The COX inhibitor screening method is based on the measurement of $PGF_{2\alpha}$ by the stannous chloride ($SnCl_2$)-catalyzed reduction of PGH_2 that is produced in the COX reaction via cELISA using a broadly specific antibody against all PGs and Ellman's reagent [67]. The assay reagent includes both ovine COX-1 and human recombinant COX-2 enzymes together with acetylcholinesterase-conjugated PG (AChE-PG) tracer and Ellman's reagent containing acetylcholine and 5,5′-dithio-bis-(2-nitrobenzoic acid). In this study, five different extracts of espresso, boiled and drip coffee (10 μL each) in duplicate were firstly incubated with the COXs reagent (10 μL) at $37\,^\circ C$ for 10 min. Secondly, the 10 μM ARA substrate solution (10 μL) was added to the mixture, and the mixture was then further incubated for exactly 2 min. Thirdly, $SnCl_2$ solution (30 μL) was added to the mixture to stop the process of enzyme catalysis. Fourthly, treated coffee extracts, standard PG (15.6–2000 pg/mL) and ELISA buffer (50 μL each) were added to mouse anti-rabbit IgG-immobilized plate wells, followed by the addition of AChE-PG tracer (50 μL) to complete the free PGs for the purpose of binding the rabbit antibody against PGs (50 μL), for which the AChE-PG-anti-PG complex was subsequently bound to the immobilized anti-rabbit IgG. The amount of AChE-PG that could bind to the anti-PG was found to be inversely proportional to the concentrations of PGs in the well. Ellman's reagent was added into each well to develop the yellow colored 5-thio-2-nitrobenzoic acid product that was determined photometrically at 405 nm using a microplate reader. The percentage of COX activity inhibition was obtained using the following forula: $100 \times (OD_{PBS} - OD_{PRP})/OD_{PBS}$.

4.6. Statistical Analysis

Experimental data were analyzed using SPSS Statistics Program (IBM SPSS® Software version 22, IBM Corporation, Armonk, NY, USA, shared license by Chiang Mai University, Chiang Mai, Thailand). Data are expressed as mean $\pm$ standard deviation (SD) values with standard error of mean (SEM). Statistical significance was determined using one-way analysis of variance (ANOVA) followed by Tukey's HSD posttest. p value < 0.05 was considered statistically significant.

5. Conclusions

Limitations of the study include the fact that only small numbers of subjects were studied; that a small range of sources of coffee product (e.g., Arabica coffee from the Royal Project Foundation) was used and the fact that only in vitro studies were performed and it is therefore not certain whether the same effect would be observed ex vivo in normal volunteers. Overall, the results strongly corroborate the hypothesis that the brewing procedure, type of coffee and potential active compounds, such as caffeine, chlorogenic acid and phenolic compounds, could influence platelet aggregation and the activity of cyclooxygenases. Notably, drip and espresso coffee brews are more effective in amelioration of cyclooxygenase-catalyzed anti-/pro-inflammation. Major bioactive compounds in the coffee beverages need to be identified using liquid chromatography/mass spectrometry together with a thorough assessment of their potential anti-thrombotic effects. Consequently, coffee beverages might indeed lower the risk of developing a hypercoagulable state and thrombosis.

Author Contributions: N.H. designed and conducted experiments and analyzed data. P.A. analyzed and approved of the accumulated coffee information, edited the experimental results and shared in the discussion. C.C. supplied green and roasted coffee beans and approved of the accumulated coffee information. S.F. advised on experiments and shared in the discussion. I.J.M. supervised the study of platelet aggregation and COX activities, shared discussion. J.B.P. shared discussion for the study of platelet aggregation and COX activities, proofread the manuscript and made the revision. S.S. conceived of the study and experiments, contributed to the discussion and wrote the manuscript. All authors have read and agreed to the published version of the manuscript.

Funding: This work has been funded by the Research and Researchers for Industries (RRI) PhD. Program (PHD60I0020), Thailand Science Research and Innovation (formerly Thailand Research Fund), the Royal Project Foundation, Chiang Mai, and the Medical Faculty Endowment Fund, Faculty of Medicine, Chiang Mai University (113/2563).

Institutional Review Board Statement: The study was conducted according to the guidelines of the Declaration of Helsinki, and approved by the Research Ethics Committee No. 4 of Faculty of Medicine, Chiang Mai University, Chiang Mai, Thailand (Research ID: 7575/Study Code: BIO-2563-07575/Date of Approval: 28 September 2020).

Informed Consent Statement: Informed consent was obtained from all subjects involved in the study.

Data Availability Statement: Data available in a publicly accessible repository.

Acknowledgments: We gratefully acknowledge the Research and Researchers for Industries (RRI) Program through Nuntouchaporn Hutachok (PHD60I0020) of Thailand Science Research and Innovation (formerly Thailand Research Fund) and the Faculty of Medicine Endowment Fund, Chiang Mai University, Thailand for supplying the research grant (113//2563). We express our appreciation to the Royal Project Foundation, Chiang Mai, Thailand for supplying green and roasted Arabica coffee beans and the Haemostasis Research Unit, Department of Haematology, UCL, London, UK for providing the laboratory facilities used in platelet aggregation and COX activity assays. John B. Porter would like to acknowledge UCL Biomedical Research Centre, Cardiometabolic Programme.

Conflicts of Interest: The authors declare that they hold no conflict of interest.

Abbreviations

AA = antioxidant activity; ABTS = 2,2′-azino-*bis* (3-ethylbenzothiazoline-6-sulfonic acid) diammonium salt; AChE = acetylcholinesterase; AChE-PG = prostaglandin-acetylcholinesterase; ADP = adenosine diphosphate sodium salt; ARA = arachidonic acid; CA = caffeic acid; cAMP = cyclic adenosine-5′-monophosphate; cELISA = competitive enzyme linked immunosorbent assay; CF = caffeine; CGA = chlorogenic acid; cGMP = cyclic guanosine-5′-monophosphate; CMA = *p*-coumaric acid; CQA = 5-*O*-caffeoylquinic acid; COX = cyclooxygenase; COX-1 = cyclooxygenase-1; COX-2 = cyclooxygenase-2; DMEM = Dulbecco's minimal essential medium; DMSO = dimethyl sulfoxide; DPPH = 2, 2- diphenyl-1-picrylhydrazyl; ELISA = enzyme linked immunosorbent assay; FA = ferulic acid; FBS = fetal bovine serum; Fe^{2+} = ferrous ion; Fe^{3+} = ferric ion; FQA = feruloylquinic; GA = gallic acid; GAE = gallic acid equivalent; HPLC = high performance liquid chromatography; HPLC/DAD = high performance liquid chromatography/diode array detection; IC_{50} = half-maximal inhibitory concentration; iNOS = inducible nitric oxide synthase; $K_2S_8O_8$ = potassium thiosulfate; Na_2CO_3 = sodium carbonate; ND = not detectable; nm = nanometer; nM = nanomolar; NO• = nitric oxide; NSAIDs = nonsteroidal anti-inflammatory drugs; OD = optical density; PBS = phosphate buffered saline; PCA = protocatechuic acid; PG = prostaglandin; PGE_3 = prostaglandin E_3; $PGF_{2}\alpha$ = prostaglandin F_{2a}; PGI2 = prostacyclin; PGs = prostaglandins; PPP = platelet-poor plasma; PRP = platelet-rich plasma; QA = quinic acid; ROS = reactive oxygen species; rpm = revolutions per minute; SD = standard deviation; SEM = standard error of the mean; TE = Trolox equivalent; TEAC

= Trolox equivalent antioxidant capacity; TPC = total phenolic content; T_R = retention time; TRAP = thrombin receptor agonist peptide; TXB_2 = thromboxane B_2; TXs = thromboxanes; U46619 = 9,11-dideoxy-11α,9α-epoxymethanoprostaglandin $F_{2\alpha}$; VA = vanillic acid; *v/v* = volume by volume.

References

1. Eira, M.T.; Silva, E.; De Castro, R.D.; Dussert, S.; Walters, C.; Bewley, J.D.; Hilhorst, H.W. Coffee seed physiology. *Braz. J. Plant. Physiol.* **2006**, *18*, 149–163. [CrossRef]
2. Higdon, J.V.; Frei, B. Coffee and health: A review of recent human research. *Crit. Rev. Food Sci. Nutr.* **2006**, *46*, 101–123. [CrossRef] [PubMed]
3. Nawrot, P.; Jordan, S.; Eastwood, J.; Rotstein, J.; Hugenholtz, A.; Feeley, M. Effects of caffeine on human health. *Food Addit. Contam.* **2003**, *20*, 1–30. [CrossRef] [PubMed]
4. Cavalcante, J.W.S.; Santos, P.R.M., Jr.; de Menezes, M.G.F.; Marques, H.O.; Cavalcante, L.P.; Pacheco, W.S. Influence of caffeine on blood pressure and platelet aggregation. *Arq. Bras. Cardiol.* **2000**, *75*, 102–105. [CrossRef] [PubMed]
5. Franks, A.; Schmidt, J.; McCain, K.R.; Fraer, M. Comparison of the effects of energy drink versus caffeine supplementation on indices of 24-hour ambulatory blood pressure. *Ann. Pharmacother.* **2012**, *46*, 192–199. [CrossRef] [PubMed]
6. De Mejia, E.G.; Ramirez-Mares, M.V. Impact of caffeine and coffee on our health. *Trends Endocrinol. Metab.* **2014**, *25*, 489–492. [CrossRef]
7. Monteiro, M.; Farah, A.; Perrone, D.; Trugo, L.C.; Donangelo, C. Chlorogenic acid compounds from coffee are differentially absorbed and metabolized in humans. *J. Nutr.* **2007**, *137*, 2196–2201. [CrossRef]
8. Farah, A.; Monteiro, M.; Donangelo, C.M.; Lafay, S. Chlorogenic acids from green coffee extract are highly bioavailable in humans. *J. Nutr.* **2008**, *138*, 2309–2315. [CrossRef]
9. Mullen, W.; Nemzer, B.; Ou, B.; Stalmach, A.; Hunter, J.; Clifford, M.N.; Combet, E. The antioxidant and chlorogenic acid profiles of whole coffee fruits are influenced by the extraction procedures. *J. Agric. Food Chem.* **2011**, *59*, 3754–3762. [CrossRef]
10. Gloess, A.N.; Schonbachler, B.; Klopprogge, B.; D'Ambrosio, L.; Chatelain, K.; Bongartz, A.; Strittmatter, A.; Rast, M.; Yeretzian, C. Comparison of nine common coffee extraction methods: Instrumental and sensory analysis. *Eur. Food Res. Technol.* **2013**, *236*, 607–627. [CrossRef]
11. Rodrigues, C.I.; Marta, L.; Maia, R.; Miranda, M.; Ribeirinho, M.; Maguas, C. Application of solid-phase extraction to brewed coffee caffeine and organic acid determination by UV/HPLC. *J. Food Compost. Anal.* **2007**, *20*, 440–448. [CrossRef]
12. Budryn, G.; Nebesny, E.; Podsedek, A.; Zyzelewicz, D.; Materska, M.; Jankowski, S.; Janda, B. Effect of different extraction methods on the recovery of chlorogenic acids, caffeine and Maillard reaction products in coffee beans. *Eur. Food Res. Technol.* **2009**, *228*, 913–922. [CrossRef]
13. Niseteo, T.; Komes, D.; Belscak-Cvitanovic, A.; Horzic, D.; Budec, M. Bioactive composition and antioxidant potential of different commonly consumed coffee brews affected by their preparation technique and milk addition. *Food Chem.* **2012**, *134*, 1870–1877. [CrossRef] [PubMed]
14. Vignoli, J.A.; Bassoli, D.G.; Benassi, M.T. Antioxidant activity, polyphenols, caffeine and melanoidins in soluble coffee: The influence of processing conditions and raw material. *Food Chem.* **2011**, *124*, 863–868. [CrossRef]
15. Taher, A.T.; Otrock, Z.K.; Uthman, I.; Cappellini, M.D. Thalassemia and hypercoagulability. *Blood Rev.* **2008**, *22*, 283–292. [CrossRef]
16. Naithani, R.; Chandra, J.; Narayan, S.; Sharma, S.; Singh, V. Thalassemia major—On the verge of bleeding or thrombosis? *Hematology* **2006**, *11*, 57–61. [CrossRef]
17. Eldor, A.; Rachmilewitz, E.A. The hypercoagulable state in thalassemia. *Blood.* **2002**, *99*, 36–43. [CrossRef]
18. Rund, D.; Rachmilewitz, E. Medical progress: Beta-thalassemia. *N. Engl. J. Med.* **2005**, *353*, 1135–1146. [CrossRef]
19. Musallam, K.M.; Taher, A.T.; Rachmilewitz, E.A. Beta-thalassemia intermedia: A clinical perspective. *Csh. Perspect. Med.* **2012**, *2*, a013482.
20. Lee, S.P.; Ataga, K.I.; Zayed, M.; Manganello, J.M.; Orringer, E.P.; Phillips, D.R.; Parise, L.V. Phase I study of eptifibatide in patients with sickle cell anaemia. *Br. J. Haematol.* **2007**, *139*, 612–620. [CrossRef]
21. Jakubowski, J.A.; Zhou, C.M.; Jurcevic, S.; Winters, K.J.; Lachno, D.R.; Frelinger, A.L.; Gupta, N.; Howard, J.; Payne, C.D.; Mant, T.G. A phase 1 study of prasugrel in patients with sickle cell disease: Effects on biomarkers of platelet activation and coagulation. *Thromb. Res.* **2014**, *133*, 190–195. [CrossRef] [PubMed]
22. Echeverri, D.; Montes, F.R.; Cabrera, M.; Galan, A.; Prieto, A. Caffeine's vascular mechanisms of action. *Int. J. Vasc. Med.* **2019**, *2019*, 834060.
23. Lippi, G.; Mattiuzzi, C.; Franchini, M. Venous thromboembolism and coffee: Critical review and meta-analysis. *Ann. Transl. Med.* **2015**, *3*, 152. [PubMed]
24. Buerkle, M.A.; Lehrer, S.; Sohn, H.Y.; Conzen, P.; Pohl, U.; Krotz, F. Selective inhibition of cyclooxygenase-2 enhances platelet adhesion in hamster arterioles in vivo. *Circulation* **2004**, *110*, 2053–2059. [CrossRef]
25. Rabausch, K.; Bretschneider, E.; Sarbia, M.; Meyer-Kirchrath, J.; Censarek, P.; Pape, R.; Fischer, J.W.; Schror, K.; Weber, A.A. Regulation of thrombomodulin expression in human vascular smooth muscle cells by COX-2—Derived prostaglandins. *Circ. Res.* **2005**, *96*, E1–E6. [CrossRef]
26. Stadler, R.H.; Markovic, J.; Turesky, R.J. In vitro anti- and pro-oxidative effects of natural polyphenols. *Biol. Trace Elem. Res.* **1995**, *47*, 299–305. [CrossRef]

27. Abbe Maleyki, M.J., Jr.; Azrina, A.; Amin, I. Assessment of antioxidant capacity and phenolic content of selected commercial beverages. *Malays. J. Nutr.* **2007**, *13*, 149–159.
28. Natella, F.; Nardini, M.; Belelli, F.; Scaccini, C. Coffee drinking induces incorporation of phenolic acids into LDL and increases the resistance of LDL to ex vivo oxidation in humans. *Am J. Clin. Nutr.* **2007**, *86*, 604–609. [CrossRef]
29. Frank, O.; Blumberg, S.; Kunert, C.; Zehentbauer, G.; Hofmann, T. Structure determination and sensory analysis of bitter-tasting 4-vinylcatechol oligomers and their identification in roasted coffee by means of LC-MS/MS. *J. Agric. Food Chem.* **2007**, *55*, 1945–1954. [CrossRef]
30. Khamitova, G.; Angeloni, S.; Fioretti, L.; Ricciutelli, M.; Sagratini, G.; Torregiani, E.; Vittori, S.; Caprioli, G. The impact of different filter baskets, heights of perforated disc and amount of ground coffee on the extraction of organics acids and the main bioactive compounds in espresso coffee. *Food Res. Int.* **2020**, *133*, 109220. [CrossRef]
31. Acidri, R.; Sawai, Y.; Sugimoto, Y.; Handa, T.; Sasagawa, D.; Masunaga, T.; Yamamoto, S.; Nishihara, E. Phytochemical profile and antioxidant capacity of coffee plant organs compared to green and roasted coffee beans. *Antioxidants* **2020**, *9*, 93. [CrossRef] [PubMed]
32. Aguilera, Y.; Rebollo-Hernanz, M.; Canas, S.; Taladrid, D.; Martin-Cabrejas, M.A. Response surface methodology to optimise the heat-assisted aqueous extraction of phenolic compounds from coffee parchment and their comprehensive analysis. *Food Funt.* **2019**, *10*, 4739–4750. [CrossRef] [PubMed]
33. Angeloni, S.; Nzekoue, F.K.; Navarini, L.; Sagratini, G.; Torregiani, E.; Vittori, S.; Caprioli, G. An analytical method for the simultaneous quantification of 30 bioactive compounds in spent coffee ground by HPLC-MS/MS. *J. Mass Spectrom.* **2020**, *55*, e4519. [CrossRef] [PubMed]
34. Alves, R.C.; Costa, A.S.G.; Jerez, M.; Casal, S.; Sineiro, J.; Nunez, M.J.; Oliveira, B. Antiradical activity, phenolics profile, and hydroxymethylfurfural in espresso coffee: Influence of technological factors. *J. Agric. Food Chem.* **2010**, *58*, 12221–12229. [CrossRef]
35. Amarowicz, R.; Pegg, R.B. Natural antioxidants of plant origin. *Adv. Food Nutr. Res.* **2019**, *90*, 1–81.
36. An, B.H.; Jeong, H.; Kim, J.H.; Park, S.; Jeong, J.H.; Kim, M.J.; Chang, M. Estrogen receptor-mediated transcriptional activities of spent coffee grounds and spent coffee grounds compost, and their phenolic acid constituents. *J. Agric. Food Chem.* **2019**, *67*, 8649–8659. [CrossRef]
37. Benoit, I.; Navarro, D.; Marnet, N.; Rakotomanomana, N.; Lesage-Meessen, L.; Sigoillot, J.C.; Asther, M.; Asther, M. Feruloyl esterases as a tool for the release of phenolic compounds from agro-industrial by-products. *Carbohydr. Res.* **2006**, *341*, 1820–1827. [CrossRef]
38. Bobkova, A.; Hudacek, M.; Jakabova, S.; Belej, L.; Capcarova, M.; Curlej, J.; Bobko, M.; Arvay, J.; Jakab, I.; Capla, J.; et al. The effect of roasting on the total polyphenols and antioxidant activity of coffee. *J. Environ. Sci. Health. B.* **2020**, *55*, 495–500. [CrossRef]
39. Sommerauer, M.; Ates, M.; Guhring, H.; Brune, K.; Amann, R.; Peskar, B.A. Ketoprofen-induced cyclooxygenase inhibition in renal medulla and platelets of rats treated with caffeine. *Pharmacology* **2001**, *63*, 234–239. [CrossRef]
40. Chang, M.C.; Chang, H.H.; Wang, T.M.; Chan, C.P.; Lin, B.R.; Yeung, S.Y.; Yeh, C.Y.; Cheng, R.H.; Jeng, J.H. Antiplatelet effect of catechol is related to inhibition of cyclooxygenase, reactive oxygen species, ERK/p38 signaling and thromboxane A2 production. *PLoS ONE.* **2014**, *9*, e104310. [CrossRef]
41. Chang, M.C.; Chang, B.E.; Pan, Y.H.; Lin, B.R.; Lian, Y.C.; Lee, M.S.; Yeung, S.Y.; Lin, L.D.; Jeng, J.H. Antiplatelet, antioxidative, and anti-inflammatory effects of hydroquinone. *J. Cell. Physiol.* **2019**, *234*, 18123–18130. [CrossRef] [PubMed]
42. Cho, H.J.; Kang, H.J.; Kim, Y.J.; Lee, D.H.; Kwon, H.W.; Kim, Y.Y.; Park, H.J. Inhibition of platelet aggregation by chlorogenic acid via cAMP and cGMP-dependent manner. *Blood Coagul. Fibrinolysis.* **2012**, *23*, 629–635. [CrossRef]
43. Nyambe-Silavwe, H.; Williamson, G. Chlorogenic and phenolic acids are only very weak inhibitors of human salivary alpha-amylase and rat intestinal maltase activities. *Food Res. Int.* **2018**, *113*, 452–455. [CrossRef] [PubMed]
44. Ludwig, I.A.; de Pena, M.P.; Cid, C.; Crozier, A. Catabolism of coffee chlorogenic acids by human colonic microbiota. *Biofactors* **2013**, *39*, 623–632. [CrossRef] [PubMed]
45. Aro, A.; Kostiainen, E.; Huttunen, J.K.; Seppala, E.; Vapaatalo, H. Effects of coffee and tea on lipoproteins and prostanoids. *Atherosclerosis* **1985**, *57*, 123–128. [CrossRef]
46. Tomac, I.; Seruga, M.; Labuda, J. Evaluation of antioxidant activity of chlorogenic acids and coffee extracts by an electrochemical DNA-based biosensor. *Food Chem.* **2020**, *325*, 126787. [CrossRef]
47. Puangpraphant, S.; de Mejia, E.G. Saponins in yerba mate tea (Ilex paraguariensis A. St.-Hil) and quercetin synergistically inhibit iNOS and COX-2 in lipopolysaccharide-induced macrophages through NFkappaB pathways. *J. Agric. Food Chem.* **2009**, *57*, 8873–8883. [CrossRef]
48. Naito, S.; Yatagai, C.; Maruyama, M.; Sumi, H. Effect of coffee extracts on plasma fibrinolysis and platelet aggregation. *Nihon Ishigaku Zasshi* **2011**, *46*, 260–269.
49. Stefanello, N.; Schmatz, R.; Pereira, L.B.; Cardoso, A.M.; Passamonti, S.; Spanevello, R.M.; Thome, G.; de Oliveira, G.M.T.; Kist, L.W.; Bogo, M.R.; et al. Effects of chlorogenic acid, caffeine and coffee on components of the purinergic system of streptozotocin-induced diabetic rats. *J. Nutr. Biochem.* **2016**, *38*, 145–153. [CrossRef]
50. Montagnana, M.; Favaloro, E.J.; Lippi, G. Coffee intake and cardiovascular disease: Virtue does not take center stage. *Semin. Thromb. Hemost.* **2012**, *38*, 164–177. [CrossRef]

51. Silverio, A.D.D.; Pereira, R.G.F.A.; Lima, A.R.; Paula, F.B.D.; Rodrigues, M.R.; Baldissera, L.; Duarte, S.M.D. The effects of the decaffeination of coffee samples on platelet aggregation in hyperlipidemic rats. *Plant Foods Hum. Nutr.* **2013**, *68*, 268–273. [CrossRef] [PubMed]

52. Kim, J.Y.; Kim, D.H.; Jeong, H.G. Inhibitory effect of the coffee diterpene kahweol on carrageenan-induced inflammation in rats. *Biofactors* **2006**, *26*, 17–28. [CrossRef] [PubMed]

53. Kim, J.Y.; Jung, K.S.; Jeong, H.G. Suppressive effects of the kahweol and cafestol on cyclooxygenase-2 expression in macrophages. *FEBS Lett.* **2004**, *569*, 321–326. [CrossRef] [PubMed]

54. Cardenas, C.; Quesada, A.R.; Medina, M.A. Anti-angiogenic and anti-inflammatory properties of kahweol, a coffee diterpene. *PLoS ONE* **2011**, *6*, e23407. [CrossRef]

55. Rebollo-Hernanz, M.; Zhang, Q.Z.; Aguilera, Y.; Martin-Cabrejas, M.A.; de Mejia, E.G. Phenolic compounds from coffee by-products modulate adipogenesis-related inflammation, mitochondrial dysfunction, and insulin resistance in adipocytes, via insulin/PI3K/AKT signaling pathways. *Food Chem. Toxicol.* **2019**, *132*, 110672. [CrossRef]

56. Soares, P.V.; Kannen, V.; Jordao, A.A.; Garcia, S.B. Coffee, but neither decaffeinated coffee nor caffeine, elicits chemoprotection against a direct carcinogen in the colon of wistar rats. *Nutr. Cancer* **2019**, *71*, 615–623. [CrossRef]

57. Chen, X.M.; Mu, K.W.; Kitts, D.D. Characterization of phytochemical mixtures with inflammatory modulation potential from coffee leaves processed by green and black tea processing methods. *Food Chem.* **2019**, *271*, 248–258. [CrossRef]

58. Rubio-Senent, F.; de Roos, B.; Duthie, G.; Fernandez-Bolanos, J.; Rodriguez-Gutierrez, G. Inhibitory and synergistic effects of natural olive phenols on human platelet aggregation and lipid peroxidation of microsomes from vitamin E-deficient rats. *Eur. J. Nutr.* **2015**, *54*, 1287–1295. [CrossRef]

59. Antolic, A.; Males, Z.; Tomicic, M.; Bojic, M. The effect of short-toothed and dalmatiansage extracts on platelet aggregation. *Food Technol. Biotechnol.* **2018**, *56*, 265–269. [CrossRef]

60. Tsoupras, A.; Lordan, R.; Harrington, J.; Pienaar, R.; Devaney, K.; Heaney, S.; Koidis, A.; Zabetakis, I. The effects of oxidation on the antithrombotic properties of tea lipids against PAF, thrombin, collagen, and ADP. *Foods* **2020**, *9*, 385. [CrossRef]

61. Caprioli, G.; Cortese, M.; Sagratini, G.; Vittori, S. The influence of different types of preparation (espresso and brew) on coffee aroma and main bioactive constituents. *Int. J. Food Sci. Nutr.* **2015**, *66*, 505–513. [CrossRef] [PubMed]

62. Aguilera-Barreiro, M.D.; Rivera-Marquez, J.A.; Trujillo-Arriaga, H.M.; Ruiz-Acosta, J.M.; Rodriguez-Garcia, M.E. Antioxidant capacity, phenolic acids and caffeine content of some commercial coffees available on the Romanian market. *Arch. Latinoam. Nutr.* **2013**, *63*, 87–94.

63. Ainsworth, E.A.; Gillespie, K.M. Estimation of total phenolic content and other oxidation substrates in plant tissues using Folin-Ciocalteu reagent. *Nat. Prot.* **2007**, *2*, 875–877. [CrossRef] [PubMed]

64. Re, R.; Pellegrini, N.; Proteggente, A.; Pannala, A.; Yang, M.; Rice-Evans, C. Antioxidant activity applying an improved ABTS radical cation decolorization assay. *Free Radic. Biol. Med.* **1999**, *26*, 1231–1237. [CrossRef]

65. Cattaneo, M.; Lecchi, A.; Zighetti, M.L.; Lussana, F. Platelet aggregation studies: Autologous platelet-poor plasma inhibits platelet aggregation when added to platelet-rich plasma to normalize platelet count. *Haematologica* **2007**, *92*, 694–697. [CrossRef] [PubMed]

66. Linnemann, B.; Schwonberg, J.; Mani, H.; Prochnow, S.; Lindhoff-Last, E. Standardization of light transmittance aggregometry for monitoring antiplatelet therapy: An adjustment for platelet count is not necessary. *J. Thromb. Haemost.* **2008**, *6*, 677–683. [CrossRef]

67. Pradelles, P.; Grassi, J.; Maclouf, J. Enzyme immunoassays of eicosanoids using acetylcholine esterase as label: An alternative to radioimmunoassay. *Anal. Chem.* **1985**, *57*, 1170–1173. [CrossRef]

 *molecules*

Article

The Anti-Cancer Effect of Linusorb B3 from Flaxseed Oil through the Promotion of Apoptosis, Inhibition of Actin Polymerization, and Suppression of Src Activity in Glioblastoma Cells

Nak Yoon Sung [1,†], Deok Jeong [1,†], Youn Young Shim [1,2,3,†], Zubair Ahmed Ratan [4], Young-Jin Jang [5], Martin J. T. Reaney [2,3], Sarah Lee [6], Byoung-Hee Lee [6], Jong-Hoon Kim [5,*], Young-Su Yi [7,*] and Jae Youl Cho [1,*]

[1] Department of Integrative Biotechnology, Biomedical Institute for Convergence at SKKU (BICS), Sungkyunkwan University, Suwon 16419, Korea; nakyoon.sung@monash.edu (N.Y.S.); jd279601@gmail.com (D.J.); younyoung.shim@usask.ca (Y.Y.S.)

[2] Department of Plant Sciences, University of Saskatchewan, Saskatoon, SK S7N 5A8, Canada; martin.reaney@usask.ca

[3] Guangdong Saskatchewan Oilseed Joint Laboratory, Department of Food Science and Engineering, Jian University, Guangzhou 510632, China

[4] Department of Biomedical Engineering, Khulna University of Engineering and Technology, Khulna 9203, Bangladesh; zubairahmed@bme.kuet.ac.bd

[5] College of Veterinary Medicine, Chonbuk National University, Iksan 54596, Korea; jyj3010@daum.net

[6] National Institute of Biological Resources, Environmental Research Complex, Incheon 22689, Korea; lsr57@korea.kr (S.L.); dpt510@korea.kr (B.-H.L.)

[7] Department of Life Sciences, Kyonggi University, Suwon 16227, Korea

*   Correspondence: jhkim1@jbnu.ac.kr (J.-H.K.); ysyi@kgu.ac.kr (Y.-S.Y.); jaecho@skku.edu (J.Y.C.)

†   These authors contributed equally to this work.

Received: 29 October 2020; Accepted: 10 December 2020; Published: 12 December 2020

Abstract: Linusorbs (LOs) are natural peptides found in flaxseed oil that exert various biological activities. Of LOs, LOB3 ([1–9-NαC]-linusorb B3) was reported to have antioxidative and anti-inflammatory activities; however, its anti-cancer activity has been poorly understood. Therefore, this study investigated the anti-cancer effect of LOB3 and its underlying mechanism in glioblastoma cells. LOB3 induced apoptosis and suppressed the proliferation of C6 cells by inhibiting the expression of anti-apoptotic genes, B cell lymphoma 2 (Bcl-2) and p53, as well as promoting the activation of pro-apoptotic caspases, caspase-3 and -9. LOB3 also retarded the migration of C6 cells, which was achieved by suppressing the formation of the actin cytoskeleton critical for the progression, invasion, and metastasis of cancer. Moreover, LOB3 inhibited the activation of the proto-oncogene, Src, and the downstream effector, signal transducer and activator of transcription 3 (STAT3), in C6 cells. Taken together, these results suggest that LOB3 plays an anti-cancer role by inducing apoptosis and inhibiting the migration of C6 cells through the regulation of apoptosis-related molecules, actin polymerization, and proto-oncogenes.

Keywords: flax seed oil; linusorb B3; anti-cancer; apoptosis; actin polymerization; Src; glioblastoma

1. Introduction

Glioma is a general term describing brain tumors and includes astrocytic tumors (astrocytomas), oligodendrogliomas, ependymomas, brain stem gliomas, optic pathway gliomas, and mixed gliomas [1]. About one-third of total brain tumors are gliomas originating in the glial cells that surround and support nerve cells in the brain, such as astrocytes, oligodendrocytes, and ependymal cells. Glioblastoma multiforme (GBM), a grade IV astrocytoma, is the most aggressive type of cancer that develops primarily in the brain and spreads into nearby brain tissue. GBM accounts for around 60% of the total primary brain tumors in adults [2]. The annual incidence rate of GBM is up to 5 per 100,000 persons worldwide, and the average survival time is 12–18 months with less than 10% 5-year survival rates after standard treatment [3]. GBM can occur at a broad range of ages but tends to occur more in older adults between the age of 45 and 70, and the mean age for death from brain cancers and other regions of the central nervous system was is 64. GBM diagnosis includes sophisticated imaging techniques, such as computer tomography and magnetic resonance imaging. GBM can be very difficult to treat and a cure is often not possible. Treatments of GBM may slow cancer progression and reduce the signs and symptoms, but there are no known methods to prevent GBM. Current standard treatment usually involves radiation and chemotherapy therapy followed by surgery [4–6]. Surgery removes as much of the tumors as possible, but GBM grows into the normal tissue, so complete removal is not possible [7]. Radiation therapy uses high-energy radiation to kill cancer cells and is usually recommended after surgery in the combination with chemotherapy [7]. Chemotherapy uses medications to kill cancer cells. Chemotherapy is also recommended after surgery and is often used during and after radiation therapy [7]. Immunotherapy of GBM is also being studied using programmed cell death protein 1 (PD-1)/PD ligand 1 (PD-L1) immune checkpoint inhibitors [8–10]. Preclinical studies in GBM mouse models showed the safety and efficacy, including significant tumor regression and longer survival rate of monoclonal antibody therapeutics targeting PD-1/PD-L1 axis [8]. Currently, monoclonal antibody therapy targeting PD-1/PD-L1 axis is being evaluated in clinical trials concerning GBM patients. However, despite recent medical and surgical advances, treatment of GBM remains very difficult, with poor prognoses and disappointingly low survival rates, and one of the critical concerns of the current chemotherapy is a toxicity issue, which raises the demand for the development of more effective and less toxic medications, such as the natural product-derived complementary and alternative medicines to treat GBM.

Flax (*Linum usitatissimum* L.), also known as linseed, is a fibrous crop and bluish-flowering plant that belongs to the family *Linaceae*. It has been cultivated as a fiber crop and food (flaxseed) in cooler regions of countries, such as Canada, China, and Russia for a long time and is also used in Ayurvedic medicines [11]. Flax is originally cultivated for its fiber, and flax fiber has long been used for manufacturing linen, fabrics, yarn, cordage in many textile industies [12]. Flax is also cultivated for flaxseed. Flaxseed contains 20–25% proteins and 40–45% fatty acids, including the major bioactive ingredients, such as polyunsaturated fatty acids, short-chain omega-3, lignan, mucilage, and linusorbs (LOs) [13]. Flaxseed has been consumed as a dietary supplement for human health and herbal medicines with the purpose of ameliorating many human diseases, including cardiovascular diseases, hypertension, renal disease, cancers, diabetes, stroke, skin disease, gastrointestinal disease, and inflammatory diseases [11,14–24]. Flaxseed has been also used for extracting flaxseed oil that is the oldest commercial oil for foods and pharmaceutical purposes, and flaxseed oil contains many bioactive ingredients such as omega-3 fatty acids, alpha-linolenic acid, lignan secoisolariciresinol diglucoside, and LOs [11,25]. LOs, whose name is derived from *L. usitatissimum*, are natural bioactive orbitide consisting of eight, nine or ten amino acid residues with a molecular weight of approximately 1 kDa and can also be found in flaxseed oil [20,26–28]. Studies have demonstrated that LOs have various biological and pharmacological activities, including immunosuppressive, anti-inflammatory, anti-malarial, and anti-cancer effects, [18,20,29,30]. LOB3 ([1–9-NαC]-linusorb B3) was the first LO to be discovered and isolated in flaxseed in 1959 [30] and is the most abundant cyclic nonapeptide. LOB3 and its analogs were reported to have pharmacological properties in disease conditions, such as antioxidative, immunosuppressive, anti-malarial, and anti-inflammatory properties [20,31–33]. A few studies have reported that LOB3 also shows cytotoxic activity against several types of cancers [29,34], but its anti-cancer activity, especially anti-GBM activity, and the underlying molecular mechanisms still remain poorly understood. Therefore, the present study investigated the anti-cancer activity of LOB3 and the underlying molecular mechanism in glioblastoma cells.

2. Results and Discussion

2.1. Cytotoxic and Anti-Proliferative Effect of LOB3 in Cancer Cells

LOs are small biologically active cyclic peptides found in flaxseed oil, and many types of LOs have been identified and named based on their structures [35]. Of LOs, LOB3 (Figure 1A, molecular weight: 1040.34) and its analogs have been demonstrated to play pivotal roles in antioxidative [31] and anti-inflammatory actions [20]. Interestingly, recent studies have reported the cytotoxic effect of LOB3 on cancer cells [29,34,36]; however, the anti-cancer effect of LOB3 and the underlying mechanisms are poorly understood. Therefore, this study investigated the anti-cancer activity of LOB3 and the underlying mechanisms in C6 glioblastoma cells, since glioblastoma multiforme is the most aggressive type of brain cancer with a high recurrence rate and a low 5-year survival rate [37].

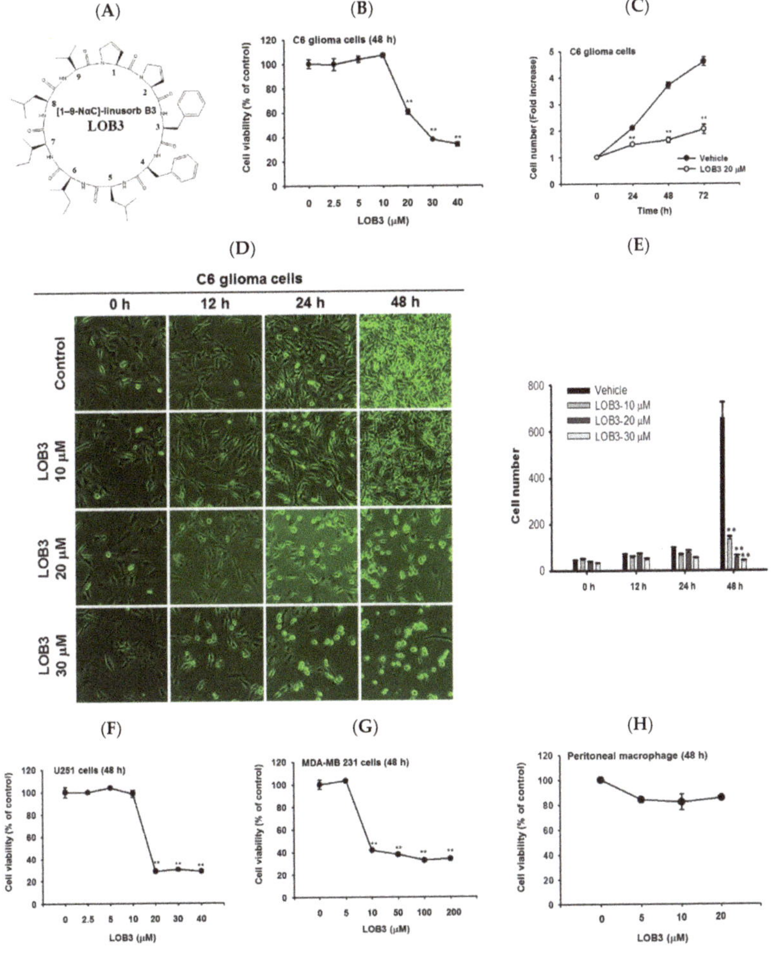

Figure 1. Cytotoxic and anti-proliferative effect of LOB3 in cancer cells. (**A**) Chemical structure of LOB3. (**B**) C6 cells were treated with the indicated doses of LOB3 for 48 h, and cell viability was determined by

a conventional 3-(4,5-dimethylthiazol-2-yl)-2,5-diphenyltetrazolium bromide (MTT) assay. (**C**) C6 cells were treated with LOB3 (20 μM) for the indicated time, and viable cell numbers were determined by a conventional MTT assay. (**D,E**) C6 cells were treated with the indicated doses of LOB3 for the indicated time, and cell numbers and shapes were observed under a light microscope. Photos of the cells were taken by a digital camera (**D**) and numbers of cells were counted by a cell counter (**E**). (**F–H**) U251, MDA-MD-231, and peritoneal macrophage cells were treated with the indicated doses of LOB3 for 48 h, and cell viability was determined by a conventional MTT assay. The data (**B,C,E,F–H**) are expressed as the means ± standard error of the mean (SEM) of three independent experiments. Statistical significance was analyzed by the Mann-Whitney U test. ** $p < 0.01$ compared to the vehicle-treated controls.

First, the cytotoxic effect of LOB3 was evaluated in C6 cells. C6 cells were treated with various doses of LOB3 for 48 h, and cell viability was examined by an MTT [3-(4,5-dimethylthiazol-2-yl)-2, 5-diphenyltetrazolium bromide] assay. LOB3 from 20 to 40 μM exerted a cytotoxic effect in C6 cells in a dose-dependent manner, while no cytotoxic effect of LOB3 was shown at doses lower than 10 μM (Figure 1B). One of the fundamental features of cancer is tumor clonality and uncontrolled proliferation. Therefore, the anti-proliferative effect of LOB3 was also evaluated in C6 cells. C6 cells treated with LOB3 (20 μM) were cultured for 72 h, and the proliferation rate of LOB3-treated C6 cells was significantly reduced compared to that of the vehicle-treated control cells (Figure 1C). These results were confirmed by observing the shape and the numbers of C6 cells after LOB3 treatment. Similar to the results depicted in Figure 1B,C, LOB3 exerted a cytotoxic effect in C6 cells by changing the cell shape and reducing cell numbers at 20 and 30 μM in a time- and dose-dependent manner (Figure 1D,E).

The cytotoxic effect of LOB3 on cancer cells was further investigated in another glioblastoma cell line, U251 cells, and a breast cancer cell line, MDA-MB-231 cells. Similar to the C6 cells, LOB3 significantly reduced the viability of U251 cells at doses of 20 μM and greater, but no cytotoxic effect was observed at doses lower than 10 μM (Figure 1F). Similar to a previous study [34], LOB3 also induced the cytotoxicity of MDA-MB-231 cells, but MDA-MB-231 cells were more sensitive to LOB3. LOB3 exerted the cytotoxic effect in the breast cancer cells at doses as low as 10 μM (Figure 1G), while the two glioblastoma cell lines were sensitive and dead at 20 μM (Figure 1B,C,F), indicating that the drug sensitivity and cytotoxic effect of LOB3 on cancer cells depend on the types of cancer cells. Furthermore, cytotoxicity of this compound was not found in non-malignant cells, peritoneal macrophages (Figure 1H). Similarly, other natural bioactive orbitides such as surfactins and beauvericin displayed anti-cancer activity in giant-cell tumors of the bone (GCTB) cells, MCF-7 breast tumor cells, and CT-26 lymphoma [38–40], implying that cytotoxic activity of LOB3 might be due to structural feature of this compound. Taken together, these results suggest that LOB3 plays an anti-cancer role by inducing cytotoxicity and reducing the proliferation of cancer cells).

2.2. Cytotoxic Effect of LOB3 on C6 Cells by Apoptosis

Apoptosis is a form of programmed cell death occurring in multicellular organisms and is characterized by many biochemical events leading to cell changes and eventually death [41,42]. surfactins and beauvericin were reported to induce apoptosis in cancer cells [36,38,43]; therefore, whether the cytotoxic effect of LOB3 on cancer cells is mediated by apoptosis was evaluated in C6 cells. One of the major characteristics of apoptosis is nuclear shrinking and fragmentation [44,45], and these events were examined in LOB3-treated C6 cells by Hoechst nuclear staining. Compared to the control, LOB3 (20 and 30 μM) had a similar effect on C6 cells as staurosporine, an apoptosis inducer [46], in that it stimulated nuclear shrinking and fragmentation (Figure 2A). Double staining of annexin V and propidium iodide (PI) is a commonly used analytical approach for detecting apoptosis of cells [47], and this method was used to examine LOB3-induced apoptosis of C6 cells. The proportions of early and late apoptosis of LOB3-treated C6 cells were quantified by flow cytometry analysis after Annexin V/PI staining, and the results indicated that LOB3 significantly induced the apoptotic population of C6 cells at doses of 30 μM, but not 10 μM (Figure 2B). This result is consistent with the result that LOB3 exhibited the cytotoxicity-inducing effect from doses of 20 μM (Figure 1B,D).

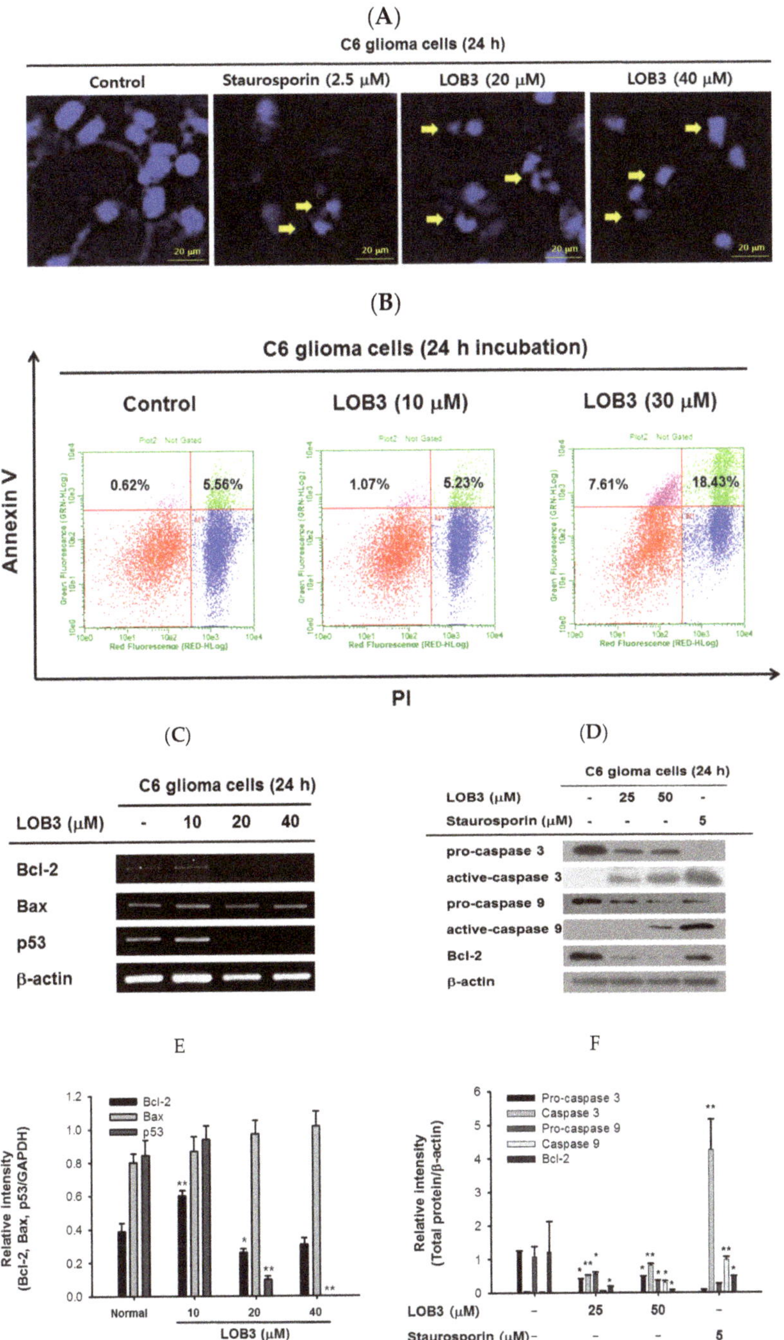

Figure 2. Cytotoxic effect of LOB3 on C6 cells by apoptosis. (**A**) The nuclei of C6 cells treated with either staurosporine (2.5 μM) or LOB3 (20 and 30 μM) were stained with Hoechst 33342 and observed

under a fluorescence microscope. Yellow arrows indicate nuclear shrinking and fragmentation. (**B**) C6 cells treated with the indicated doses of LOB3 for 24 h were stained with PI and annexin V-FITC, and the cell population was determined by flow cytometry analysis. (**C**) C6 cells were treated with the indicated doses of LOB3 for 24 h, and mRNA levels of Bcl-2, BAX, and p53 were analyzed by semiquantitative RT-PCR. (**D**) C6 cells were treated with either staurosporine (5 µM) or LOB3 (25 and 50 µM) for 24 h, and protein levels of pro-caspase-3, caspase-3, pro-caspase-9, and caspase-9 were determined by Western blot analysis. The data (**E,F**) are expressed as the means ± standard deviation (SD) of three experiments. Statistical significance was analyzed by the Mann-Whitney U test. Results (**A,B**). Data of band intensity (**E,F**) were measured and quantified using ImageJ. * $p < 0.05$ and ** $p < 0.01$ compared to the vehicle-treated controls.

The mechanism by which LOB3 exhibited an apoptotic effect on C6 cells was next evaluated at a molecular level. Bcl-2 family members play pivotal roles in the regulation of apoptosis and are categorized into two major groups: anti-apoptotic members, including Bcl-2, Bcl-X_L, Bcl-W, MCL-1, and BFL-1/A1, and pro-apoptotic members, including BAX, BAK, BOK, and BAD [48]. The effect of LOB3 on the mRNA expression of both anti-apoptotic and pro-apoptotic Bcl-2 family members was examined in C6 cells. LOB3 decreased mRNA expression of the anti-apoptotic member Bcl-2 at doses of 20 and 40 µM (Figure 2C–F) but showed no marked effect on mRNA expression of the pro-apoptotic member BAX at all doses in C6 cells (Figure 2C,E), suggesting that LOB3 induced apoptotic death of C6 cells by inhibiting the expression of the anti-apoptotic member, Bcl-2 rather than increasing the expression of the pro-apoptotic member, BAX. p53 is a tumor suppressor, but strong evidence has accumulated to indicate that p53 plays an anti-apoptotic role by transcriptionally activating many genes whose products efficiently suppress apoptosis [49]. Therefore, the effect of LOB3 on mRNA expression of p53 was examined, and mRNA expression of p53 was markedly decreased in the LOB3-treated C6 cells at doses of 20 and 40 µM (Figure 2C,E).

Caspases are a family of endonucleases that act as critical links in the molecular networks that control apoptosis and play critical roles in the induction of apoptosis as both initiators (caspase-2, -8, -9, -10, 20, and -22) and executioners (caspase-3, -6, -7, and 21) [50,51]. Caspases are initially expressed as inactive procaspases that are activated by pro-apoptotic signals via proteolytic cleavage. Therefore, the effect of LOB3 on the proteolytic activation of caspases was examined in C6 cells. LOB3 activated both an apoptosis initiator, caspase-9, as well as an executioner, caspase-3, by promoting the proteolytic cleavage of these caspases at doses of 25 and 50 µM in C6 cells (Figure 2D,F), indicating that LOB3 induces apoptosis of C6 cells by activating the apoptosis initiators and executioners. Moreover, similar to the semi-quantitative RT-PCR result (Figure 2C,E), LOB3 markedly reduced the protein expression of the anti-apoptotic Bcl family member, Bcl-2, at doses of 25 and 50 µM in C6 cells (Figure 2D,F). Taken together, these results suggest that the cytotoxic effect of LOB3 on C6 cells is mediated by the induction of apoptosis through inhibiting the expression of anti-apoptotic genes, such as Bcl-2 and p53, as well as activating the proteolytic processing of both apoptosis initiator, caspase-9, and executioner, caspase-3.

2.3. Anti-Migratory Effect of LOB3 in C6 Cells

Another fundamental feature of cancer is the migration of tumor cells from the original location where tumors arise to other parts of the body. It was reported that surfactin can reduce the 12-*O*-tetradecanoylphorbol-13-acetate (TPA)-induced metastatic potentials, including invasion and migration of human breast carcinoma cells [52]. Therefore, the anti-migratory effect of LOB3 was evaluated in C6 cells. C6 cells treated with LOB3 (20 and 30 µM) were cultured for 6 h, and the degree of C6 cell migration was examined. Compared to the control, LOB3 markedly suppressed the migration of C6 cells at a dose of 30 µM (Figure 3A,B). Interestingly, although LOB3 exerted a cytotoxic and anti-proliferative effect on C6 cells (Figure 1B–D) by facilitating apoptosis at doses as low as 20 µM (Figure 2), LOB3 did not suppress the migration of C6 cells at 20 µM, but 30 µM. This result indicates that the doses of LOB3 required to inhibit the proliferation and migration of C6 cells might not be same, since the molecules critical for cell proliferation and migration are different, and the inhibitory effect of LOB3 on the biological actions of these molecules might also be different in C6 cells. To identify the effective and optimal doses targeting both sets of these molecules and thereby inhibiting both the proliferation and migration of C6 cells, further molecular mechanism studies using various doses of LOB3 will be required. Taken together, these results suggest that LOB3 plays an anti-cancer role by not only inducing cytotoxicity but also suppressing the migration of C6 cells.

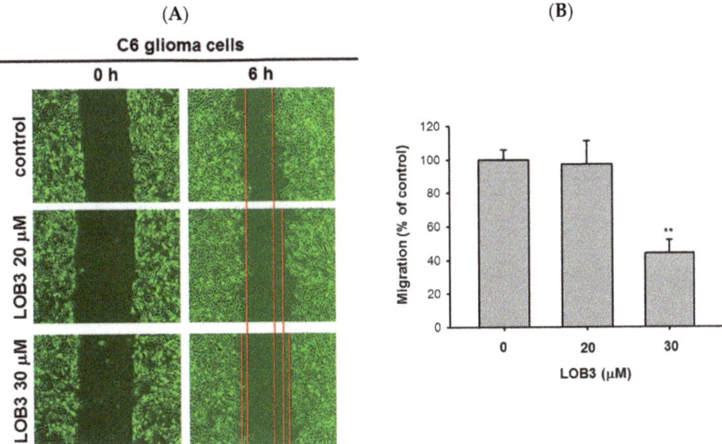

Figure 3. Anti-migratory effect of LOB3 in C6 cells. (**A**) Migration of C6 cells treated with the indicated doses of LOB3 for 6 h was analyzed under a light microscope. (**B**) Migration areas (% of control) were calculated and plotted with Figure 3A. Statistical significance was analyzed by the Mann-Whitney U test. Result (**A**) is a representative of three experiments. The data (**B**) are expressed as the means ± SD of three experiments. ** $p < 0.01$ compared to the vehicle-treated controls.

2.4. Inhibitory Effect of LOB3 on Actin Polymerization in C6 Cells through the Targeting of Src and STAT3

The cytoskeleton is a dynamic and complex intracellular network of protein filaments interlinking in the cytoplasm of cells, and its primary function is to provide the cells with their shape and with mechanical resistance to deformation stresses. Of the three main components of the cytoskeleton, the actin cytoskeleton is essential to enabling cell motility by maintaining the shape and integrity of the cell. In addition, the actin cytoskeleton plays a critical role in the migration, invasion, and metastasis of cancer cells, as well as overall cancer progression [53–56]; therefore, the selective and effective targeting of actin in cancer cells is a worthwhile strategy in the development of anti-cancer therapeutics [56–60]. The effect of LOB3 on the actin cytoskeleton was examined in C6 cells. C6 cells treated with LOB3 or cytochalasin B (CytoB), an actin polymerization inhibitor [61], were incubated with phalloidin to visualize filamentous actin (F-actin) [62]. Although CytoB moderately inhibited the formation of the actin cytoskeleton, LOB3 (30 µM) dramatically inhibited the formation of the actin cytoskeleton in C6 cells (Figure 4A). Since LOB3 almost completely inhibited the formation of the actin cytoskeleton in C6 cells at 30 µM (Figure 4A), the inhibitory effect of LOB3 on the formation of the actin cytoskeleton in C6 cells was further examined at a lower dose (20 µM) for different time periods. LOB3 (20 µM) also markedly inhibited the formation of the actin cytoskeleton at 12 h and 24 h after treatment, while LOB3 (20 µM) moderately inhibited the formation of the actin cytoskeleton at 6 h after treatment (Figure 4B). This result does not necessarily mean LOB3 needs at least 6 h to inhibit the formation of the actin cytoskeleton in C6 cells, and the time required to inhibit actin cytoskeleton formation might vary depending on LOB3 doses. As discussed earlier, caspase-3 is an executioner of apoptosis [35,36], and similar to the previous result (Figure 2D), LOB3 (20 µM) induced proteolysis of pro-caspase-3 and produced active caspase-3 at 24 h in C6 cells (Figure 4B).

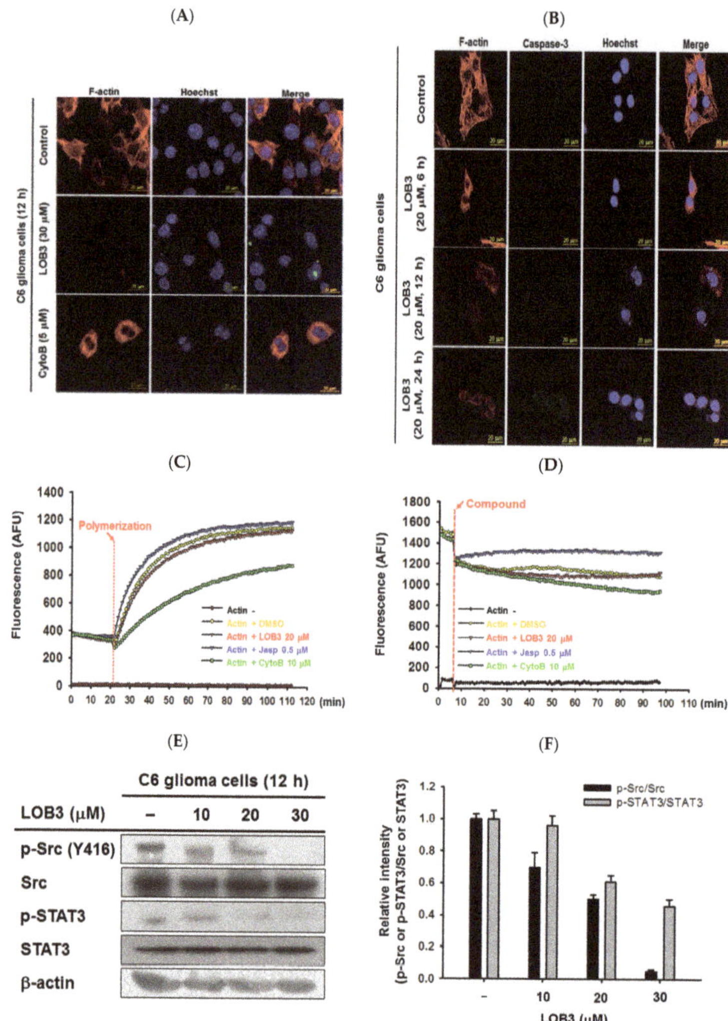

Figure 4. Inhibitory effect of LOB3 on actin polymerization by targeting Src and STAT3 in C6 cells. (**A**) Actin filaments (F-actin) and the nuclei of C6 cells treated with either LOB3 (30 μM) or CytoB (5 μM) for 12 h were stained with phalloidin and Hoechst 33342, respectively, and visualized under a confocal microscope. (**B**) Actin filaments (F-actin), caspase-3, and the nuclei of C6 cells treated with LOB3 (20 μM) for the indicated time were stained with phalloidin, caspase-3 antibody, and Hoechst 33342, respectively, and visualized under a confocal microscope. (**C**) Actin monomers were incubated with the indicated compounds for the indicated time, and actin polymerization was analyzed by an in vitro actin polymerization assay. (**D**) Actin filaments were incubated with the indicated compounds for the indicated time, and actin depolymerization was analyzed by an in vitro actin de-polymerization assay. (**E**) C6 cells were treated with the indicated doses of LOB3 for 12 h, and the protein levels of the phosphor and total forms of Src and STAT3 were determined by Western blot analysis. Results (**A–D**) are representative of three independent experiments. Statistical significance was analyzed by the Mann-Whitney U test. Data of band intensity (**F**) were measured and quantified using ImageJ. * $p < 0.05$ and ** $p < 0.01$ compared to the vehicle-treated controls.

The effect of LOB3 on the formation of the actin cytoskeleton was further evaluated by in vitro actin polymerization and depolymerization assays. LOB3 (20 µM) suppressed actin polymerization, while jasplakinolide (Jasp, 0.5 µM), an actin polymerization promoting peptide [63], and CytoB (10 µM) induced and suppressed actin polymerization, respectively (Figure 4C). Similarly, LOB3 (20 µM) promoted actin de-polymerization, while Jasp (0.5 µM) and CytoB (10 µM) inhibited and promoted actin depolymerization, respectively (Figure 4D). LOB3 suppressed the migratory ability of C6 cells (Figure 3), and this suppressive effect of LOB3 on C6 cell migration could be achieved by the suppression of actin polymerization. Taken together, these results suggested that LOB3 played an anti-cancer role by suppressing actin polymerization as well as promoting actin depolymerization.

Src is a proto-oncogene that is strongly implicated in the growth, progression, invasion, and metastasis of many types of human cancers [64,65]. Interestingly, actin polymerization induces Src activation with delivery to the cell membrane [66]. Since LOB3 played an inhibitory role in actin polymerization (Figure 4A–D), whether LOB3 suppresses Src activation was next examined in C6 cells. LOB3 suppressed Src activation in C6 cells in a dose-dependent manner (Figure 4E,F), thereby indicating that LOB3 suppresses the activation of a proto-oncogene, Src, by inhibiting actin polymerization in C6 cells. Src regulates various downstream signaling pathways in cancer cells, leading to the development and progression of cancers, and one of the most critical downstream target molecules is STAT3 [67–69]. Therefore, whether LOB3 also suppresses the activation of STAT3 was examined, and as expected, LOB3 was found to suppress STAT3 activation in C6 cells (Figure 4E,F). Taken together, LOB3-induced anti-cancer activity in C6 cells is mediated by the inhibition of actin polymerization and the subsequent suppression of Src and the downstream molecule, STAT3.

3. Materials and Methods

3.1. Materials

LOB3 (Figure 1A) was provided as a generous contribution from Prairie Tide Diversified Inc. (Saskatoon, SK, Canada). The C6 human glioblastoma cell line, U251 human glioblastoma cell line, and MDA-MB-231 human breast cancer cell line were purchased from the American Type Culture Collection (Rockville, MD, USA). Dulbecco's modified Eagle's medium (DMEM), fetal bovine serum (FBS), phosphate-buffered saline (PBS), penicillin, streptomycin, L-glutamine, bovine serum albumin (BSA), apoptosis analysis kit (Dead Cell Apoptosis Kit with Annexin V FITC and PI), MuLV reverse transcriptase (RT), and Lipofectamine® 2000 reagent were purchased from Thermo Fisher Scientific (Waltham, MA, USA). MTT, Hoechst 33342, staurosporine, cytochalasin B (CytoB), and jasplakinolide (Jasp) were purchased from Sigma-Aldrich (St. Louis, MO, USA). TRI reagent® was purchased from Molecular Research Center, Inc. (Cincinnati, OH, USA). Primers for semi-quantitative RT polymerase chain reaction (PCR) were synthesized, and PCR premix was purchased from Bioneer, Inc. (Daejeon, Korea). An enhanced chemiluminescence system was purchased from AbFrontier (Seoul, Korea). Antibodies specific to each target used for Western blot analysis and immunofluorescence staining were purchased from Cell Signaling Technology (Beverly, MA, USA) and Santa Cruz Biotechnology (Santa Cruz, CA, USA). The Actin Polymerization Biochem kit was purchased from Cytoskeleton (Denver, CO, USA).

3.2. Preparation of Peritoneal Macrophages

Peritoneal exudates were isolated from ICR mice (6-week-old, 17 to 21 g) by lavage 4 days after intraperitoneal treatment with 4% thioglycolate broth (Difco Laboratories, Detroit, MI, USA). After the blood was removed from the exudates using RBC lysis buffer (Sigma Chemical Co., St. Louis, MO, USA), the extracted peritoneal macrophages (1×10^6 cells/mL) were plated in a 100 mm tissue culture plate and incubated for 4 h at 37 °C in a 5% CO_2 humidified atmosphere. The ICR male mice were purchased from Daehan Bio Link Co., Ltd. (Chungbuk, Korea) and housed at seven mice per group under a 12-h light/dark cycle (lights on at 6 a.m.). Water and a pellet diet (Samyang, Daejeon,

Korea) were supplied ad libitum. Animal care followed guidelines issued by the National Institutes of Health for the Care and Use of Laboratory Animals (NIH Publication 80–23, revised in 1996) and the Institutional Animal Care and Use Committee at Sungkyunkwan University (Suwon, Korea).

3.3. Cell Culture

C6 cells, U251 cells, and MDA-MB-231 cells as well as peritoneal macrophages were cultured or maintained in DMEM supplemented with 10% FBS, penicillin (100 U/mL), streptomycin (100 mg/mL), and L-glutamine (2 mM) at 37 °C in a humidified incubator with 5% CO_2. Cells were kept fresh by splitting them 2–3 times per week.

3.4. Cell Proliferation and Viability Assay

C6, U251, and MDA-MB-231 cells as well as peritoneal macrophages were treated with LOB3 at the indicated doses and time periods, and cell viability was determined by a conventional MTT assay [70,71]. For an MTT assay, the MTT solution was incubated with the cells at 37 °C for 4 h, and then stop solution (10% SDS in 0.01 N HCl) was added to the cells. After incubation at 37 °C for 24 h, the optical density was determined at 540 nm using a microplate reader (BioTek, Winooski, VT, USA).

3.5. Cell Death Assays and Flow Cytometry Analysis

The death of C6 cells was also evaluated by observing their shapes. C6 cells were treated with LOB3 (0, 10, 20, and 30 µM) for 0, 12, 24, and 48 h, and the shapes of the cells were observed and evaluated under a light microscope. The death of C6 cells was also analyzed by Hoechst 33342 (10 µg/mL in PBS) staining. C6 cells were stained with Hoechst 33342 at room temperature for 30 min and washed with PBS three times. Hoechst 33342-stained nuclei were observed and analyzed under a fluorescence microscope. The apoptotic death of C6 cells was evaluated by flow cytometry analysis using an apoptosis analysis kit (see Materials) according to the manufacturer's instructions. Briefly, C6 cells pretreated with LOB3 (0, 10, and 30 µM) for 24 h were incubated with propidium iodide (PI) and annexin V-FITC in a binding buffer (50 mM HEPES, 700 mM NaCl, 12.5 mM $CaCl_2$, pH 7.4) at room temperature for 15 min. After washing the cells with cold PBS three times, the population of fluorescent cells was determined by flow cytometry analysis.

3.6. Semi-Quantitative RT-PCR

Total RNA was extracted from the C6 cells treated with LOB3 (0, 10, 20, and 40 µM) for 24 h using TRI reagent®, followed immediately by the synthesis of cDNA from total RNA (1 µg) using MuLV RT according to the manufacturer's instructions. Semi-quantitative RT-PCR was conducted using the cDNA to determine the mRNA expression levels of Bcl-2, Bax, and p53. The experimental conditions and the primer sequences used for the semi-quantitative RT-PCR in this study are listed in Tables 1 and 2, respectively.

Table 1. The experimental conditions of the semi-quantitative RT-PCR in this study.

Targets	Annealing Temp.	Cycle No.	Fragment Size (Base Pair)
Bcl-2	60	30	304
Bax	60	30	240
p53	60	30	560
GAPDH	60	25	350

Table 2. Primer sequences used for the semi-quantitative RT-PCR in this study.

Targets		Sequences (5′ to 3′)
Bcl-2	Forward	CACCCCTGGCATCTTCTCCTT
	Reverse	CACAATCCTCCCCCAGTTCACC
Bax	Forward	ATGGCTGGGGAGACACCTGAG
	Reverse	CTAGCAAAGTAGAAAAGGGCAAC
p53	Forward	CTCTGTCATCTTCCGTCCCTTC
	Reverse	AGGACAGGCACAAACACGAAC
GAPDH	Forward	CACTCACGGCAAATTCAACGGCAC
	Reverse	GACTCCACGACATACTCAGCAC

3.7. Western Blot Analysis

Whole lysates of C6 cells treated with the indicated concentration of LOB3 or staurosporine (5 μM) for the indicated time were prepared by lysing the cells using radioimmunoprecipitation assay (RIPA) buffer (50 mM Tris-HCl pH 8.0, 150 mM NaCl, 1% Nonidet P-40, 0.5% sodium deoxycholate, and 0.1% sodium dodecyl sulfate [SDS]) containing proteinase inhibitors (1 mM sodium orthovanadate, 10 μg/mL aprotinin, 10 μg/mL pepstatin, 1 mM benzamide, and 2 mM PMSF) in ice for 30 min, followed by sonication for 10 sec. Whole lysates of HEK293 cells treated with LOB3 (0, 20, and 40 μM) and transfected with either empty plasmids (pcDNA) or HA-Src plasmids for 12 h were also prepared according to the same method as was used with the C6 cells. Sample buffer (62.5 mM Tris-HCl pH 6.8, 2.5% SDS, 0.002% bromophenol blue, 5% β-mercaptoethanol, 10% glycerol) was added to the whole cell lysates, followed by boiling for 10 min. For Western blot analysis, the whole cell lysates were subjected to SDS-polyacrylamide electrophoresis and transferred to polyvinylidene difluoride membrane (250 mA for 1 h) in transfer buffer (25 mM Tris, 192 mM glycine, pH 8.3, 20% methanol (*v/v*)). Targets (pro-caspase-3 and -9, caspase-3 and -9, Bcl-2, p-Src [Y416], p-Src [Y527], Src, p-STAT3, STAT-3, and HA) were detected with their specific primary (1:1000) and secondary (1:15,000) antibodies, and the immune complexes were visualized using an enhanced chemiluminescence system according to the manufacturer's instructions as reported previously [71].

3.8. In Vitro Cell Migration Assay

C6 cells grown to a confluent monolayer were treated with LOB3 (0, 20, and 30 μM) and scratched using a pipette tip as previously described with slight modification [72,73]. After 6 h, the migrated cells to the scratched regions were observed and taken pictures under a light microscope (Olympus, Japan). The migrated C6 cells were measured using ImageJ software (NIH, Bethesda, MD, USA), and compared by plotting (% control).

3.9. Confocal Microscopy

C6 cells were treated with either LOB3 (0, 20, and 30 μM) or CytoB (5 μM) for the indicated time. For confocal microscopic analysis, the cells were fixed with 4% paraformaldehyde in PBS for 10 min and permeabilized with 0.5% Triton X-100 in PBS for 10 min. The cells were next blocked with 1% BSA in PBS for 1 h, followed by incubation with Rhodamine Phalloidin reagent or the antibodies specific for F-actin and cleaved caspase-3 at 4°C overnight. The cells were then incubated with Alexa Fluor 488-or 568-conjugated secondary antibodies for 1 h. The DNA of these cells was stained with Hoechst 33342 (10 μg/mL in PBS) for 30 min, followed by washing with PBS for 5 min three times. The cells were mounted on the glass slides and imaged using a laser-scanning confocal microscope (Zeiss LSM 710 META, Oberkochen, Germany) with a 63× oil-immersion objective lens.

3.10. In Vitro Actin Polymerization and Depolymerization Assays

Actin polymerization and depolymerization assays were conducted in the presence and absence of either LOB3 (20 mM), Jasp (0.5 mM), or CytoB (10 mM), using the Actin Polymerization Biochem kit according to the manufacturer's instructions. For actin polymerization assay, LOB3, Jasp, or CytoB were mixed with pyrene-labeled globular actin (G-actin) in actin polymerization buffer, and the fluorescence was measured for 90 min using a fluorescence microplate reader (BioTek, Winooski, VT, USA). For actin depolymerization assay, LOB3, Jasp, or CytoB were incubated with pyrene-labeled F-actin in depolymerization buffer, and the fluorescence was measured for 90 min using a fluorescence microplate reader. The effect of each compound on actin polymerization and de-polymerization was determined by fluorescence of pyrene-labeled actin, measured with an excitation wavelength of 350 nm and an emission wavelength of 407 nm at 25 °C every 60 s for 1 h.

3.11. Statistical Analysis

Data (Figure 1B,C,E–G) are presented as the mean ± standard error of the mean (SEM) of three independent sets of experimental data performed with at least three samples. The data (Figure 1D right panel, bottom panels of Figure 2C,D, Figure 3B, and Figure 4E right panel) are expressed as the means ± standard deviation (SD) of three experiments. Statistical differences between groups in these data were analyzed by the Mann-Whitney U test, and p-value < 0.05 was considered to indicate a statistically significant difference. All statistical analyses were conducted using SPSS software (SPSS Inc., Chicago, IL, USA). Other results are representative of at least two of the data sets.

4. Conclusions

The current study investigated the anti-cancer effect of LOB3 and the underlying molecular mechanism in glioblastoma C6 cells. LOB3 induces the cytotoxicity of C6 cells by promoting apoptosis through modulating the expression of apoptosis-related genes and molecules. LOB3 also suppressed the motility of C6 cells, which is critical for cancer cell migration, invasion, and metastasis, by inhibiting actin polymerization, and LOB3 suppressed the activation of Src and STAT3, which are proto-oncogenic factors activated by actin polymerization in cancer cells. Despite these results, this study was limited to in vitro experiments using cancer cell lines, and further ex vivo studies using tumor cells from cancer animal models or human patients as well as in vivo studies using animal xenograft or orthotopic models are required to support and confirm the results of this study. In addition, the anti-cancer effect of LOB3 needs to be expanded to other types of cancers to confirm the general anti-cancer effect of LOB3 in a broad range of cancers. Moreover, the comparison studies for the anti-cancer effect of various LOs also need to be further investigated. In conclusion, LOB3 plays an anti-cancer role by facilitating the apoptotic death of C6 cells as well as inhibiting the migratory activity of C6 cells by modulating multiple factors associated with apoptosis, motility, cytoskeleton formation, and proto-oncogenic functions, as described in Figure 5. Given this evidence, this study proposes an anti-cancer role of a cyclic peptide, LOB3, which is present in flaxseed oil, in glioblastoma cells with a new understanding of the underlying molecular mechanisms, which could provide insight into the development of effective and safer LO-based therapeutics to prevent and treat glioblastoma and even other types of cancers.

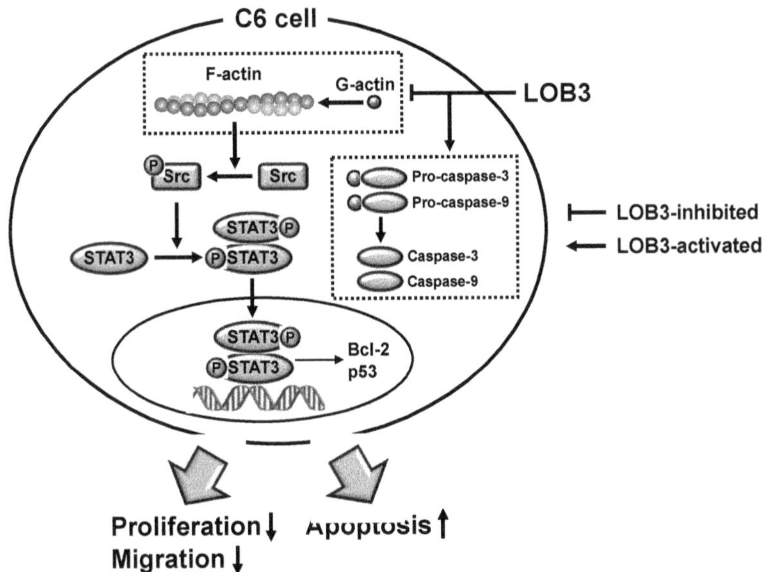

Figure 5. The proposed model to illustrate the anti-cancer activity of LOB3 in a C6 cell.

Author Contributions: Conceptualization, Z.A.R., D.J., Y.-J.J., M.J.T.R., Y.S.Y., and J.Y.C.; formal analysis, N.Y.S., Z.A.R., S.L., B.-H.L., M.J.T.R., Y.S.Y., and J.Y.C.; investigation, Z.A.R., D.J., N.Y.S., and Y.Y.S.; writing—original draft preparation, Z.A.R., D.J., and Y.-S.Y.; writing—review and editing, Y.Y.S., J.-H.K., M.J.T.R., Y.-S.Y., and J.Y.C.; funding acquisition, Y.Y.S., and J.Y.C. All authors have read and agreed to the published version of the manuscript.

Funding: This research was supported by Basic Science Research (2017R1A6A1A03015642 and 2015K2A1A2070737) and Brain Pool Program (2019H1D3A2A01102248) through the National Research Foundation of Korea (NRF) funded by the Ministry of Education and the Ministry of Science and ICT, respectably.

Acknowledgments: The authors acknowledge the kind contribution of LOB3 from Prairie Tide Diversified Inc. (Saskatoon, SK, Canada).

Conflicts of Interest: The authors declare no conflict of interest.

References

1. Agnihotri, S.; Burrell, K.E.; Wolf, A.; Jalali, S.; Hawkins, C.; Rutka, J.T.; Zadeh, G. Glioblastoma, a brief review of history, molecular genetics, animal models and novel therapeutic strategies. *Arch. Immunol. Ther. Exp.* **2013**, *61*, 25–41. [CrossRef] [PubMed]

2. Rock, K.; McArdle, O.; Forde, P.; Dunne, M.; Fitzpatrick, D.; O'Neill, B.; Faul, C. A clinical review of treatment outcomes in glioblastoma multiforme—The validation in a non-trial population of the results of a randomised Phase III clinical trial: Has a more radical approach improved survival? *Br. J. Radiol.* **2012**, *85*, e729–733. [CrossRef] [PubMed]

3. Grech, N.; Dalli, T.; Mizzi, S.; Meilak, L.; Calleja, N.; Zrinzo, A. Rising Incidence of Glioblastoma Multiforme in a Well-Defined Population. *Cureus* **2020**, *12*, e8195. [CrossRef] [PubMed]

4. Gallego, O. Nonsurgical treatment of recurrent glioblastoma. *Curr. Oncol.* **2015**, *22*, e273–281. [CrossRef] [PubMed]

5. Khosla, D. Concurrent therapy to enhance radiotherapeutic outcomes in glioblastoma. *Ann. Transl. Med.* **2016**, *4*, 54. [CrossRef] [PubMed]

6. Hirsch, S.; Roggia, C.; Biskup, S.; Bender, B.; Gepfner-Tuma, I.; Eckert, F.; Zips, D.; Malek, N.P.; Wilhelm, H.; Renovanz, M.; et al. Depatux-M and temozolomide in advanced high-grade glioma. *Neuro Oncol. Adv.* **2020**, *2*, vdaa063. [CrossRef] [PubMed]

7. Bahadur, S.; Sahu, A.K.; Baghel, P.; Saha, S. Current promising treatment strategy for glioblastoma multiform: A review. *Oncol. Rev.* **2019**, *13*, 417. [CrossRef]

8. Litak, J.; Mazurek, M.; Grochowski, C.; Kamieniak, P.; Rolinski, J. PD-L1/PD-1 Axis in Glioblastoma Multiforme. *Int. J. Mol. Sci.* **2019**, *20*, 5347. [CrossRef]

9. Caccese, M.; Indraccolo, S.; Zagonel, V.; Lombardi, G. PD-1/PD-L1 immune-checkpoint inhibitors in glioblastoma: A concise review. *Crit. Rev. Oncol. Hematol.* **2019**, *135*, 128–134. [CrossRef]

10. Wang, X.; Guo, G.; Guan, H.; Yu, Y.; Lu, J.; Yu, J. Challenges and potential of PD-1/PD-L1 checkpoint blockade immunotherapy for glioblastoma. *J. Exp. Clin. Cancer Res.* **2019**, *38*, 87. [CrossRef]

11. Parikh, M.; Maddaford, T.G.; Austria, J.A.; Aliani, M.; Netticadan, T.; Pierce, G.N. Dietary Flaxseed as a Strategy for Improving Human Health. *Nutrients* **2019**, *11*, 1171. [CrossRef] [PubMed]

12. Saleem, M.H.; Fahad, S.; Khan, S.U.; Din, M.; Ullah, A.; Sabagh, A.E.; Hossain, A.; Llanes, A.; Liu, L. Copper-induced oxidative stress, initiation of antioxidants and phytoremediation potential of flax (*Linum usitatissimum* L.) seedlings grown under the mixing of two different soils of China. *Environ. Sci. Pollut. Res. Int.* **2020**, *27*, 5211–5221. [CrossRef] [PubMed]

13. Saleem, M.H.; Ali, S.; Hussain, S.; Kamran, M.; Chattha, M.S.; Ahmad, S.; Aqeel, M.; Rizwan, M.; Aljarba, N.H.; Alkahtani, S.; et al. Flax (*Linum usitatissimum* L.): A Potential Candidate for Phytoremediation? Biological and Economical Points of View. *Plants* **2020**, *9*, 496. [CrossRef] [PubMed]

14. Sharma, J.; Singh, R.; Goyal, P.K. Chemomodulatory Potential of Flaxseed Oil Against DMBA/Croton Oil-Induced Skin Carcinogenesis in Mice. *Integr. Cancer Ther.* **2016**, *15*, 358–367. [CrossRef]

15. Zhang, X.; Wang, H.; Yin, P.; Fan, H.; Sun, L.; Liu, Y. Flaxseed oil ameliorates alcoholic liver disease via anti-inflammation and modulating gut microbiota in mice. *Lipids Health Dis.* **2017**, *16*, 44. [CrossRef]

16. Zhu, H.; Wang, H.; Wang, S.; Tu, Z.; Zhang, L.; Wang, X.; Hou, Y.; Wang, C.; Chen, J.; Liu, Y. Flaxseed Oil Attenuates Intestinal Damage and Inflammation by Regulating Necroptosis and TLR4/NOD Signaling Pathways Following Lipopolysaccharide Challenge in a Piglet Model. *Mol. Nutr. Food Res.* **2018**, *62*, e1700814. [CrossRef]

17. Bashir, S.; Sharma, Y.; Jairajpuri, D.; Rashid, F.; Nematullah, M.; Khan, F. Alteration of adipose tissue immune cell milieu towards the suppression of inflammation in high fat diet fed mice by flaxseed oil supplementation. *PLoS ONE* **2019**, *14*, e0223070. [CrossRef]

18. De Silva, S.F.; Alcorn, J. Flaxseed Lignans as Important Dietary Polyphenols for Cancer Prevention and Treatment: Chemistry, Pharmacokinetics, and Molecular Targets. *Pharmaceuticals* **2019**, *12*, 68. [CrossRef]

19. Parikh, M.; Pierce, G.N. Dietary flaxseed: What we know and don't know about its effects on cardiovascular disease. *Can. J. Physiol. Pharmacol.* **2019**, *97*, 75–81. [CrossRef]

20. Ratan, Z.A.; Jeong, D.; Sung, N.Y.; Shim, Y.Y.; Reaney, M.J.T.; Yi, Y.S.; Cho, J.Y. LOMIX, a Mixture of Flaxseed Linusorbs, Exerts Anti-Inflammatory Effects through Src and Syk in the NF-kappaB Pathway. *Biomolecules* **2020**, *10*, 859. [CrossRef]

21. Watanabe, Y.; Ohata, K.; Fukanoki, A.; Fujimoto, N.; Matsumoto, M.; Nessa, N.; Toba, H.; Kobara, M.; Nakata, T. Antihypertensive and Renoprotective Effects of Dietary Flaxseed and its Mechanism of Action in Deoxycorticosterone Acetate-Salt Hypertensive Rats. *Pharmacology* **2020**, *105*, 54–62. [CrossRef] [PubMed]

22. Zhu, L.; Sha, L.; Li, K.; Wang, Z.; Wang, T.; Li, Y.; Liu, P.; Dong, X.; Dong, Y.; Zhang, X.; et al. Dietary flaxseed oil rich in omega-3 suppresses severity of type 2 diabetes mellitus via anti-inflammation and modulating gut microbiota in rats. *Lipids Health Dis.* **2020**, *19*, 20. [CrossRef] [PubMed]

23. Saleem, M.H.; Kamran, M.; Zhou, Y.; Parveen, A.; Rehman, M.; Ahmar, S.; Malik, Z.; Mustafa, A.; Ahmad Anjum, R.M.; Wang, B.; et al. Appraising growth, oxidative stress and copper phytoextraction potential of flax (*Linum usitatissimum* L.) grown in soil differentially spiked with copper. *J. Environ. Manag.* **2020**, *257*, 109994. [CrossRef] [PubMed]

24. Imran, M.; Sun, X.; Hussain, S.; Ali, U.; Rana, M.S.; Rasul, F.; Saleem, M.H.; Moussa, M.G.; Bhantana, P.; Afzal, J.; et al. Molybdenum-Induced Effects on Nitrogen Metabolism Enzymes and Elemental Profile of Winter Wheat (Triticum aestivum L.) Under Different Nitrogen Sources. *Int. J. Mol. Sci.* **2019**, *2*, 3009. [CrossRef] [PubMed]

25. Kajla, P.; Sharma, A.; Sood, D.R. Flaxseed-a potential functional food source. *J. Food Sci. Technol.* **2015**, *52*, 1857–1871. [CrossRef] [PubMed]

26. Bruhl, L.; Matthaus, B.; Fehling, E.; Wiege, B.; Lehmann, B.; Luftmann, H.; Bergander, K.; Quiroga, K.; Scheipers, A.; Frank, O.; et al. Identification of bitter off-taste compounds in the stored cold pressed linseed oil. *J. Agric. Food Chem.* **2007**, *55*, 7864–7868. [CrossRef] [PubMed]

27. Burnett, P.G.; Jadhav, P.D.; Okinyo-Owiti, D.P.; Poth, A.G.; Reaney, M.J. Glycine-containing flaxseed orbitides. *J. Nat. Prod.* **2015**, *78*, 681–688. [CrossRef]

28. Burnett, P.G.; Olivia, C.M.; Okinyo-Owiti, D.P.; Reaney, M.J. Orbitide Composition of the Flax Core Collection (FCC). *J. Agric. Food Chem.* **2016**, *64*, 5197–5206. [CrossRef]

29. Okinyo-Owiti, D.P.; Dong, Q.; Ling, B.; Jadhav, P.D.; Bauer, R.; Maley, J.M.; Reaney, M.J.T.; Yang, J.; Sammynaiken, R. Evaluating the cytotoxicity of flaxseed orbitides for potential cancer treatment. *Toxicol. Rep.* **2015**, *2*, 1014–1018. [CrossRef]

30. Okinyo-Owiti, D.P.; Young, L.; Burnett, P.G.; Reaney, M.J. New flaxseed orbitides: Detection, sequencing, and (^{15})N incorporation. *Biopolymers* **2014**, *102*, 168–175. [CrossRef]

31. Sharav, O.; Shim, Y.Y.; Okinyo-Owiti, D.P.; Sammynaiken, R.; Reaney, M.J. Effect of cyclolinopeptides on the oxidative stability of flaxseed oil. *J. Agric. Food Chem.* **2014**, *62*, 88–96. [CrossRef] [PubMed]

32. Drygala, P.; Olejnik, J.; Mazur, A.; Kierus, K.; Jankowski, S.; Zimecki, M.; Zabrocki, J. Synthesis and immunosuppressive activity of cyclolinopeptide A analogues containing homophenylalanine. *Eur. J. Med. Chem.* **2009**, *44*, 3731–3738. [CrossRef] [PubMed]

33. Bell, A.; McSteen, P.M.; Cebrat, M.; Picur, B.; Siemion, I.Z. Antimalarial activity of cyclolinopeptide A and its analogues. *Acta Pol. Pharm.* **2000**, *57*, 134–136. [PubMed]

34. Yang, J.; Jadhav, P.D.; Reaney, M.J.T.; Sammynaiken, R.; Yang, J. A novel formulation significantly increases the cytotoxicity of flaxseed orbitides (linusorbs) LOB3 and LOB2 towards human breast cancer MDA-MB-231 cells. *Die Pharm.* **2019**, *74*, 520–522. [CrossRef]

35. Shim, Y.Y.; Young, L.W.; Arnison, P.G.; Gilding, E.; Reaney, M.J. Proposed systematic nomenclature for orbitides. *J. Nat. Prod.* **2015**, *78*, 645–652. [CrossRef] [PubMed]

36. Zou, X.G.; Li, J.; Sun, P.L.; Fan, Y.W.; Yang, J.Y.; Deng, Z.Y. Orbitides isolated from flaxseed induce apoptosis against SGC-7901 adenocarcinoma cells. *Int. J. Food Sci. Nutr.* **2020**, *71*, 929–939. [CrossRef] [PubMed]

37. Ostrom, Q.T.; Cioffi, G.; Gittleman, H.; Patil, N.; Waite, K.; Kruchko, C.; Barnholtz-Sloan, J.S. CBTRUS Statistical Report: Primary Brain and Other Central Nervous System Tumors Diagnosed in the United States in 2012–2016. *Neuro Oncol.* **2019**, *21*, v1–v100. [CrossRef]

38. Heilos, D.; Rodriguez-Carrasco, Y.; Englinger, B.; Timelthaler, G.; van Schoonhoven, S.; Sulyok, M.; Boecker, S.; Sussmuth, R.D.; Heffeter, P.; Lemmens-Gruber, R.; et al. The Natural Fungal Metabolite Beauvericin Exerts Anticancer Activity In Vivo: A Pre-Clinical Pilot Study. *Toxins* **2017**, *9*, 258. [CrossRef]

39. Cao, X.H.; Wang, A.H.; Wang, C.L.; Mao, D.Z.; Lu, M.F.; Cui, Y.Q.; Jiao, R.Z. Surfactin induces apoptosis in human breast cancer MCF-7 cells through a ROS/JNK-mediated mitochondrial/caspase pathway. *Chem. Biol. Interact.* **2010**, *183*, 357–362. [CrossRef]

40. Taniguchi, Y.; Yamamoto, N.; Hayashi, K.; Takeuchi, A.; Miwa, S.; Igarashi, K.; Higuchi, T.; Abe, K.; Yonezawa, H.; Araki, Y.; et al. Anti-tumor Effects of Cyclolinopeptide on Giant-cell Tumor of the Bone. *Anticancer Res.* **2019**, *39*, 6145–6153. [CrossRef]

41. Elmore, S. Apoptosis: A review of programmed cell death. *Toxicol. Pathol.* **2007**, *35*, 495–516. [CrossRef] [PubMed]

42. He, B.; Lu, N.; Zhou, Z. Cellular and nuclear degradation during apoptosis. *Curr. Opin. Cell Biol.* **2009**, *21*, 900–912. [CrossRef] [PubMed]

43. Wang, C.L.; Liu, C.; Niu, L.L.; Wang, L.R.; Hou, L.H.; Cao, X.H. Surfactin-induced apoptosis through ROS-ERS-Ca2$^+$-ERK pathways in HepG2 cells. *Cell Biochem. Biophys.* **2013**, *67*, 1433–1439. [CrossRef] [PubMed]

44. Saraste, A.; Pulkki, K. Morphologic and biochemical hallmarks of apoptosis. *Cardiovasc. Res.* **2000**, *45*, 528–537. [CrossRef]

45. Zhang, J.H.; Xu, M. DNA fragmentation in apoptosis. *Cell Res.* **2000**, *10*, 205–211. [CrossRef] [PubMed]

46. Belmokhtar, C.A.; Hillion, J.; Segal-Bendirdjian, E. Staurosporine induces apoptosis through both caspase-dependent and caspase-independent mechanisms. *Oncogene* **2001**, *20*, 3354–3362. [CrossRef]

47. Cornelissen, M.; Philippe, J.; De Sitter, S.; De Ridder, L. Annexin V expression in apoptotic peripheral blood lymphocytes: An electron microscopic evaluation. *Apoptosis Int. J. Program. Cell Death* **2002**, *7*, 41–47. [CrossRef]

48. Kale, J.; Osterlund, E.J.; Andrews, D.W. BCL-2 family proteins: Changing partners in the dance towards death. *Cell Death Differ.* **2018**, *25*, 65–80. [CrossRef]

49. Janicke, R.U.; Sohn, D.; Schulze-Osthoff, K. The dark side of a tumor suppressor: Anti-apoptotic p53. *Cell Death Differ.* **2008**, *15*, 959–976. [CrossRef]

50. Espinosa-Oliva, A.M.; Garcia-Revilla, J.; Alonso-Bellido, I.M.; Burguillos, M.A. Brainiac Caspases: Beyond the Wall of Apoptosis. *Front. Cell. Neurosci.* **2019**, *13*, 500. [CrossRef]

51. McIlwain, D.R.; Berger, T.; Mak, T.W. Caspase functions in cell death and disease. *Cold Spring Harb. Perspect. Biol.* **2015**, *7*, a026716. [CrossRef] [PubMed]

52. Park, S.Y.; Kim, J.H.; Lee, Y.J.; Lee, S.J.; Kim, Y. Surfactin suppresses TPA-induced breast cancer cell invasion through the inhibition of MMP-9 expression. *Int. J. Oncol.* **2013**, *42*, 287–296. [CrossRef] [PubMed]

53. Yamaguchi, H.; Condeelis, J. Regulation of the actin cytoskeleton in cancer cell migration and invasion. *Biochim. Biophys. Acta* **2007**, *1773*, 642–652. [CrossRef] [PubMed]

54. Olson, M.F.; Sahai, E. The actin cytoskeleton in cancer cell motility. *Clin. Exp. Metastasis* **2009**, *26*, 273–287. [CrossRef]

55. Izdebska, M.; Zielinska, W.; Grzanka, D.; Gagat, M. The Role of Actin Dynamics and Actin-Binding Proteins Expression in Epithelial-to-Mesenchymal Transition and Its Association with Cancer Progression and Evaluation of Possible Therapeutic Targets. *BioMed Res. Int.* **2018**, *2018*, 4578373. [CrossRef]

56. Stehn, J.R.; Haass, N.K.; Bonello, T.; Desouza, M.; Kottyan, G.; Treutlein, H.; Zeng, J.; Nascimento, P.R.; Sequeira, V.B.; Butler, T.L.; et al. A novel class of anticancer compounds targets the actin cytoskeleton in tumor cells. *Cancer Res.* **2013**, *73*, 5169–5182. [CrossRef]

57. Foerster, F.; Braig, S.; Moser, C.; Kubisch, R.; Busse, J.; Wagner, E.; Schmoeckel, E.; Mayr, D.; Schmitt, S.; Huettel, S.; et al. Targeting the actin cytoskeleton: Selective antitumor action via trapping PKCvarepsilon. *Cell Death Dis.* **2014**, *5*, e1398. [CrossRef]

58. Gandalovicova, A.; Rosel, D.; Fernandes, M.; Vesely, P.; Heneberg, P.; Cermak, V.; Petruzelka, L.; Kumar, S.; Sanz-Moreno, V.; Brabek, J. Migrastatics-Anti-metastatic and Anti-invasion Drugs: Promises and Challenges. *Trends Cancer* **2017**, *3*, 391–406. [CrossRef]

59. Xuan, B.; Ghosh, D.; Cheney, E.M.; Clifton, E.M.; Dawson, M.R. Dysregulation in Actin Cytoskeletal Organization Drives Increased Stiffness and Migratory Persistence in Polyploidal Giant Cancer Cells. *Sci. Rep.* **2018**, *8*, 11935. [CrossRef]

60. Kim, M.Y.; Kim, J.H.; Cho, J.Y. Cytochalasin B modulates macrophage-mediated inflammatory responses. *Biomol. Ther.* **2014**, *22*, 295–300. [CrossRef]

61. Craig, E.W.; Avasthi, P. Visualizing Filamentous Actin Using Phalloidin in Chlamydomonas reinhardtii. *Bio-Protocol* **2019**, *9*, e3274. [CrossRef] [PubMed]

62. Cheng, Y.; Feng, Y.; Jansson, L.; Sato, Y.; Deguchi, M.; Kawamura, K.; Hsueh, A.J. Actin polymerization-enhancing drugs promote ovarian follicle growth mediated by the Hippo signaling effector YAP. *FASEB J. Off. Publ. Fed. Am. Soc. Exp. Biol.* **2015**, *29*, 2423–2430. [CrossRef] [PubMed]

63. Dehm, S.M.; Bonham, K. SRC gene expression in human cancer: The role of transcriptional activation. *Biochem. Cell Biol. Biochim. Biol. Cell.* **2004**, *82*, 263–274. [CrossRef] [PubMed]

64. Sen, B.; Johnson, F.M. Regulation of SRC family kinases in human cancers. *J. Signal Transduct.* **2011**, *2011*, 865819. [CrossRef]

65. Sandilands, E.; Cans, C.; Fincham, V.J.; Brunton, V.G.; Mellor, H.; Prendergast, G.C.; Norman, J.C.; Superti-Furga, G.; Frame, M.C. RhoB and actin polymerization coordinate Src activation with endosome-mediated delivery to the membrane. *Dev. Cell* **2004**, *7*, 855–869. [CrossRef]

66. Garcia, R.; Bowman, T.L.; Niu, G.; Yu, H.; Minton, S.; Muro-Cacho, C.A.; Cox, C.E.; Falcone, R.; Fairclough, R.; Parsons, S.; et al. Constitutive activation of Stat3 by the Src and JAK tyrosine kinases participates in growth regulation of human breast carcinoma cells. *Oncogene* **2001**, *20*, 2499–2513. [CrossRef]

67. Bjorge, J.D.; Pang, A.S.; Funnell, M.; Chen, K.Y.; Diaz, R.; Magliocco, A.M.; Fujita, D.J. Simultaneous siRNA targeting of Src and downstream signaling molecules inhibit tumor formation and metastasis of a human model breast cancer cell line. *PLoS ONE* **2011**, *6*, e19309. [CrossRef]

68. Harada, D.; Takigawa, N.; Kiura, K. The Role of STAT3 in Non-Small Cell Lung Cancer. *Cancers* **2014**, *6*, 708–722. [CrossRef]

69. Mosmann, T. Rapid colorimetric assay for cellular growth and survival: Application to proliferation and cytotoxicity assays. *J. Immunol. Methods* **1983**, *65*, 55–63. [CrossRef]

70. Lee, J.O.; Choi, E.; Shin, K.K.; Hong, Y.H.; Kim, H.G.; Jeong, D.; Hossain, M.A.; Kim, H.S.; Yi, Y.S.; Kim, D.; et al. Compound K, a ginsenoside metabolite, plays an antiinflammatory role in macrophages by targeting the AKT1-mediated signaling pathway. *J. Ginseng Res.* **2019**, *43*, 154–160. [CrossRef]
71. Lee, J.O.; Kim, J.H.; Kim, S.; Kim, M.Y.; Hong, Y.H.; Kim, H.G.; Cho, J.Y. Gastroprotective effects of the nonsaponin fraction of Korean Red Ginseng through cyclooxygenase-1 upregulation. *J. Ginseng Res.* **2020**, *44*, 655–663. [CrossRef] [PubMed]
72. Zhao, L.; Xu, G.; Zhou, J.; Xing, H.; Wang, S.; Wu, M.; Lu, Y.P.; Ma, D. The effect of RhoA on human umbilical vein endothelial cell migration and angiogenesis in vitro. *Oncol. Rep.* **2006**, *15*, 1147–1152. [CrossRef] [PubMed]
73. Yi, Y.S.; Baek, K.S.; Cho, J.Y. L1 cell adhesion molecule induces melanoma cell motility by activation of mitogen-activated protein kinase pathways. *Die Pharm.* **2014**, *69*, 461–467.

Sample Availability: Sample of the compound linusorb B3 (LOB3), is available from the authors.

Publisher's Note: MDPI stays neutral with regard to jurisdictional claims in published maps and institutional affiliations.

 *molecules*

Review

Bergenia Genus: Traditional Uses, Phytochemistry and Pharmacology

Bhupendra Koul [1,*], Arvind Kumar [2], Dhananjay Yadav [3,*] and Jun-O. Jin [3,4,*]

1   School of Bioengineering and Biosciences, Lovely Professional University, Phagwara 144411, India
2   Research Center for Chromatography and Mass Spectrometry, CROM-MASS, CIBIMOL-CENIVAM, Industrial University of Santander, Carrera 27, Calle 9, Edificio 45, Bucaramanga 680002, Colombia; arvindtomer81@gmail.com
3   Department of Medical Biotechnology, Yeungnam University, Gyeongsan 38541, Korea
4   Research Institute of Cell Culture, Yeungnam University, Gyeongsan 38541, Korea
*   Correspondence: bhupendra.18673@lpu.co.in (B.K.); dhanyadav16481@gmail.com (D.Y.); jinjo@yu.ac.kr (J.-O.J.); Tel.: +91-9454320518 (B.K.); +82-1022021191 (D.Y.); +82-53-810-3033 (J.-O.J.); Fax: +82-53-810-4769 (J.-O.J.)

Academic Editors: Raffaele Pezzani and Sara Vitalini
Received: 23 October 2020; Accepted: 14 November 2020; Published: 26 November 2020

Abstract: *Bergenia* (Saxifragaceae) genus is native to central Asia and encompasses 32 known species. Among these, nine are of pharmacological relevance. In the Indian system of traditional medicine (Ayurveda), "Pashanabheda" (stone breaker) is an elite drug formulation obtained from the rhizomes of *B. ligulata*. *Bergenia* species also possess several other biological activities like diuretic, antidiabetic, antitussive, insecticidal, anti-inflammatory, antipyretic, anti-bradykinin, antiviral, antibacterial, antimalarial, hepatoprotective, antiulcer, anticancer, antioxidant, antiobesity, and adaptogenic. This review provides explicit information on the traditional uses, phytochemistry, and pharmacological significance of the genus *Bergenia*. The extant literature concerned was systematically collected from various databases, weblinks, blogs, books, and theses to select 174 references for detailed analysis. To date, 152 chemical constituents have been identified and characterized from the genus *Bergenia* that belong to the chemical classes of polyphenols, phenolic-glycosides, lactones, quinones, sterols, tannins, terpenes, and others. *B. crassifolia* alone possesses 104 bioactive compounds. Meticulous pharmacological and phytochemical studies on *Bergenia* species and its conservation could yield more reliable compounds and products of pharmacological significance for better healthcare.

Keywords: *Bergenia* species; botanical description; traditional uses; phytochemistry; pharmacology; anti-urolithiatic activity; bergenin

1. Introduction

The use of herbs for healing diseases and disorders can be dated back to at least 1500 BC [1]. The traditional system of medicine (TCM) is a source of >60% of the commercialized drugs and is still used by the population in lower income countries for the cure of chronic diseases [2]. As far as primary healthcare is concerned, approximately 75% of Indians rely on Ayurvedic formulations [3,4]. Many medicinal plants containing various phytochemicals have been successfully used to cure diabetes, cancers, gastrointestinal disorders, cardiovascular, and urological disorders [1].

Among the urological disorders, "urolithiasis" is the third most common disorder with a high relapse rate [5–8]. The invasive treatments of urolithiasis are costly and precarious, so the search for natural anti-urolithiatic drugs is of immense importance [9,10].

The Ayurvedic preparations have used *Bergenia* species down the centuries to dissolve bladder and kidney stones and to treat piles, abnormal leucorrhea, and pulmonary infections [11–13].

These pharmacological properties can be attributed to a wide-range polyphenols, flavonoids, and quinones present in *Bergenia* species. The polyphenols constitute a major share of the active ingredients, and the elite among them are "arbutin" and "bergenin" [14–19]. Bergenin alone possesses burn-wound healing, antiulcer, anti-arrhythmic, antihepatotoxic, neuroprotective, antifungal, antidiabetic, antilithiatic, anti-inflammatory, anti-nociceptive, anti-HIV, and immunomodulatory properties [20–22]. *Bergenia ligulata* Wall. Engl. [synonym of *B. pacumbis*] is an essential ingredient of an Ayurvedic formulation, "Pashanbheda" (Paashan = rockstone, bheda = piercing), which is used as a kidney stone dissolver in the indigenous system of medicine [23,24]. This drug has been listed in ancient Indian chronicles of medicine including *"Charak Samhita"*, *"Sushruta Samhita"* and *"Ashtang-Hridaya"*. *B. ligulata* is reputedly known by other names such as "Pashana", "Ashmabhid", "Ashmabhed", "Asmaribheda", "Nagabhid", "Parwatbhed", "Upalbhedak", and "Shilabhed" [25].

The unavailability of a compendious review on bioactive molecules present in *Bergenia* genus prompted us to compile the same. The present review provides explicit knowledge on the traditional and medicinal importance and phytochemistry of the *Bergenia* species.

2. Review Methodology

The extant literature (abstracts, blogs, full-text articles, PhD theses, and books) on the *Bergenia* species was reviewed systematically to generate concise and resourceful information regarding their distribution, phytochemistry, traditional medicinal uses, and pharmacological activities. For this purpose, different bibliographic search engines and online databases (Google Scholar, WoS, PubMed, CAB abstracts, INMEDPLAN, Scopus, NATTS, EMBASE, SciFinder, MEDLINE) and websites (www.sciencedirect.com; eflora.org; jstor.org; pfaf.org) were referred, to select 174 references for detailed analysis. Each botanical name has been validated through www.theplantlist.org and https://www.catalogueoflife.org/ online repositories. ChemDraw software (version 12.0) was used to draw the structures of the chemical compounds.

3. Distribution

The plant family Saxifragaceae encompasses 48 genera and 775 species, which are mostly distributed in South East Asia. The name "Bergenia" was coined by Conrad Moench in 1794, in the memory of Karl August von Bergen (German botanist and physician). Genus *Bergenia* harbors 32 species of flowering plants, including highly valued ornamental, rhizomatous, and temperate medicinal herbs [16]. Central Asia is the native place for genus *Bergenia* [26,27]. The geographical distribution of 32 species of genus *Bergenia* are detailed in Figure 1, which depicts the worldwide distribution through the map. In China, seven species are reported from three provinces and two autonomous regions: Shanxi, Sichuan, and Sanxi and Tibet and Xinjiang, respectively. Among the seven species, four (*B. yunnanensis*, *B. scopulosa*, *B. emeiensis*, and *B. tianquanensis*) are endemic to China [28–30].

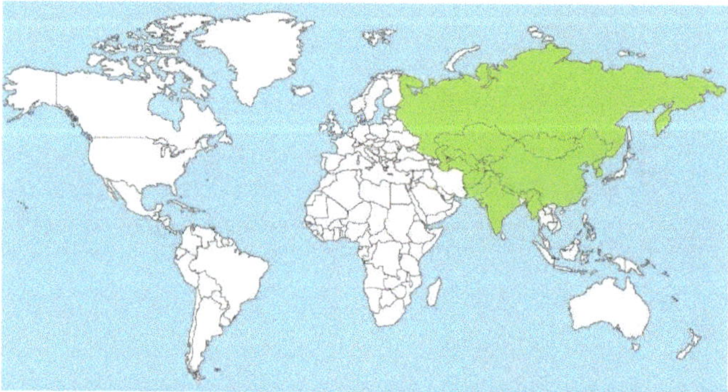

Figure 1. A world map showing the geographical distribution of *Bergenia* species (in green).

4. Botanical Description

Bergenia(s) are evergreen, perennial, drought-resistant, herbaceous plants that bear pink flowers produced in a cyme. Due to the leaf shape and leathery texture, *Bergenia*(s) have earned some interesting nicknames such as "pigsqueak", "elephant-ear", "heartleaf", "leather cabbage", or "picnic plates". The plants should be planted about two feet apart as they spread horizontally up to 45–60 cm. The botanical description of *Bergenia* species [31–34] is described in Supplementary Table S1.

5. Traditional Medicinal Uses

Bergenia species have been used in the traditional medicines for a long time. In Unani and Ayurvedic systems of medicine, *Bergenia* spp. rhizomes and roots have been used for curing kidney and, bladder diseases, dysuria, heart diseases, lung and liver diseases, spleen enlargement, tumors, ulcers, piles, dysentery, menorrhagia, hydrophobia, biliousness, eyesores, cough, and fever [35–37]. The burns or wounds may be treated with rhizome paste for three to four days [38–40]. The paste can be applied on dislocated bones after setting, or consumed to treat diarrhea or along with honey in fevers [41,42].

The leaf extract of *B. ciliata* possesses antimalarial property [43]. Its leaves are revered to as "Pashanabheda", which designates the litholytic property [44]. In Nepal, 1:1 mixture (one teaspoon) of the dried B. ciliata rhizome-juice and honey is administered to post-partum women 2–3 times a day as a tonic and remedy for digestive disorders (carminative) [38]. The rhizome-decoction may also be consumed orally as antipyretic and antihelmintic [45].

Since ancient times, consumption of water-extract of *B. ligulata* has cured urogenital and kidney-stone complaints [23,35,46,47]. In Nepal, the rhizome paste of *B. ligulata* is consumed for treating many diseases including diarrhea, ulcer, dysuria, spleen enlargement, pulmonary infusion, cold, cough, and fever [45]. The intestinal worms can also be removed by consuming rhizomes along with molasses (two times/day, 3–4 days) [38]. The Indians use the dried roots of *B. ligulata* for treating burns, boils, wounds, and ophthalmia [46,48]. The dried leaf powder of *B. pacumbis* may be inhaled to bring relief from heavy sneezing [49]. In Lahul (Punjab), the locals use *B. stratecheyi* plants to prepare a poultice, which is applied to heal the joint-stiffness [50]. *Bergenia* species are also used for the treatment of boils and even blisters [19].

In Russian tradition, *B. crassifolia* leaves are commonly used to prepare a health drink. Buryats and Mongols used *B. crassifolia*-young leaves of to prepare tea. Interestingly, in Altai, tea is prepared from old blackened leaves (chagirsky tea having lesser amounts of tannins) [51]. The rhizome infusions can treat fevers, cold, headache, gastritis, dysentery, and enterocolitis [52]. They are also used to treat oral diseases (bleeding gums, periodontitis, gingivitis, and stomatitis) and also possess adaptogenic properties [51,53–55]. Mongols used the extracts for treating typhoid, gastro-intestinal ailments, diarrhoea, and lung inflammation. The rhizome extract is also used to strengthen capillary walls to stop bleeding after abortions, alleviate excessive menstruation, and cervical erosion. Therefore, the roots and rhizomes of *B. crassifolia* are claimed as antimicrobial, anti-inflammatory, haemostatic, and as astringent in the officinal medicine of Mongolia [54].

Tibetans apply fresh leaf-paste on their skin to protect them from harmful ultraviolet radiations [56]. The chewing of leaf helps in relieving constipation and the leaf-juice can treat earaches [11,38,42]. The bullocks and cows are fed on a mixture of *Bergenia* inflorescence and barley-flour to treat hematuria [38]. *Bergenia* roots are also effective in preventing venereal diseases [57]. Thick leaves of Bergenias are used in Chinese Medicine to stop bleeding, treat cough, dizziness, hemoptysis, and asthma, and to strengthen immunity [27,58].

6. Phytochemistry

Nowadays, HPLC and HPTLC have become routine analytical techniques due to their reliability in quantitation of analytes at the micro or even nanogram levels plus the cost effectiveness. Phytochemical investigation of nine *Bergenia* species (*B. ciliata*, *B. crassifolia*, *B. emeiensis*, *B. ligulata*, *B. scopulosa*, *B. stracheyi*, *B. hissarica*, *B. purpurascens*, and *B. tianquanesis*) led to the characterization of several chemical constituents [16,59–63]. The review of the extant literature reveals the presence of 152 chemical compounds (volatile: 47 and non-volatile: 105) (Table 1) as shown in Supplementary Figure S1. The constituents have been categorized into polyphenols, flavonoids, quinones, sterols, terpenes, tannins, lactones, and others [16,26,64–67]. The major bioactive compounds are bergenin (**1**), (+)-catechin (**2**), gallic acid (**3**), β-sitosterol (**4**), catechin-7-*O*-β-D-glucoside (**5**), (+)-afzelechin (**6**), *arbutin (10)*,4-*O*-galloylbergenin (**12**), 11-*O*-galloylbergenin (**13**), caffeoylquinic acid (**21**), pashaanolactone (**26**), 3,11-di-*O*-galloylbergenin (**64**), bergapten (**66**), kaempferol-3-*O*-rutinoside (**70**), quercetin-3-*O*-rutinoside (**79**), (+)-catechin-3-*O*-gallate (**83**), 2-*O*-caffeoylarbutin (**86**), leucocyanidin (**124**), methyl gallate (gallicin) (**125**), sitoinoside I (**126**), β-sitosterol-D-glucoside (**127**), avicularin (**128**), reynoutrin (**129**), procyanidin B1 (**135**), afzelin (**140**), and aloe-emodin (**146**).

Arbutin (**10**) inhibits tyrosinase, prevents the formation of melanin and thus prevents skin darkening [68]. Bergenin (**1**) is a pharmaceutically important molecule that has hepatoprotective and immunomodulatory potential [69]. It is used clinically for eliminating phlegm, relieving cough, inflammation, etc. [20,70,71]. (+)-catechin (**2**) possesses antioxidant, glucosidase, renoprotective, matrix-metalloproteinase inhibitory, and cancer preventive activity. Gallicin (**125**) exhibits antioxidant, anti-tumor, antimicrobial, anti-inflammatory, and cyclooxygenase-2/5-lipoxygenase inhibitory activity [72]. Gallic acid (**3**) possesses anti-inflammatory, antioxidant, cytotoxic, bactericidal, gastroprotective, and antiangiogenic activity. β-sitosterol (**4**) is well-known for its antioxidant, anti-inflammatory, analgesic, and anti-helminthic effects. It is also efficient in the curing prostate enlargement [73].

Recently, bergenicin (**151**) and bergelin (**152**) have been isolated from leaves of *B. himalaica* Boriss [71]. The chemistry of *B. tianquanesis* plant has not been reported to date. Although several bioactive compounds have been isolated and characterized from *Bergenia* species, there is still scope for extended research on their efficacy and versatility.

Table 1. Bioactive compounds and medicinal properties of different *Bergenia* species.

Bergenia Species	Distribution	Medicinal Property	Part Used	Chemical Constituents (Structure Number)	Reference(s)
Bergenia ciliata (Haw.) Sternb.	Central Asia, Afghanistan to China, Himalayan region. Altitude range (1800–3000 m)	Analgesic, Antiarrhythmic, Antiwrinkle, Antiasthma, Antibacterial, Anticancer, Antidiabetic, Antidiarrheal, Antidotary, Antiepileptic, Antiflatulent, Antifungal, Anti-haemorrhoidal, Antiviral, Anti-inflammatory, Antilithiatic, Antimalaria; Antimenorrhagic, Antiobesity, Antiophthalmia, Antioxidant, Antipyretic, Antispasmodic, Antiulcer, Burn wound healing, Deobstruent, Cerebroprotective, Diuretic, Ecbolic, Emmenagogue, Expectorant, Hepatoprotective, Immunomodulatory, Pulmonary affection	Whole plant	Bergenin (**1**) [a] Catechin (**2**) [a] Gallic acid (**3**) [a] β-Sitosterol (**4**) [d] Catechin-7-*O*-glucoside (**5**) [a] Afzelechin (**6**) [a] Quercetin-3-*O*-β-D-xylopyranoside (**7**) [a] Quercetin-3-*O*-α-L-arbinofuranoxide (**8**) [a] Eryodictiol-7-*O*-β-D-glucopyranoside (**9**) [a] Arbutin (**10**) [c] 6-*O*-*p*-Hydroxybenzoyl arbutin (**11**) [a] 4-*O*-Galloylbergenin (**12**) [a] 11-*O*-Galloylbergenin (**13**) [a] *p*-Hydroxybenzoic acid (**14**) [f] Protocatechuic acid (**15**) [a] 6-*O*-Protocatechuoyl arbutin (**16**) [a] 11-*O*-*p*-Hydroxybenzoyl bergenin (**17**) [a] 11-*O*-Protocatechuoyl bergenin (**18**) [a] 6-*O*-*p*-Hydroxybenzoyl parasorboside (**19**) [a]	[11,16,31,43, 72–91]
				Ellagitannins (**20**) [a] Gallic acid (**3**) [a] Arbutin (**10**) [c] Bergenin (**1**) [a] Caffeoylquinic acid (**21**) [c] Monogalloylquinic acid (**22**) [c] 2,4,6-Tri-*O*-galloyl-β-D-glucose (**23**) [a] Pedunculagin (**24**) [a] Tellimagrandin I (**25**) [a] Catechin-7-*O*-β-D-glucoside (**5**) [a] Paashanolactone (**26**) [b] Catechin (**2**) [a] β-Sitosterol (**4**) [b] 2,4-Heptadienal (**27**) [f] Benzaldehyde (**28**) [f] Benzeneacetaldehyde (**29**) [f] Decadienal (**30**) [f]	

Table 1. *Cont.*

Bergenia Species	Distribution	Medicinal Property	Part Used	Chemical Constituents (Structure Number)	Reference(s)
Bergenia crassifolia (L.) Fritsch [Synonym: *Bergenia cordifolia* (Haw.) Sternb.]	North Eastern Asia. Altitude range (200–2000 m)	Antihypertensive, Anti-inflammatory, Antilithiatic, Antiobesity, Antioxidant, Antipyretic, Cerebroprotective, Diuretic, Hepatoprotective, Immunomodulatory	Whole plant	Decanal (**31**) [f] Dimethylcyclohexene acetaldehyde (**32**) [f] (*E*)-2-Decenal (**33**) [f] (*E*)-2-Nonenal (**34**) [f] Nonanal (**35**) [f] *p*-Menthenal (**36**) [f] (*E*)-β-Damascenone (**37**) [e] (*E*)-β-Damascone (**38**) [e] 3-Thujen-2-one (**39**) [e] Caryophyllene (**40**) [e] Cedranol (**41**) [e] (*E*)-2-Decenol (**42**) [e] Farnesol (**43**) [e] Farnesyl acetone (**44**) [e] Geraniol (**45**) [e] Geranyl acetone (**46**) [e] Hexahydrofarnesyl acetone (**47**) [e] Ionone (**48**) [e] Linalool (**49**) [e] *m*-Mymene (**50**) [e] Nerolidol (**51**) [e] Phytol (**52**) [e] *p*-Menth-1-en-4-ol (**53**) [e] Prenol (**54**) [e] Thymol (**55**) [e] α-Bisabolol (**56**) [e] α-Bisabololoxide B (**57**) [e] α-Cadinol (**58**) [e] α-Terpineol (**59**) [e] β-Elemene (**60**) [e] β-Eudesmol (**61**) [e] δ-Cadinene (**62**) [e] 11-*O*-(*p*-Hydroxybezoyl)bergenin (**63**) [a] 3,11-Di-*O*-galloylbergenin (**64**) [a] 4,11-Di-*O*-galloylbergenin (**65**) [a] Bergapten (**66**) [a] Kaempferol-3-*O*-xylosylgalactoside (**67**) [a] Kaempferol-3-*O*-xylosylglucoside (**68**) [a] Kaempferol-3-*O*-arabinoside (**69**) [a] Kaempferol-3-*O*-rutinoside (**70**) [a] Norathyriol (**71**) [a] Norbergenin (**72**) [a] Quercetin-3-*O*-xylosylgalactoside (**73**) [a] Quercetin-3-*O*-xylosylglucoside (**74**) [a] Quercetin-3-*O*-arabinoside (**75**) [a] Quercetin-3-*O*-galactoside (**76**) [a] Quercetin-3-*O*-glucoside (**77**) [a] Quercetin-3-*O*-rhamnoside (**78**) [a] Quercetin-3-*O*-rutinoside (**79**) [a] Quercetin-3-*O*-xyloside (**80**) [a] Trihydroxycoumarin (**81**) [a] (+)-Catechin-3,5-di-*O*-gallate (**82**) [a] (+)-Catechin-3-*O*-gallate (**83**) [a] 1,2,4,6-Tetra-*O*-galloy-β-ᴅ-glucopyranose (**84**) [a] 1-*O*-Galloylglucose (**85**) [a] 2-*O*-Caffeoylarbutin (**86**) [a] 6-*O*-Galloylarbutin (**87**) [a] Ellagic acid (**88**) [a] Hydroquinone (**89**) [a] *p*-Galloyloxyphenyl-β-ᴅ-glucoside (**90**) [a] Pyrogallol (**91**) [a] Acetylsalicylic acid (**92**) [f] Fumaric acid (**93**) [f] Furancarboxylic acid (**94**) [f] Protocatechuic acid (**15**) [f] Malic acid (**95**) [f] Quinic acid (**96**) [f] 4-Methoxystyrene (**97**) [f] 9,12-Octadecadienoic acid (**98**) [f] 9-Octadecenoic acid (**99**) [f] Decanoic acid (**100**) [f] Dodecanoic acid (**101**) [f] Hexadecanoic acid (**102**) [f] *n*-Cetyl alcohol (**103**) [f] *n*-Eicosanol (**104**) [f] *n*-Hentriacontane (**105**) [f] *n*-Heptacosane (**106**) [f] *n*-Nonacosane (**107**) [f] Nonanoic acid (**108**) [f] *n*-pentacosane (**109**) [f] Pentadecanoic acid (**110**) [f] Rhododendrin (**111**) [f] Stearic acid (**112**) [f] Tetradecanoic acid (**113**) [f] Tetramethyl hexadecenol (**114**) [f] Trimethyl dihydronaphthalene (**115**) [f] Trimethyl-3-methylene hexadecatetraene (**116**) [f]	[14,28,31,64, 79,85,92–104]

Table 1. *Cont.*

Bergenia Species	Distribution	Medicinal Property	Part Used	Chemical Constituents (Structure Number)	Reference(s)
Bergenia emeiensis C.Y. Wu ex J.T. Pan.	China. Altitude range (1600–4200 m)	Antiwrinkle, Anti-inflammatory, Antiobesity, Antioxidant	Whole plant	Bergenin (**1**) [a] Tannic acid (**117**) [a] Arbutin (**10**) [c]	[31,65,85,105]
Bergenia ligulata Wall. Engl. [Accepted name: *Bergenia pacumbis* (Buch.-Ham. Ex D. Don.) C.Y. Wu & J.T. Pan]	Temperate Himalayas. Altitude range (2134–3048 m)	Analgesic, Antiarrhythmic, Anticancer, Antidiabetic, Antifungal, Anti-haemorrhoidal, Antiviral, Anti-inflammatory, Antilithiatic, Antiprotozoal, Antipyretic, Antiscorbutic, Antispasmodic, Antitumor, Antiulcer, Astringent, burn wound healing, Cerebroprotective, Diuretic, Expectorant, Hepatoprotective, Immunomodulatory	Root, Rhizome	Bergenin (**1**) [a] Gallic acid (**3**) [a] Tannic acid (**117**) [a] Arbutin (**10**) [c] Catechin (**2**) [a] β-Sitosterol (**4**) [b] Stigmasterol (**118**) [d] Afzelechin (**6**) [a] 1,8-Cineole (**119**) [e] Isovaleric acid (**120**) [f] (+)-(6S)-Parasorbic acid (**121**) [b] Terpinen-4-ol (**122**) [e] (Z)-asarone (**123**) [f] Leucocyanidin (**124**) [a] Methyl gallate (**125**) [a] Sitoinoside I (**126**) [d] β-Sitosterol-D-glucoside (**127**) [d] Avicularin (**128**) [a] Eriodictyol-7-O-β-D-glucopyranoside (**9**) [a] Reynoutrin (**129**) [a] 11-O-Galloylbergenin (**13**) [a] Pashaanolactone (**26**) [b] Catechin-7-O-glucoside (**5**) [a] Coumarin (**130**) [b] 11-O-p-Hydroxybenzoyl bergenin (**17**) [a] 11-O-Protocatechuoyl bergenin (**18**) [a] 4-O-Galloylbergenin (**12**) [a] 6-O-p-Hydroxybenzoyl arbutin (**11**) [a] Hexan-5-olide (**131**) [b] Quercetin (**132**) [a] β-Sitosterol-D-glucoside (**127**) [d]	[3,16,23,24,26,37,61,73,78,83,85–88,90,95,106–127]
Bergenia purpurascens (Hook.f. & Thomson) Engl.	Eastern Himalayas. Altitude range (2800–4800 m)	Antibacterial, Anti-inflammatory, Antilithiatic, Antipyretic	Rhizome	Catechin (**2**) [a] Gallic acid (**3**) [a] Bergenin (**1**) [a] Arbutin (**10**) [c] 1,2,3,4,6-Penta-O-galloyl-β-D-glucose (**133**) [a] 4,6-Di-O-galloyl-β-D-glucose (**134**) [a] 6-O-Galloylarbutin (**87**) [a] 11-O-Galloylbergenin (**13**) [a] 4-O-Galloylbergenin (**12**) [a] 2,3,4,6-Tetra-O-galloyl-β-D-glucose (**84**) [a] Procyanidin B1 (**135**) [a] 2,4,6-Tri-O-galloyl-β-D-glucose (**23**) [a] Procyanidin B3 (**136**) [a]	[31,44,56,62,85,95,100,128–133]
Bergenia scopulosa (T.P. Wang)	China. Altitude range (2400–3600 m)	Anti-hypertensive, Anti-inflammatory, Antiobesity, Antioxidant, Antitussive, Cerebroprotective, Diuretic, Hepatoprotective, Immunomodulatory	Leaf, Root, Rhizome	Bergenin (**1**) [a] Arbutin (**10**) [c] Catechin (**2**) [a] β-Sitosterol (**4**) [d] 6-O-Galloylarbutin (**87**) [a] Catechin-7-O-β-D-glucopyranoside (**5**) [a] Phenylalanine (**137**) [f] Succinic acid (**138**) [f] Protocatechuic acid (**15**) [a] Gallic acid (**3**) [a] Methyl gallate (**125**) [a] Quercetin (**133**) [a] Hyperoside (**139**) [a] Quercetin-3-O-rutinoside (**79**) [a] Afzelin (**140**) [a] Chrysophanol-8-O-β-D-glucopyranoside (**141**) [c] 11-O-Galloylbergenin (**13**) [a]	[31,59,66,67,85,134–138]
Bergenia stracheyi (Hook. f. & Thomas) Engl.	Afghanistan, Pakistan, Nepal. Altitude range (3000–4600 m)	Antifungal, Anti-haemorrhoidal, Anti-inflammatory, Antilithiatic, Antiobesity, Antioxidant, Poultice to treat the stiff joints	Rhizome	Bergenin (**1**) [a] (+)-Catechin-3-O-gallate (**83**) [a] Gallic acid (**3**) [a] Tannic acid (**117**) [a] Phytol (**142**) [e] Caryophyllene (**40**) [e] Damascenone (**143**) [f] β-Eudesmol (**144**) [e] 3-Methyl-2-buten-1-ol (**145**) [e]	[13,83,85,86,88,90,116,139–141]

Table 1. *Cont.*

Bergenia Species	Distribution	Medicinal Property	Part Used	Chemical Constituents (Structure Number)	Reference(s)
Bergenia hissarica (A. Boriss.)	Central Asia, Uzbekistan, Hissar. Altitude range (1200–1600 m)	Stimulantlaxative, Neuroprotective, Antioxidant	Root, Rhizome	Aloe emodin (**146**) [c] Aloeemodin-8-*O*-β-D-glucoside (**147**) [c] Chrysophanein (**148**) [c] Emodin-1-*O*-β-D-glucoside (**149**) [c] Physeion (**150**) [d]	[142,143]
Bergenia tianquanesis (J.T. Pan)	China. Altitude range (2200–3400 m)	Not reported			[29,32]

[a] Polyphenols; [b] Lactones; [c] Quinones; [d] Sterols; [e] Terpenes; [f] Others. Number beside each bioactive compound represents the structure number as shown in Supplementary Figure S1.

7. Pharmacological Activities

The pharmaceutical importance of *Bergenia* species has been known since ancient times. Therefore, numerous biopharmaceutical products encompassing leaf or stem extracts are available in the markets and are being used to cure specific ailments (Figure 2).

Figure 2. Pharmacological significance of *Bergenia* species.

7.1. Antilithiatic Activity

The major contribution of *B. ligulata* towards pharmaceutical applications is that of an antilithiatic agent. Lower dose (0.5 mg/kg) of the EtOH extract of *B. ligulata* rhizome encourages diuresis in rats and is effective in dissolving preformed stones [144]. The MeOH extracts of the rhizome also possess an antilitihiatic property that has been tested both in vitro and in vivo. In male Wistar rats, 5–10 mg/kg of the extract inhibited calcium oxalate crystal ($CaC_2O_4 \bullet_x$) aggregation in the renal tubes. There are several other reports that state that *Bergenia* extracts exerts its antilithiactic effect by diuresis, inhibition of $CaC_2O_4 \bullet_x$ crystal formation and aggregation, and hypermagnesemic and antioxidant activity [106,145,146].

7.2. Diuretic Activity

Bergenia species are also known to possess diuretic properties. The EtOH extracts of *B. ligulata* roots were tested for their diuretic activity in rats. The Na^+, K^+, and Cl^- ion concentrations and the volume of urine excreted was measured after an interval of 5 h. It was observed that the EtOH extract showed significant diuretic activity [107]. *Bergenia crassifolia* (L.) Fritsch. leaf extract contains 15–20% arbutin, which has the potential to treat genitourinary diseases. In a 14 day experiment, the rats were injected with arbutin (**10**) and hydroquinone (**89**), 5 mg/kg (seven days) and 15 mg/kg (seven days). During the experiment, the arbutin (**10**) treatment increased the urine output (diuresis) along with creatinine and potassium, while hydroquinone (**89**) did not [147].

7.3. Antidiabetic Activity

After rigorus researches on animal models, it has now been proved that *B. ciliata*, *B. ligulata*, and *B. himalaica* possess an antidiabetic property [71]. The EtOH extracts of *B. ligulata* roots exhibit a remarkable hypoglycaemic effect in diabetic rats [108]. Saijyo et al., (2008) isolated the antidiabetic principle (α-glucosidase inhibitor) from *B. ligulata* rhizome extract by column chromatography, which was characterized as (+)-afzelechin (**6**), by NMR technique [61]. The antidiabetic property of *B. ligulata* can be useful in developing nutraceuticals (value-added food products) for diabetics [61,71,108].

7.4. Antitussive Activity

Bergenia species possesses the potential antitussive property. Different concentrations of arbutin (**10**) were administered to cough-induced mice, and it was observed that a dose of 200 mg/kg had the similar effect as that of 30 mg/kg antitussive drug codeine phosphate [138].

7.5. Insecticidal Activity

It has been recently discovered that *B. ligulata* exhibit an insecticidal property. The volatile oil from roots of *B. ligulata* containing 1,8-cineole (**119**) [4.24%], (+)-(6S)-isovaleric acid (**120**) [6.25%], (+)-(6S)-parasorbic acid (**121**) [47.45%], terpinen-4-ol (**122**) [2.96%], and (Z)-asarone (**123**) [3.50%] was tested for its insecticidal activity against *Drosophila melanogaster*, which was found to be significant [109]. Thus, volatile oil from *Bergenia* species or its specific component could be deployed as a natural insecticidal agent [24,109].

7.6. Anti-Inflammatory Activity

Bergenia species do have anti-inflammatory potential. The aqueous and EtOH (50%) extract of the rhizomes were introduced to animal model (rats) to demonstrate the anti-inflammatory activity. The succinate dehydrogenase (SDH) activity level (represented higher in inflammation) reduced in the rats that received the therapy. The attenuation of inflammatory response was confirmed through pharmacological and biochemical measurements [148]. Different concentrations of the MeOH extract of *B. ciliata* rhizomes have also been tested on a rat model with 100 mg/kg phenylbutazone (an anti-inflammatory agent) as a standard. Maximum inhibition of the inflammatory response was recorded at a dose of 300 mg/kg [74]. In a study by Churin et al. (2005), the dry extract of *B. crassifolia* leaves was administered to DBA/2 mice to study the effect on immune response. The extract declined the inflammatory process by preventing T-lymphocyte accumulation and cytokine production in the inflammatory region [149].

In another study, the delayed type hypersensitivity reaction was significantly elevated in mice administered with 100 μg/mL of bergenan BC (pectic polysaccharide) extracted from *B. crassifolia* leaves. It enhanced the uptake volume of neutrophils and mediated oxygen radicals' production by mouse peritoneal macrophages [150]. In mice model (balb/c mice), the increasing dose of bergenin (**1**) extracted from the rhizomes of *B. stracheyi* exhibited anti-arthritic property in a dose-dependent manner up to a dose of 40 mg/kg, while a higher dose of 80 mg/kg caused a reduction in the same [151]. These studies along with several others explain the anti-inflammatory activity of *Bergenia* species [92,110,151].

7.7. Antipyretic Activity

B. ligulata possess a significant antipyretic property. In a study by Singh et al. (2009b), the EtOH (95%) and aqueous extract of *B. ligulata* prepared in 2% gum acacia was administered to Wistar rats (300 and 500 mg/kg body weight) having pyrexia [107]. The antipyretic activity was observed using 200 mg/kg paracetamol (standard antipyretic drug) as positive control. The rectal temperature of the rats was documented after the 1 h time interval. A significant lowering in the body temperature was observed with EtOH extract (500 mg/kg). This study along with others justify that *B. ligulata* possesses significant antipyretic potential [111].

7.8. Anti-Bradykinin Activity

The anti-bradykinin activity of *B. crassifolia* leaf extract (per oral dose/treatment: 50 mg/kg for 14 days) has been studied in *spontaneously hypertensive* (SHR) rats. The reduction in the systolic blood pressure was observed after 3–6 h (by 20–25 mmHg), while a lowering of diastolic blood pressure with similar values was observed after 1 h of treatment [112,152]. The angiotensin-I-converting enzyme converts the hormone angiotensin I to the active form (vasoconstrictor: angiotensin II) and thus indirectly elevates the blood pressure by causing the blood vessels to constrict. The EtOH (70%) extract of *B. crassifolia* rhizomes significantly inhibits the angiotensin-I-converting enzyme (IC_{50} = 0.128 mg/mL), in vitro, and thus exhibits anti-bradykinin activity [153].

7.9. Antiviral Activity

The MeOH-water extract from rhizomes of *B. ligulata* have been reported to impede the in vitro replication of influenza A virus. Pre-treatment of cells with *B. ligulata* extract was effective in the preventing virus-mediated cell-destruction by repressing viral RNA and protein synthesis. The aqueous extract of *B. crassifolia* leaf supplemented with lectins reduced the virus-induced (HSV strain L2) cytopathogenic effect up to 95% [55]. The bioactive compound 1,2,3,4,6-penta-*O*-galloyl-β-D-glucose (**133**) present in the EtOH extract of *Saxifraga melanocentra* Franch. has been tested for its antiviral activity against HCV NS3 serine protease, through ELISA. The IC_{50} values of penta- (**133**), tetra- (**84**) and 2,4,6-tri-galloyl-β-D-glucose (**23**), were estimated to be 0.68–1.01 μM and exhibited 98.7–94.7% inhibition [113,128]. 1,2,3,4,6-penta-*O*-galloyl-β-D-glucose and its derivatives are also reported in *Bergenia* species. Thus, the aforementioned results support the antiviral potential in *Bergenia* species also.

7.10. Antibacterial Activity

Almost all of the aforementioned nine *Bergenia* species possess antibacterial activity. In a study by Sajad et al. (2010), the antibacterial activity of *B. ligulata* whole plant extract was analyzed based on the diffusion method. Different concentrations (10, 25, or 50 mg/mL) of the aqueous, EtOH and MeOH extracts of *B. ligulata* rhizomes exhibited antibacterial activity against *E. coli*, *B. subtilis*, and *S. Aureus* [110]. The extract concentration of 50 mg/mL was found to be most effective and was similar to that of the ciprofloxacin-antibiotic (25 μg/mL). These results show that *B. ligulata* possess significant antibacterial activity [110]. It is reported for *B. ciliata* that compared to leaf extracts, the root and rhizome extracts exhibit much higher antibacterial activity. The MeOH rhizome extracts of *B. scopulosa* were tested on eight different bacteria(s) using the agar-well diffusion assay method. It was concluded from the bacterial susceptibility test that both Gram-ve and +ve bacteria are susceptible as evident from the zone of inhibition that ranged from 13 to 15 mm. However, *E. coli*, *P. aeruginosa*, *K. pneumoniae*, and *S. aureus* were found to be vulnerable, as they were considerably inhibited at a concentration of 12.5 mg/mL [129]. In a similar study, the *B. scopulosa* MeOH extract was tested for its inhibitory effect on *S. aureus*, *P. aeruginosa*, and *E. coli*, through zone-inhibition assay. It was interesting to note that the inhibitory impact on *S. aureus* was stronger than that on *P. aeruginosa* and *E. coli* [134].

7.11. Antimalarial Activity

Malaria is a notorious disease and one of the main causes of high morbidity and mortality in many tropical and subtropical areas. The ethnopharmacological relevance of the *Bergenia* species for treating fever has been time-tested. EtOH leaf extracts *B. ciliata* (ELEBC) has been tested for its antiplasmodial (*Plasmodium berghei*) activity using a rodent-malaria model, along with chloroquine (10 μM) as a positive control. The IC_{50} of ELEBC was found to be less than 10 μg/mL. Thus, both the in vitro and in vivo experiments have confirmed the antimalarial activity of ELEBC [43].

7.12. Hepatoprotective Activity

Bergenia species do possess hepatoprotective potential. In a study, the EtOH root-extract of *B. ligulata* was evaluated for its hepatoprotective activity in CCl_4 treated (toxicant) albino rats. The estimation of hepatoprotective activity was confirmed by measuring the decline in the elevated levels of serum marker-enzymes such as SGPT, SGOT, ALP, and total bilirubin levels [107]. In another study conducted by Mansoor et al. (2015), the *B. ligulata* leaf extract (dose of 500 mg/kg) fully restored the carbon tetrachloride (potent hepatotoxicant)-induced variations in carbon tetrachloride intoxicated rats [154]. Moreover, the histopathological examination of the liver tissue further confirmed the hepatoprotective effect [154]. *B. crassifolia* dry extract has also been reported to exhibit hepatoprotective property in rats intoxicated with 4-pentenoic acid, thus confirming its hepatoprotective potential [155].

7.13. Antiulcer Activity

In some areas of South East Asia, *B. ciliata* has been used in the treatment of stomach disorders as a folkloric medicine. An experiment was performed to assess the gastro-protective activity of *B. ciliata* extracts on stomach ulcer-induced rats. Different doses (15, 30, and 60 mg/kg) of the aqueous and MeOH rhizome extracts were administered 1 h after the ulcerogenic treatment. Among the two treatments, the aqueous extract reduced the stomach-ulcer lesions to a better degree. It was concluded that the rhizome extract exhibited its cytoprotective effect (anti-ulcer activity) by facilitating the improvement of gastric mucosal barrier [75].

7.14. Anticancer Activity

Bergenia ciliata rhizome extracts (MeOH and aqueous) were tested for their cytotoxicity on human breast, liver, and prostate cancer cell-lines by XTT assay, respectively. Both the extracts exhibited concentration-dependent toxicity in each of three cell lines [156]. The IC_{50} value of both extracts fell within the acceptable range in all cell-lines (except Hep 3B cell-lines). Thus, Bergenias possess potential antineoplastic activity that may have probable clinical use as preventive medicine [76,77].

7.15. Antioxidant Activity

Undoubtedly, *Bergenia* species are an excellent source of antioxidants. *B. ciliata* MeOH leaf extract has been reported to be a potent free-radical scavenger (EC_{50} of 36.24 μg/mL), as confirmed through DPPH assay [78,157]. *B. ligulata* also possess considerable antioxidant activity, as confirmed by DPPH assay

(IC$_{50}$ value: 50 μg/mL) [93]. Ivanov et al. (2011) reported that the antioxidant properties of *B. crassifolia* is due to the presence of two compounds, (+)-catechin-3,5-di-*O*-gallate (**82**) and (+)-catechin-3-*O*-gallate (**83**). They were isolated from its aqueous EtOH leaf extract and exhibited strong antioxidant properties, as determined by DPPH assay, with SC$_{50}$ = 1.04 and 1.33 g/mL, respectively [72].

Shilova et al. (2006) performed a study using green and black leaves EtOH extracts of *B. crassifolia* and examined the oxygen uptake rate in a gasometric system with 2,2′-azobisisobutyronitrile-initiated oxidation of isopropylbenzene. The green leaves showed the most pronounced antioxidant effect [158]. In another study, the separation of main phenolic compounds of *B. crassifolia* followed by their DPPH assay with the post-chromatographic derivatization of TLC plates. The increasing order of the free-radical scavenging activity was found to be gallic acid > arbutin > ellagic acid > hydroquinone > ascorbic acid [94]. A comparative assessment of the antioxidant activity, free radical scavenging activity, and inhibition of lipid-peroxidation using MeOH and aqueous extracts of *B. ciliata* rhizomes was performed. The MeOH extract exhibited a better antioxidant activity [76].

7.16. Antiobesity Activity

It was reported by Ivanov et al. (2011) that crude extracts of *B. crassifolia* rhizomes can efficiently suppress the human pancreatic lipase activity (IC$_{50}$ = 3.4 g/mL) in vitro [72]. The *B. crassifolia* leaf extracts are known to suppress the appetite as well as energy intake in rats suffering from high-calorie diet-induced obesity. Compared to controls, a 40% reduction in the daily dietary consumption of the rats tested with 50 mg/kg *Bergenia* aqueous leaf extract (seven days of oral treatment) was observed. Moreover, a reasonable reduction (45%) in the triglyceride level was also observed after seven-day therapy [159]. 3,11-Di-*O*-galloylbergenin (**64**), a galloylbergenin from *B. crassifolia* roots has been reported (using MC3T3-G2/PA6 murine preadipocytes) to exhibit a moderate anti-lipid accumulation activity [160].

7.17. Adaptogenic Activity

An adaptogen increases the resistance power against various stresses such as physical, chemical, or biological stress and has a stabilizing effect on the body functions [161]. *B. crassifolia* can also be considered as a promising phytoadaptogen [53,55]. In a treadmill test, the running-time of rats fed (for 10 days) on 300 mg/kg *Bergenia* black leaves extract was elevated by 30% more than the control group. The running-time was similar to that of rats administered with 5 mL/kg of extract of *Eleutherococcus senticosus* [162]. Similarly, the swimming capacity of the mice treated with infusions prepared from *B. crassifolia* fermented leaves was observed to significantly increase by 2.2-fold, compared to the control. The swimming capacity was increased with a simultaneous increase in glucose utilization and without changing the body weight [163]. A similar study revealed that the endurance capability of rats exposed to a very low temperature of −15 °C (3 h, for 21 days) was significantly ameliorated after treatment with extracts of *Bergenia* black-leaves. Moreover, the floating-time of the rats supplemented with 100 mg/kg extract was considerably augmented after 21 days of treatment, whereas in the other group treated with liposome-encapsulated-extract the swimming-time was increased after seven days of treatment, under extreme circumstances (e.g., hypoxia) [164], because, under hypoxic conditions, the adaptive response of an organism activates mitoK$_{ATP}$ channel and increases the ATP-dependent potassium transport in mitochondria. Mironova et al. explored the the activation ability of mitoK$_{ATP}$ channel through water-soluble flavonoid-containing plant preparations of *Bergenia* (*Bergenia crassifolia*) in a rat model [165].

8. Other Benefits of *Bergenia* Species

Bergenias are a reservoir of nutrients and are therefore used in culinary preparations [63]. Furthermore, the arbutin (**10**) content of Bergenias inhibits the degradation of insulin and is useful for diuresis and can work as a urinary disinfectant [56]. Bergenias are also being used in the field of cosmetics, owing to the presence of arbutin [166]. The arbutin can make skin whiten because it can prevent tyrosinase activity and can reduce the skin's melanin (pigment) production [14,167]. *B. ligulata* is used for manufacturing cosmetic brightening agents and under-eye creams [23]. *B. emeinensis* extracts have also been used to treat skin wrinkles [168].

9. Conclusions and Future Perspectives

It is quite evident from this review that the *Bergenia* species contains a wide range of bioactive compounds of therapeutic value. The safety and efficacy of *Bergenia* leaves and rhizomes has been time-tested and documented during the long-period of traditional use. However, there is still a scope of research on the mechanism of action of several other aforementioned therapeutic activities. Moreover, among the 32 species, only nine species have been experimentally reported to possess the pharmacological properties. There is a scope for phytochemical analysis and clinical efficacy trials with the rest of the 23 species. To date, 152 compounds have been isolated and characterized from the genus *Bergenia*.

The studies done so far on Bergenias have focused on investigation and assessment of germplasm resources, functional credentials of extracts and isolation of bioactive components, but the reports on cytological and molecular researches and standardization of plant-extracted drugs for product-development are still fragmentary. *B. hissarica* and *B. tianquanensis* are extremely rare species with very few reports on their biological activities. Therefore, the conservation of the *Bergenia* species is of immense concern from a biodiversity, ethnobotanical, and pharmacological perspective. Although the research is progressing on *Bergenia* species, their robust tissue culture protocols are yet to be discovered, as the publications [97,169–174] on tissue culture and germplasm maintenance activities are fragmentary (Supplementary Table S2). The present study proposes a wide scope for multiple benefits of *Bergenia* in the field of floriculture, health foods, pharmaceuticals, cosmetics, and many other industrial and economic ventures. To conclude, *Bergenia* species have huge potential to act as a panacea to numerous health-related maladies, and therefore their conservation is necessary.

Supplementary Materials: The following are available online: Figure S1: Chemical structures of isolated and characterized phytochemicals from *Bergenia* species, Table S1: Botanical description of *Bergenia* species, Table S2: Tissue culture reports of *Bergenia* species.

Author Contributions: Conceptualization, B.K. and D.Y.; methodology, B.K.; software, A.K.; writing—original draft preparation, B.K., D.Y.; writing—review and editing, B.K.; D.Y., J.-O.J. All authors have read and agreed to the published version of the manuscript.

Funding: This work was supported by the National Research Foundation of Korea (NRF) funded by the Ministry of Education (NRF2019R1G1A1008566) and (NRF-2020R1A6A1A03044512).

Acknowledgments: The authors are thankful to Lovely Professional University (LPU), Punjab, India for the infrastructural support. Authors are also thankful to the CROM-MASS, CIBIMOL-CENIVAM, Industrial University of Santander, Colombia, for awarding the postdoctoral fellowship (Apoyo a estancias postdoctorales-UIS) to Arvind Kumar.

Conflicts of Interest: The authors declare that there is no conflict of interest regarding publication.

References

1. Koul, B. *Herbs for Cancer Treatment*, 1st ed.; Springer: New York, NY, USA, 2020.
2. Cragg, G.M.; Newman, D.J. Natural products: A continuing source of novel drug leads. *Biochim. Biophys. Acta Gen. Subj.* **2013**, *1830*, 3670–3695. [CrossRef]
3. Pandey, M.; Rastogi, S.; Rawat, A. Indian traditional ayurvedic system of medicine and nutritional supplementation. *Evid. Based Complement. Alternat. Med.* **2013**, *2013*, 1–12. [CrossRef]
4. Sen, S.; Chakraborty, R. Toward the integration and advancement of herbal medicine: A focus on traditional Indian medicine. *Bot. Target Ther.* **2015**, *5*, 33–44. [CrossRef]
5. Kasote, D.M.; Jagtap, S.D.; Thapa, D.; Khyade, M.S.; Russell, W.R. Herbal remedies for urinary stones used in India and China: A review. *J. Ethnopharmacol.* **2017**, *203*, 55–68. [CrossRef]
6. Liu, Y.; Chen, Y.; Liao, B.; Luo, D.; Wang, K.; Li, H.; Zeng, G. Epidemiology of urolithiasis in Asia. *Asian J. Urol.* **2018**, *5*, 205–214. [CrossRef]

7. Vitale, C.; Croppi, E.; Marangella, M. Biochemical evaluation in renal stone disease. *Clin. Cases Miner. Bone Metab.* **2008**, *5*, 127.

8. Ramello, A.; Vitale, C.; Marangella, M. Epidemiology of nephrolithiasis. *J. Nephrol.* **2000**, *13* (Suppl. S3), S45–S50.

9. Sharma, I.; Khan, W.; Parveen, R.; Alam, M.; Ahmad, I.; Ansari, M.H.R.; Ahmad, S. Antiurolithiasis activity of bioactivity guided fraction of *Bergenia ligulata* against ethylene glycol induced renal calculi in rat. *Biomed. Res. Int.* **2017**, *2017*, 1–11.

10. Wadkar, K.A.; Kondawar, M.S.; Lokapure, S.G. Standardization of marketed cystone tablet: A herbal formulation. *J. Pharmacogn. Phytochem.* **2017**, *6*, 10–16.

11. Ahmad, M.; Butt, M.A.; Zhang, G.; Sultana, S.; Tariq, A.; Zafar, M. *Bergenia ciliata*: A comprehensive review of its traditional uses, phytochemistry, pharmacology and safety. *Biomed. Pharmacother.* **2018**, *97*, 708–721. [CrossRef]

12. Ruby, K.; Chauhan, R.; Dwivedi, J. Himalayan bergenia a comprehensive review. *Int. J. Pharm. Sci.* **2012**, *14*, 139–141.

13. Srivastava, S.; Rawat, A.K.S. Botanical and phytochemical comparison of three bergenia species. *J. Sci. Ind. Res.* **2008**, *67*, 65–72.

14. Árok, R.; Végh, K.; Alberti, Á.; Kéry, Á. Phytochemical comparison and analysis of *Bergenia crassifolia* l.(fritsch.) and *Bergenia cordifolia* sternb. *Eur. Chem. Bull.* **2012**, *1*, 31–34.

15. de Oliveira, C.M.; Nonato, F.R.; de Lima, F.O.; Couto, R.D.; David, J.P.; David, J.M.; Soares, M.B.P.; Villarreal, C.F. Antinociceptive properties of bergenin. *J. Nat. Prod.* **2011**, *74*, 2062–2068. [CrossRef]

16. Dhalwal, K.; Shinde, V.; Biradar, Y.; Mahadik, K. Simultaneous quantification of bergenin, catechin, and gallic acid from *Bergenia ciliata* and *Bergenia ligulata* by using thin-layer chromatography. *J. Food Compos. Anal.* **2008**, *21*, 496–500. [CrossRef]

17. Li, F.; Zhou, D.; Qin, X.; Zhang, Z.-R.; Huang, Y. Studies on the physicochemical properties of bergenin. *Chin. Pharm. J.* **2009**, *44*, 92–95.

18. Rastogi, S.; Rawat, A. A comprehensive review on bergenin, a potential hepatoprotective and antioxidative phytoconstituent. *Herba Polonica* **2008**, *54*, 66–79.

19. Singh, D.P.; Srivastava, S.K.; Govindarajan, R.; Rawat, A.K.S. High-performance liquid chromatographic determination of bergenin in different *bergenia* species. *Acta Chromatogr.* **2007**, *19*, 246–252.

20. Nazir, N.; Koul, S.; Qurishi, M.A.; Najar, M.H.; Zargar, M.I. Evaluation of antioxidant and antimicrobial activities of bergenin and its derivatives obtained by chemoenzymatic synthesis. *Eur. J. Med. Chem.* **2011**, *46*, 2415–2420. [CrossRef]

21. Rousseau, C.; Martin, O.R. Synthesis of bergenin-related natural products by way of an intramolecular c-glycosylation reaction. *Tetrahedron: Asymmetry* **2000**, *11*, 409–412. [CrossRef]

22. Suh, K.S.; Chon, S.; Jung, W.W.; Choi, E.M. Effect of bergenin on rankl-induced osteoclast differentiation in the presence of methylglyoxal. *Toxicol. In Vitro* **2019**, *61*, 104613. [CrossRef]

23. Gurav, S.; Gurav, N. A comprehensive review: *Bergenia ligulata* wall-a controversial clinical candidate. *Int. J. Pharm. Sci. Rev. Res.* **2014**, *5*, 1630–1642.

24. Singh, N.; Gupta, A.; Juyal, V. A review on *Bergenia ligulata* wall. *IJCAS* **2010**, *1*, 71–73.

25. Chitme, H.R.; Alok, S.; Jain, S.; Sabharwal, M. Herbal treatment for urinary stones. *Int. J. Pharm. Sci. Res.* **2010**, *1*, 24–31.

26. Chandrareddy, U.D.; Chawla, A.S.; Mundkinajeddu, D.; Maurya, R.; Handa, S.S. Paashaanolactone from *Bergenia ligulata*. *Phytochemistry* **1998**, *47*, 907–909. [CrossRef]

27. Khan, M.Y.; Vimal, K.V. Phytopharmacological and chemical profile of *Bergenia ciliate*. *Int. J. Phytopharm.* **2016**, *6*, 90–98.

28. Hendrychová, H.; Tůmová, L. Bergenia genus-content matters and biological activity. *Ceska a Slovenska farmacie Casopis Ceske farmaceuticke spolecnosti a Slovenske farmaceuticke spolecnosti* **2012**, *61*, 203–209.

29. Liu, S.J.; Yu, B.; Hu, C.H. In The variation of pod activities in *Bergenla tianquanensis* in tissue culture progress. In *Advanced Materials Research*; Trans Tech Publications Ltd.: Stafa-Zurich, Switzerland, 2011; pp. 196–200.

30. Wu, Z.-Y.; Raven, P.H. *Flora of China*; Science Press (Beijing) & Missouri Botanical Garden Press: St. Louis, MO, USA, 2001; Volume 8.

31. Zhang, Y.; Liao, C.; Liu, X.; Li, J.; Fang, S.; Li, Y.; He, D. Biological advances in bergenia genus plant. *Afr. J. Biotechnol.* **2011**, *10*, 8166–8169.

32. Jin-tang, P. New taxa of the genus bergenia from Hengduan mountains. *Acta Phytotax. Sin.* **1994**, *32*, 571–573.

33. Jin-tang, P.; Soltis, D.E. Flora China. *Bergenia* **2001**, *8*, 278–280.

34. Zhou, G.Y.; Li, W.C.; Guo, F.G. Resource investigation and observation of biological characteristics of *Bergenia purpurascens* (Hook. f. et. Thoms.). Engl. *Chin. Agric. Sci. Bull.* **2007**, *23*, 390–392.

35. Alok, S.; Jain, S.K.; Verma, A.; Kumar, M.; Sabharwal, M. Pathophysiology of kidney, gallbladder and urinary stones treatment with herbal and allopathic medicine: A review. *Asian Pac. J. Trop. Dis.* **2013**, *3*, 496–504. [CrossRef]

36. Chowdhary, S.; Verma, D.; Kumar, H. Biodiversity and traditional knowledge of *Bergenia* spp. In kumaun himalaya. *Sci. J.* **2009**, *2*, 105–108.

37. Rajbhandari, M.; Mentel, R.; Jha, P.; Chaudhary, R.; Bhattarai, S.; Gewali, M.; Karmacharya, N.; Hipper, M.; Lindequist, U. Antiviral activity of some plants used in nepalese traditional medicine. *Evid. Based Complement. Alternat. Med.* **2009**, *6*, 517–522. [CrossRef]

38. Kumar, V.; Tyagi, D. Review on phytochemical, ethnomedical and biological studies of medically useful genus bergenia. *Int. J. Curr. Microbiol. App. Sci* **2013**, *2*, 328–334.

39. Patel, A.M.; Kurbetti, S.; Savadi, R.; Thorat, V.; Takale, V.; Horkeri, S. Preparation and evaluation of wound healing activity of new polyherbal formulations in rats. *Am. J. Phytomed. Clin. Ther.* **2013**, *1*, 498–506.

40. Raina, R.; Prawez, S.; Verma, P.; Pankaj, N. Medicinal plants and their role in wound healing. *Vet. Scan.* **2008**, *3*, 1–7.

41. Shakya, A.K. Medicinal plants: Future source of new drugs. *Int. J. Herb. Med.* **2016**, *4*, 59–64.

42. Singh, K.J.; Thakur, A.K. Medicinal plants of the shimla hills, himachal pradesh: A survey. *Int. J. Herbal Med.* **2014**, *2*, 118–127.

43. Walter, N.S.; Bagai, U.; Kalia, S. Antimalarial activity of *Bergenia ciliata* (haw.) sternb. against *Plasmodium berghei*. *Parasitol. Res.* **2013**, *112*, 3123–3128. [CrossRef]

44. Bahu, C.P.; Seshadri, R.T. *Advances in Research in "Indian Medicine; "Pashanbedi" Drugs for Urinary Calculus*; Udupa, K.N., Ed.; Banaras Hindu University: Varanasi, India, 1970; pp. 77–98.

45. Manandhar, N.P. A survey of medicinal plants of jajarkot district, Nepal. *J. Ethnopharmacol.* **1995**, *48*, 1–6. [CrossRef]

46. Kapur, S. Ethno-medico plants of kangra valley (Himachal Pradesh). *J. Econ. Taxon. Bot.* **1993**, *17*, 395–408.

47. Mukerjee, T.; Bhalla, N.; Singh, A.; Jain, H. Herbal drugs for urinary stones. *Indian Drugs* **1984**, *21*, 224–228.

48. Shah, N.; Jain, S. Ethnomedico-botany of the kumaon himalaya, india. *Soc. Pharmacol.* **1988**, *2*, 359–380.

49. Rani, S.; Rana, J.C. Ethnobotanical uses of some plants of bhattiyat block in district chamba, Himachal Pradesh (Western Himalaya). *Ethnobot. Res. Appl.* **2014**, *12*, 407–414. [CrossRef]

50. Koelz, W.N. Notes on the ethnobotany of lahul, a province of the Punjab. *Q. J. Crude Drug Res.* **1979**, *17*, 1–56. [CrossRef]

51. Vereschagin, V.; Sobolevskaya, K.; Yakubova, A. *Useful Plants of West Siberia*; Publishing of Academy of Science of USSR: Moscow-Leningrad, Russia, 1959.

52. Gammerman, A.; Kadaev, G.; Yacenko-Khmelevsky, A. *Medicinal Plants (Herbs-Healers)*; High School: Moscow, Russia, 1984.

53. Panossian, A.G. Adaptogens: Tonic herbs for fatigue and stress. *Altern. Complement. Ther.* **2003**, *9*, 327–331. [CrossRef]

54. Sokolov, S.Y. *Phytotherapy and Phytopharmacology: The Manual for Doctors*; Medical News Agency: Moscow, Russia, 2000; pp. 197–199.

55. Suslov, N.; Churin, A.; Skurikhin, E.; Provalova, N.; Stal'bovskiĭ, A.; Litvinenko, V.; Dygaĭ, A. Effect of natural nootropic and adaptogen preparations on the cortex bioelectrical activity in rats. *Eksp. Klin. Farmakol.* **2002**, *65*, 7–10.

56. Li, W.-C.; Gou, F.-G.; Zhang, L.-M.; Yu, H.-M.; Li, X.; Lin, C. The situation and prospect of research on *Bergenia purpurascens*. *J. -Yunnan Agric. Univ.* **2006**, *21*, 845.

57. Pokhrel, P.; Parajuli, R.R.; Tiwari, A.K.; Banerjee, J. A short glimpse on promising pharmacological effects of *Begenia ciliata*. *J. Appl. Pharm. Res.* **2014**, *2*, 1–6.

58. Xie, G.; Zhou, J.; Yan, X. *Encyclopedia of Traditional Chinese Medicines: Molecular Structures, Pharmacological Activities, Natural Sources and Applications*; Springer: Berlin/Heidelberg, Germany, 2011; Volume 2.

59. Chen, Y.; Jia, X.; Zhang, Y. Studies on chemical compositions of *Bergenia scopulosa* T. P. Wang. *J. Chin. Med. Mater.* **2008**, *31*, 1006–1007.

60. Hasan, A.; Husain, A.; Khan, M.A. Flavonol glycosides from leaves of *Bergenia himalaica*. *Asian J. Chem.* **2005**, *17*, 822.

61. Saijyo, J.; Suzuki, Y.; Okuno, Y.; Yamaki, H.; Suzuki, T.; Miyazawa, M. A-glucosidase inhibitor from *Bergenia ligulata*. *J. Oleo Sci.* **2008**, *57*, 431–435. [CrossRef] [PubMed]

62. Xin-Min, C.; Yoshida, T.; Hatano, T.; Fukushima, M.; Okuda, T. Galloylarbutin and other polyphenols from *Bergenia purpurascens*. *Phytochemistry* **1987**, *26*, 515–517. [CrossRef]

63. Yang, X.; Wang, Z.; Wang, Z.; Li, R. Analysis of nutritive components and mineral element of *Bergenae pacumbis* intibet. *J. Chang. Veg.* **2009**, *22*, 57–58.

64. Carmen, P.; Vlase, L.; Tamas, M. Natural resources containing arbutin. Determination of arbutin in the leaves of *Bergenia crassifolia* (L.) fritsch. Acclimated in romania. *Not. Bot. Horti Agrobot. Cluj-Napoca* **2009**, *37*, 129–132.

65. Chen, J.; Li, Y.; Cai, L. Determination of total flavonoids in *Bergenia emeiensis* leaf and rhizome by spectrophotometry. *J. China West Norm. Univ. (Nat. Sci.)* **2008**, *29*, 141–143.

66. Lu, X. Studies on chemical compositions of *Bergenia scopulosa* TP Wang. *Zhong Yao Cai* **2003**, *26*, 791–792.

67. Wang, J.; Lu, X. Studies on chemical compositions of *Bergenia scopulosa* T. P. Wang. *J. Chin. Med. Mater.* **2005**, *28*, 23–24. [CrossRef]

68. Lim, Y.-J.; Lee, E.H.; Kang, T.H.; Ha, S.K.; Oh, M.S.; Kim, S.M.; Yoon, T.-J.; Kang, C.; Park, J.-H.; Kim, S.Y. Inhibitory effects of arbutin on melanin biosynthesis of α-melanocyte stimulating hormone-induced hyperpigmentation in cultured brownish guinea pig skin tissues. *Arch. Pharm. Res* **2009**, *32*, 367–373. [CrossRef]

69. Samant, S.; Pant, S. Diversity, distribution pattern and conservation status of the plants used in liver diseases/ailments in Indian himalayan region. *J. Mt. Sci.* **2006**, *3*, 28–47. [CrossRef]

70. Jiang, H.; Guo, F.; Zhang, L.; Chen, Y.; Yang, S. Comparison of bergenin contents of *Bergenia purpurascens* among different regions in yunnan province. *J. Yunnan Agric. Univ.* **2010**, *25*, 895–898.

71. Siddiqui, B.S.; Hasan, M.; Mairaj, F.; Mehmood, I.; Hafizur, R.M.; Hameed, A.; Shinwari, Z.K. Two new compounds from the aerial parts of *Bergenia himalaica* boriss and their anti-hyperglycemic effect in streptozotocin-nicotinamide induced diabetic rats. *J. Ethnopharmacol.* **2014**, *152*, 561–567. [CrossRef] [PubMed]

72. Ivanov, S.A.; Nomura, K.; Malfanov, I.L.; Sklyar, I.V.; Ptitsyn, L.R. Isolation of a novel catechin from bergenia rhizomes that has pronounced lipase-inhibiting and antioxidative properties. *Fitoterapia* **2011**, *82*, 212–218. [CrossRef] [PubMed]

73. Dharmender, R.; Madhavi, T.; Reena, A.; Sheetal, A. Simultaneous quantification of bergenin,(+)-catechin, gallicin and gallic acid; and quantification of β-sitosterol using hptlc from *Bergenia ciliata* (haw.) sternb. Forma ligulata yeo (pasanbheda). *Pharm. Anal. Acta* **2010**, *1*, 104. [CrossRef]

74. Sinha, S.; Murugesan, T.; Maiti, K.; Gayen, J.R.; Pal, M.; Saha, B. Evaluation of anti-inflammatory potential of *Bergenia ciliata* sternb. Rhizome extract in rats. *J. Pharm. Pharmacol.* **2001**, *53*, 193–196. [CrossRef] [PubMed]

75. Kakub, G.; Gulfraz, M. Cytoprotective effects of *Bergenia ciliata* sternb, extract on gastric ulcer in rats. *Phytother. Res.* **2007**, *21*, 1217–1220. [CrossRef]

76. Bhandari, M.R.; Jong-Anurakkun, N.; Hong, G.; Kawabata, J. α-glucosidase and α-amylase inhibitory activities of nepalese medicinal herb pakhanbhed (*Bergenia ciliata*, haw.). *Food Chem.* **2008**, *106*, 247–252. [CrossRef]

77. Mazhar-Ul-Islam, I.A.; Mazhar, F.; Usmanghani, K.; Gill, M.A. Evaluation of antibacterial activity of *Bergenia ciliata*. *Pak. J. Pharm. Sci.* **2002**, *15*, 21–27.

78. Bagul, M.S.; Ravishankara, M.; Padh, H.; Rajani, M. Phytochemical evaluation and free radical scavenging properties of rhizome of *Bergenia ciliata* (haw.) sternb. Forma ligulata yeo. *J. Nat. Remedies* **2003**, *3*, 83–89.

79. Sticher, O.; Soldati, F.; Lehmann, D. High-performance liquid chromatographic separation and quantitative determination of arbutin, methylarbutin, hydroquinone and hydroquinone-monomethylether in arctostaphylos, bergenia, calluna and vaccinium species [blueberry]. *Planta Med.* **1979**, *35*, 253–261. [CrossRef] [PubMed]

80. Fujii, M.; Miyaichi, Y.; Tomimori, T. Studies on nepalese crude drugs. XXII: On the phenolic constituents of the rhizome of *Bergenia ciliata* (haw.) sternb. *Nat. Med.* **1996**, *50*, 404–407.

81. Sinha, S.; Murugesan, T.; Maiti, K.; Gayen, J.; Pal, B.; Pal, M.; Saha, B. Antibacterial activity of *Bergenia ciliata* rhizome. *Fitoterapia* **2001**, *72*, 550–552. [CrossRef]

82. Mazhar-Ul-Islam, I.A.; Usmanghani, K.; Shahab-ud-Din, A. Antifungal activity evaluation of *Bergenia ciliata*. *Pak. J. Pharmacol.* **2002**, *19*, 1–6.

83. Chowdhary, S.; Verma, K. Some peculiar structures in bergenia species growing in western himalaya. *Nat. Sci.* **2010**, *8*, 1–4.

84. Rajkumar, V.; Guha, G.; Kumar, R.A.; Mathew, L. Evaluation of antioxidant activities of *Bergenia ciliata* rhizome. *Rec. Nat. Prod.* **2010**, *4*, 38–48.

85. Zhang, Y.; Liao, C.; Li, J.; Liu, X. A review on resource status, bioactive ingredients, clinical applications and biological progress in bergenia. *J. Med. Plant Res.* **2011**, *5*, 4396–4399.

86. Chauhan, R.; Ruby, K.; Dwivedi, J. *Bergenia ciliata* mine of medicinal properties: A review. *Int. J. Pharm. Sci. Rev. Res* **2012**, *15*, 20–23.

87. Chauhan, R.; Ruby, K.; Dwivedi, J. Golden herbs used in piles treatment: A concise report. *Int. J. Drug Dev. Res.* **2012**, *4*, 50–68.

88. Ruby, K.; Chauhan, R.; Sharma, S.; Dwivedi, J. Polypharmacological activities of bergenia species. *Int. J. Pharm. Sci. Rev. Res.* **2012**, *13*, 100–110.

89. Patel, A.M.; Savadi, R.V. Pharmacognostic and phytochemical evaluation of *Bergenia ciliata* rhizome. *Int. J. Pharm. Rev. Res.* **2014**, *4*, 52–55.

90. Ruby, K.; Sharma, S.; Chauhan, R.; Dwivedi, J. In-vitro antioxidant and hemorrhoidal potential of hydroethanolic leaf extracts of *Bergenia ciliata*, *Bergenia ligulata* and *Bergenia stracheyi*. *Asian J. Plant Sci. Res.* **2015**, *5*, 34–46.

91. Srivastava, N.; Srivastava, A.; Srivastava, S.; Rawat, A.K.S.; Khan, A.R. Simultaneous quantification of syringic acid and kaempferol in extracts of bergenia species using validated high-performance thin-layer chromatographic-densitometric method. *J. Chromatogr. Sci.* **2015**, *54*, 460–465. [PubMed]

92. Shikov, A.N.; Pozharitskaya, O.N.; Makarova, M.N.; Makarov, V.G.; Wagner, H. *Bergenia crassifolia* (l.) Fritsch-pharmacology and phytochemistry. *Phytomedicine* **2014**, *21*, 1534–1542. [CrossRef] [PubMed]

93. Vaishali, A.S.; Vikas, M.D.; Krishnapriya, M.; Sanjeevani, G. Identification of potential antioxidants by in-vitro activity guided fractionation of *Bergenia ligulata*. *Pharmacogn. Mag.* **2008**, *4*, 79–84.

94. Pozharitskaya, O.N.; Ivanova, S.A.; Shikov, A.N.; Makarov, V.G. Separation and evaluation of free radical-scavenging activity of phenol components of *Emblica officinalis* extract by using an HPTLC–DPPH method. *J. Sep. Sci.* **2007**, *30*, 1250–1254. [CrossRef]

95. Vasi, I.; Kalintha, V. Chemical analysis of *Bergenia lingulata* roots. *Comp. Physiol. Ecol.* **1981**, *6*, 127–128.

96. Ostrowska, B.; Gorecki, P.; Wolska, D. Investigation on possibility of utilization of bergenia leaves to therapeutics in place of arbutin and tannin raw materials deficiency. Pt. 2. Isolation of bergenin and a method of its quantitative determination [*Bergenia crassifolia*, *Bergenia cordifolia*]. *Herba Polonica (Poland)* **1989**, *35*, 117–122.

97. Furmanowa, M.; Rapczewska, L. Bergenia crassifolia (l.) fritsch (bergenia): Micropropagation and arbutin contents. In *Medicinal and Aromatic Plants IV*; Springer: Berlin/Heidelberg, Germany, 1993; pp. 18–33.

98. Golovchenko, V.; Bushneva, O.; Ovodova, R.; Shashkov, A.; Chizhov, A.; Ovodov, Y.S. Structural study of bergenan, a pectin from *Bergenia crassifolia*. *Russ. J. Bioorg. Chem.* **2007**, *33*, 47–56. [CrossRef]

99. Roselli, M.; Lentini, G.; Habtemariam, S. Phytochemical, antioxidant and anti-α-glucosidase activity evaluations of *Bergenia cordifolia*. *Phytother. Res.* **2012**, *26*, 908–914. [CrossRef]

100. Sun, X.; Huang, W.; Ma, M.; Guo, B.; Wang, G. Comparative studies on content of arbutin, bergenin and catechin in different part of *Bergenia purpurascens* and *B. crassifolia*. *China J. Chin. Mater. Med.* **2010**, *35*, 2079–2082.

101. Chernetsova, E.S.; Crawford, E.A.; Shikov, A.N.; Pozharitskaya, O.N.; Makarov, V.G.; Morlock, G.E. ID-CUBE direct analysis in real time high-resolution mass spectrometry and its capabilities in the identification of phenolic components from the green leaves of *Bergenia crassifolia* L. *Rapid Commun. Mass Spectrom.* **2012**, *26*, 1329–1337. [CrossRef] [PubMed]

102. Habtemariam, S. The hidden treasure in europe's garden plants: Case examples; *Berberis darwinni* and *Bergenia cordifolia*. *Med. Aromat. Plants* **2013**, *2*, 1–5.

103. Chernetsova, E.S.; Shikov, A.N.; Crawford, E.A.; Grashorn, S.; Laakso, I.; Pozharitskaya, O.N.; Makarov, V.G.; Hiltunen, R.; Galambosi, B.; Morlock, G.E. Characterization of volatile and semi-volatile compounds in green and fermented leaves of *Bergenia crassifolia* L. By gas chromatography-mass spectrometry and id-cube direct analysis in real time-high resolution mass spectrometry. *Eur. J. Mass Spectrom* **2014**, *20*, 199–205. [CrossRef] [PubMed]

104. Salminen, J.-P.; Shikov, A.N.; Karonen, M.; Pozharitskaya, O.N.; Kim, J.; Makarov, V.G.; Hiltunen, R.; Galambosi, B. Rapid profiling of phenolic compounds of green and fermented *Bergenia crassifolia* l. Leaves by UPLC-DAD-QqQ-MS AND HPLC-DAD-ESI-QTOF-MS. *Nat. Prod. Res.* **2014**, *28*, 1530–1533. [CrossRef] [PubMed]

105. Ogisu, M.; Rix, M. 572. Bergenia emeiensis: Saxifragaceae. *Curtis's Bot. Mag.* **2007**, *24*, 2–6. [CrossRef]

106. Bashir, S.; Gilani, A.H. Antiurolithic effect of *Bergenia ligulata* rhizome: An explanation of the underlying mechanisms. *J. Ethnopharmacol.* **2009**, *122*, 106–116. [CrossRef]

107. Singh, N.; Juyal, V.; Gupta, A.K.; Gahlot, M. Evaluation of ethanolic extract of root of *Bergenia ligulata* for hepatoprotective, diuretic and antipyretic activities. *J. Pharm. Res.* **2009**, *2*, 958–960.

108. Singh, N.; Juyal, V.; Gupta, A.; Gahlot, M.; Prashant, U. Antidiabetic activity of ethanolic extract of root of *Bergenia ligulata* in alloxan diabetic rats. *Indian Drugs* **2009**, *46*, 247–249.

109. Kashima, Y.; Yamaki, H.; Suzuki, T.; Miyazawa, M. Insecticidal effect and chemical composition of the volatile oil from *Bergenia ligulata*. *J. Agric. Food Chem.* **2011**, *59*, 7114–7119. [CrossRef]

110. Sajad, T.; Zargar, A.; Ahmad, T.; Bader, G.; Naime, M.; Ali, S. Antibacterial and anti-inflammatory potential *Bergenia ligulata*. *Am. J. Biomed. Sci.* **2010**, *2*, 313–321. [CrossRef]

111. Nardev, S.; Gupta, A.; Vijay, J.; Renu, C. Study on antipyretic activity of extracts of *Bergenia ligulata* wall. *Int. J. Pharma Bio Sci.* **2010**, *1*, 1–5.

112. Chauhan, R.; Saini, R.; Dwivedi, J. Secondary metabolites found in bergenia species: A compendious review. *Int. J. Pharm. Pharm. Sci.* **2013**, *5*, 9–16.

113. Rajbhandari, M.; Wegner, U.; Schoepke, T.; Lindequist, U.; Mentel, R. Inhibitory effect of *Bergenia ligulata* on influenza virus A. *Pharmazie* **2003**, *58*, 268–271.

114. Jain, M.; Gupta, K. Isolation of bergenin from *Saxifraga ligulata* wall. *J. Ind. J. Chem. Soc.* **1962**, *39*, 559–560.

115. Tucci, A.P.; Delle, F.M.; Marini-Bettolo, G.B. The occurrence of (+) afzelechin in *Saxifraga ligulata* wall. *Ann. Ist Super Sanita* **1969**, *5*, 555–556.

116. Bahl, C.; Murari, R.; Parthasarathy, M.; Seshadri, T. Components of *Bergenia strecheyi* & *Bergenia ligulata*. *Indian J. Chem.* **1974**, *12*, 1038–1039.

117. Gehlot, N.; Sharma, V.; Vyas, D. Some pharmacological studies on ethanolic extract of *Bergenia ligulata*. *Indian J. Pharmacol.* **1976**, *8*, 92–94.

118. Dix, B.; Srivastava, S. Tannin constituents of *Bergenia ligulata* roots. *Ind. J. Nat. Prod.* **1989**, *5*, 24–25.

119. Reddy, U.D.C.; Chawla, A.S.; Deepak, M.; Singh, D.; Handa, S.S. High pressure liquid chromatographic determination of bergenin and (+)-afzelechin from different parts of paashaanbhed (*Bergenia ligulata* yeo). *Phytochem. Anal.* **1999**, *10*, 44–47. [CrossRef]

120. Chauhan, S.K.; Singh, B.; Agrawal, S. Simultaneous determination of bergenin and gallic acid in *Bergenia ligulata* wall by high-performance thin-layer chromatography. *J. AOAC Int.* **2000**, *83*, 1480–1483. [CrossRef] [PubMed]

121. Joshi, V.S.; Parekh, B.B.; Joshi, M.J.; Vaidya, A.D. Inhibition of the growth of urinary calcium hydrogen phosphate dihydrate crystals with aqueous extracts of tribulus terrestris and *Bergenia ligulata*. *Urol. Res.* **2005**, *33*, 80–86. [CrossRef] [PubMed]

122. Kumar, S. Herbaceous flora of Jaunsar-Bawar (Uttarkhand), India: Enumerations. *Phytotaxonomy* **2012**, *12*, 33–56.

123. Goswami, P.K.; Samant, M.; Srivastava, R.S. Multi faceted *Saxifraga ligulata*. *Int. J. Res. Ayurveda Pharm.* **2013**, *4*, 608–611. [CrossRef]

124. Jani, S.; Shukla, V.J.; Harisha, C. Comparative pharmacognostical and phytochemical study on *Bergenia ligulata* wall. and *Ammania buccifera* linn. *Ayu* **2013**, *34*, 406–410. [CrossRef]

125. Agnihotri, V.; Sati, P.; Jantwal, A.; Pandey, A. Antimicrobial and antioxidant phytochemicals in leaf extracts of *Bergenia ligulata*: A himalayan herb of medicinal value. *Nat. Prod. Res.* **2015**, *29*, 1074–1077. [CrossRef]

126. Messaoudi, D.; Bouriche, H.; Demirtas, I.; Senator, A. Phytochemical analysis and hepatoprotective activity of Algerian *Santolina chamaecyparissus*. Extracts. *Annu. Res. Rev. Biol.* **2018**, *25*, 1–12. [CrossRef]

127. Pushpalatha, H.B.; Pramod, K.; Devanathan, R.; Sundaram, R. Use of bergenin as an analytical marker for standardization of the polyherbal formulation containing *Saxifraga ligulata*. *Pharmacogn. Mag.* **2015**, *11*, S60. [CrossRef]

128. Zuo, G.-Y.; Li, Z.-Q.; Chen, L.-R.; Xu, X.-J. In vitro anti-hcv activities of *Saxifraga melanocentra* and its related polyphenolic compounds. *Antivir. Chem. Chemother.* **2005**, *16*, 393–398. [CrossRef]

129. Bajracharya, G.B.; Maharjan, R.; Maharjan, B.L. Potential antibacterial activity of *Bergenia purpurascens*. *Nepal J. Sci. Technol.* **2011**, *12*, 157–162. [CrossRef]

130. Chen, W.; Nie, M. HPLC determination of bergenin in *Astilbe chinensis* (maxim.) franch. Et sav. And *Bergenia purpurascens* (hook. F. Et thoms.) engl. *Acta Pharm. Sin.* **1988**, *23*, 606–609.

131. Li, B.-H.; Wu, J.-D.; Li, X.-L. LC–MS/MS determination and pharmacokinetic study of bergenin, the main bioactive component of *Bergenia purpurascens* after oral administration in rats. *J. Pharm. Anal.* **2013**, *3*, 229–234. [CrossRef] [PubMed]

132. Ren, Y.; Cao, L.; Chang, L.; Zhi, X.; Yuan, L.; Sheng, N.; Zhang, L.-T. Simultaneous determination of nine compounds in *Bergenia purpurascens* by HPLC-MS. *Chin. Pharm. J.* **2013**, *6*, 477–481.

133. Shi, X.; Li, X.; He, J.; Han, Y.; Li, S.; Zou, M. Study on the antibacterialactivity of *Bergenia purpurascens* extract. *Afr. J. Tradit. Complement. Altern. Med.* **2014**, *11*, 464–468. [CrossRef]

134. Ma, L. The antibacterial activity and antibacterial mechanism of *Bergenia scopulosa* TP Wang extract. *Adv. J. Food Sci. Technol.* **2014**, *6*, 994–997. [CrossRef]

135. Cui, Y. Chemical constituents from rhizomes of *Bergenia scopulosa* (ii). *Chin. Tradi. Herb. Drugs* **2012**, *43*, 1704–1707.

136. Yao-yuan, W. Chemical constituents from Bergenia scopulosa (I). *Chin. J. Exp. Tradit. Med Formulae* **2012**, *9*, 154–156.

137. Wei, L.; Si, M.; Long, M.; Zhu, L.; Li, C.; Shen, X.; Wang, Y.; Zhao, L.; Zhang, L. *Rhizobacter bergeniae* sp. Nov., isolated from the root of *Bergenia scopulosa*. *Int. J. Syst. Evol. Microbiol.* **2015**, *65*, 479–484. [CrossRef]

138. Li, S.; Liu, G.; Zhang, Y.; Xu, J. Experimental study on antitussive effect or arbutin. *Yaoxue Tongbao* **1982**, *17*, 720–722.

139. Kumar, V.; Tyagi, D. Antifungal activity evaluation of different extracts of *Bergenia stracheyi*. *Int. J. Curr. Microbiol. App. Sci.* **2013**, *2*, 69–78.

140. Kumar, V.; Tyagi, D. Phytochemical screening and free-radical scavenging activity of *Bergenia stracheyi*. *J. Pharmacogn. Phytochem.* **2013**, *2*, 175–180.

141. Ali, I.; Bibi, S.; Hussain, H.; Bano, F.; Ali, S.; Khan, S.W.; Ahmad, V.U.; Al-Harrasi, A. Biological activities of *Suaeda heterophylla* and *Bergenia stracheyi*. *Asian Pac. J. Trop. Dis.* **2014**, *4*, S885–S889. [CrossRef]

142. Yuldashev, M.; Batirov, È.K.; Malikov, V. Anthraquinones of *Bergenia hissarica*. *Chem. Nat. Compd.* **1993**, *29*, 543–544. [CrossRef]

143. Izhaki, I. Emodin-a secondary metabolite with multiple ecological functions in higher plants. *New Phytol.* **2002**, *155*, 205–217. [CrossRef]

144. Garimella, T.; Jolly, C.; Narayanan, S. In vitro studies on antilithiatic activity of seeds of *Dolichos biflorus* linn. and rhizomes of *Bergenia ligulata* wall. *Phytother. Res.* **2001**, *15*, 351–355. [CrossRef]

145. Nagal, A.; Singla, R.K. Herbal resources with antiurolithiatic effects: A review. *Indo Glob. J. Pharm. Sci.* **2013**, *3*, 6–14.

146. Satish, H.; Umashankar, D. Comparative study of methanolic extract of *Bergenia ligulata* yeo., with isolated constituent bergenin in urolithiatic rats. *Biomed* **2006**, *1*, 80–86.

147. Voloboy, N.; Smirnov, I.; Bondarev, A. Features of diuretic activity of arbutin and hydroquinone. *Sib. Med. J.* **2012**, *27*, 131–134.

148. Naik, S.; Kalyanpur, S.; Sheth, U. Effects of anti-inflammatory drugs on glutathione levels and liver succinic dehydrogenase activity in carrageenin edema and cotton pellet granuloma in rats. *Biochem. Pharmacol.* **1972**, *21*, 511–516. [CrossRef]

149. Churin, A.; Masnaia, N.; Sherstoboev, E.Y.; Suslov, N. Effect of *Bergenia crassifolia* extract on specific immune response parameters under extremal conditions. *Eksp. Klin. Farmakol.* **2005**, *68*, 51–54.

150. Popov, S.; Popova, G.Y.; Nikolaeva, S.Y.; Golovchenko, V.; Ovodova, R. Immunostimulating activity of pectic polysaccharide from *Bergenia crassifolia* (L.) fritsch. *Phytother. Res.* **2005**, *19*, 1052–1056. [CrossRef] [PubMed]

151. Nazir, N.; Koul, S.; Qurishi, M.A.; Taneja, S.C.; Ahmad, S.F.; Bani, S.; Qazi, G.N. Immunomodulatory effect of bergenin and norbergenin against adjuvant-induced arthritis-a flow cytometric study. *J. Ethnopharmacol.* **2007**, *112*, 401–405. [CrossRef] [PubMed]

152. Makarova, M.; Makarov, V. *Molecular Biology of Flavonoids (Chemistry, Biochemistry, Pharmacology): Manual for Doctors*; Lema Publishing: St-Petersburg, Russia, 2010; pp. 272–290.

153. Ivanov, S.; Garbuz, S.; Malfanov, I.; Ptitsyn, L. Screening of Russian medicinal and edible plant extracts for angiotensin I-converting enzyme (ACE I) inhibitory activity. *Russ. J. Bioorganic Chem.* **2013**, *39*, 743–749. [CrossRef]

154. Mansoor, M.; Bhagyarao, D.; Srinivasa Rao, D. Photochemical analysis and hepatoprotective activity of *Saxifraga ligulata* leaves extract. *J. Sci. Res. Pharm.* **2015**, *4*, 93–97.

155. Shutov, D.V. Hepatoprotective effect of *Bergenia crassifolia* extract and silymarin at experimental inhibition of (3-oxidation of fatty acids caused by 4-pentenioc acid. *Bull. Sib. Med.* **2007**, *7*, 64–70.

156. Rajkumar, V.; Guha, G.; Kumar, R.A. Anti-neoplastic activities of *Bergenia ciliata* rhizome. *J. Pharm. Res.* **2011**, *4*, 443–445.

157. Zafar, R.; Ullah, H.; Zahoor, M.; Sadiq, A. Isolation of bioactive compounds from *Bergenia ciliata* (haw.) sternb rhizome and their antioxidant and anticholinesterase activities. *BMC Complement. Altern. Med.* **2019**, *19*, 296. [CrossRef]

158. Shilova, I.; Pisareva, S.; Krasnov, E.; Bruzhes, M.; Pyak, A. Antioxidant properties of *Bergenia crassifolia* extract. *Pharm. Chem. J.* **2006**, *40*, 620–623. [CrossRef]

159. Shikov, A.N.; Pozharitskaya, O.N.; Makarova, M.N.; Kovaleva, M.A.; Laakso, I.; Dorman, H.D.; Hiltunen, R.; Makarov, V.G.; Galambosi, B. Effect of *Bergenia crassifolia* L. Extracts on weight gain and feeding behavior of rats with high-caloric diet-induced obesity. *Phytomedicine* **2012**, *19*, 1250–1255. [CrossRef]

160. Janar, J.; Fang, L.; Wong, C.P.; Kaneda, T.; Hirasawa, Y.; Morita, H.; Shahmanovna, B.; Abduahitovich, A. A new galloylbergenin from *Bergenia crassifolia* with anti-lipid droplet accumulation activity. *Heterocycles* **2012**, *86*, 1591–1595.

161. Panossian, A.; Wikman, G.; Wagner, H. Plant adaptogens III. Earlier and more recent aspects and concepts on their mode of action. *Phytomedicine* **1999**, *6*, 287–300. [CrossRef]

162. Tsyrenzhapova, O.D.; Lubsandorzhieva, P.B.; Bryzgalov, G.Y. *Conservation of Biological Diversity in the Baikal Region: Problems, Approaches, Practice*; Korsunov, V.M., Ed.; Baikal Scientific Center SB RAS: Ulan-Ude, Russia, 1996; pp. 157–158. (In Russian)

163. Shikov, A.N.; Pozharitskaya, O.N.; Makarova, M.N.; Dorman, H.D.; Makarov, V.G.; Hiltunen, R.; Galambosi, B. Adaptogenic effect of black and fermented leaves of *Bergenia crassifolia* L. In mice. *J. Funct. Foods* **2010**, *2*, 71–76. [CrossRef]

164. Bolshunova, E.; Lamazhapova, G.; Zhamsaranova, S. Research of liposomal form of *Bergenia crassifolia* (L.) fritsch influence on formation of adaptation potencial of the body. *ESSUTM Bull.* **2010**, *4*, 83–88.

165. Mironova, G.; Shigaeva, M.; Belosludtseva, N.; Gritsenko, E.; Belosludtsev, K.; Germanova, E.; Lukyanova, L. Effect of several flavonoid-containing plant preparations on activity of mitochondrial ATP-dependent potassium channel. *Bull. Exp. Biol. Med.* **2008**, *146*, 229–233. [CrossRef] [PubMed]

166. Yaginuma, A.; Murata, K.; Matsuda, H. Beta-gulcan and *Bergenia ligulata* as cosmetics ingredient. *Fragrance J.* **2003**, *31*, 114–119.

167. Guo, H.; Song, K.; Chen, Q. The synthesis of two arbutin derivatives and inhibitory effect of them on tyrosinase. *J. Xiamen Univ. (Nat. Sci.)* **2004**, *43*, 1–4.

168. Lee, K.-T.; Lee, S.-y.; Lee, K.-S.; Jeong, J.-H. Cosmetic Composition for Remedying Skin Wrinkles Comprising Bergenia emeiensis Extract as Active Ingredient. Google Patents US20040115286A1, 2004.

169. Shrestha, U.K.; Pant, B. Production of bergenin, an active chemical constituent in the callus of *Bergenia ciliate* (Haw.) Sternb. Botanica Orientalis. *J. Plant Sci.* **2011**, *8*, 40–44.

170. Verma, R.; Parkash, V.; Kumar, D. Ethnomedicinal uses of some plants of Kanag Hill in Shimla, Himachal Pradesh, India. *Int. J. Res. Ayurveda Pharm.* **2012**, *3*, 319–322.

171. Rafi, S.; Kamili, A.N.; Ganai, B.A.; Mir, M.Y.; Parray, J.A. In vitro Culture and Biochemical Attributes of *Bergenia ciliate* (Haw.) Sternb. *Proc. Natl. Acad. Sci. India Sect. B Biol. Sci.* **2018**, *88*, 609–619. [CrossRef]

172. Liu, M.; Hao, X.Y.; Xu, Q.; Bo, L.T.; Kang, X.L.; Wang, X.J. Tissue culture of wild flower *Bergenia crassifolia* and establishment of its regeneration system. *J. Anhui. Agric. Sci.* **2009**, *37*, 3455–3456.

173. Parveen, S.; Kamili, A.N. In vitro shoot multiplication response from shoot tips of *Bergenia ligulata* Engl. on different nutrient media-A comparative study. *Int. J. Innov. Res. Dev.* **2013**, *2*, 65–67.

174. Lu, X.M.; Wang, J.X. Research advancement on *Bergenia* genus plants. *Chin. Med. Mat.* **2003**, *26*, 58–60.

MDPI

St. Alban-Anlage 66

4052 Basel

Switzerland

Tel. +41 61 683 77 34

Fax +41 61 302 89 18

www.mdpi.com

Molecules Editorial Office

E-mail: molecules@mdpi.com

www.mdpi.com/journal/molecules